Introductory Algebra

FIFTH EDITION

FIFTH EDITION

Introductory Algebra

Mervin L. Keedy
PURDUE UNIVERSITY

Marvin L. Bittinger
INDIANA UNIVERSITY—
PURDUE UNIVERSITY AT INDIANAPOLIS

ADDISON-WESLEY PUBLISHING COMPANY

Reading, Massachusetts · Menlo Park, California · Don Mills, Ontario
Wokingham, England · Amsterdam · Sydney
Singapore · Tokyo · Madrid · Bogotá · Santiago · San Juan

Sponsoring Editor	•	Susan Zorn
Production Supervisor	•	Susanah H. Michener
Design, Editorial, and Production Services	•	Quadrata, Inc.
Illustrator	•	Textbook Art Associates
Manufacturing Supervisor	•	Ann DeLacey
Cover Designer	•	Marshall Henrichs

PHOTO CREDITS   1, © Ross Lewis   65, © Hiroji Kubota, Magnum Photos, Inc.
115, AP/Wide World Photos   167, Frank Siteman, Taurus Photos
207, © Ross Lewis   269, Catherine Ursillo, Photo Researchers, Inc.
307, © Bill Ross 1983, Woodfin Camp and Associates   351, Ken Karp
375, © Richard Wood, Taurus Photos   445, Rick Haston, Latent Images
489, © Bohdan Hrynewych, Stock/Boston

Library of Congress Cataloging-in-Publication Data

Keedy, Mervin Laverne.
   Introductory algebra.

   Includes index.
    1. Algebra.   I. Bittinger, Marvin L.   II. Title.
QA152.2.K43   1986   512.9   86-14164
ISBN 0-201-15431-5

CDEFGHIJ-MU-898

Preface

A NOTE FROM THE PUBLISHER

The Keedy/Bittinger texts are well known as the first mathematics books to incorporate the interactive worktext pedagogy that has proven so successful in helping students both to learn and retain mathematical concepts. These texts have been revised and improved with each edition to meet the everchanging needs of the developmental-mathematics marketplace.

Introductory Algebra, Fifth Edition, is a significant revision of the Fourth Edition. It focuses on new skill-maintenance features, more applications, exercises, and examples, and the inclusion of extension exercises in every exercise set. The presentation of key concepts has been refined and improved in response to extensive user feedback. An expanded supplements package includes a computerized testbank with extensive graphics capabilities, interactive software, an Instructor's Resource Guide with numerous helpful pedagogical features, and a specially bound Teacher's Edition with answers provided in color.

INTERACTIVE WORKTEXT PEDAGOGY carefully designed to help students learn and retain mathematical concepts.

- *Objectives.* The objectives for each section of material are stated in the margin, and are identified with a domino symbol: ▪▪. The same symbol is placed in the text beside the corresponding expository material, in the exercise sets, and in the answers to the summary—reviews and cumulative reviews. Students can immediately relate material they study to specific objectives, and can easily find appropriate review material if they are unable to do an exercise.

- *Margin Exercises.* This important feature is designed to actively involve students in the development of material and to provide immediate reinforcement of concepts covered in each section. The

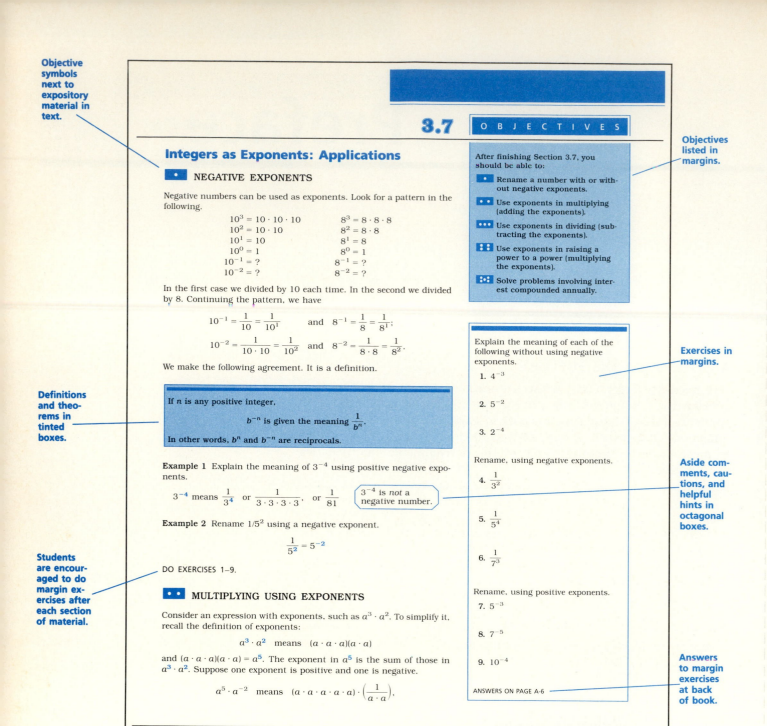

Objective symbols next to expository material in text.

O B J E C T I V E S

Integers as Exponents: Applications

NEGATIVE EXPONENTS

Negative numbers can be used as exponents. Look for a pattern in the following.

$$10^3 = 10 \cdot 10 \cdot 10 \qquad 8^3 = 8 \cdot 8 \cdot 8$$
$$10^2 = 10 \cdot 10 \qquad 8^2 = 8 \cdot 8$$
$$10^1 = 10 \qquad 8^1 = 8$$
$$10^0 = 1 \qquad 8^0 = 1$$
$$10^{-1} = ? \qquad 8^{-1} = ?$$
$$10^{-2} = ? \qquad 8^{-2} = ?$$

In the first case we divided by 10 each time. In the second we divided by 8. Continuing the pattern, we have

$$10^{-1} = \frac{1}{10} = \frac{1}{10^1} \quad \text{and} \quad 8^{-1} = \frac{1}{8} = \frac{1}{8^1};$$

$$10^{-2} = \frac{1}{10 \cdot 10} = \frac{1}{10^2} \quad \text{and} \quad 8^{-2} = \frac{1}{8 \cdot 8} = \frac{1}{8^2}.$$

We make the following agreement. It is a definition.

Definitions and theorems in tinted boxes.

> If n is any positive integer,
>
> $$b^{-n} \text{ is given the meaning } \frac{1}{b^n}.$$
>
> In other words, b^n and b^{-n} are reciprocals.

Example 1 Explain the meaning of 3^{-4} using positive negative exponents.

$$3^{-4} \text{ means } \frac{1}{3^4} \text{ or } \frac{1}{3 \cdot 3 \cdot 3 \cdot 3}, \text{ or } \frac{1}{81}$$

3^{-4} is *not* a negative number.

Aside comments, cautions, and helpful hints in octagonal boxes.

Example 2 Rename $1/5^2$ using a negative exponent.

$$\frac{1}{5^2} = 5^{-2}$$

Students are encouraged to do margin exercises after each section of material.

DO EXERCISES 1–9.

MULTIPLYING USING EXPONENTS

Consider an expression with exponents, such as $a^3 \cdot a^2$. To simplify it, recall the definition of exponents:

$$a^3 \cdot a^2 \text{ means } (a \cdot a \cdot a)(a \cdot a)$$

and $(a \cdot a \cdot a)(a \cdot a) = a^5$. The exponent in a^5 is the sum of those in $a^3 \cdot a^2$. Suppose one exponent is positive and one is negative.

$$a^5 \cdot a^{-2} \text{ means } (a \cdot a \cdot a \cdot a \cdot a) \cdot \left(\frac{1}{a \cdot a}\right),$$

Objectives listed in margins.

After finishing Section 3.7, you should be able to:

- Rename a number with or without negative exponents.
- Use exponents in multiplying (adding the exponents).
- Use exponents in dividing (subtracting the exponents).
- Use exponents in raising a power to a power (multiplying the exponents).
- Solve problems involving interest compounded annually.

Exercises in margins.

Explain the meaning of each of the following without using negative exponents.

1. 4^{-3}

2. 5^{-2}

3. 2^{-4}

Rename, using negative exponents.

4. $\frac{1}{3^2}$

5. $\frac{1}{5^4}$

6. $\frac{1}{7^3}$

Rename, using positive exponents.

7. 5^{-3}

8. 7^{-5}

9. 10^{-4}

ANSWERS ON PAGE A-6

Answers to margin exercises at back of book.

margin exercises are similar to examples in the text and homework exercises, providing students with plenty of practice.

- *Exercise Sets.* Exercise sets at the end of each section are provided on tearout sheets. They are carefully graded by difficulty, with easier exercises at the beginning. The exercises themselves are indexed to objectives, and identified by domino symbols. They also are paired, so that each even-numbered exercise is very much like the

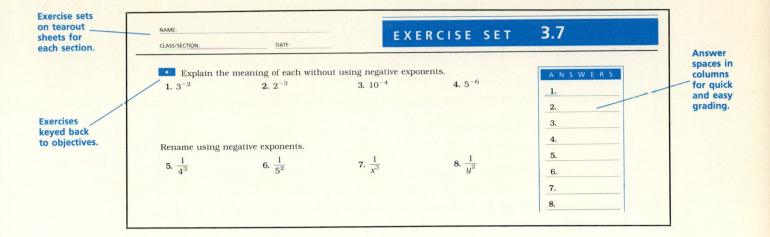

Exercise sets on tearout sheets for each section.

Exercises keyed back to objectives.

Answer spaces in columns for quick and easy grading.

NAME:

CLASS/SECTION: DATE:

EXERCISE SET   3.7

• Explain the meaning of each without using negative exponents.

1. 3^{-2} 2. 2^{-3} 3. 10^{-4} 4. 5^{-6}

Rename using negative exponents.

5. $\dfrac{1}{4^3}$ 6. $\dfrac{1}{5^2}$ 7. $\dfrac{1}{x^3}$ 8. $\dfrac{1}{y^2}$

ANSWERS

1. _____
2. _____
3. _____
4. _____
5. _____
6. _____
7. _____
8. _____

odd-numbered exercise preceding it, allowing instructors to assign odd and even exercises separately. Answers to odd-numbered exercises are at the back of the book. Optional calculator exercises are interspersed throughout the exercise sets and marked with the symbol ▦.

SKILL MAINTENANCE FEATURES continually reinforce students' mastery of skills learned earlier in the text.

- *Skill Maintenance Exercises.* At the end of most exercise sets are several skill maintenance exercises designated by ✔. These exercises may review *any* skill covered earlier in the text. The test for each chapter except Chapter 1 contains four skill maintenance questions covering specific sections from earlier chapters. These sections are reviewed in the summary–review sections preceding the chapter tests.

- *Extension Exercises.* At the end of most exercise sets is a set of extension exercises marked with the symbol ☆. These are designed to challenge students, and may require them to synthesize the objectives of several sections.

- *Summary–Reviews.* At the end of each chapter is a Summary– Review section. The objectives of each chapter are stated in boldface type and followed by review exercises. Answers are at the back of the book, together with section and domino objective references so that students can easily find the appropriate material to restudy when necessary.

- *Chapter Tests.* This revision includes new versions of tests for each chapter and a new final examination. Answers to the tests and final exam are not in the book so that students can be tested and test themselves without having access to answers. Six alternative test forms for both the chapter tests and final exam can be found in the *Instructor's Resource Guide.*

- *Cumulative Reviews.* An important new feature, these reviews are located every three or four chapters, and contain exercises covering

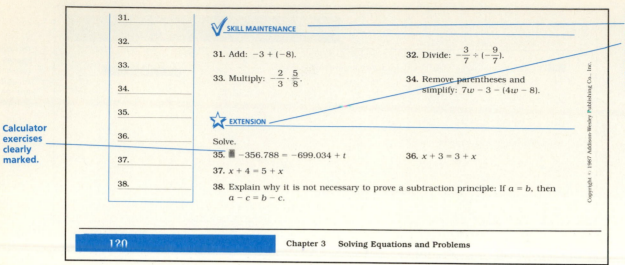

Calculator exercises clearly marked.

31.

32.

33.

34.

35.

36.

37.

38.

Skill maintenance and extension exercises at the end of most exercise sets.

✓ SKILL MAINTENANCE

31. Add: $-3 + (-8)$.

32. Divide: $-\frac{3}{7} \div (-\frac{9}{7})$.

33. Multiply: $-\frac{2}{3} \cdot \frac{5}{8}$.

34. Remove parentheses and simplify: $7w - 3 - (4w - 8)$.

⭐ EXTENSION

Solve.

35. 🖩 $-356.788 = -699.034 + t$

36. $x + 3 = 3 + x$

37. $x + 4 = 5 + x$

38. Explain why it is not necessary to prove a subtraction principle: If $a = b$, then $a - c = b - c$.

120   Chapter 3   Solving Equations and Problems

Summary and review section for each chapter.

CHAPTER **3**

Summary and Review

The following contains a summary of what you should be able to do after completing this chapter. The review exercises are for practice. Answers are at the back of the book. If you miss an exercise, restudy the section and objective indicated alongside the answer.

The review sections to be tested in addition to the material in this chapter are Sections 1.3, 1.7, 2.5, 2.6, and 2.8.

Practice for chapter test review sections.

You should be able to:

Objectives listed in boldface type.

Solve equations using the addition principle, the multiplication principle, the addition and multiplication principles together, and the distributive laws to collect like terms and to remove parentheses.

Solve.

1. $x + 5 = -17$

2. $-8x = -56$

Chapter tests on tearout sheets.

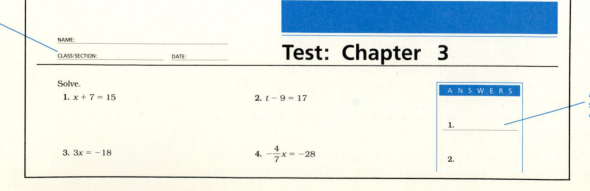

NAME:

CLASS/SECTION: DATE:

Test: Chapter 3

Solve.

1. $x + 7 = 15$

2. $t - 9 = 17$

3. $3x = -18$

4. $-\frac{4}{7}x = -28$

ANSWERS

1.

2.

Answer spaces in columns.

CHAPTERS **1-3**

Cumulative Review

Evaluate.

1. $\dfrac{y - x}{4}$ for $y = 12$ and $x = 6$

2. $\dfrac{3x}{y}$ for $x = 5$ and $y = 4$

3. $x^3 - 3$ for $x = 3$

4. Find the perimeter of a rectangle when l is 12 cm and w is 14 cm.

Translate to an algebraic expression.

5. Four less than twice w

6. Thirty-four percent of some number

the preceding three or four chapters. Cumulative review answers together with section and domino objective references are at the back of the book. Additional cumulative review tests for every chapter except Chapter 1 are located in the *Instructor's Resource Guide*. These cover all preceding material. The exercises for the cumulative reviews and cumulative review tests are not arranged by objective, so that students must decide themselves what concepts are involved in solving a particular problem.

Content Changes

In this edition, the presentation of key concepts has been polished in response to user feedback. Following are specific changes.

Chapter 1. An introduction to algebra and more applications have been added.

Chapter 2. The first section has been rewritten. Section 2.3 has been restructured to emphasize the use of $a - b = a + (-b)$ for subtraction. Many more examples have been added. A new section has been added on the order of operations.

Chapter 3. New examples, applications, and problems have been added. The material on properties of exponents and scientific notation has been moved from Chapter 2 to the end of this chapter in order to reduce Chapter 2's content load and to place the material in a more advantageous position prior to the discussion of polynomials in Chapter 4.

Chapter 4. Section 4.1 has been revised and reorganized. A general strategy for the multiplication of polynomials is now included in Section 4.7.

Chapter 5. The material on factoring of trinomials has been broken down into more steps. Many applications and problems have been added. In order to minimize the time demands of the text, "Polynomials in Several Variables" is covered in this chapter in Section 5.9; in the previous edition it was in a separate chapter. There are many exercises for this section, some of which may be cut to save time.

Chapter 6. Many users felt that Chapter 6 in the Fourth Edition was too long. In response to this and to allow for the inclusion of optional material on slope and equations of lines, it has been divided into two chapters in this edition. This is the first of the two new chapters, and includes the optional material on slope and equations of lines. For applications, the topic of direct variation is also included in this chapter.

Chapter 7. This chapter contains the remaining topics from Chapter 6 in the previous edition. Sections 7.1 and 7.2 were each condensed from two sections in the previous edition. Two sections of problems at the end of the chapter in the previous edition have been combined so that students can practice solving a mixture of different types of problems.

Chapter 8. This was Chapter 11 in the Fourth Edition.

Chapter 9. Material on solving equations and problem solving has been moved so that students cover this material earlier than in the Fourth Edition. Sections that are more optional have been moved to the end of the chapter.

Chapter 10. Many more applications and problems have been added, and the first two sections have been reorganized to introduce real numbers earlier.

Chapter 11. A new method of graphing quadratic equations has been added to this chapter.

Intended Audience and Scope

Introductory Algebra, Fifth Edition, is one of a series of basic mathematics texts that includes the following:

Basic Mathematics
Introductory Algebra
Intermediate Algebra
Essential Mathematics: combined arithmetic, introductory algebra, geometry, intermediate algebra, and trigonometry
Developmental Mathematics: combined arithmetic and introductory algebra
College Algebra (or *Algebra and Trigonometry*)

Introductory Algebra is the second book in this series, and is for students who have not studied algebra. It is a beginner's introduction, although it may also be appropriate for those needing a review. The first book in the series, *Basic Mathematics*, covers arithmetical skills as a prelude to an introductory algebra course. *Intermediate Algebra*, the third book in the series, is intended for use by students who have finished a first course in algebra in high school or have taken a comparable course, such as that provided by *Introductory Algebra*. *Intermediate Algebra* "starts at the beginning," but proceeds faster than an introductory book does, and covers the usual topics of intermediate algebra, including quadratics, exponential functions, and logarithms. It provides adequate preparation for a course in finite mathematics or brief calculus. The last book, *College Algebra*, provides the level of algebraic maturity usually desired as a prerequisite to the mainstream calculus.

A Placement Test is available to help place students in the appropriate book in the series.

Teaching Modes

The flexible worktext format of *Introductory Algebra* allows the book to be used in many ways.

- *As an ordinary textbook.* To use the book this way, the instructor and students simply ignore the margin exercises.

- *For a modified lecture.* To bring student-centered activity into the class, the instructor stops lecturing and has the students do margin exercises at appropriate times.

- *For a no-lecture class.* The instructor makes assignments that students do on their own, including working exercise sets. During the class period following the assignment, the instructor answers questions, and students have an extra day or two to polish their work before handing it in. In the meantime, they are working on the next assignment. This method provides individualization while keeping a class together. It minimizes the number of instructor hours required and has been found to work well with large classes.

- *In a learning laboratory.* This book has a low reading level and is easy to understand, so that it can be used in a learning laboratory or any other self-study situation.

Acknowledgments

We wish to thank the following people who reviewed the manuscript: Laverne Blagmon, *University of the District of Columbia*; Mary Jean Brod, *University of Montana*; Ira Lansing, *Marin Community College*; Sharon MacKendrick, *Grants Branch of New Mexico State University*; and Dwain Small, *Valencia Community College*.

In addition, Kathleen Berver and her staff at *New Mexico State University* provided the author with many creative suggestions for improvement.

We are also grateful to the following focus group participants:

Sandy Ajose, *Essex County College*
Marcelle Bessman, *University of Tampa*
Laverne Blagmon, *University of the District of Columbia*
Paul Boltz, *Harrisburg Community College*
Sue Boyer, *University of Maryland, Baltimore County*
Evelyn Brodbeck, *Suffolk Community College*
Brenda Brown, *University of the District of Columbia*
Anita Buker, *Miami-Dade Community College, North Campus*
Laura Cameron, *University of New Mexico*
Ben Cheatham, *Valencia Community College*
Doug Dawson, *Arizona State University*

Dennis Ebersole, *Northampton Area Community College*
Dennis Epley, *Community College of Baltimore*
Ronald Epperlein, *Arizona State University*
Betty Field, *Phoenix College*
Catherine Folio, *Brookdale Community College*
Virginia Hamilton, *Ball State University*
Ed Hoff, *Glendale Community College*
Giselle Icore, *Community College of Baltimore*
Don Johnson, *Scottsdale Community College*
Glenn Johnston, *Morehead State University*
Judy Jones, *Valencia Community College*
Rosemary Karr, *Eastern Kentucky University*
Linda Knepp, *Cincinnati Technical College*
Linda Kyle, *Tarrant County Junior College*
Jo Lane, *Eastern Kentucky University*
Eric Lubot, *Bergen Community College*
Robert Lutz, *Harrisburg Community College*
Shirley Markus, *University of Louisville*
Mary Mentzer, *University of Louisville*
Don Poulson, *Mesa Community College*
Charles Robinson, *Eastern Kentucky University*
Sharon Ross, *Dekalb Community College*
Winona Sathre, *Valencia Community College*
Karen Skulder, *University of Maryland, Baltimore County*
Dwain Small, *Valencia Community College*
John Taylor, *Hillsborough Community College*
Arlene Watkins, *University of Arizona*
Paul Welsh, *Peabody Community College—East Campus*
Pat Wilkinson, *Borough of Manhattan Community College.*

We would like to give special thanks to the many people without whose committed efforts our work could not have been completed. Judy Penna, Karen Anderson, Gloria Schnippel, Michael Dagg, Sharon Mac-Kendrick, and Sue Haberhern did a thorough and conscientious job of checking the manuscript. Judy Penna and Judy Beecher coordinated the *Instructor's Resource Guide* to perfection. Virginia Hamilton, Linda Kyle, and Jim Magliano did a superb job on the computerized testbank, and Julie Stephenson and Patsy Hammond made the typing of the supplements smooth sailing.

M.L.K.
M.L.B.

- *Teacher's Edition.* A specially bound version of the student text with answers in color.

- *Instructor's Resource Guide.* The elaborate *Instructor's Resource Guide* is a significant new supplement containing the following:

  a) Six alternative forms of the chapter tests and final exam with answers. Three of the final exams contain questions arranged by chapter and three contain questions categorized by type.

  b) Two versions with answers of a cumulative review test for every chapter except Chapter 1.

  c) Extra practice sheets covering the most difficult topics in the text. Each sheet has several examples and 15–20 basic exercises.

  d) All even-numbered answers to the exercise sets in the text. These can be copied and used to supplement odd answers at the back of the students' books.

  e) Videotape indexes listing all topics covered on the videotapes with counter numbers for easy access.

  f) A page of grids for making up tests and transparency masters of different types of grids for lectures on graphing.

- *Computerized Testbank, Testgen II.* Test items can be selected from a printed bank of test questions, and a test can be generated with graphics by an Apple II series computer and printer or by the IBM PC. There are roughly 10 questions at two levels of difficulty for every terminal objective. Questions are numbered and coded by section, objective, and level of difficulty, and can be selected by any one of these four factors. The test questions are in multiple-choice format, but most are written so that they can also be generated in open-ended format. Questions can be scrambled within a given test, and answer options can also be scrambled, enabling instructors to generate many permutations of the same test.

- *Printed Testbank.* This is a printed version of the computerized testbank, and can be used by those who wish to generate tests without using a computer.

- *Student's Solutions Manual.* Complete worked-out solutions to all odd-numbered exercises in the exercise sets are provided in this booklet. This is an improvement over the Fourth Edition, in which only 70% of the odd-numbered exercises were worked out.

- *Instructional Software.* This is an interactive software package covering 12 topics in introductory algebra. It generates random exercises, help screens, messages, and hints for the student, and both tutorial and quiz modes, and is available for the Apple II series.

- *Videotape Cassettes.* A lecturer speaks to students and works out examples with lucid explanations. These video cassettes can be used to supplement lectures or in individualized learning situations.

- *Placement Test.* This short test is designed to place students properly in one of the texts in the Keedy/Bittinger series. It has been thoroughly evaluated statistically and is known to give a good screening with respect to placement into one of the four categories into which the books of this series fall.

Once specific placement within a text is achieved, chapter placement can be determined by using one of the final exams with questions arranged by chapter in the *Instructor's Resource Guide.*

Texts covered by the placement test are:

I. *Basic Mathematics,* Fifth Edition

II. *Introductory Algebra,* Fifth Edition
   A Problem-Solving Approach to Introductory Algebra, Second Edition (hardback)
   Introductory Algebra with Problem Solving

III. *Intermediate Algebra,* Fifth Edition
   A Problem-Solving Approach to Intermediate Algebra, Second Edition (hardback)
   Intermediate Algebra with Problem Solving

IV. *College Algebra: A Functions Approach,* Fourth Edition
   Fundamental College Algebra, Second Edition (hardback)
   Algebra and Trigonometry: A Functions Approach, Fourth Edition
   Fundamental Algebra and Trigonometry, Second Edition (hardback)
   Trigonometry: Triangles and Functions, Fourth Edition

For more information on the Keedy/Bittinger supplements, please contact your regional sales office:

EASTERN REGIONAL OFFICE
Addison-Wesley Publishing Co.
Jacob Way
Reading, MA   01867
(617) 246-5100

WESTERN REGIONAL OFFICE
Addison-Wesley Publishing Co.
390 Bridge Parkway, Suite 200
Redwood City, CA   94065
(415) 594-4400

MIDWESTERN REGIONAL OFFICE
Addison-Wesley Publishing Co.
1843 Hicks Road
Rolling Meadows, IL   60008
(312) 991-7878

SOUTHERN REGIONAL OFFICE
Addison-Wesley Publishing Co.
1100 Ashwood Parkway, Suite 145
Atlanta, GA   30338
(404) 394-7268

Contents

1

Introduction to Algebra; Review of Arithmetic

Contents

Integers and Rational Numbers

3

Solving Equations and Problems

Polynomials

5

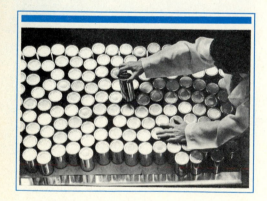

Polynomials and Factoring

Graphs, Linear Equations, and Slope

7

Systems of Equations

Inequalities and Sets

Fractional Expressions and Equations

Contents

10

Radical Expressions and Equations

Contents

11

Quadratic Equations

Introductory Algebra

FIFTH EDITION

AN APPLICATION
The length *l* of a standard-sized football field in the United States is 360 ft. The width *w* is 160 ft. Find the perimeter of such a football field.

THE MATHEMATICS
The perimeter of a rectangle is given by the formula $2l + 2w$. We substitute 360 for *l* and 160 for *w*:

$$2l + 2w = 2 \cdot 360 + 2 \cdot 160 = 720 + 320$$
$$= 1040 \text{ ft.}$$

↑————————This is an algebraic expression.

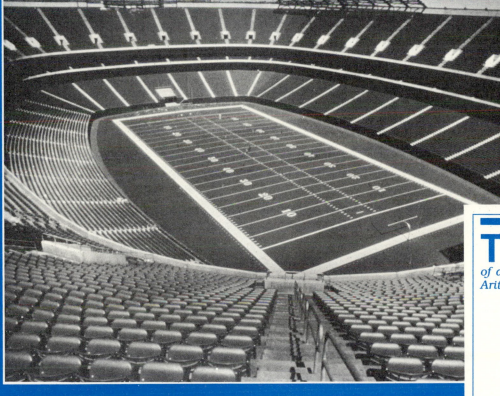

This chapter presents algebraic expressions as part of an introduction to algebra. Arithmetic is also reviewed.

Introduction to Algebra; Review of Arithmetic

1.1

Translate this problem to an equation. Use the chart at the right.

1. How many more flights are there on the Dallas–Houston route than on the New York–Boston route?

Introduction to Algebra and Expressions

This section will introduce you to algebra. We will study evaluating expressions and translating to expressions of the type used in algebra.

• ALGEBRAIC EXPRESSIONS

In arithmetic you have worked with expressions such as

$$23 + 54, \quad 4 \times 5, \quad 16 - 7, \quad \text{and} \quad \frac{5}{8}.$$

In algebra we use certain letters, or *variables*, for numbers and work with *algebraic expressions* such as

$$23 + x, \quad 4 \times t, \quad 16 - y, \quad \text{and} \quad \frac{a}{8}.$$

How do these expressions arise? Most often they arise in problem solving. For example, consider the following chart, which you might see in a magazine.

Taking flight in traffic

Here's how many flights are made monthly on the busiest air routes:

Dallas–Houston	2,866
Los Angeles–San Francisco	2,822
New York–Chicago	2,710
New York–Washington	2,442
New York–Boston	2,128

Suppose we want to know how many more flights there are on the Los Angeles–San Francisco route than on the New York–Washington route.

In algebra we translate this problem to an equation. It might be done as follows:

(Flights NY–Wash) plus (How many?) is (Flights LA–SF)

$$2442 \quad + \quad x \quad = \quad 2882.$$

Note that we have an algebraic expression on the left. To find the number x, we can subtract 2442 on both sides of the equation:

$$x = 2822 - 2442.$$

Then we carry out the subtraction and obtain the answer, 380.

In arithmetic, you probably would do this subtraction right away without considering an equation. In algebra, you will find most problems extremely difficult to solve without first translating to an equation.

DO EXERCISE 1 IN THE MARGIN AT THE LEFT.

ANSWER ON PAGE A-1

An *algebraic expression* consists of variables, numerals, and operation signs. When we replace a variable by a number, we say that we are *substituting* for the variable. When we calculate the results, we get a number. This process is called *evaluating the expression.*

Example 1 Evaluate $x + y$ for $x = 23$ and $y = 54$.

We substitute 23 for x and 54 for y and carry out the addition:

$$x + y = 23 + 54 = 77.$$

The number 77 is called the *value* of the expression.

Example 2 Evaluate a/b for $a = 56$ and $b = 7$.

We substitute 56 for a and 7 for b and carry out the division:

$$\frac{a}{b} = \frac{56}{7} = 8.$$

DO EXERCISES 2–4.

In arithmetic, we often use $\times$ for a multiplication. We also use a dot.

$$4 \cdot 9 \quad \text{means} \quad 4 \times 9.$$

In algebra, when two letters or a number and a letter are written together, that also means that they are to be multiplied. Numbers to be multiplied are called *factors.* We usually write a factor that is a number before any factor named by a letter. For example,

$$3y \quad \text{means} \quad 3 \cdot y \quad \text{or} \quad 3 \times y,$$

and

$$ab \quad \text{means} \quad a \cdot b \quad \text{or} \quad a \times b.$$

Now we evaluate expressions involving products.

Example 3 Evaluate $3y$ for $y = 14$.

We substitute 14 for y and carry out the multiplication:

$$3y = 3 \cdot 14 = 42.$$

Example 4 The area of a rectangle of length l and width w is lw. Find the area when l is 24.5 in. and w is 16 in.

We substitute 24.5 for l and 16 for w and carry out the multiplication:

$$lw = 24.5 \times 16 = 392 \text{ square inches.}$$

DO EXERCISES 5 AND 6.

Example 5 Evaluate $12m/n$ for $m = 8$ and $n = 16$.

We substitute 8 for m and 16 for n and carry out the calculation:

$$\frac{12m}{n} = \frac{12 \cdot 8}{16} = \frac{96}{16} = 6.$$

DO EXERCISE 7.

2. Evaluate $a + b$ for $a = 38$ and $b = 26$.

3. Evaluate $x - y$ for $x = 57$ and $y = 29$.

4 Evaluate a/b for $a = 200$ and $b = 8$.

5. Evaluate $4t$ for $t = 15$.

6. Find the area of a rectangle when l is 24 ft and w is 8 ft.

7. Evaluate $10p/q$ when $p = 40$ and $q = 25$.

ANSWERS ON PAGE A-1

Translate to an algebraic expression.

8. Twelve less than some number

Translate to an algebraic expression.

9. Twelve more than some number

Translate to an algebraic expression.

10. Four less than some number

11. Half of some number

12. Six more than eight times some number

13. The difference of two numbers

14. Fifty-nine percent of some number

15. Two hundred less than the product of two numbers

16. The sum of two numbers

ANSWERS ON PAGE A-1

•• TRANSLATING TO ALGEBRAIC EXPRESSIONS

In algebra we translate problems to equations. The parts of equations are translations of phrases to algebraic expressions. To help you become familiar with algebra and to make such translations easier when we do them later, we now practice translating.

Example 6 Translate to an algebraic expression: seven less than some number.

Although we can use any variable we wish, such as x, y, m, or n, we let t represent the number. If we knew the number to be 23, then the translation would be $23 - 7$. If we knew the number to be 345, then the translation would be $345 - 7$. Since we are using a variable for the number, the translation is

$$t - 7.$$

DO EXERCISE 8.

Example 7 Translate to an algebraic expression: fifteen more than some number.

This time we let y represent the number. If we knew the number to be 47, then the translation would be $47 + 15$, or $15 + 47$. If we knew the number to be 1.688, then the translation would be $1.688 + 15$, or $15 + 1.688$. Since we are using a variable, the translation is

$$y + 15, \quad \text{or} \quad 15 + y.$$

DO EXERCISE 9.

Example 8 Translate each of the following to an algebraic expression.

Phrase	Algebraic Expression
Eight more than some number	$m + 8$, or $8 + m$
Eight less than some number	$a - 8$
The difference of two numbers	$a - b$, or $b - a$
One-fourth of a number	$\frac{1}{4}x$, or $x/4$
Three more than five times a number	$3 + 5t$, or $5t + 3$
Seven less than the product of two numbers	$pq - 7$
Ninety-four percent of some number	$94\%z$, or $0.94z$

DO EXERCISES 10–16.

• Substitute to find values of the expressions.

1. Chris is 4 years younger than her brother Lowell. Suppose the variable x stands for Lowell's age. Then $x - 4$ stands for Chris's age. How old is Chris when Lowell is 14? 29? 52?

2. Employee A took five times as long to do a job as employee B. Suppose t stands for the time it takes B to do the job. Then $5t$ stands for the time it takes A. How long did it take A if B took 30 sec? 90 sec? 2 min?

3. The area of a triangle with base b and height h is $\frac{1}{2}bh$. Find the area when $b = 16$ yd and $h = 9$ yd.

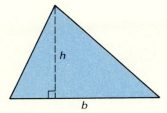

4. The area of a parallelogram with base b and height h is bh. Find the area of a parallelogram with a height of 15.4 cm (centimeters) and a base of 6.5 cm.

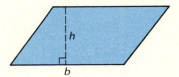

Evaluate.

5. $7y$ for $y = 8$

6. $6x$ for $x = 9$

7. $\dfrac{x}{y}$ for $x = 9$ and $y = 3$

8. $\dfrac{m}{n}$ for $m = 14$ and $n = 2$

9. $\dfrac{m - n}{8}$ for $m = 20$ and $n = 4$

10. $\dfrac{p + q}{5}$ for $p = 10$ and $q = 20$

11. $\dfrac{5z}{y}$ for $z = 8$ and $y = 2$

12. $\dfrac{18m}{n}$ for $m = 4$ and $n = 18$

•• Translate to an algebraic expression.

13. 6 more than m

14. 8 more than t

15. 9 less than c

16. 4 less than d

17. 6 greater than q

18. 11 greater than z

ANSWERS
1.
2.
3.
4.
5.
6.
7.
8.
9.
10.
11.
12.
13.
14.
15.
16.
17.
18.

19. b more than a

20. c more than d

21. x less than y

22. c less than b

23. 98% of x

24. Forty-five percent of m

25. The sum of r and s

26. The sum of d and f

27. Twice x

28. Four times p

29. 5 times t

30. 9 times d

31. The difference of 3 and b

32. The difference of p and q

33. Six more than some number

34. One more than some number

35. Four less than some number

36. Forty-three less than some number

37. A number x plus three times y

38. A number a plus 2 plus b

⭐ **EXTENSION**

This symbol means that the exercises that follow are more challenging, requiring you to put together the objectives of this section or of other sections of the book.

39. Translate to an algebraic expression: the perimeter of a rectangle of length l and width w.

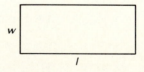

40. You invest P dollars at 10% simple interest. Write an algebraic expression for the number of dollars in the account a year from now. Evaluate for $P = \$2000$.

Evaluate.

41. $\dfrac{256y}{32x}$ for $y = 1$ and $x = 4$

42. $\dfrac{y + x}{2} + \dfrac{3y}{x}$ for $x = 2$ and $y = 4$

Review of Arithmetic: Factoring and LCMs

In algebra the notion of factoring is quite important. Eventually, you will learn to factor algebraic expressions. For now, we review factoring of numbers in order to review addition and subtraction using fractional notation.

• FACTORING

In this section we consider the set of *natural numbers*: 1, 2, 3, 4, 5, and so on.

To *factor* a number means to name it as a product.

Example 1 Factor the number 8.

There are several ways to name the number 8 as a product.

$$2 \cdot 4, \quad 1 \cdot 8, \quad 2 \cdot 2 \cdot 2$$

DO EXERCISES 1 AND 2.

A symbol that names a number as a product is called a *factorization* of the number.

Example 2 Write several factorizations of the number 12.

$$1 \cdot 12, \quad 2 \cdot 6, \quad 3 \cdot 4, \quad 2 \cdot 2 \cdot 3$$

DO EXERCISES 3 AND 4.

•• PRIME NUMBERS AND PRIME FACTORIZATIONS

Some numbers have exactly two factors, themselves and 1. Such numbers are called *prime*.

A *prime* number is a natural number that has exactly two different factors.

Example 3 Which of these numbers are prime? 7, 16, 1, 4, 11

7 is prime. It has exactly two different factors, 7 and 1.

16 is not prime. It has factors 1, 2, 4, 8, and 16.

1 is not prime. It has only one factor, itself.

4 is not prime. It has three different factors, 1, 2, and 4.

11 is prime. It has exactly two different factors, 11 and 1.

After finishing Section 1.2, you should be able to:

• Factor numbers.

•• Find prime factorizations.

••• Make lists of multiples and from them find common multiples.

:: Find the LCM of two or more numbers using prime factorizations.

Factor the number.

1. 9

2. 16

Write several factorizations for the number.

3. 18

4. 20

ANSWERS ON PAGE A-1

Which of these numbers is or are prime?

5. 8, 6, 13, 14, 1

Find the prime factorization.

6. 48

7. 50

8. 770

The following is a table of those primes that occur from 2 to 101. There are more extensive tables, but these will be the most helpful to you in this book.

A TABLE OF PRIMES
2, 3, 5, 7, 11, 13, 17, 19, 23, 29, 31, 37, 41, 43, 47, 53, 59, 61, 67, 71, 73, 79, 83, 89, 97, 101

DO EXERCISE 5.

If a natural number, other than 1, is not prime, we call it *composite*. Every composite number can be factored into prime numbers. Such a factorization is called a *prime factorization*.

Example 4 Find the prime factorization of 36.

We begin by factoring 36 any way we can. One way is like this:

$$36 = 4 \cdot 9.$$

The factors 4 and 9 are not prime, so we factor them.

$$36 = 4 \cdot 9 = \underbrace{2 \cdot 2}_{4} \cdot \underbrace{3 \cdot 3}_{9}.$$

The factors in the last factorization are all prime, so we have the *prime factorization*.

Example 5 Find the prime factorization of 60.

This time, we use our list of primes from the table. We go through the table until we find a prime that divides 60. The first such prime is 2.

$$60 = 2 \cdot 30$$

Keep dividing by 2 until it is not possible to do so.

$$60 = 2 \cdot 30 = 2 \cdot 2 \cdot 15$$

Now go to the next prime in the table, 3.

$$60 = 2 \cdot 15 = 2 \cdot 2 \cdot 3 \cdot 5$$

We now have the prime factorization, $2 \cdot 2 \cdot 3 \cdot 5$.

DO EXERCISES 6–8.

••• MULTIPLES

The *multiples* of a number all have that number as a factor. For example, the multiples of 2 are

2, 4, 6, 8, 10, 12, 14, 16, 18, 20, ...

We can show that each has 2 as a factor. For example, $14 = 2 \cdot 7$.

The multiples of 3 all have 3 as a factor. They are

3, 6, 9, 12, 15, 18, 21, 24, ...

Chapter 1   Introduction to Algebra; Review of Arithmetic

Two or more numbers always have a great many multiples in common. From lists of multiples, we can find common multiples.

Example 6 Find some of the multiples that 2 and 3 have in common.

We make lists of their multiples and circle the multiples that appear in both lists.

2, 4, ⑥, 8, 10, ⑫, 14, 16, ⑱, 20, 22, ㉔, 26, 28, ㉚, 32, 34, ㊱, 38,. . .

3, ⑥, 9, ⑫, 15, ⑱, 21, ㉔, 27, ㉚, 33, ㊱, 39, 42, . . .

The common multiples of 2 and 3 are

$$6, 12, 18, 24, 30, 36, \ldots .$$

DO EXERCISE 9.

∷ LEAST COMMON MULTIPLES

In Example 6 we found many common multiples of 2 and 3. The *least*, or smallest, of those common multiples is 6. We abbreviate *least common multiple* as LCM.

There are several methods that work well for finding the LCMs of natural numbers. Some of these do not work well in algebra. We now learn, or review, a method that will work in arithmetic *and in algebra as well*. The method uses factorizations. To see how it works, let's look at the prime factorizations of 9 and 15 in order to find the LCM:

$$9 = 3 \cdot 3,$$
$$15 = 3 \cdot 5.$$

Any multiple of 9 must have *two* 3's as factors. Any multiple of 15 must have *one* 3 and *one* 5 as factors. The smallest number satisfying these conditions is

Two 3's; 9 is a factor.

$3 \cdot 3 \cdot 5$, 45.

One 3, one 5; 15 is a factor.

The LCM must have all the factors of 9 and all the factors of 15, but the factors do not have to be repeated when they are common to both numbers.

> **To find the LCM of several factors:**
>
> 1. Write the prime factorization of each number.
> 2. Write each factor the greatest number of times it appears in any one factorization.

Example 7 Find the LCM of 24 and 36.

a) We find the factorizations.

$$24 = 2 \cdot 2 \cdot 2 \cdot 3$$
$$36 = 2 \cdot 2 \cdot 3 \cdot 3$$

9. Find several of the common multiples of 3 and 5 by making lists of multiples.

ANSWER ON PAGE A-1

Find the LCM by factoring.

10. 8 and 10

11. 18 and 40

12. Find the LCM of 24, 35, and 45.

Find the LCM.

13. 3, 18

14. 12, 24

Find the LCM.

15. 4, 9

16. 5, 6, 7

ANSWERS ON PAGE A-1

b) We write 2 as a factor 3 times (the greatest number of times it occurs). We write 3 as a factor 2 times (the greatest number of times it occurs).

The LCM is $2 \cdot 2 \cdot 2 \cdot 3 \cdot 3$, or 72.

DO EXERCISES 10 AND 11.

Example 8 Find the LCM of 27, 84, and 90.

a) We find the factorizations.

$$27 = 3 \cdot 3 \cdot 3$$
$$84 = 2 \cdot 2 \cdot 3 \cdot 7$$
$$90 = 2 \cdot 3 \cdot 3 \cdot 5$$

b) We write 2 as a factor 2 times, 3 as a factor 3 times, 5 one time, and 7 one time.

The LCM is $2 \cdot 2 \cdot 3 \cdot 3 \cdot 3 \cdot 5 \cdot 7$, or 3780.

DO EXERCISE 12.

Example 9 Find the LCM of 7 and 21.

Since 7 is a prime, it has no factorization. We still need it as a factor:

$$7 = \phantom{3 \cdot{}} 7,$$
$$21 = 3 \cdot 7.$$

The LCM is $3 \cdot 7$, or 21.

> If one number is a factor of another, the LCM is the larger of the numbers.

DO EXERCISES 13 AND 14.

Example 10. Find the LCM of 8 and 9.

$$8 = 2 \cdot 2 \cdot 2,$$
$$9 = 3 \cdot 3$$

The LCM is $2 \cdot 2 \cdot 2 \cdot 3 \cdot 3$, or 72.

> If two or more numbers have no common factor, the LCM is the product of the numbers.

DO EXERCISES 15 AND 16.

Chapter 1   Introduction to Algebra; Review of Arithmetic

- Write at least one factorization of each number.

1. 21   **2.** 30   **3.** 144   **4.** 102

•• Find the prime factorization of each number.

5. 14   **6.** 15   **7.** 33   **8.** 55

9. 9   **10.** 25   **11.** 49   **12.** 121

13. 18   **14.** 24   **15.** 40   **16.** 56

17. 90   **18.** 120   **19.** 210   **20.** 330

21. 91   **22.** 143   **23.** 119   **24.** 221

••• Make lists of multiples and from them find the first three common multiples.

25. 3, 7   **26.** 8, 12   **27.** 12, 16   **28.** 24, 36

∷ Find the prime factorization of each number. Then find the LCM.

29. 12, 18   **30.** 18, 30   **31.** 45, 72   **32.** 30, 36

A N S W E R S

1. _____
2. _____
3. _____
4. _____
5. _____
6. _____
7. _____
8. _____
9. _____
10. _____
11. _____
12. _____
13. _____
14. _____
15. _____
16. _____
17. _____
18. _____
19. _____
20. _____
21. _____
22. _____
23. _____
24. _____
25. _____
26. _____
27. _____
28. _____
29. _____
30. _____
31. _____
32. _____

33. 30, 50   **34.** 24, 30   **35.** 30, 40   **36.** 13, 23

33. _____

34. _____

35. _____

36. _____

37. _____

38. _____

39. _____

40. _____

41. _____

42. _____

43. _____

44. _____

45. _____

46. _____

47. _____

48. _____

49. _____

50. _____

51. _____

52. _____

53. _____

37. 18, 24   **38.** 12, 28   **39.** 35, 45   **40.** 60, 70

41. 2, 3, 5   **42.** 3, 5, 7   **43.** 24, 36, 12   **44.** 8, 16, 22

45. 5, 12, 15   **46.** 12, 18, 40   **47.** 6, 12, 18   **48.** 18, 24, 30

☑ SKILL MAINTENANCE

This symbol indicates that the exercises that follow are *Skill Maintenance Exercises,* which review skills previously studied in the text. You can expect such exercises in almost every exercise set.

49. Evaluate xy when $x = 7$ and $y = 42$.   **50.** Translate to an algebraic expression: twenty-seven percent of x.

⭐ EXTENSION

Application: Planet Orbits and LCMs. The earth, Jupiter, Saturn, and Uranus all revolve around the sun. The earth takes 1 year, Jupiter 12 years, Saturn 30 years, and Uranus 84 years to make a complete revolution. On a certain night you look at all the planets and wonder how many years it will take before they have the same position again. (*Hint:* To find out, you find the LCM of 12, 30, and 84. It will be that number of years.)

51. How often will Jupiter and Saturn appear in the same direction?   **52.** How often will Saturn and Uranus appear in the same direction?

53. How often will Jupiter, Saturn, and Uranus appear in the same direction?

Review of Arithmetic: Basic Operations

We now review the numbers of arithmetic and such procedures as simplifying fractional notation, as well as adding, subtracting, and multiplying using fractional notation.

After finishing Section 1.3, you should be able to:

 • Write several numerals for any number of arithmetic by multiplying by 1.

 • • Find simplest fractional notation for a number of arithmetic.

 • • • Add, subtract, and multiply with fractional notation.

• NUMBERS OF ORDINARY ARITHMETIC

The numbers used for counting are called *natural numbers.* They are

$$1, 2, 3, 4, 5, 6, 7, 8, 9, 10, 11, \text{ and so on.}$$

The *whole numbers* consist of the natural numbers and zero. They are

$$0, 1, 2, 3, 4, 5, 6, 7, 8, 9, 10, 11, \text{ and so on.}$$

The *numbers of ordinary arithmetic* consist of the whole numbers and the fractions, such as $\frac{2}{3}$ and $\frac{9}{5}$. These numbers are sometimes called the *numbers of arithmetic.* All the numbers of arithmetic can be named by *fractional notation a/b, b ≠ 0* (*b* is not zero).

Example 1 Write three fractional numerals for the number *three fourths.*

$$\frac{3}{4}, \quad \frac{6}{8}, \quad \frac{600}{800}$$

Example 2 Write three fractional numerals for the number *one third.*

$$\frac{1}{3}, \quad \frac{2}{6}, \quad \frac{10}{30}$$

Numerals are *names* for numbers. A number has many names.

Any fractional numeral with a denominator three times the numerator names the number *one third.*

DO EXERCISES 1–3.

Whole numbers can also be named with fractional notation.

Examples

3. $0 = \dfrac{0}{1} = \dfrac{0}{4} = \dfrac{0}{56}$, and so on.

4. $2 = \dfrac{2}{1} = \dfrac{6}{3} = \dfrac{128}{64}$, and so on.

1. Write three fractional numerals for $\frac{2}{3}$.

2. Write three fractional numerals for $\frac{1}{2}$.

3. Write three fractional numerals for $\frac{3}{5}$.

ANSWERS ON PAGE A-1

4. Write three fractional numerals for 1.

5. $1 = \dfrac{2}{2} = \dfrac{5}{5} = \dfrac{642}{642}$, and so on.

DO EXERCISES 4 AND 5.

To find different names, we can use the idea of multiplying by 1.

Examples

6. $\dfrac{2}{3} = \dfrac{2}{3} \times 1$

$= \dfrac{2}{3} \times \dfrac{5}{5}$ Using $\dfrac{5}{5}$ for 1

$= \dfrac{10}{15}$

How did we choose $\frac{5}{5}$? We just picked a number—in this case, 5—and multiplied by $\frac{5}{5}$.

5. Write three fractional numerals for 4.

7. $\dfrac{2}{3} = \dfrac{2}{3} \times 1$

$= \dfrac{2}{3} \times \dfrac{13}{13}$ This time we use $\dfrac{13}{13}$ for 1.

$= \dfrac{26}{39}$

DO EXERCISES 6 AND 7.

6. Multiply by 1 to find three different names for $\frac{4}{5}$.

●● SIMPLEST FRACTIONAL NOTATION

The simplest fractional notation for a number has the smallest possible numerator and denominator. To simplify, we reverse the process used in Examples 6 and 7.

Examples Simplify.

8. $\dfrac{10}{15} = \dfrac{2 \times 5}{3 \times 5}$ Factoring numerator and denominator.
Factoring means writing as a product.

$= \dfrac{2}{3} \times \dfrac{5}{5}$ Factoring the fraction

$= \dfrac{2}{3}$ We "removed" a factor of 1.

We have now simplified. We have the smallest possible numerator and denominator.

7. Multiply by 1 to find three different names for $\frac{8}{7}$.

Chapter 1   Introduction to Algebra; Review of Arithmetic

9. $\dfrac{36}{24} = \dfrac{6 \times 6}{4 \times 6} = \dfrac{3 \times 2 \times 6}{2 \times 2 \times 6} = \dfrac{3}{2} \times \dfrac{2 \times 6}{2 \times 6} = \dfrac{3}{2}$

You may be in the habit of canceling. For example, you might have done Example 9 as follows. This is a shortcut for the procedure used in Examples 8 and 9.

$$\begin{array}{c} 3 \\ \cancel{18} \\ \dfrac{\cancel{36}}{\cancel{24}} \\ \cancel{12} \\ 2 \end{array} \quad \text{or} \quad \dfrac{36}{24} = \dfrac{3 \times \cancel{12}}{2 \times \cancel{12}} = \dfrac{3}{2}$$

> *Caution:* The difficulty with canceling is that it is applied incorrectly in situations such as the following:
>
> $$\dfrac{\cancel{2}+3}{\cancel{2}} = 3, \qquad \dfrac{\cancel{4}+1}{\cancel{4}+2} = \dfrac{1}{2}, \qquad \dfrac{1\cancel{5}}{\cancel{5}4} = \dfrac{1}{4}.$$
>
> Wrong!   Wrong!   Wrong!
>
> In each of these situations the expressions canceled were *not* factors. Factors are parts of products. For example, in $2 \cdot 3$, 2 and 3 are factors, but in $2 + 3$, 2 and 3 are *not* factors. If you can't factor, you can't cancel! If in doubt, don't cancel!

DO EXERCISES 8–10.

We can put a factor of 1 in the numerator or denominator.

Examples Simplify.

10. $\dfrac{18}{72} = \dfrac{1 \times 18}{4 \times 18}$ Putting a 1 in the numerator

$\qquad = \dfrac{1}{4} \times \dfrac{18}{18}$

$\qquad = \dfrac{1}{4}$

11. $\dfrac{72}{9} = \dfrac{8 \times 9}{1 \times 9} = \dfrac{8}{1} \times \dfrac{9}{9} = \dfrac{8}{1} = 8$

DO EXERCISES 11 AND 12.

Simplify.

8. $\dfrac{18}{27}$

9. $\dfrac{38}{18}$

10. $\dfrac{56}{49}$

Simplify.

11. $\dfrac{27}{54}$

12. $\dfrac{48}{12}$

ANSWERS ON PAGE A-1

Multiply and simplify.

13. $\dfrac{6}{5} \cdot \dfrac{25}{12}$

14. $\dfrac{3}{8} \cdot \dfrac{5}{3} \cdot \dfrac{7}{2}$

Add and simplify.

15. $\dfrac{5}{6} + \dfrac{7}{10}$

16. $\dfrac{1}{4} + \dfrac{1}{3}$

Example 12 Multiply and simplify: $\frac{5}{6} \cdot \frac{9}{25}$.

$$\dfrac{5}{6} \cdot \dfrac{9}{25} = \dfrac{5 \cdot 9}{6 \cdot 25}$$ Multiplying numerators and denominators

$$= \dfrac{1 \cdot 5 \cdot 3 \cdot 3}{2 \cdot 3 \cdot 5 \cdot 5}$$ Factoring numerator and denominator

$$= \dfrac{3 \cdot 5 \cdot 1 \cdot 3}{3 \cdot 5 \cdot 2 \cdot 5}$$

$$= \dfrac{3 \cdot 5}{3 \cdot 5} \cdot \dfrac{1 \cdot 3}{2 \cdot 5}$$ Factoring the fraction

$$= \dfrac{3}{10}$$ "Removing" a factor of 1

DO EXERCISES 13 AND 14.

We can multiply by 1 to find common denominators.

Example 13 Add and simplify: $\frac{3}{8} + \frac{5}{12}$.

The LCM of the denominators is 24. We multiply by 1 to obtain the least common multiple for each denominator.

$$\dfrac{3}{8} + \dfrac{5}{12} = \dfrac{3}{8} \cdot \dfrac{3}{3} + \dfrac{5}{12} \cdot \dfrac{2}{2}$$ Multiplying by 1. Since $3 \cdot 8 = 24$, we multiply the first number by $\frac{3}{3}$. Since $2 \cdot 12 = 24$, we multiply the second number by $\frac{2}{2}$.

$$= \dfrac{9}{24} + \dfrac{10}{24}$$

$$= \dfrac{19}{24}$$

DO EXERCISES 15 AND 16.

Chapter 1   Introduction to Algebra; Review of Arithmetic

Example 14 Subtract and simplify: $\frac{9}{8} - \frac{4}{5}$.

The denominators, 8 and 5, have no common prime factor. Thus the LCM of the denominators is the product $8 \cdot 5$, or 40.

$$\frac{9}{8} - \frac{4}{5} = \frac{9}{8} \cdot \frac{5}{5} - \frac{4}{5} \cdot \frac{8}{8} = \frac{45}{40} - \frac{32}{40} = \frac{13}{40}$$

Example 15 Subtract and simplify: $\frac{7}{10} - \frac{1}{5}$.

The denominator 5 is a factor of 10. Thus the LCM is the larger number, 10.

$$\frac{7}{10} - \frac{1}{5} = \frac{7}{10} - \frac{1}{5} \cdot \frac{2}{2} = \frac{7}{10} - \frac{2}{10} = \frac{5}{10} = \frac{1}{2} \cdot \frac{5}{5} = \frac{1}{2}$$

DO EXERCISES 17 AND 18.

> In arithmetic you usually write $1\frac{1}{8}$ rather than $\frac{9}{8}$. In algebra you will find that the so-called *improper* symbols such as $\frac{9}{8}$ are more useful.

Subtract and simplify.

17. $\dfrac{4}{5} - \dfrac{4}{6}$

18. $\dfrac{5}{12} - \dfrac{2}{9}$

ANSWERS ON PAGE A-1

FACTORS AND SUMS

In the table, the top number has been factored in such a way that the sum of the factors is the bottom number. For example, in the first column 56 has been factored as $7 \cdot 8$, and $7 + 8 = 15$, the bottom number.

Product	56	63	36	72	140	96		168	110			
Factor	7										9	24	3
Factor	8						8	8			10	18	
Sum	15	16	20	38	24	20	14		21			24

EXERCISE

Find the missing numbers in the table.

EXERCISE SET 1.3

- Write four different fractional numerals for each number.

1. $\dfrac{4}{3}$

2. $\dfrac{5}{9}$

3. $\dfrac{6}{11}$

4. $\dfrac{15}{7}$

5. $\dfrac{2}{11}$

6. 1

7. 5

8. 0

•• Simplify.

9. $\dfrac{8}{6}$

10. $\dfrac{15}{25}$

11. $\dfrac{17}{34}$

12. $\dfrac{35}{25}$

13. $\dfrac{100}{50}$

14. $\dfrac{13}{39}$

15. $\dfrac{250}{75}$

16. $\dfrac{12}{18}$

••• Compute and simplify.

17. $\dfrac{1}{4} \cdot \dfrac{1}{2}$

18. $\dfrac{11}{10} \cdot \dfrac{8}{5}$

19. $\dfrac{17}{2} \cdot \dfrac{3}{4}$

20. $\dfrac{11}{12} \cdot \dfrac{12}{11}$

ANSWERS
1.
2.
3.
4.
5.
6.
7.
8.
9.
10.
11.
12.
13.
14.
15.
16.
17.
18.
19.
20.

21. $\dfrac{1}{2} + \dfrac{1}{2}$ 22. $\dfrac{1}{2} + \dfrac{1}{4}$ 23. $\dfrac{4}{9} + \dfrac{13}{18}$ 24. $\dfrac{4}{5} + \dfrac{8}{15}$

25. $\dfrac{3}{10} + \dfrac{8}{15}$ 26. $\dfrac{9}{8} + \dfrac{7}{12}$ 27. $\dfrac{5}{4} - \dfrac{3}{4}$ 28. $\dfrac{12}{5} - \dfrac{2}{5}$

29. $\dfrac{13}{18} - \dfrac{4}{9}$ 30. $\dfrac{13}{15} - \dfrac{8}{45}$ 31. $\dfrac{11}{12} - \dfrac{2}{5}$ 32. $\dfrac{15}{16} - \dfrac{2}{3}$

 SKILL MAINTENANCE

33. Translate to an algebraic expression: the sum of p and q.

34. Evaluate $\dfrac{y}{x}$ when $x = 72$ and $y = 9$.

35. Find the prime factorization of 48.

36. Find the LCM of 12, 24, and 56.

⭐ EXTENSION

Simplify.

37. $\dfrac{3x}{4x}$ 38. $\dfrac{xy}{5y}$ 39. $\dfrac{4 \cdot 9 \cdot 16}{2 \cdot 8 \cdot 15}$ 40. $\dfrac{pqrs}{qrst}$

Decimal and Exponential Notation

DECIMAL NOTATION

Another notation for the numbers of arithmetic is *decimal notation*. It is based on ten. Standard notation, such as 9345 and 20,867, is decimal notation for whole numbers. Decimal notation for fractions, such as 0.9 and 34.248, contains *decimal points*.

• FROM DECIMAL TO FRACTIONAL NOTATION

To convert from decimal to fractional notation, we can multiply by 1, as shown in the following examples.

Examples Write fractional notation.

1. $34.2 = \dfrac{34.2}{1} \cdot \dfrac{10}{10}$ Multiplying by 1

     $= \dfrac{342}{10}$ 34.2 is 34 and two tenths, or 342 tenths.

> Why did we multiply by $\frac{10}{10}$? We did not want a decimal point in the answer, and we knew that multiplying 34.2 by 10 would give us a whole number, 342. Thus we multiplied by $\frac{10}{10}$ to rename the number.

2. $16.453 = \dfrac{16.453}{1} \cdot \dfrac{1000}{1000}$ Multiplying by 1

     $= \dfrac{16,453}{1000}$ 16.453 is 16,453 thousandths.

DO EXERCISES 1 AND 2.

•• FROM FRACTIONAL TO DECIMAL NOTATION

To convert from fractional to decimal notation we divide.

Example 3 $\frac{3}{8}$ means $3 \div 8$, so we divide.

$$
\begin{array}{r}
0.375 \\
8\overline{)3.000} \\
\underline{2\ 400} \\
600 \\
\underline{560} \\
40 \\
\underline{40}
\end{array}
$$

Write fractional notation. You need not simplify.

1. 1.62

2. 35.431

ANSWERS ON PAGE A-1

Sometimes we get a repeating decimal when we divide.

Example 4 $\frac{4}{11}$ means $4 \div 11$, so we divide.

$$
\begin{array}{r}
0.3636\ldots \\
11\overline{)\ 4\ .00} \\
\underline{3\ 3} \\
70 \\
\underline{66} \\
4\ 0 \\
\underline{3\ 3} \\
70 \\
\underline{66} \\
4
\end{array}
$$

Thus $\frac{4}{11} = 0.363636\ldots$. Such decimals are often abbreviated by putting a bar over the repeating part, as follows.

$$\frac{4}{11} = 0.\overline{36}$$

DO EXERCISES 3–6.

••• EXPONENTIAL NOTATION

Shorthand notation for $10 \cdot 10 \cdot 10$ is called *exponential notation*.

For $\underline{10 \cdot 10 \cdot 10}$ we write 10^3.

3 factors

This is read "ten cubed" or "ten to the third power." We call the number 3 an *exponent* and we say that 10 is the *base*. For $10 \cdot 10$ we write 10^2, read "ten squared," or "ten to the second power."

Example 5 Write exponential notation for $10 \cdot 10 \cdot 10 \cdot 10 \cdot 10$.

$$10 \cdot 10 \cdot 10 \cdot 10 \cdot 10 = 10^5$$

DO EXERCISE 7.

An exponent of 2 or greater tells us how many times the base is used as a factor.

Examples

6. 3^5 means $3 \cdot 3 \cdot 3 \cdot 3 \cdot 3$.

7. 7^4 means $7 \cdot 7 \cdot 7 \cdot 7$.

8. If we use n to stand for a number, n^4 means $n \cdot n \cdot n \cdot n$, or $nnnn$.

DO EXERCISES 8 AND 9.

Write decimal notation. Use a bar for repeating decimals.

3. $\dfrac{7}{8}$

4. $\dfrac{4}{5}$

5. $\dfrac{9}{11}$

6. $\dfrac{23}{9}$

7. Write exponential notation for $10 \cdot 10 \cdot 10 \cdot 10$.

8. What is the meaning of 5^4?

9. What is the meaning of x^5?

Chapter 1   Introduction to Algebra; Review of Arithmetic

:: 1 AND 0 AS EXPONENTS

Look for a pattern.

$$10 \cdot 10 \cdot 10 \cdot 10 = 10^4$$
$$10 \cdot 10 \cdot 10 = 10^3$$
$$10 \cdot 10 = 10^2$$
$$10 = 10^?$$
$$1 = 10^?$$

We are dividing by 10 each time. To continue the pattern we would say that

$$10 = 10^1 \quad \text{and} \quad 1 = 10^0.$$

We make the following agreement. It is a definition.

> For any number n,
>
> $$n^1 \quad \text{means} \quad n,$$
>
> and if n is not 0, then
>
> $$n^0 \quad \text{means} \quad 1.$$

If n is 0, then n^0 is meaningless. We explain why later.

Examples What is the meaning of each of the following?

9. 5^1 *Answer*: 5   10. 8^1 *Answer*: 8

11. 3^0 *Answer*: 1   12. 5^0 *Answer*: 1

DO EXERCISES 10 AND 11.

In summary, note that there are many kinds of names, or notations, for a number. For example,

      0.64   (decimal notation),
      $\frac{16}{25}$ (fractional notation),

and

    0.8^2, or 0.8×0.8 (exponential notation)

are all names for the *same* number. Later, we review *percent notation*, which gives us another name, 64%.

:: EVALUATING EXPRESSIONS CONTAINING EXPONENTIAL NOTATION

We now consider evaluating expressions containing exponential notation.

Example 13 Evaluate x^4 for $x = 2$.

We substitute 2 for x:

$$x^4 = 2^4 \quad \text{Substituting}$$
$$= 2 \cdot 2 \cdot 2 \cdot 2$$
$$= 16.$$

10. What is 4^1?

11. What is 6^0?

ANSWERS ON PAGE A-1

12. Evaluate t^3 for $t = 5$.

13. Find the area of a circle when $r = 22$ cm. Use 3.14 for π.

14. Evaluate $200 - n^4$ for $n = 3$.

15. a) Evaluate $5y^3$ for $y = 2$.

b) Evaluate $(5y)^3$ for $y = 2$.

c) Determine whether $5y^3$ and $(5y)^3$ are equivalent.

ANSWERS ON PAGE A-2

Example 14 The area of a circle is given by $A = \pi r^2$, where r is the radius. Find the area of a circle with a radius of 10 cm. Use 3.14 for π.

$$\begin{aligned} A = \pi r^2 &\approx 3.14 \times (10 \text{ cm})^2 \\ &= 3.14 \times 10 \text{ cm} \times 10 \text{ cm} \\ &= 3.14 \times 100 \text{ cm}^2 \\ &= 314 \text{ cm}^2 \end{aligned}$$

Here "cm^2" means "square centimeters," and "$\approx$" means "approximately equal to."

DO EXERCISES 12 AND 13.

Example 15 Evaluate $m^3 + 5$ for $m = 4$.

We agree to evaluate $m^3 + 5$ by evaluating m^3 first and then adding 5:

$$\begin{aligned} m^3 + 5 &= 4^3 + 5 \qquad \text{Substituting} \\ &= (4 \cdot 4 \cdot 4) + 5 = 64 + 5 \\ &= 69. \end{aligned}$$

DO EXERCISE 14.

Example 16 Evaluate $3x^2$ and $(3x)^2$ for $x = 5$.

We agree to evaluate $3x^2$ by evaluating x^2 first and then multiplying by 3:

$$\begin{aligned} 3x^2 &= 3 \cdot 5^2 \qquad \text{Substituting} \\ &= 3 \cdot 25 \\ &= 75. \end{aligned}$$

To evaluate $(3x)^2$ we agree to do what is in parentheses first. Then we raise the answer to the power 2.

$$\begin{aligned} (3x)^2 &= (3 \cdot 5)^2 \qquad \text{Substituting} \\ &= (15)^2 \\ &= 225 \end{aligned}$$

Note in Example 16 that the expressions do not have the same values.

> **Expressions that have the same values for all replacements are called *equivalent expressions*.**

The expressions $3x^2$ and $(3x)^2$ are *not* equivalent. The expressions y^4 and $yyyy$ are equivalent.

DO EXERCISE 15.

ANSWERS ON PAGE A-2

Chapter 1 Introduction to Algebra; Review of Arithmetic

⬛ Write fractional notation. You need not simplify.

1. 29.1 **2.** 16.33 **3.** 4.67 **4.** 3.1415

5. 3.62 **6.** 13.617 **7.** 18.789 **8.** 309.62

⬛⬛ Write decimal notation.

9. $\dfrac{1}{2}$ **10.** $\dfrac{1}{4}$ **11.** $\dfrac{3}{5}$ **12.** $\dfrac{6}{5}$

13. $\dfrac{2}{9}$ **14.** $\dfrac{4}{9}$ **15.** $\dfrac{1}{8}$ **16.** $\dfrac{5}{8}$

17. $\dfrac{5}{11}$ **18.** $\dfrac{7}{11}$ **19.** $\dfrac{1}{12}$ **20.** $\dfrac{23}{12}$

⬛⬛⬛ What is the meaning of each of the following?

21. 5^2 **22.** 8^3 **23.** m^3 **24.** t^2

What is the meaning of each of the following? Do not use $\times$ or $\cdot$.

25. x^3 **26.** m^2 **27.** y^4 **28.** p^5

⬛⬛ What is the meaning of each of the following?

29. $y^0,\ y \neq 0$ **30.** $h^0,\ h \neq 0$ **31.** p^1 **32.** z^1

33. M^1 **34.** 9.68^0 **35.** $\left(\dfrac{a}{b}\right)^0,\ a \neq 0,$ **36.** $(ab)^1$
$b \neq 0$

⬛⬛ Evaluate each expression.

37. m^3 for $m = 3$ **38.** x^6 for $x = 2$ **39.** p^1 for $p = 19$

ANSWERS
1.
2.
3.
4.
5.
6.
7.
8.
9.
10.
11.
12.
13.
14.
15.
16.
17.
18.
19.
20.
21.
22.
23.
24.
25.
26.
27.
28.
29.
30.
31.
32.
33.
34.
35.
36.
37.
38.
39.

40. x^{19} for $x = 0$ **41.** x^4 for $x = 4$ **42.** y^{15} for $y = 1$

43. n^0 for $n = 5$ **44.** $y^2 - 7$ for $y = 10$ **45.** z^5 for $z = 2$

46. Find the area of a circle when $r = 34$ ft. Use 3.14 for π.

47. The area of a square with sides of length s is given by $A = s^2$. Find the area of a square with sides of length 24 m (meters).

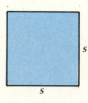

48. Evaluate Pr^2 when $P = 1000$ and $r = 0.16$.

49. Evaluate $(2y)^4$ and $2y^4$ when $y = 2$.

50. Evaluate $5p^2$ and $(5p)^2$ when $p = 4$.

 SKILL MAINTENANCE

51. Translate to an algebraic expression: x minus y.

52. Add and simplify: $\dfrac{11}{12} + \dfrac{15}{16}$.

53. Find the LCM of 18, 27, and 54.

54. Subtract and simplify: $\dfrac{7}{8} - \dfrac{2}{3}$.

⭐ **EXTENSION**

55. Evaluate x^3y^2z for $x = 2$, $y = 1$, and $z = 3$.

56. Evaluate $x^2 + 2xy + y^2$ for $x = 7$ and $y = 8$.

57. Write exponential notation for $xxxyyyy$.

58. Find $x^{149}y$ for $x = 0$ and $y = 13$.

Properties of Numbers of Arithmetic

Some properties of the numbers of arithmetic are so simple that they may seem unimportant. However, they are important, especially in algebra, because they allow us to manipulate algebraic expressions. Such manipulation may allow us to find a simpler expression after we have translated a problem to an equation.

• PARENTHESES. SYMBOLS OF GROUPING

What does $5 \times 2 + 4$ mean? If we multiply 5 by 2 and add 4, we get 14. If we add 2 and 4 and multiply by 5, we get 30. To tell which operation to do first, we use parentheses. For example,

$$(3 \times 5) + 6 \text{ means } 15 + 6, \text{ or } 21;$$
$$3 \times (5 + 6) \text{ means } 3 \times 11, \text{ or } 33.$$

DO EXERCISES 1–5.

•• GROUPING AND ORDER

What does $3 + 5 + 4$ mean? Does it mean $(3 + 5) + 4$ or $3 + (5 + 4)$? Either way the sum is 12, so it doesn't matter. In fact, if we are doing addition only, we can group numbers in any manner. This means that we really don't need parentheses if we are doing only addition.

Another basic property of numbers is that they can be added or multiplied in any order. For example, $3 + 2$ and $2 + 3$ are the same. Also, $5 \cdot 7$ and $7 \cdot 5$ are the same. The properties we have illustrated are the following.

For any numbers a, b, and c,

$(a + b) + c = a + (b + c)$.   (The associative law of addition)
$a + b = b + a$;   (The commutative law of addition)
$a \cdot b = b \cdot a$.   (The commutative law of multiplication)

DO EXERCISES 6–9.

The commutative and associative laws together help make addition easier.

Example 1 Add $3 + 4 + 7 + 6 + 8$. Look for combinations that make ten.

Add 3 and 7 to make 10 and 4 and 6 to make 10, and then add the two 10's. Finally add the 8. The sum is 28.

DO EXERCISES 10 AND 11.

Do these calculations.

1. $(5 \times 4) + 2$

2. $5 \times (4 + 2)$

3. $(4 \times 6) + 2$

4. $5 \times (2 \times 3)$

5. $(6 \times 2) + (3 \times 5)$

Do these calculations.

6. $17 + 10$

7. $10 + 17$

8. 26×70

9. 70×26

Add. Look for combinations that make ten.

10. $5 + 2 + 3 + 5 + 8$

11. $1 + 5 + 6 + 9 + 4$

ANSWERS ON PAGE A-2

Which laws are illustrated by these sentences?

12. $61 \times 56 = 56 \times 61$

13. $(3 + 5) + 2 = 3 + (5 + 2)$

14. $4 + (2 + 5) = (4 + 2) + 5$

15. $7 \cdot (9 \cdot 8) = (7 \cdot 9) \cdot 8$

Do the calculations as shown.

16. a) $(2 + 5) \cdot 4$

b) $(2 \cdot 4) + (5 \cdot 4)$

17. a) $(7 + 4) \cdot 7$

b) $(7 \cdot 7) + (4 \cdot 7)$

Compute.

18. $3 \cdot 5 + 2 \cdot 4$

19. $4 \cdot 2 + 7 \cdot 1$

ANSWERS ON PAGE A-2

For multiplication, does the grouping matter? For example, is $2 \cdot (5 \cdot 3) = (2 \cdot 5) \cdot 3$? Our experience with arithmetic tells us that multiplication is also associative.

> For any numbers a, b, and c,
>
> $\quad a \cdot (b \cdot c) = (a \cdot b) \cdot c$.　　(The associative law of multiplication)

Examples Which laws are illustrated by these sentences?

2. $3 + 5 = 5 + 3$ 　Commutative law of addition (order changed)

3. $(2 + 3) + 5 = 2 + (3 + 5)$ 　Associative law of addition (grouping changed)

4. $(3 \cdot 5) \cdot 2 = 3 \cdot (5 \cdot 2)$ 　Associative law of multiplication

Note that the commutative and associative laws tell us that certain expressions are equivalent. For example, $x + 2$ and $2 + x$ are equivalent, as are $4 \cdot (b \cdot c)$ and $(4 \cdot b) \cdot c$.

DO EXERCISES 12–15.

••• THE DISTRIBUTIVE LAW

If we wish to multiply a number by a sum of several numbers, we can either add and then multiply or multiply and then add.

Example 5 Compute in two ways: $(4 + 8) \cdot 5$.

$$\left.\begin{array}{r} (4 + 8) \cdot 5 \\ 12 \cdot 5 \\ 60 \end{array}\right\} \quad \text{Adding and then multiplying}$$

$$\left.\begin{array}{r} (4 \cdot 5) + (8 \cdot 5) \\ 20 + 40 \\ 60 \end{array}\right\} \quad \text{Multiplying and then adding}$$

DO EXERCISES 16 AND 17.

The property we are investigating is the *distributive law of multiplication over addition*. Before we state it formally, we need to make an agreement about parentheses. We agree that in an expression like $(4 \cdot 5) + (3 \cdot 7)$, we can omit the parentheses. Thus $4 \cdot 5 + 3 \cdot 7$ means $(4 \cdot 5) + (3 \cdot 7)$. In other words, we do the multiplications first.

DO EXERCISES 18 AND 19.

> In an expression such as $ab + cd$, it is understood that parentheses belong around ab and cd. In other words, the multiplications are to be done first.

Using our agreement about parentheses, we now state the distributive law.

For any numbers a, b, and c,

$$a(b + c) = ab + ac. \qquad \text{(The distributive law)}$$

Note that there are *two* operations involved in the distributive law: addition and multiplication.

We cannot omit parentheses on the left above. If we did we would have $ab + c$, which by our agreement means $(ab) + c$.

Note that the distributive law can be extended to more than two numbers inside the parentheses:

$$a(b + c + d) = ab + ac + ad.$$

The distributive law can also be applied on the right. That is, $(b + c)a = ba + ca$. For example, $(2 + 3)x = 2x + 3x$.

The distributive law would apply to the following situation. Someone decides to invest $1000 in one bank at 8% simple interest and $2000 in another bank at 8% simple interest. At the end of one year the total interest from the two investments would be

$$(8\% \cdot 1000) + (8\% \cdot 2000).$$

The same interest would also have been made by investing the entire $3000 in just one bank. The interest is

$$8\% \cdot (1000 + 2000), \quad \text{or} \quad 8\% \cdot 3000.$$

DO EXERCISE 20.

▮▮ FACTORING

Any equation can be reversed. Thus for the distributive law we could also write $ab + ac = a(b + c)$. The distributive law is the basis for *factoring* algebraic expressions.

To *factor* an expression is to find an equivalent expression that is a product.

Example 6 Factor: $3x + 3y$.

By the distributive law,

$$3x + 3y = 3(x + y).$$

When we write $3(x + y)$, we say that we have *factored* $3x + 3y$. That is, we have an equivalent expression that is a product.

Example 7 Factor: $5x + 5y + 5z$.

$$5x + 5y + 5z = 5(x + y + z)$$

Note that we manipulated the expression on the left according to the distributive law. We did not have to know what the unknowns x, y, and z were.

DO EXERCISES 21–23.

Do these calculations.

20. a) $(0.08 \times 1000) + (0.08 \times 2000)$

b) $0.08 \times (1000 + 2000)$

Factor.

21. $4x + 4y$

22. $5a + 5b$

23. $7p + 7q + 7r$

ANSWERS ON PAGE A-2

24. A standard tennis court has a length of 78 ft and a width of 36 ft. Find the perimeter two ways using the two expressions for perimeter given in Example 10.

Factor. Then evaluate both the original and factored expressions when $x = 4$ and $y = 3$.

25. $5x + 5y$

26. $7x + 7y$

:·: FACTORING AND EVALUATING

It is important to realize that when we factor an expression like $2l + 2w$, the factored expression is equivalent to the original one.

Example 8 The *perimeter* of a rectangle is the distance around it. The perimeter of a rectangle can be found by adding the lengths of the four sides, or by doubling the length and width and then adding. The perimeter is then given by the algebraic expression

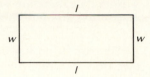

$$2l + 2w,$$

where l is the length and w is the width.

Factor $2l + 2w$. Then evaluate both the original and factored expressions when l is 360 ft and w is 160 ft. This will give the perimeter of a standard-sized football field in the United States.

First, factor $2l + 2w$: $2l + 2w = 2(l + w)$.
Evaluate $2l + 2w$:

$$2l + 2w = 2 \cdot 360 + 2 \cdot 160 \qquad \text{Substituting 360 for } l \text{ and 160 for } w$$
$$= 720 + 320 = 1040.$$

Evaluate $2(l + w)$:

$$2(l + w) = 2(360 + 160) = 2 \cdot 520 = 1040.$$

The perimeter of the football field is 1040 ft. Note that for all replacements of l and w by arithmetic numbers, $2l + 2w = 2(l + w)$.

DO EXERCISE 24.

Example 9 Factor $6x + 6y$. Then evaluate both the original and factored expressions when $x = 3$ and $y = 8$.

a) $6x + 6y = 6(x + y)$ Factoring

b) $6x + 6y = 6 \cdot 3 + 6 \cdot 8$ Evaluating $6x + 6y$
$$= 18 + 48 = 66$$

c) $6(x + y) = 6(3 + 8)$ Evaluating $6(x + y)$
$$= 6 \cdot 11 = 66$$

DO EXERCISES 25 AND 26.

In summary, we list the five properties considered in this section:

> **The associative laws:** For any numbers a, b, and c,
> $$a + (b + c) = (a + b) + c, \qquad a(bc) = (ab)c.$$
> **The commutative laws:** For any numbers a and b,
> $$a + b = b + a, \qquad ab = ba.$$
> **The distributive law:** For any numbers a, b, and c,
> $$a(b + c) = ab + ac, \quad \text{or} \quad (b + c)a = ba + ca.$$

• Do these calculations.

1. $(10 + 4) + 8$

2. $10 \times (9 + 4)$

3. $(10 \cdot 7) + 19$

4. $(10 \cdot 7) + (20 \cdot 14)$

•• Do these calculations. Choose grouping and ordering to make the work easy.

5. $8 + 4 + 5 + 2 + 6 + 15 + 1$

6. $9 + 6 + 3 + 4 + 1 + 7 + 11$

7. $14 + 3 + 12 + 7 + 8 + 6 + 9$

8. $17 + 7 + 16 + 3 + 4 + 3 + 8$

Which laws are illustrated by these sentences?

9. $67 + 3 = 3 + 67$

10. $15 \cdot 44 = 44 \cdot 15$

11. $6 + (9 + 5) = (6 + 9) + 5$

12. $8 \cdot (7 \cdot 6) = (8 \cdot 7) \cdot 6$

••• Compute in two ways.

13. $(6 + 7) \cdot 4$

14. $(8 + 10) \cdot 2$

⦂⦂ Factor.

15. $9x + 9y$

16. $7w + 7u$

17. $\frac{1}{2}a + \frac{1}{2}b$

18. $\frac{3}{4}x + \frac{3}{4}y$

19. $1.5x + 1.5z$

20. $0.7a + 0.7b$

ANSWERS
1.
2.
3.
4.
5.
6.
7.
8.
9.
10.
11.
12.
13.
14.
15.
16.
17.
18.
19.
20.

21. $4x + 4y + 4z$

22. $10a + 10b + 10c$

23. $\frac{4}{7}a + \frac{4}{7}b + \frac{4}{7}c + \frac{4}{7}d$

24. $\frac{3}{5}x + \frac{3}{5}y + \frac{3}{5}z + \frac{3}{5}w$

 Factor. Then evaluate both expressions when $x = 5$ and $y = 10$.

25. $9x + 9y$ 26. $8x + 8y$ 27. $10x + 10y$ 28. $2x + 2y$

Factor. Then evaluate both expressions when $a = 0$ and $b = 9$.

29. $5a + 5b$ 30. $7a + 7b$ 31. $20a + 20b$ 32. $14a + 14b$

33. A standard badminton court is 40 ft by 20 ft. Find the perimeter of a badminton court using the two expressions for perimeter.

34. A batter's box in baseball is 6 ft by 4 ft. Find the perimeter of a batter's box using the two expressions for perimeter.

 SKILL MAINTENANCE

35. Evaluate $2x^3$ and $(2x)^3$ when $x = 2$.

36. Subtract and simplify: $\frac{11}{12} - \frac{3}{8}$.

37. Find the prime factorization of 96.

38. A number t is 7 less than a larger number. Write an expression for the larger number.

⭐ **EXTENSION**

39. Which of the following expressions represent the same number?

   a) $3 \cdot 4 + 2$

   b) $3(4 + 2)$

   c) $3 \cdot 4 + 3 \cdot 2$

   d) $3(2 + 4)$

40. Name the *two* laws illustrated by each sentence.

   a) $(2 + 3)x = x(2 + 3) = 2x + 3x$

   b) $5 \cdot (4 \cdot a) = (4 \cdot 5) \cdot a$

41. Which of the following are equivalent?

   a) $5x + 5y$

   b) $5(x + y)$

   c) $5y + 5x$

   d) $5x + y$

42. Which of the following are equivalent?

   a) $7y + 9y$

   b) $9y + 7y$

   c) $(7 + 9)y$

   d) $16y$

Using the Distributive Law

The distributive law is the basis of many procedures in both arithmetic and algebra. Below are some examples of other procedures based on this property and some further examples of factoring.

After finishing Section 1.6, you should be able to:

: Use the distributive law to multiply expressions like $5(x + 3)$.

: : Factor expressions like $5x + 10$ by using the distributive law.

: : : Collect like terms in expressions such as $3x + 4y + 5x + 3y$.

• MULTIPLYING

In the expression $x + y + z$, the parts separated by plus signs are called *terms*. Thus x, y, and z are terms in $x + y + z$. The distributive law is the basis of a procedure called "multiplying." For example, consider

$$8(a + b).$$

Using the distributive law, we multiply each term of $(a + b)$ by 8:

$$8 \cdot (a + b) = 8 \cdot a + 8 \cdot b.$$

Examples Multiply.

1. $3(x + 2) = 3 \cdot x + 3 \cdot 2$ Using the distributive law
$= 3x + 6$

2. $6(s + 2t + 5w) = 6 \cdot s + 6 \cdot 2t + 6 \cdot 5w$ Using the distributive law
$= 6s + 12t + 30w$

We multiplied each term inside the parentheses by the factor outside.

DO EXERCISES 1–3.

• • FACTORING

In the expression $6x + 3y + 9z$ the terms are $6x$, $3y$, and $9z$. In this case, the terms are products. To factor, look for a factor common to all the terms. Then "remove" it, so to speak, using the distributive law.

Example 3 Factor: $6x + 3y + 9z$.

$6x + 3y + 9z = 3 \cdot 2x + 3 \cdot y + 3 \cdot 3z$ The common factor is 3.
$= 3(2x + y + 3z)$ Using the distributive law

Note that factoring is the reverse of multiplying. In fact, you can check factoring by multiplying:

$$3(2x + y + 3z) = 3(2x) + 3y + 3(3z) = 6x + 3y + 9z.$$

Example 4 Factor: $7y + 21z + 7$.

$7y + 21z + 7 = 7 \cdot y + 7 \cdot 3z + 7 \cdot 1$ The common factor is 7.
$= 7(y + 3z + 1)$

Caution! Be sure not to omit the common factor 7.

Caution! Be sure not to omit the 1.

DO EXERCISES 4–7.

Multiply.

1. $5(y + 3)$

2. $4(x + 2y + 5)$

3. $8(m + 3n + 4p)$

Factor.

4. $5x + 10$

5. $12 + 3x$

6. $6x + 12 + 9y$

7. $5x + 10y + 25$

ANSWERS ON PAGE A-2

8. Factor: $Q + Qab$.

Collect like terms.

9. $6y + 2y$

10. $4x + x$

11. $x + 0.03x$

12. $10p + 8p + 4q + 5q$

13. $7x + 3y + 4x + 5y$

Collect like terms.

14. $4y + 12y$

15. $3s + 4s + 6w + 7w$

16. $5x + 4y + 4x + 6y$

17. $5a + b + a + 0.07b$

Example 5 *Application: Simple Interest.* Simple interest on a principal of P dollars invested at interest rate r for t years is given by the expression Prt. In t years, principal P will grow to the amount

$$(\text{Principal}) + (\text{Interest}) = P + Prt.$$

Factor this expression.

$$P + Prt = P \cdot 1 + Prt = P(1 + rt)$$

DO EXERCISE 8.

••• COLLECTING LIKE TERMS

If two terms have the same letters, they are called *like* terms, or *similar* terms.* We can often simplify expressions by *collecting* or *combining like terms.* (We could also say *collecting* or *combining similar terms.*)

Examples Collect like terms.

6. $3x + 4x = (3 + 4)x$ Using the distributive law
$$= 7x$$

7. $2x + 3y + 5x + 8y = 2x + 5x + 3\;y\; + 8\;y$ Regrouping and reordering using the associative and commutative laws

$$= (2 + 5)x + (3 + 8)\;y \quad \text{Factoring}$$
$$= 7x + 11y$$

8. $7x + x = 7 \cdot x + 1 \cdot x = (7 + 1)x = 8x$

9. $x + 0.05x = 1 \cdot x + 0.05\;x = (1 + 0.05)\;x = 1.05x$

DO EXERCISES 9–13.

With practice we can leave out some steps, collecting like terms mentally, Note that constants like 4 and 7 are considered like terms.

Examples Collect like terms.

10. $5y + 2y + 4y = 11y$

11. $3x + 7x + 2y = 10x + 2y$

12. $3a + 5a + 8t + 2t = 8a + 10t$

13. $8p + q + p + 0.3q = 9p + 1.3q$

14. $4 + x + 7 = x + 11$

15. $3x + 25 + 7y + 8x + 11 = 11x + 7y + 36$

DO EXERCISES 14–17.

*Later we will amend this definition.

· Multiply.

1. $3(x + 1)$

2. $2(x + 2)$

3. $4(1 + y)$

4. $9(s + 1)$

5. $9(4t + 3z)$

6. $8(5x + 3y)$

7. $7(x + 4 + 6y)$

8. $8(9x + 5y + 8)$

9. $5(3x + 9 + 7y)$

10. $4(5x + 8 + 3z)$

· · Factor. Check by multiplying.

11. $2x + 4$

12. $9x + 27$

13. $6x + 24$

14. $5y + 20$

15. $9x + 3y$

16. $15x + 5y$

17. $14x + 21y$

18. $18x + 24y$

19. $5 + 10x + 15y$

20. $7 + 14b + 56w$

21. $8a + 16b + 64$

22. $9x + 27y + 81$

23. $3x + 18y + 15z$

24. $4r + 28s + 16t$

ANSWERS
1.
2.
3.
4.
5.
6.
7.
8.
9.
10.
11.
12.
13.
14.
15.
16.
17.
18.
19.
20.
21.
22.
23.
24.

●●● Collect like terms.

25. $2x + 3 + 3x + 9$

26. $7y + 9 + 8 + 20y$

27. $10a + a$

28. $16x + x$

29. $2x + 9z + 6x$

30. $3a + 5b + 7a$

31. $41a + 90c + 60c + 2a$

32. $42x + 6b + 4x + 2b$

33. $x + 0.09x + 0.2t + t$

34. $0.01a + 0.23b + a + b$

35. $8u + 3t + 10u + 6u + 2t$

36. $5t + 6h + t + 8t + 9h$

37. $23 + 5t + 7y + t + y + 27$

38. $45 + 90d + 87 + 9d + 3 + 7d$

39. $\frac{1}{2}b + \frac{2}{3} + \frac{1}{2}b + \frac{2}{3}$

40. $\frac{2}{3}x + \frac{5}{8} + \frac{1}{3}x + \frac{3}{8}$

41. $2y + \frac{1}{4}y + y$

42. $\frac{1}{2}a + a + 5a$

✓ **SKILL MAINTENANCE**

43. Multiply and simplify: $\frac{15}{16} \cdot \frac{8}{9}$.

44. Find the area of a rectangle when the length l is 24 yd and the width w is 35 yd.

 EXTENSION

Collect like terms and then factor.

45. $3x + 10y + 2x + 5y + 25$

46. $\frac{1}{3}x + \frac{1}{4} + \frac{1}{6}y + \frac{1}{4}x + \frac{1}{6} + \frac{3}{4}y$

1.7

The Number 1 and Reciprocals

The number 1 has some very special properties important in both arithmetic and algebra. When we multiply any number by 1, we get that same number.

For any number n,

$$n \cdot 1 = n.$$

The number 1 is also called the *multiplicative identity*.

When we divide a number by 1, we get the same number with which we started. Since a/b means $a \div b$, $n/1 = n \div 1 = n$.

For any number n,

$$\frac{n}{1} = n.$$

When we divide a number by itself, the result is the number 1. This is true for any number except zero. We will see later why we do not divide by zero.

For any number n except zero,

$$\frac{n}{n} = 1.$$

Thus the expression $\dfrac{n}{n}$ is equivalent to 1.

Simplify.

1. $\dfrac{4}{7} \cdot \dfrac{11}{11}$

2. $\dfrac{67}{67}$

3. $\dfrac{\frac{2}{3}}{1}$

4. $\dfrac{\frac{7}{5}}{\frac{7}{5}}$

Examples Simplify.

1. $\frac{3}{5} \cdot \frac{7}{7} = \frac{3}{5} \cdot 1 = \frac{3}{5}$ 2. $\frac{\frac{3}{5}}{1} = \frac{3}{5}$ 3. $\frac{\frac{4}{3}}{\frac{4}{3}} = 1$

DO EXERCISES 1–4.

• RECIPROCALS

Two numbers whose product is 1 are called *reciprocals* of each other. All the numbers of arithmetic, except zero, have reciprocals.

Examples

4. The reciprocal of $\frac{2}{3}$ is $\frac{3}{2}$ because $\frac{2}{3} \cdot \frac{3}{2} = \frac{6}{6} = 1$.

5. The reciprocal of 9 is $\frac{1}{9}$ because $9 \cdot \frac{1}{9} = \frac{9}{9} = 1$.

6. The reciprocal of $\frac{1}{4}$ is 4 because $\frac{1}{4} \cdot 4 = 1$.

DO EXERCISES 5–8.

Find the reciprocal.

5. $\dfrac{4}{11}$

6. $\dfrac{15}{7}$

7. 5

8. $\dfrac{1}{3}$

ANSWERS ON PAGE A-2

Divide by multiplying by 1.

9. $\dfrac{\dfrac{3}{5}}{\dfrac{4}{7}}$

10. $\dfrac{\dfrac{5}{11}}{\dfrac{3}{2}}$

11. $\dfrac{\dfrac{9}{7}}{\dfrac{4}{5}}$

Divide by multiplying by the reciprocal of the divisor.

12. $\dfrac{4}{3} \div \dfrac{7}{2}$

13. $\dfrac{5}{4} \div \dfrac{3}{2}$

14. $\dfrac{\dfrac{2}{9}}{\dfrac{5}{7}}$

•• RECIPROCALS AND DIVISION

The number 1 and reciprocals can be used to explain division of numbers of arithmetic. To divide, we can multiply by 1, choosing carefully the symbol for 1.

Example 7 Divide $\frac{2}{3}$ by $\frac{7}{5}$.

$$\frac{\frac{2}{3}}{\frac{7}{5}} = \frac{\frac{2}{3}}{\frac{7}{5}} \times \frac{\frac{5}{7}}{\frac{5}{7}} \qquad \text{Multiplying by } \frac{\frac{5}{7}}{\frac{5}{7}}$$

We use $\frac{5}{7}$ because it is the reciprocal of $\frac{7}{5}$.

$$= \frac{\frac{2}{3} \times \frac{5}{7}}{\frac{7}{5} \times \frac{5}{7}} \qquad \text{Multiplying numerators and denominators}$$

$$= \frac{\frac{2}{3} \times \frac{5}{7}}{\frac{35}{35}} = \frac{\frac{2}{3} \times \frac{5}{7}}{1} = \frac{2}{3} \times \frac{5}{7}$$

$$= \frac{10}{21}$$

After multiplying we got 1 for a denominator. The numerator (in color) shows multiplication by the reciprocal.

DO EXERCISES 9–11.

When multiplying by 1 to divide, we get a denominator of 1. What do we get in the numerator? In Example 7, we got $\frac{2}{3} \times \frac{5}{7}$. This is the product of $\frac{2}{3}$, the dividend, and $\frac{5}{7}$, the reciprocal of the divisor.

> **To divide, multiply by the reciprocal of the divisor:**
> $$\frac{a}{b} \div \frac{c}{d} = \frac{a}{b} \cdot \frac{d}{c}.$$

Example 8 Divide by multiplying by the reciprocal of the divisor.

$$\frac{1}{2} \div \frac{3}{5} = \frac{1}{2} \cdot \frac{5}{3} = \frac{5}{6} \qquad \frac{5}{3} \text{ is the reciprocal of } \frac{3}{5}$$

After dividing, simplification is often possible and should be done.

Example 9 Divide.

$$\frac{2}{3} \div \frac{4}{9} = \frac{2}{3} \cdot \frac{9}{4} = \frac{18}{12} = \frac{3 \cdot 6}{2 \cdot 6} = \frac{3}{2} \cdot \frac{6}{6} = \frac{3}{2}$$

Simplifying

DO EXERCISES 12–14.

••• THE NUMBER LINE AND ORDER

The order of numbers of arithmetic can be shown on a number line.

DO EXERCISES 15 AND 16.

Note that $\frac{1}{2}$ is less than 1, and $\frac{1}{2}$ is to the left of 1 on the number line. To say that $\frac{1}{2}$ is less than 1, we write $\frac{1}{2} < 1$.

For any numbers a and b,

$$a < b \quad \text{(read "a is less than b")}$$

means that a is to the left of b on the number line.

The number $\frac{5}{2}$ is to the right of $\frac{1}{2}$. This means that $\frac{5}{2}$ is greater than $\frac{1}{2}$. To say that $\frac{5}{2}$ is greater than $\frac{1}{2}$, we write $\frac{5}{2} > \frac{1}{2}$.

For any numbers a and b,

$$a > b \quad \text{(read "a is greater than b")}$$

means that a is to the right of b on the number line.

Sentences such as $\frac{5}{2} > \frac{1}{2}$ and $x < 2$ are called *inequalities*.

Examples Determine whether true or false. Use the number line above.

10. $1.5 > 1$; true   1.5 is to the right of 1

11. $\frac{5}{2} < \frac{3}{4}$; false $\frac{5}{2}$ is not to the left of $\frac{3}{4}$

12. $\frac{1}{2} < 6$; true $\frac{1}{2}$ is to the left of 6

DO EXERCISES 17–19.

⠿ COMPARING NUMBERS

We want to develop a more efficient way of comparing numbers of arithmetic. Consider $\frac{4}{5}$ and $\frac{3}{5}$. The number line shows that $\frac{4}{5} > \frac{3}{5}$.

Note also that $4 > 3$. When denominators are the same, we just compare numerators.

Graph the number on a number line.

15. $\dfrac{6}{5}$

16. $\dfrac{17}{18}$

Determine whether true or false.

17. $\dfrac{1}{2} < \dfrac{3}{4}$

18. $\dfrac{3}{4} > \dfrac{1}{2}$

19. $\dfrac{13}{4} < \dfrac{5}{2}$

ANSWERS ON PAGE A-2

Use the proper symbol, $>$, $<$, or $=$, between the pair of numerals.

20. $\dfrac{3}{4}$ $\dfrac{4}{4}$

21. $\dfrac{1}{2}$ $\dfrac{1}{2}$

22. $\dfrac{22}{19}$ $\dfrac{21}{19}$

Use the proper symbol $>$, $<$, or $=$.

23. $\dfrac{9}{5}$ $\dfrac{11}{7}$

24. $\dfrac{16}{12}$ $\dfrac{13}{8}$

25. $\dfrac{23}{7}$ $\dfrac{29}{9}$

26. $\dfrac{23}{13}$ $\dfrac{18}{11}$

For any numbers of arithmetic $\dfrac{a}{b}$ and $\dfrac{c}{b}$,

$$\dfrac{a}{b} > \dfrac{c}{b} \text{ when } a > c,$$

$$\dfrac{a}{b} < \dfrac{c}{b} \text{ when } a < c,$$

$$\text{and } \dfrac{a}{b} = \dfrac{c}{b} \text{ when } a = c.$$

DO EXERCISES 20–22.

It is not so easy to tell which of $\frac{5}{7}$ and $\frac{2}{3}$ is larger. Let's find a common denominator and compare numerators.

Example 13 Insert the proper symbol $>$, $<$, or $=$ between $\frac{5}{7}$ and $\frac{2}{3}$.

We multiply by 1 to find a common denominator:

$$\frac{5}{7} = \frac{5}{7} \cdot \frac{3}{3} = \frac{15}{21} \quad \text{and} \quad \frac{2}{3} = \frac{2}{3} \cdot \frac{7}{7} = \frac{14}{21}.$$

Since $15 > 14$, it follows that $\frac{15}{21} > \frac{14}{21}$, so $\frac{5}{7} > \frac{2}{3}$.

Example 14 Insert the proper symbol $>$, $<$, or $=$ between $\frac{7}{10}$ and $\frac{8}{9}$.

$$\frac{7}{10} = \frac{7}{10} \cdot \frac{9}{9} = \frac{63}{90} \quad \text{and} \quad \frac{8}{9} = \frac{8}{9} \cdot \frac{10}{10} = \frac{80}{90}.$$

Since $63 < 80$, it follows that $\frac{63}{90} < \frac{80}{90}$, so $\frac{7}{10} < \frac{8}{9}$.

Example 15 Insert the proper symbol $>$, $<$, or $=$ between $\frac{6}{27}$ and $\frac{2}{9}$.

Here we need only multiply $\frac{2}{9}$ by $\frac{3}{3}$ and we obtain the other fraction.

$$\frac{2}{9} = \frac{2}{9} \cdot \frac{3}{3} = \frac{6}{27}$$

Since the numerators (and denominators) are the same, $\frac{6}{27} = \frac{2}{9}$.

We could also do these problems using a calculator. We find decimal notation and compare.

Example 16 Insert the proper symbol between $\frac{5}{7}$ and $\frac{2}{3}$.

$\frac{5}{7} \approx 0.7142857143$ 📟, rounded to ten decimal places
$\frac{2}{3} \approx 0.6666666667$

Thus $\frac{5}{7} > \frac{2}{3}$.

DO EXERCISES 23–26.

EXERCISE SET   1.7

· Find the reciprocal of each number.

1. $\dfrac{3}{4}$

2. $\dfrac{5}{8}$

3. $\dfrac{1}{8}$

4. $\dfrac{1}{10}$

5. 1

6. 9

·· Divide by multiplying by the reciprocal of the divisor.

7. $\dfrac{7}{6} \div \dfrac{3}{5}$

8. $\dfrac{7}{5} \div \dfrac{3}{4}$

9. $\dfrac{8}{9} \div \dfrac{4}{15}$

10. $\dfrac{3}{4} \div \dfrac{3}{7}$

11. $\dfrac{1}{4} \div \dfrac{1}{2}$

12. $\dfrac{1}{10} \div \dfrac{1}{5}$

13. $\dfrac{\frac{13}{12}}{\frac{39}{5}}$

14. $\dfrac{\frac{17}{6}}{\frac{3}{8}}$

15. $100 \div \dfrac{1}{5}$

16. $78 \div \dfrac{1}{6}$

17. $\dfrac{3}{4} \div 10$

18. $\dfrac{5}{6} \div 15$

··· Graph each number on the number line.

19. $\dfrac{5}{4}$

20. $\dfrac{7}{6}$

21. $\dfrac{15}{16}$

22. $\dfrac{11}{12}$

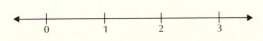

ANSWERS
1.
2.
3.
4.
5.
6.
7.
8.
9.
10.
11.
12.
13.
14.
15.
16.
17.
18.
19.
20.
21.
22.

Use the proper symbol >, <, or = between each pair of numerals.

23. $\dfrac{1}{2}$ $\dfrac{2}{4}$

24. $\dfrac{9}{12}$ $\dfrac{3}{4}$

25. $\dfrac{11}{15}$ $\dfrac{13}{24}$

26. $\dfrac{19}{16}$ $\dfrac{5}{4}$

27. $\dfrac{13}{8}$ $\dfrac{8}{5}$

28. $\dfrac{21}{16}$ $\dfrac{5}{4}$

29. $\dfrac{4}{5}$ $\dfrac{8}{10}$

30. $\dfrac{8}{9}$ $\dfrac{16}{18}$

31. $\dfrac{7}{22}$ $\dfrac{1}{3}$

32. $\dfrac{8}{23}$ $\dfrac{1}{3}$

SKILL MAINTENANCE

33. Find the area of a triangle when the base b is 48 cm and the height h is 17 cm.

34. Factor: $7x + 14y + 21z$.

35. Find the LCM of 45, 55, and 75.

36. Multiply: $10(8x + 5y + 7z)$.

EXTENSION

The symbol means that the exercise is to be done with a calculator.

37. Find the decimal notation for the reciprocal of 0.3125.

38. Find decimal notation for the reciprocal of 0.24.

39. Use the proper symbol >, <, or =.

$\dfrac{1439}{2007}$ $\dfrac{2359}{2876}$

40. Find decimal notation for each. Use this to order from smallest to largest.

$\dfrac{11}{13}, \ \dfrac{17}{20}, \ \dfrac{2}{3}, \ \dfrac{35}{37}, \ \dfrac{5}{6}, \ \dfrac{23}{25}$

1.8

Solving Equations

• EQUATIONS AND SOLUTIONS

An *equation* is a number sentence with = for its verb. For example,

$$3 + 2 = 5, \quad 7 - 2 = 8, \quad \text{and} \quad x + 6 = 13.$$

Some equations are true. Some are false. Some are neither true nor false.

Examples

1. The equation $3 + 2 = 5$ is true.
2. The equation $7 - 2 = 8$ is false.
3. The equation $x + 6 = 13$ is neither true nor false because we don't know what x stands for.

An equation says that the symbols on either side of the equals sign stand for, or name, the same number. For example, $5 - 4 = 3 + 7$ says that $5 - 4$ names the same number as $3 + 7$. This equation is false.

DO EXERCISES 1–3.

The letter x in $x + 6 = 13$ is called a *variable*. Some replacements for the variable make the equation true. Some make it false.

> The replacements that make an equation true are called *solutions*. To *solve* an equation means to find all of its solutions.

There are several ways to solve equations.

Example 4 Solve $x + 6 = 13$ by trial.

If we replace x by 2 we get a false equation: $2 + 6 = 13$.
If we replace x by 8 we get a false equation: $8 + 6 = 13$.
If we replace x by 7 we get a true equation: $7 + 6 = 13$.

No other replacement makes the equation true, so the only solution is 7.

DO EXERCISES 4–8.

•• SOLVING EQUATIONS $x + a = b$

Solving equations is an important part of algebra. Here we consider enough equation-solving techniques to give you a feel for this part of algebra. Later, we study equation solving in more detail. Consider the true sentence

$$2 + 3 = 5.$$

1. Write three true equations.

2. Write three false equations.

3. Write three equations that are neither true nor false.

4. Find three replacements that make $x + 5 = 12$ false.

5. Find the replacement that makes $x + 5 = 12$ true.

Solve by trial.

6. $x + 4 = 10$

7. $3x = 12$

8. $3y + 1 = 16$

ANSWERS ON PAGE A-2

If we subtract 3 on both sides, we still get a true equation.

$$
\begin{array}{r|r}
2 + 3 = & 5 \\
-3 & -3 \\
\hline
2 & 2
\end{array}
$$
    Two names of the same number
    Subtract 3 on both sides.
    We get a true equation.

Suppose we want to solve $x + 3 = 5$. We subtract 3 on both sides.

$$
\begin{array}{r|r}
x + 3 = & 5 \\
-3 & -3 \\
\hline
x & 2
\end{array}
$$

We subtract 3 to "undo" the addition $x + 3$. This gets x by itself. This equation is easy to solve. It is true when x is replaced by 2. We check to see if $x + 3 = 5$ is true when x is replaced by 2.

$$
\begin{array}{c|c}
\multicolumn{2}{c}{x + 3 = 5} \\
\hline
2 + 3 & 5 \\
5 &
\end{array}
$$

Thus 2 is a solution.

> **To solve $x + a = b$, subtract a on both sides.**

Example 5 Solve: $x + 6 = 13$.

$$
\begin{aligned}
x + 6 &= 13 \\
x + 6 - 6 &= 13 - 6 \qquad \text{Subtracting 6. Subtraction "undoes" addition.} \\
x + 0 &= 7 \\
x &= 7
\end{aligned}
$$

The solution to $x = 7$ is obviously the number 7. It is also the solution of the original equation. We check to find out. To do this we replace x by 7. When we simplify, we get 13 on both sides.

Check:
$$
\begin{array}{c|c}
\multicolumn{2}{c}{x + 6 = 13} \\
\hline
7 + 6 & 13 \\
13 &
\end{array}
$$

The solution is 7.

In the preceding discussion you may have questioned using the technique when you could "see" the answers. In the next example it is not so easy to "see" the answer.

Example 6 Solve: $x + 1.64 = 9.08$.

$$
\begin{aligned}
x + 1.64 &= 9.08 \\
x + 1.64 - 1.64 &= 9.08 - 1.64 \qquad \text{Subtracting 1.64} \\
x + 0 &= 7.44 \\
x &= 7.44
\end{aligned}
$$

Chapter 1   Introduction to Algebra; Review of Arithmetic

Check:

$$\begin{array}{r} x + 1.64 = 9.08 \\ \hline 7.44 + 1.64 \quad\bigm|\quad 9.08 \\ 9.08 \quad\bigm| \end{array}$$

The solution is 7.44.

DO EXERCISES 9–11.

Example 7 Solve: $x + \dfrac{3}{4} = \dfrac{5}{6}$.

$$x + \frac{3}{4} - \frac{3}{4} = \frac{5}{6} - \frac{3}{4} \qquad \text{Subtracting } \frac{3}{4}$$

$$x = \frac{5}{6} \cdot \frac{2}{2} - \frac{3}{4} \cdot \frac{3}{3} \qquad \begin{array}{l}\text{The LCM is 12. We multiply by 1}\\ \text{to obtain the LCM of the denominators.}\end{array}$$

$$x = \frac{10}{12} - \frac{9}{12}$$

$$x = \frac{1}{12}$$

Check:

$$\begin{array}{r} x + \frac{3}{4} = \frac{5}{6} \\ \hline \frac{1}{12} + \frac{3}{4} \quad\bigm|\quad \frac{5}{6} \\ \frac{1}{12} + \frac{9}{12} \quad\bigm| \\ \frac{10}{12} \quad\bigm| \\ \frac{5}{6} \quad\bigm| \end{array}$$

The solution is $\frac{1}{12}$.

DO EXERCISE 12.

••• SOLVING EQUATIONS $ax = b$

Consider the true equation

$$5 \cdot 9 = 45.$$

If we divide by 5 on both sides, we still get a true equation:

$$\frac{5 \cdot 9}{5} = \frac{45}{5}, \quad \text{or} \quad 9 = 9.$$

Suppose we want to solve $5x = 45$. We divide by 5 on both sides to undo the multiplication $5x$.

$$\frac{5x}{5} = \frac{45}{5} \qquad \boxed{\text{We can also think of this as multiplying by } \tfrac{1}{5}.}$$

or $\qquad x = 9.$

The solution of $x = 9$ is 9. We check to see if 9 is a solution of $5x = 45$.

$$\begin{array}{r} 5x = 45 \\ \hline 5 \cdot 9 \quad\bigm|\quad 45 \\ 45 \quad\bigm| \end{array}$$

Thus 9 is a solution.

Solve. Be sure to check.

9. $x + 9 = 17$

10. $x + 234 = 507$

11. $x + 2.78 = 7.65$

12. Solve: $x + \dfrac{11}{8} = \dfrac{5}{2}$.

ANSWERS ON PAGE A-2

Solve. Be sure to check.

13. $5x = 35$

14. $8x = 36$

15. $1.2x = 6.72$

Determine whether each division is possible.

16. $\frac{21}{6}$

17. $\frac{13}{0}$

18. $\frac{0}{8}$

19. $\dfrac{11}{10 - 10}$

20. $\dfrac{9}{12 - (4 \cdot 3)}$

21. $\dfrac{P}{x - x}$

ANSWERS ON PAGE A-2

To solve $ax = b$ when a is nonzero, divide on both sides by a (or multiply on both sides by $1/a$).

Example 8 Solve: $5x = 16$.

$$5x = 16$$
$$\frac{1}{5} \cdot 5x = \frac{16}{5} \qquad \text{Dividing by 5, or multiplying by } \tfrac{1}{5}$$
$$1 \cdot x = 3.2$$
$$x = 3.2$$

We divide by 5 to "undo" the multiplication and get x by itself. This equation is easy to solve. It is true when x is replaced by 3.2.

Check:

$$
\begin{array}{c|c}
5x = 16 & \\
\hline
5(3.2) & 16 \\
16 &
\end{array}
$$

The solution is 3.2. (It could also be left as $\frac{16}{5}$.)

Example 9 Solve: $4.7x = 40.42$.

$$4.7x = 40.42$$
$$\frac{1}{47} \times 4.7x = \frac{40.42}{4.7} \qquad \text{Dividing by 4.7, or multiplying by } \frac{1}{4.7}$$
$$1 \cdot x = 8.6$$
$$x = 8.6$$

Check:

$$
\begin{array}{c|c}
4.7x = 40.42 & \\
\hline
4.7(8.6) & 40.42 \\
40.42 &
\end{array}
$$

The solution is 8.6.

DO EXERCISES 13–15.

⠿ DIVISION BY ZERO

We cannot use the preceding method to solve an equation $ax = b$ when $a = 0$, because this would result in division by zero. But why? The division $a \div b$ or a/b is defined to be some number c such that $a = bc$. Thus $a/0$ would be some number c such that $0 \cdot c = a$. But $0 \cdot c = 0$, so the only possible number a that could be divided by 0 is 0. Look for a pattern.

a) $\frac{0}{0} = 5$ because $0 = 0 \cdot 5$

b) $\frac{0}{0} = 789$ because $0 = 0 \cdot 789$

c) $\frac{0}{0} = 17$ because $0 = 0 \cdot 17$

d) $\frac{0}{0} = \frac{1}{2}$ because $0 = 0 \cdot \frac{1}{2}$

It looks as if $\frac{0}{0}$ could be any number at all. This would be very confusing, getting any answer we want when we divide 0 by 0. Thus we agree to exclude division by zero.

We never divide by zero.

DO EXERCISES 16–21.

Chapter 1   Introduction to Algebra; Review of Arithmetic

• Solve by trial.

1. $x + 8 = 10$ **2.** $x + 5 = 13$

3. $x - 4 = 5$ **4.** $x - 6 = 12$

5. $5x = 25$ **6.** $7x = 42$

7. $5y + 7 = 107$ **8.** $9x + 5 = 86$

9. $7x - 1 = 48$ **10.** $4y - 2 = 10$

•• Solve. Be sure to check.

11. $x + 17 = 22$ **12.** $x + 18 = 32$

13. $x + 56 = 75$ **14.** $x + 47 = 83$

15. $x + 2.78 = 8.44$ **16.** $x + 3.04 = 4.69$

17. $x + 5064 = 7882$ **18.** $x + 4112 = 8007$

19. $x + \dfrac{1}{4} = \dfrac{2}{3}$ **20.** $x + \dfrac{1}{3} = \dfrac{4}{5}$

21. $x + \dfrac{2}{3} = \dfrac{5}{6}$ **22.** $x + \dfrac{3}{4} = \dfrac{7}{8}$

ANSWERS
1.
2.
3.
4.
5.
6.
7.
8.
9.
10.
11.
12.
13.
14.
15.
16.
17.
18.
19.
20.
21.
22.

••• Solve. Be sure to check.

23. $6x = 24$

24. $4x = 32$

25. $4x = 5$

26. $6x = 27$

27. $10y = 2.4$

28. $9x = 3.6$

29. $2.9y = 8.99$

30. $5.5y = 34.1$

31. $6.2x = 52.7$

32. $0.4x = 23.5$

33. $\frac{3}{4}x = 35$

34. $\frac{4}{5}x = 27$

⠿ Determine whether each division is possible.

35. $\dfrac{6}{0}$

36. $\dfrac{0}{5}$

37. $\dfrac{0}{0}$

38. $\dfrac{8}{7-7}$

39. $\dfrac{2-2}{4}$

40. $\dfrac{1}{x-x}$

41. $\dfrac{8-8}{t-t}$

✓ SKILL MAINTENANCE

42. Find the area of a circle with radius $r = 12$ ft. Use 3.14 for π.

43. Divide and simplify: $\frac{7}{8} \div \frac{14}{15}$.

44. Factor: $16a + 8b + 64c$.

45. Collect like terms: $9u + 4t + 11u + 7u + 28$.

☆ EXTENSION

Solve.

46. $x + 506{,}233 = 976{,}421$

47. ▦ $0.1265x = 1065.636$

48. Write an equation that has *no* whole-number solution.

49. Write an equation for which *every* whole number is a solution.

Solving Problems

1.9

Solving Problems

■   Why learn to solve equations? Because applied problems can be solved using equations. To do this, we first translate the problem situation to an equation and then solve the equation. Then we check the solution of the equation to see if it is a solution of the original problem. The problem situation may be explained in words (as in a textbook) or mlay come from a real-world situation.

The problems we solve here are simple. You might use other methods to solve them, but it is better if you do not. Remember, you are learning how algebra works.

Example 1 *Application: Population Growth.* The bar graph shows the population growth of three states since 1980. How much more has the population of Texas grown than the population of California?

We reword the problem and translate it to an equation as follows:

(What number)   plus   (growth of California)   is   (growth of Texas)?

x +   2,148,688   =   2,155,609

The translation gives us the equation

$$x + 2,148,688 = 2,155,609.$$

We solve it:

$x = 2,155,609 - 2,148,688$ Subtracting 2,148,688
$x = 6921$.

To check, we find out whether $6921 + 2,148,688$ is $2,155,609$:

$$6921 + 2,148,688 = 2,155,609.$$

We have the answer: 6921.

DO EXERCISE 1.

Example 2 Solve this problem.

Three-fourths   of   what number   is   thirty-five?

$\frac{3}{4}$ · x =   35

The translation gives us the equation

$$\frac{3}{4}x = 35.$$

We solve it:

$$x = \frac{35}{\frac{3}{4}}$$ Dividing by $\frac{3}{4}$

$$x = 35 \cdot \frac{4}{3}$$

$$x = \frac{140}{3}.$$

O B J E C T I V E S

After finishing Section 1.9, you should be able to:

■ Solve applied problems by translating to equations and solving.

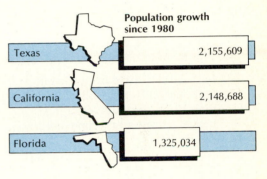

Population growth since 1980

Texas	2,155,609
California	2,148,688
Florida	1,325,034

Translate to an equation. Then solve and check.

1. Refer to the bar graph in Example 1. How much more did the population of California grow than the population of Florida?

ANSWER ON PAGE A-2

1.9 Solving Problems   **49**

Translate to an equation. Then solve and check.

2. Two-thirds of what number is forty-four?

Translate to an equation. Then solve and check.

3. In Australia there are 145 million sheep. This is 132 million more than the number of people. How many people are in Australia?

Translate to an equation. Then solve and check.

4. An investment was made that grew to $14,500 after one year. This was 1.16 times what was originally invested. How much was originally invested?

To check, we find out whether $\frac{3}{4}$ of this number is 35:

$$\frac{3}{4} \cdot \frac{140}{3} = \frac{3 \cdot 140}{4 \cdot 3} = \frac{3 \cdot 35 \cdot 4}{4 \cdot 3} = 35.$$

The number 140/3 is the answer.

> Note that in translating, *is* translates to =. The word *of* translates to ·, and the unknown number translates to a variable.

DO EXERCISE 2.

Sometimes it helps to reword a problem before translating.

Example 3 One year a person earned a salary of $23,400. This was $1700 more than last year. What was last year's salary?

Rewording: Last year's salary plus $1700 is this year's salary.

Translating: x +   1700   =   23,400

The translation gives us the equation $x + 1700 = 23,400$. We solve it:

$$x = 23,400 - 1700 \qquad \text{Subtracting 1700}$$
$$x = 21,700.$$

To check, we add $1700 to $21,700:

$$\$1700 + \$21,700 = \$23,400.$$

The answer is $21,700.

DO EXERCISE 3.

Example 4 A solid-state color television set uses about 420 kilowatt hours (kWh) of electricity in a year. This is 3.5 times that used by a solid-state black-and-white set. How many kilowatt hours does the black-and-white set use each year?

Rewording: 3.5 times energy of the black-and-white set is 420.

Translating: 3.5   × y =   420

The translation gives us the equation

$$3.5y = 420.$$

We solve it:

$$y = \frac{420}{3.5} \qquad \text{Dividing by 3.5}$$
$$y = 120.$$

To check, we multiply 120 by 3.5: 3.5(120) = 420. The answer is 120 kWh.

DO EXERCISE 4.

• Translate to equations. Then solve and check.

1. Two thirds of what number is forty-eight?

2. One-eighth of what number is fifty-six?

3. What number plus five is twenty-two?

4. What number plus eight is sixty-three?

5. What number is four more than five?

6. What number is seven less than ten?

7. The area of Lake Superior is four times the area of Lake Ontario. The area of Lake Superior is 78,114 km². What is the area of Lake Ontario?

8. The area of Alaska is about 483 times the area of Rhode Island. The area of Alaska is 1,519,202 km². What is the area of Rhode Island?

9. Izzi Zlow's typing speed is 35 words per minute. This is two-fifths of Ty Preitter's speed. What is Preitter's speed?

10. Walter Logged's body contains 57 kg of water. This is two-thirds of his weight. What is Logged's weight?

11. The boiling point of ethyl alcohol is 78.3°C. This is 13.5°C more than the boiling point of methyl alcohol. What is the boiling point of methyl alcohol?

12. The height of the Eiffel Tower is 295 m. This is about 203 m more than the height of the Statue of Liberty. What is the height of the Statue of Liberty?

ANSWERS
1. _____
2. _____
3. _____
4. _____
5. _____
6. _____
7. _____
8. _____
9. _____
10. _____
11. _____
12. _____

13.

14.

15.

16.

17.

18.

19.

20.

13. A color television set with tubes uses about 640 kWh of electricity in a year. This is 1.6 times that used by a solid-state color set. How many kilowatt hours does the solid-state color model use each year?

14. The distance from the earth to the sun is about 150,000,000 km. This is about 391 times the distance from the earth to the moon. What is the distance from the earth to the moon?

15. Recently, the average cost of having a baby in one Ohio hospital was $1175. This was about 1.8 times the average cost for a certain California hospital. What was the cost of having a baby in the California hospital?

16. It takes a 60-watt bulb about 16.6 hours to use one kilowatt hour of electricity. This is about 2.5 times as long as it takes a 150-watt bulb to use one kilowatt hour. How long does it take a 150-watt bulb to use one kilowatt hour?

⭐ **EXTENSION**

17. One inch = 2.54 cm. A meter is 100 cm. Find the number of inches in a meter.

18. Sound travels at a speed of 1087 feet per second. How long does it take the sound an airplane makes to travel to your ear if the airplane is 10,000 ft overhead?

These problems are impossible to solve because some piece of information is missing. Tell what you would need to know to solve the problem.

19. A person makes three times the salary of ten years ago. What was the salary ten years ago?

20. Records were on sale for 75¢ off the marked price. After buying four records, a person has $8.72 left. How much was there to begin with?

Percent Notation

• CONVERTING TO DECIMAL NOTATION

There are other ways to name numbers of arithmetic besides using fractional and decimal notation. One other kind of notation uses the percent symbol %, which means "per hundred." We can regard the percent symbol as part of a numeral. For example,

$$37\% \quad \text{is defined to mean} \quad 37 \times 0.01 \quad \text{or} \quad 37 \times \frac{1}{100}$$

In general,

> $n\%$ means $n \times 0.01$ or $n \times \frac{1}{100}$.

Example 1 Find decimal notation for 78.5%.

$$78.5\% = 78.5 \times 0.01 \qquad \text{Replacing \% by} \times 0.01$$
$$= 0.785$$

DO EXERCISES 1 AND 2.

•• CONVERTING TO FRACTIONAL NOTATION

Example 2 Find fractional notation for 88%.

$$88\% = 88 \times \frac{1}{100} \qquad \text{Replacing \% by} \times \frac{1}{100}$$

$$= \frac{88}{100} \qquad \text{You need not simplify.}$$

Example 3 Find fractional notation for 34.7%.

$$34.7\% = 34.7 \times \frac{1}{100} \qquad \text{Replacing \% by} \times \frac{1}{100}$$

$$= \frac{34.7}{100}$$

$$= \frac{34.7}{100} \cdot \frac{10}{10} \qquad \text{Multiplying by 1 to get a whole number in the numerator}$$

$$= \frac{347}{1000}$$

> There is a table of decimal and percent equivalents on the last page of this book. If you do not already know these facts, you should memorize them.

DO EXERCISES 3–5.

After finishing Section 1.10, you should be able to:

• Convert from percent notation to decimal notation.

•• Convert from percent notation to fractional notation.

••• Convert from decimal notation to percent notation.

∷ Convert from fractional notation to percent notation.

∷• Solve applied problems involving percents.

Find decimal notation.

1. 46.2%

2. 100%

Find fractional notation.

3. 67%

4. 45.6%

5. $\frac{1}{4}\%$

ANSWERS ON PAGE A-2

Find percent notation.

6. 6.77

7. 0.9944

Find percent notation.

8. $\frac{1}{4}$

9. $\frac{3}{8}$

10. $\frac{2}{3}$

ANSWERS ON PAGE A-2

CONVERTING FROM DECIMAL TO PERCENT NOTATION

By applying the definition of % in reverse, we can convert from decimal notation to percent notation. We multiply by 1, naming it 100×0.01. Then we use the associative law of multiplication. We multiply 100 by the number, and replace "$\times 0.01$" by "%."

Example 4 Find percent notation for 0.93.

$$
\begin{aligned}
0.93 &= 0.93 \times 1 \\
&= 0.93 \times (100 \times 0.01) \qquad \text{Replacing 1 by } 100 \times 0.01 \\
&= (0.93 \times 100) \times 0.01 \qquad \text{Using associativity} \\
&= 93 \times 0.01 \\
&= 93\,\% \qquad \text{Replacing } \times 0.01 \text{ by } \%
\end{aligned}
$$

Example 5 Find percent notation for 0.002.

$$
\begin{aligned}
0.002 &= 0.002 \times (100 \times 0.01) \\
&= (0.002 \times 100) \times 0.01 \\
&= 0.2 \times 0.01 \\
&= 0.2\,\% \qquad \text{Replacing } \times 0.01 \text{ by } \%
\end{aligned}
$$

DO EXERCISES 6 AND 7.

CONVERTING FROM FRACTIONAL TO PERCENT NOTATION

We can also convert from fractional to percent notation. Again, we multiply by 1, but this time we use $100 \times \frac{1}{100}$.

Example 6 Find percent notation for $\frac{5}{8}$.

$$
\begin{aligned}
\frac{5}{8} &= \frac{5}{8} \times \left(100 \times \frac{1}{100}\right) \\
&= \left(\frac{5}{8} \times 100\right) \times \frac{1}{100} \\
&= \frac{500}{8} \times \frac{1}{100} \\
&= \frac{500}{8}\,\%, \quad \text{or} \quad 62.5\,\%
\end{aligned}
$$

DO EXERCISES 8–10.

SOLVING PROBLEMS INVOLVING PERCENTS

Let's solve some applied problems involving percents. Again, it is helpful to translate the problem situation to an equation and then solve the equation.

Example 7 What is 12% of 59?

$$x = 12\% \cdot 59 \qquad \text{Translating}$$

Solve:
$$x = (12 \times 0.01) \times 59 \qquad \text{Replacing 12\% by 12} \times 0.01$$
$$x = 0.12 \times 59 \qquad \text{This eliminates the \% sign.}$$
$$x = 7.08$$

The solution is 7.08, so 7.08 is 12% of 59.

DO EXERCISES 11 AND 12.

Example 8 What percent of 45 is 15?

$$x \quad \% \quad \cdot \quad 45 = 15 \qquad \text{Translating}$$

Solve:
$$(x \times 0.01) \times 45 = 15$$
$$x(0.45) = 15$$
$$x = \frac{15}{0.45} \qquad \text{Dividing by 0.45}$$
$$x = \frac{15}{0.45} \times \frac{100}{100} = \frac{1500}{45} = 33\tfrac{1}{3}$$

The solution is $33\tfrac{1}{3}$, so $33\tfrac{1}{3}\%$ of 45 is 15.

DO EXERCISES 13 AND 14.

Example 9 3 is 16 percent of what?

$$3 = 16 \quad \% \quad \cdot \quad y \qquad \text{Translating}$$

Solve:
$$3 = 16 \times 0.01 \times y$$
$$3 = 0.16y$$
$$0.16y = 3$$
$$y = \frac{3}{0.16} \qquad \text{Dividing by 0.16}$$
$$y = \frac{3}{0.16} \cdot \frac{100}{100} = \frac{300}{16} = 18.75$$

16% of 18.75 is 3, so the solution is 18.75.

> Perhaps you noticed in Examples 7–9 that to handle percents in such problems, you must first convert to decimal notation, and then proceed.

DO EXERCISES 15 AND 16.

Translate and solve.

11. What is 23% of 48?

12. 25% of 40 is what?

Translate and solve.

13. What percent of 50 is 16?

14. 15 is what percent of 60?

Translate and solve.

15. 45 is 20 percent of what?

16. 120 percent of what is 60?

ANSWERS ON PAGE A-3

1.10   Percent Notation

55

Translate to an equation. Then solve.

17. The area of Arizona is 19% of the area of Alaska. The area of Alaska is 586,400 mi². What is the area of Arizona?

18. An investment is made at 7% simple interest for one year. It grows to $8988. How much was originally invested (the principal)?

Translate to an equation. Then solve.

19. A person's salary increased 12% to an amount of $20,608. What was the former salary?

Sometimes it is helpful to reword the problem before translating.

Example 10 Blood is 90% water. The average adult has 5 quarts of blood. How much water is in the average adult's blood?

Rewording: 90% of 5 is what?

Translating: $90\% \cdot 5 = x$

Solve: $90 \times 0.01 \times 5 = x$

$0.90 \times 5 = x$ Converting 90% to decimal notation

$4.5 = x$

The number 4.5 checks in the problem. Thus there are 4.5 quarts of water in the average adult's blood.

Example 11 An investment is made at 8% simple interest for one year. It grows to $783. How much was originally invested (the principal)?

Rewording: (Principal) + (Interest) = Amount

Translating: $x + 8\%x = 783$ Interest is 8% of the principal

Solve:

$x + 8\%x = 783$

$x + 0.08x = 783$ Converting

$1.08x = 783$ Collecting like terms

$x = \dfrac{783}{1.08}$ Dividing by 1.08

$x = 725$

The original investment (principal) was $725.

DO EXERCISES 17 AND 18.

Example 12 The price of an automobile rose 16% to a value of $10,393.60. What was the former price?

Rewording: (Former price) + (Increase) = New price

Translating: $x + 16\%x = \$10{,}393.60$

Solve:

$x + 16\%x = 10{,}393.60$

$x + 0.16x = 10{,}393.60$ Converting

$1.16x = 10{,}393.60$ Collecting like terms

$x = \dfrac{10{,}393.60}{1.16}$ Dividing by 1.16

$x = 8960$

The former price was $8960.

A common error in a problem like this is to take 16% of the new price and subtract. This problem is easy with algebra. Without algebra it is not.

DO EXERCISE 19.

Note: If you need extra practice on percent, do Exercise Set 1.10A on pp. 59–60.

⚀ Find decimal notation.

1. 76%   2. 54%   3. 54.7%   4. 96.2%

⚁ Find fractional notation.

5. 20%   6. 80%   7. 78.6%   8. 12.5%

⚂ Find percent notation.

9. 4.54   10. 1   11. 0.998   12. 0.751

⚃ Find percent notation.

13. $\frac{1}{8}$ 14. $\frac{1}{3}$ 15. $\frac{17}{25}$ 16. $\frac{11}{20}$

⚄ Translate and solve.

17. What is 65% of 840?   18. 34% of 560 is what?

19. 24 percent of what is 20.4?   20. 45 is 30 percent of what?

21. What percent of 80 is 100?   22. 30 is what percent of 125?

Translate to an equation. Then solve.

23. On a test of 88 items, a student got 76 correct. What percent were correct?

24. A baseball player got 13 hits in 25 times at bat. What percent were hits?

ANSWERS
1.
2.
3.
4.
5.
6.
7.
8.
9.
10.
11.
12.
13.
14.
15.
16.
17.
18.
19.
20.
21.
22.
23.
24.

25. _____

26. _____

27. _____

28. _____

29. _____

30. _____

31. _____

32. _____

33. _____

34. _____

35. _____

36. _____

25. A family spent $208 one month for food. This was 26% of its income. What was their monthly income?

26. The weight of the human brain is 2.7% of the body weight. A human's brain weighs 2.1 kg (kilograms). What is its body weight?

27. The sales tax rate in New York City is 8%. How much is charged on a purchase of $428.86? How much is the total cost of the purchase?

28. Water volume increases 9% when water freezes. If 400 cm³ (cubic centimeters) of water is frozen, how much would its volume increase? What would be the volume of the ice?

29. An investment is made at 9% simple interest for one year. It grows to $8502. How much was originally invested?

30. An investment is made at 8% simple interest for one year. It grows to $7776. How much was originally invested?

31. A person earned $9600 one year. An 8% increase in salary was received, but the cost of living rose 7.4%. How much additional earning power was actually received?

32. Due to inflation the price of an item rose 8%, which was 12¢. What was the old price? The new price?

⭐ **EXTENSION**

Simplify.

33. $1 - 76\%$

34. $56\% + 28\%$

Solve.

35. It is estimated that there are 300,000 words n the English language. The average person knows 10,000 of them. What percent of the words does the average person know?

36. If x is 160% of y, then y is what percent of x?

This exercise set supplements Exercise Set 1.10 for those who need it.

⬤ Find decimal notation.

1. 38%

2. 72.1%

3. 65.4%

4. 3.25%

5. 8.24%

6. 0.61%

7. 0.012%

8. 0.023%

9. 0.73%

10. 0.0035%

11. 125%

12. 240%

⬤⬤ Find fractional notation.

13. 30%

14. 70%

15. 13.5%

16. 73.4%

17. 3.2%

18. 8.4%

19. 120%

20. 250%

21. 0.35%

22. 0.48%

23. 0.042%

24. 0.083%

⬤⬤⬤ Find percent notation.

25. 0.62

26. 0.73

27. 0.623

28. 0.812

ANSWERS	
1.	
2.	
3.	
4.	
5.	
6.	
7.	
8.	
9.	
10.	
11.	
12.	
13.	
14.	
15.	
16.	
17.	
18.	
19.	
20.	
21.	
22.	
23.	
24.	
25.	
26.	
27.	
28.	

29. 7.2

30. 3.5

31. 2

32. 5

33. 0.072

34. 0.013

35. 0.0057

36. 0.0068

⠿ Find percent notation.

37. $\dfrac{17}{100}$

38. $\dfrac{119}{100}$

39. $\dfrac{7}{10}$

40. $\dfrac{8}{10}$

41. $\dfrac{7}{20}$

42. $\dfrac{7}{25}$

43. $\dfrac{1}{2}$

44. $\dfrac{3}{4}$

45. $\dfrac{3}{5}$

46. $\dfrac{17}{50}$

47. $\dfrac{1}{3}$

48. $\dfrac{3}{8}$

⠿ Translate and solve.

49. What is 38% of 250?

50. 37.2% of 85 is what?

51. What percent of 80 is 20?

52. 35 is 20 percent of what?

53. 25 percent of what is 16?

54. 20 is what percent of 60?

Summary and Review

The following contains a summary of what you should be able to do after completing this chapter. The reveiw exercises are for practice. Answers are at the back of the book. If you miss an exercise, restudy the section and objective indicated alongside the answer.

You should be able to:

Evaluate simple algebraic expressions.

Evaluate.

1. $4a$ for $a = 5$

2. $\dfrac{x}{y}$ for $x = 12$ and $y = 2$

3. $\dfrac{2p}{q}$ for $p = 20$ and $q = 8$

4. $\dfrac{x - y}{3}$ for $x = 17$ and $y = 5$

5. $n^3 + 1$ for $n = 2$

6. x^0 for $x = 6$

7. Find the area of a circle when $t = 32$ mi. Use 3.14 for π.

Translate phrases to algebraic expressions.

Translate to an algebraic expression.

8. 8 less than z

9. Three times x

10. Nineteen percent of some number

11. A number x is 1 more than a smaller number. Write an expression for the smaller number.

Find the prime factorization of a number and find the least common multiple of two or more numbers using prime factorizations.

Find the prime factorization of each number.

12. 92

13. 1400

Find the LCM.

14. 12, 32

15. 5, 18, 45

Find simplest fractional notation for a number of arithmetic and add, subtract, multiply, and divide with fractional notation.

Simplify.

16. $\dfrac{20}{48}$

17. $\dfrac{1800}{1000}$

Compute and simplify.

18. $\dfrac{4}{9} + \dfrac{5}{12}$

19. $\dfrac{3}{4} \div 3$

20. $\dfrac{2}{3} - \dfrac{1}{15}$

21. $\dfrac{9}{10} \cdot \dfrac{16}{5}$

Use the proper symbol $>$, $<$, or $=$ between two fractional numerals.

Use the proper symbol $>$, $<$, or $=$.

22. $\dfrac{4}{9} \quad \dfrac{5}{12}$

23. $\dfrac{19}{20} \quad \dfrac{29}{30}$

Convert from decimal to fractional notation and from fractional to decimal notation.

Write fractional notation. You need not simplify.

24. 17.98

25. 0.347

Write decimal notation.

26. $\frac{13}{16}$

27. $\frac{11}{7}$

Tell the meaning of exponential notation.

What is the meaning of each of the following?

28. t^5

29. t^0, $t \neq 0$

30. t^1

Do calculations as shown by parentheses, and tell what law is illustrated by a given sentence.

Calculate.

31. $(6 \times 5) + 2$

32. $(7 + 8) \cdot 5$

Tell what law is illustrated by each sentence.

33. $5 \cdot (7 + 8) = 5 \cdot 7 + 5 \cdot 8$

34. $68 + 4 = 4 + 68$

35. $68 \cdot (4 \cdot 9) = (68 \cdot 4) \cdot 9$

36. $79 \cdot 8 = 8 \cdot 79$

Use the distributive law to multiply and factor algebraic expressions and to simplify expressions by collecting like terms.

Multiply.

37. $6(3x + 5y)$

38. $8(5x + 3y + 2)$

Factor.

39. $18x + 6y$

40. $36x + 16 + 4y$

Collect like terms.

41. $26y + 8a + 4y + 2a$

42. $54x + 3b + 9b + 6x$

Convert percent notation to decimal and fractional notation and convert decimal and fractional notation to percent notation.

43. Find decimal notation: 4.7%.

44. Find fractional notation: 60%

Find percent notation.

45. 0.886

46. $\frac{5}{8}$

47. $\frac{29}{25}$

Solve equations of the type $x + a = b$ and $ax = b$.

Solve.

48. $x + 11 = 50$

49. $x + \frac{3}{8} = \frac{1}{2}$

50. $7n = 56$

51. $1.3z = 33.8$

Solve problems.

52. What number added to 35 is 102?

53. Three-fifths of what number is thirty?

54. An artist charges $75 for each drawing. How many drawings did the artist complete in order to earn $2400?

55. 25 is 10% of what number?

56. 40% of 75 is what number?

57. A government employee received an 8% raise. The new salary is $15,336. What was the original salary?

⭐ **EXTENSION**

58. Simplify: $\frac{10ac}{18bc}$.

59. Evaluate $x^{253}y^1$ for $x = 1$ and $y = 37$.

60. Which of the following are equivalent?

    a) $7a + 7b$

    b) $7(a + b)$

    c) $(a + b)^7$

    d) $7b + 7$

61. There must be 6 parts per billion (ppb) of chlorine in a swimming pool. What percent of total volume is this?

Test: Chapter 1

Evaluate.

1. $\dfrac{3x}{y}$ for $x - 10$ and $y = 5$

2. $x^2 - 5$ for $x = 8$

3. y^1 for $y = 8$

4. Write an algebraic expression: 8 less than n.

5. Find the area of a triangle when the height h is 30 ft and the base b is 16 ft.

Compute and simplify.

6. $\dfrac{7}{8} \cdot \dfrac{16}{21}$

7. $\dfrac{7}{10} + \dfrac{5}{6}$

8. $\dfrac{3}{4} - \dfrac{15}{32}$

9. $\dfrac{3}{7} \div \dfrac{9}{28}$

10. What is the meaning of y^4?

11. Write fractional notation for 5.69.

12. Write decimal notation for $\frac{3}{8}$.

13. Calculate: $(8 \times 7) + 6$.

14. What law is illustrated by this sentence?

    $6 \cdot (4 \cdot 3) = (6 \cdot 4) \cdot 3$

15. Factor $5x + 5y$. Then evaluate both the original and factored expressions when $x = 4$ and $y = 6$.

Factor.

16. $15y + 5$

17. $24x + 16 + 8y$

18. Find the prime factorization of 300.

19. Find the LCM: 15, 24, 60.

20. Simplify: $\dfrac{16}{24}$.

Multiply.

21. $10(9x + 3y)$

22. $7(9m + 2x + 1)$

Collect like terms.

23. $21x + 96 + 5x + 6$

24. $18y + 30a + 9a + 4y$

25. Find decimal notation: 0.7%.

26. Find fractional notation: 91%.

ANSWERS
1.
2.
3.
4.
5.
6.
7.
8.
9.
10.
11.
12.
13.
14.
15.
16.
17.
18.
19.
20.
21.
22.
23.
24.
25.
26.

27. Find percent notation: $\dfrac{11}{25}$.

27. _____

28. _____

29. _____

30. _____

31. _____

32. _____

33. _____

34. _____

35. _____

36. _____

37. _____

38. _____

39. _____

40. _____

Solve.

28. $x + 3 = 11.4$

29. $5.4x = 32.4$

30. $y + \dfrac{3}{10} = \dfrac{9}{10}$

31. What number added to 43 is 60?

32. Four times what number is 56?

33. Three-fifths of what number is 18?

34. What percent of 75 is 5?

35. 16 is 25% of what number?

36. 62% of 125 is what number?

37. An investment is made at 12% simple interest for 1 year. It grows to $28,000. How much was originally invested?

38. Use the proper symbol $>$, $<$, or $=$: $\dfrac{5}{8}$ $\dfrac{6}{7}$.

⭐ **EXTENSION**

39. Simplify: $\dfrac{13,860}{42,000}$.

40. Evaluate $\dfrac{5y - x}{4}$ when $x = 20$ and $y = 16$.

2

AN APPLICATION
Death Valley is 280 ft below sea level.
What integer corresponds to this situation?

THE MATHEMATICS
The integer that corresponds to this
situation is

−280.
└─── This is a negative integer.

In this chapter the numbers of arithmetic are expanded to larger sets of numbers: the set of integers and the set of rational numbers. This sets the stage for Chapter 3, which will present solving equations that have no solution in the set of numbers of arithmetic. We will learn to add, subtract, multiply, and divide with these new numbers and carry out other algebraic manipulations.

The review sections to be tested in addition to the material in this chapter are 1.1, 1.2, 1.6, and 1.10.

Integers and Rational Numbers

2.1

After finishing Section 2.1, you should be able to:

- Tell which integer corresponds to a real-world situation.
- Write a true sentence using $<$ or $>$.
- Find the absolute value of any integer.
- Find the additive inverse of any integer.

Integers and the Number Line

The numbers of arithmetic can be represented on a number line.

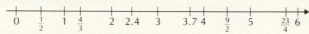

In this chapter we will use the number line to learn about other sets of numbers. The numbers of arithmetic can be used to solve many kinds of problems. However, we can solve many other kinds of problems using numbers called *integers*.

THE SET OF INTEGERS

To create the set of integers, we begin with the set of *whole* numbers, 0, 1, 2, 3, and so on. For each natural number, 1, 2, 3, and so on, we invent a new number.

For the number 1 there will be a new number named -1 (negative 1);

For the number 2 there will be a new number named -2 (negative 2);

For the number 3 there will be a new number named -3 (negative 3); and so on.

The *integers* consist of the whole numbers and these new numbers. We picture them on a number line as follows.

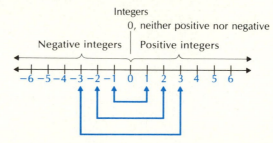

We call the newly invented numbers *negative* integers. The natural numbers are called *positive* integers. Zero is neither positive nor negative.

• INTEGERS AND REAL-WORLD SITUATIONS

Integers can be associated with many real-world problems and situations. The following examples will help you get ready to translate problem situations to mathematical language.

Example 1 Tell which integer corresponds to this situation: The temperature is 3 degrees below zero.

$$3° \text{ below zero is } -3°$$

Example 2 Tell which integer corresponds to this situation: getting set 21 points in a card game.

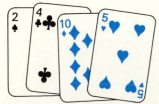

3° below zero is $-3°$.

Getting set 21 points in a card game gives you -21 points.

Getting set 21 points in a card game gives you −21 points. "Getting set" means not making a bid in a certain game such as *Hearts.* That is, you lose points.

Example 3 Tell which integer corresponds to this situation: Death Valley is 280 ft below sea level.

The integer −280 corresponds to the situation. The elevation is −280 ft.

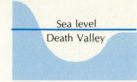

Death Valley is 280 ft below sea level. The elevation is −280 ft.

Example 4 Tell which integers correspond to this situation: A business made $234 on Monday, but lost $350 on Tuesday.

The integers 234 and −350 correspond to the situation. The integer $234 corresponds to the profit on Monday and −$350 corresponds to the loss on Tuesday.

DO EXERCISES 1–4.

•• ORDER ON THE NUMBER LINE

Numbers are named in order on the number line, with larger numbers further to the right. For any two numbers on the line, the one to the left is less than the one to the right.

We use the symbol < to mean "is less than." The sentence −5 < 9 means "−5 is less than 9." The symbol > means "is greater than." The sentence −4 > −8 means "−4 is greater than −8."

Examples Use either < or > to write true sentences.

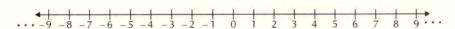

5. 3   8   Since 8 is to the right of 3, 8 is greater than 3, or 3 is less than 8. The answer is 3 < 8.

6. −9   2   Since −9 is to the left of 2, we have −9 < 2.

7. 7   −13   Since 7 is to the right of −13, we have 7 > −13.

8. −19   −6   Since −19 is to the left of −6, we have −19 < −6.

DO EXERCISES 5–8.

••• ABSOLUTE VALUE

From the number line we see that numbers like 5 and −5 are the same distance from zero.

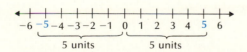

How far is 5 from 0? How far is −5 from 0? Since distance is always considered to be a nonnegative number (positive or zero), it follows that 5 is 5 units from 0 and −5 is 5 units from 0.

Tell which integers correspond to the situation.

1. The halfback gained 8 yd on first down. The quarterback was sacked for a 5-yd loss on second down.

2. The highest temperature ever recorded in the United States was 134° in Death Valley on July 10, 1913. The coldest temperature ever recorded in the United States was 76° below zero in Tanana, Alaska, in January of 1886.

3. At 10 sec before liftoff, ignition occurs. At 148 sec after liftoff, the first stage is detached from the rocket.

4. A student owes $137 to the bookstore. The student has $289 in a savings account.

Use either < or > to write true sentences.

5. 15   7   6. 12   −3

7. −13   −3   8. −4   −20

ANSWERS ON PAGE A-3

Find the absolute value.

9. |18|

10. |−9|

11. |−29|

12. |0|

The *absolute value* of a number is its distance from 0 on a number line. We use the symbol |x| to represent the absolute value of a number x.

To find absolute value:

1. If a number is negative, make it positive.
2. If a number is positive or zero, leave it alone.

Examples Find the absolute value.

9. |−3| The distance of −3 from 0 is 3, so |−3| = 3.
10. |25| The distance of 25 from 0 is 25, so |25| = 25.
11. |0| The distance of 0 from 0 is 0, so |0| = 0.

DO EXERCISES 9–12.

:: ADDITIVE INVERSES

The set of integers is shown below on a number line.

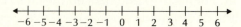

Given a number on one side of 0, we can get a number on the other side by *reflecting*. The *reflection* of 2 is −2.

We can read −2 as "negative 2," "the opposite of 2," "the additive inverse of 2," or simply "the inverse of 2."

The inverse of any number x is named −x (the inverse of x).

Example 12 Find −x when x is −3.

To find −x we reflect x to the other side of 0.

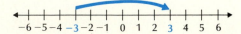

−(−3) = 3 (the inverse of negative 3 is 3, or the inverse of the inverse of 3 is 3).

ANSWERS ON PAGE A-3

Example 13 Find $-x$ when x is 0.

When we try to reflect 0 "to the other side of 0" we go nowhere:

$$-x = 0 \quad \text{when} \quad x \text{ is } 0.$$

DO EXERCISES 13–15.

Example 14 Find $-(-x)$ when x is 2.

We replace x by 2. We wish to find $-(-2)$.

We see from the figure that $-(-2) = 2$.

Example 15 Find $-(-x)$ when x is -3.

We replace x by -3. We wish to find $-(-(-3))$.

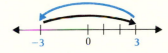

Reflecting -3 to the other side of 0 and then reflecting back gives us -3. Thus $-(-x) = -3$ when x is -3.

DO EXERCISES 16–19.

Replacing a number by its additive inverse is sometimes called *changing the sign*.

Examples Change the sign. (Find the additive inverse.)

16. -3 $-(-3) = 3$
17. -10 $-(-10) = 10$
18. 0 $-0 = 0$
19. 14 $-(14) = -14$

DO EXERCISES 20–23.

To the student and instructor: The *Skill Maintenance Exercises,* which occur at the end of most exercise sets and are indicated by the symbol ✓, review, as before, any skill that has been studied before in the text. Beginning with this chapter, however, four particular sections, along with the material of the chapter, will be tested on the Chapter Test. For this chapter the review sections to be tested are Sections 1.1, 1.2, 1.6, and 1.10.

In each case draw a number line, if necessary.

13. Find $-x$ when x is 1.

14. Find $-x$ when x is -2.

15. Find $-x$ when x is 0.

16. Find $-(-x)$ when x is 4.

17. Find $-(-x)$ when x is 1.

18. Find $-(-x)$ when x is -2.

19. Find $-(-x)$ when x is -5.

Change the sign. (Find the additive inverse.)

20. -4 **21.** -13

22. 28 **23.** 0

ANSWERS ON PAGE A-3

CALCULATOR CORNER

1. *A number pattern.* Find each of the following using your calculator. Look for a pattern.

$$1^3$$
$$1^3 + 2^3$$
$$1^3 + 2^3 + 3^3$$
$$1^3 + 2^3 + 3^3 + 4^3$$

Use the pattern to find each of the following without using your calculator.

$$1^3 + 2^3 + 3^3 + 4^3 + 5^3$$
$$1^3 + 2^3 + 3^3 + 4^3 + 5^3 + 6^3$$

2. Complete this table. Find decimal notation, rounding to four decimal places.

n	2^n	$\dfrac{1}{2^n}$
1		
2		
3		
4		
5		
6		
7		

• Tell which integers correspond to each situation.

1. The Dead Sea, between Jordan and Israel, is 1286 ft below sea level, whereas Mt. Everest is 29,028 ft above sea level.

2. During a certain time period, the United States had a deficit of $3 million in foreign trade.

3. 3 seconds before the liftoff of a rocket. 128 seconds after the liftoff of a rocket.

4. A student deposited $850 in a savings account. Two weeks later the student withdrew $432.

•• Write a true sentence using < or >.

5. 6   0

6. 0   9

7. -9 5

8. 8 -8

9. -6 6

10. 0 -7

11. -8 -5

12. -5 -3

13. -5 -11

14. -3 -4

15. -6 -5

16. -10 -14

••• Find the absolute value.

17. $|-3|$

18. $|-7|$

19. $|10|$

20. $|11|$

21. $|0|$

22. $|-4|$

23. $|-24|$

24. $|-36|$

25. $|53|$

26. $|54|$

27. $|-8|$

28. $|0|$

ANSWERS

1. ___
2. ___
3. ___
4. ___
5. ___
6. ___
7. ___
8. ___
9. ___
10. ___
11. ___
12. ___
13. ___
14. ___
15. ___
16. ___
17. ___
18. ___
19. ___
20. ___
21. ___
22. ___
23. ___
24. ___
25. ___
26. ___
27. ___
28. ___

Find $-x$ when x is each of the following.

29. -6 **30.** -7 **31.** 6 **32.** 0

Find $-(-x)$ when x is each of the following.

33. -7 **34.** -8 **35.** 1 **36.** 2

Find $-x$ when x is each of the following.

37. -12 **38.** -19 **39.** 70 **40.** 80

Find $-(-x)$ when x is each of the following.

41. 0 **42.** -2 **43.** -34 **44.** -23

Change the sign. (Find the additive inverse.)

45. -1 **46.** -7 **47.** 7 **48.** 10

49. -14 **50.** -22 **51.** 0 **52.** 1

 SKILL MAINTENANCE

53. Multiply and simplify: $\frac{21}{5} \cdot \frac{1}{7}$.

54. What is the meaning of w^4?

55. What law is illustrated by this sentence?

$$6 \cdot (4 + 8) = 6 \cdot 4 + 6 \cdot 8$$

56. Factor: $3x + 9 + 12y$.

EXTENSION

Solve. Consider integer replacements.

57. $|x| = 7$

58. $|x| < 2$

59. Simplify $-(-x)$, $-(-(-x))$, and $-(-(-(-x)))$.

60. List this set of integers in order from least to greatest.

$7^1, -5, |-6|, 4, |3|, -100, 0, 2^7, 8^0, 10^2$

Addition of Integers

• ADDITION

To explain addition of integers we can use the number line.

> To do the addition $a + b$, we start at a, and move according to b.
>
> a) If b is positive, we move to the right.
> b) If b is negative, we move to the left.
> c) If b is 0, we stay at a. The number 0 is called the *additive identity*.

Example 1 Add $3 + (-5)$.

$3 + (-5) = -2$

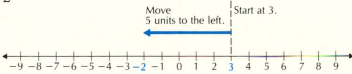

Example 2 Add $-4 + (-3)$.

$-4 + (-3) = -7$

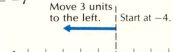

Example 3 Add $-4 + 9$.

$-4 + 9 = 5$

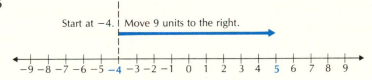

DO EXERCISES 1–7.

You may have noticed some patterns in the preceding examples and exercises. When we add positive integers and zero, the additions are, of course, additions of whole numbers. When we add two negative integers, the result is negative. We generalize as follows.

> To add two negative integers, we add their absolute values and change the sign (making the answer negative).

Examples Add.

4. $-5 + (-7) = -12$ *Think:* Add the absolute values: $5 + 7 = 12$. Then change the sign.

5. $-8 + (-2) = -10$

DO EXERCISES 8–11.

Add, using a number line.

1. $1 + (-4)$

2. $-3 + (-5)$

3. $-3 + 7$

4. $-5 + 5$

Write an addition sentence.

5.

6.

7.

Add. Do not use a number line except as a check.

8. $-5 + (-6)$

9. $-9 + (-3)$

10. $-20 + (-14)$

11. $-11 + (-11)$

ANSWERS ON PAGE A-3

Add. Do not use a number line except as a check.

12. $-4 + 6$

13. $-7 + 3$

14. $5 + (-7)$

15. $10 + (-7)$

Add. Do not use a number line except as a check.

16. $5 + (-5)$

17. $-6 + 6$

18. $-10 + 10$

19. $89 + (-89)$

Add.

20.
$$
\begin{array}{r}
-35 \\
17 \\
14 \\
-27 \\
31 \\
-12 \\
\hline
\end{array}
$$

ANSWERS ON PAGE A-3

What happens when we add a positive integer and a negative integer with different absolute values? The answer is sometimes negative and sometimes positive, depending on which number has the greater absolute value (the greater distance from 0). We can generalize as follows.

> To add a positive and a negative integer, find the difference of their absolute values.
>
> a)   If the negative integer has the greater absolute value, the answer is negative.
> b)   If the positive integer has the greater absolute value, the answer is positive.

Examples Add.

6. $3 + (-5) = -2$ *Think:* Find the difference of the absolute values. $5 - 3 = 2$. Since -5 has the greater absolute value and it is negative, make the answer negative.

7. $11 + (-8) = 3$ *Think:* Find the difference of the absolute values: $11 - 8 = 3$. Since 11 has the greater absolute value and it is positive, leave the answer positive.

8. $1 + (-5) = -4$ **9.** $-7 + 4 = -3$

10. $5 + (-10) = -5$ **11.** $7 + (-3) = 4$

12. $-6 + 10 = 4$ **13.** $5 + (-2) = 3$

DO EXERCISES 12–15.

We call $-a$ the *additive inverse of a* because adding any number to its additive inverse always gives 0—for example,

$$-8 + 8 = 0, \qquad 14 + (-14) = 0, \quad \text{and} \quad 0 + 0 = 0.$$

> For any integer a,
> $$a + (-a) = -a + a = 0.$$

DO EXERCISES 16–19.

To add several integers, some positive and some negative, we add the positive ones. We add the negative ones. Then we add the results.

Example 14

$$
\begin{array}{r}
25 \\
-14 \\
-127 \\
45 \\
32 \\
-118 \\
87 \\
\hline
\end{array}
$$

First, add the positives.
$$
\begin{array}{r}
25 \\
45 \\
32 \\
87 \\
\hline
189
\end{array}
$$

Next, add the negatives.
$$
\begin{array}{r}
-14 \\
-127 \\
-118 \\
\hline
-259
\end{array}
$$

Now add the results.
$$
\begin{array}{r}
-259 \\
189 \\
\hline
-70
\end{array}
$$

DO EXERCISE 20.

Chapter 2   Integers and Rational Numbers

• Add. Do not use a number line except as a check.

1. $-9 + 2$ 2. $2 + (-5)$ 3. $-9 + 5$ 4. $8 + (-3)$

5. $-6 + 6$ 6. $-7 + 7$ 7. $-8 + (-5)$ 8. $-3 + (-1)$

9. $-5 + (-11)$ 10. $-3 + (-4)$ 11. $-6 + (-5)$ 12. $-10 + (-14)$

13. $9 + (-9)$ 14. $10 + (-10)$ 15. $-2 + 2$ 16. $-3 + 3$

17. $0 + 8$ 18. $7 + 0$ 19. $0 + (-8)$ 20. $-7 + 0$

21. $-25 + 0$ 22. $-43 + 0$ 23. $0 + (-17)$ 24. $0 + (-19)$

25. $17 + (-17)$ 26. $-13 + 13$ 27. $-18 + 18$ 28. $11 + (-11)$

29. $8 + (-5)$ 30. $-7 + 8$ 31. $-4 + (-5)$ 32. $0 + (-3)$

33. $0 + (-5)$ 34. $10 + (-12)$ 35. $13 + (-6)$ 36. $-3 + 14$

ANSWERS
1.
2.
3.
4.
5.
6.
7.
8.
9.
10.
11.
12.
13.
14.
15.
16.
17.
18.
19.
20.
21.
22.
23.
24.
25.
26.
27.
28.
29.
30.
31.
32.
33.
34.
35.
36.

ANSWERS

37.
38.
39.
40.
41.
42.
43.
44.
45.
46.
47.
48.
49.
50.
51.
52.
53.
54.
55.
56.
57.
58.
59.
60.
61.
62.
63.
64.
65.
66.
67.
68.

37. $-10 + 7$ **38.** $0 + (-9)$ **39.** $-6 + 6$ **40.** $-8 + 1$

41. $11 + (-16)$ **42.** $-7 + 15$ **43.** $-15 + (-6)$ **44.** $-8 + 8$

45. $11 + (-9)$ **46.** $-14 + (-19)$ **47.** $-20 + (-6)$ **48.** $17 + (-17)$

49. $-15 + (-7)$ **50.** $23 + (-5)$ **51.** $40 + (-8)$ **52.** $-23 + (-9)$

53. $-25 + 25$ **54.** $40 + (-40)$ **55.** $63 + (-18)$ **56.** $85 + (-65)$

57. -35
          -63
          -27
          -14
          $\underline{-59}$

58. -24
          -37
          -19
          -45
          $\underline{-35}$

59. 27
          -54
          -32
           65
          $\underline{46}$

60. -62
           53
          -87
           14
          $\underline{-28}$

✓ SKILL MAINTENANCE

61. Multiply: $7(3z + y + 2)$.   **62.** Divide and simplify: $\frac{7}{2} \div \frac{3}{8}$.

63. What is 25% of 14?   **64.** Find the LCM: 36, 24, 48.

⭐ EXTENSION

Add.

65. -64374
          -27159
           53690
           39087
           41646
          $\underline{-11953}$

66. -49752
           28351
          -92265
          -40870
          $\underline{35649}$

67. For what numbers x is $-x$ negative?   **68.** For what numbers x is $-x$ positive?

Subtraction of Integers

• SUBTRACTION

Subtraction is defined as follows.

> The difference $a - b$ is the number which when added to b gives a.

Example 1 Subtract: $5 - 8$.

Think: $5 - 8$ is the number which when added to 8 gives 5. What number can we add to 8 to get 5? The number has to be negative. The number is -3; $5 - 8 = -3$.

DO EXERCISES 1–3.

The definition above is *not* the most efficient way to do subtraction. From that definition a faster way can be developed. Look for a pattern in the following examples. The subtractions on the left were done in either Example 1 above or in Margin Exercises 1–3.

Subtractions	*Adding an Inverse*
$5 - 8 = -3$	$5 + (-8) = -3$
$-6 - 4 = -10$	$-6 + (-4) = -10$
$-7 - (-10) = 3$	$-7 + 10 = 3$
$-7 - (-2) = -5$	$-7 + 2 = -5$

DO EXERCISES 4–7.

Perhaps you have noticed in the preceding examples and exercises that we can subtract by adding an additive inverse. This can always be done.

> For any integers a and b,
> $$a - b = a + (-b).$$
> (To subtract, we can add the additive inverse of the subtrahend.)

The preceding is the method one normally uses for quick subtraction of integers.

Example 2 Subtract: $2 - 6$. Check by addition.

$2 - 6 = 2 + (-6)$ Changing the sign of 6 to -6, and
$\quad\quad = -4$ changing the subtraction to addition

Check: $6 + (-4) = 2$; $2 - 6$, or -4, is that number which when added to 6 gives 2.

DO EXERCISE 8.

Subtract.

1. $-6 - 4$ *Think:* $-6 - 4$ is the number which when added to 4 gives -6. What number can be added to 4 to get -6?

2. $-7 - (-10)$ *Think:* $-7 - (-10)$ is the number which when added to -10 gives -7. What number can be added to -10 to get -7?

3. $-7 - (-2)$ *Think:* $-7 - (-2)$ is the number which when added to -2 gives -7. What number can be added to -2 to get -7?

Complete each addition and compare with the subtraction.

4. $4 + (-6) = $ _____ ;
$4 - 6 = -2$

5. $-3 + (-8) = $ _____ ;
$-3 - 8 = -11$

6. $-5 + 9 = $ _____ ;
$-5 - (-9) = 4$

7. $-5 + 3 = $ _____ ;
$-5 - (-3) = -2$

8. Subtract: $2 - 8$. Check by addition.

ANSWERS ON PAGE A-3

9. Subtract: $-6 - 10$. Check by addition.

10. Subtract: $12 - 5$. Check by addition.

11. Subtract: $-8 - (-11)$. Check by addition.

12. Subtract: $-8 - (-2)$. Check by addition.

Read each of the following. Then subtract by adding the inverse of the subtrahend.

13. $3 - 11$

14. $12 - 5$

15. $-12 - (-9)$

16. $-12 - 10$

17. $-14 - (-14)$

ANSWERS ON PAGE A-3

Example 3 Subtract: $-3 - 8$. Check by addition.

$$-3 - 8 = -3 + (-8)$$ Changing the sign of 8 to -8, and changing the subtraction to addition
$$= -11$$

Check: $8 + (-11) = -3$; $-3 - 8$, or -11, is that number which when added to 8 gives -3.

DO EXERCISE 9.

Example 4 Subtract: $10 - 7$. Check by addition.

$$10 - 7 = 10 + (-7)$$ Changing the sign of 7 to -7, and changing the subtraction to addition
$$= 3$$

Check: $7 + 3 = 10$; $10 - 7$, or 3, is that number which when added to 7 gives 10.

DO EXERCISE 10.

Example 5 Subtract: $-4 - (-9)$. Check by addition.

$$-4 - (-9) = -4 + 9$$ Changing the sign of -9 to 9, and changing the subtraction to addition
$$= 5$$

Check: $-9 + 5 = -4$; $-4 - (-9)$, or 5, is that number which when added to -9 gives -4.

DO EXERCISE 11.

Example 6 Subtract: $-4 - (-3)$. Check by addition.

$$-4 - (-3) = -4 + 3$$ Changing the sign of -3 to 3, and changing the subtraction to addition
$$= -1$$

Check: $-3 + (-1) = -4$; $-4 - (-3)$, or -1, is that number which when added to -3 gives -4.

DO EXERCISE 12.

Examples Read each of the following. Then subtract by adding the inverse of the subtrahend.

7. $18 - 20$;   Read "eighteen minus twenty"
   $18 - 20 = 18 + (-20) = -2$

8. $3 - 7$;   Read "three minus seven"
   $3 - 7 = 3 + (-7) = -4$

9. $-14 - (-24)$;   Read "negative fourteen minus negative twenty-four"
   $-14 - (-24) = -14 + 24 = 10$

10. $-54 - 30$;   Read "negative fifty-four minus thirty"
    $-54 - 30 = -54 + (-30) = -84$

11. $-10 - (-10)$;   Read "negative ten minus negative ten"
    $-10 - (-10) = -10 + 10 = 0$

DO EXERCISES 13–17.

When several additions and subtractions occur together, we can convert the subtractions to additions.

Example 12 Simplify: $8 - (-4) - 2 - (-9) + 2$.

$$8 - (-4) - 2 - (-9) + 2 = 8 + 4 + (-2) + 9 + 2$$
$$= 21$$

DO EXERCISE 18.

•• SOLVING PROBLEMS

We can use addition and subtraction of integers to solve problems.

Example 13 A small business made a profit of $18 on Monday. There was a loss of $7 on Tuesday. On Wednesday there was a loss of $5, and on Thursday there was a profit of $11. Find the total profit or loss.

We can represent a loss with a negative number and a profit with a positive number. Thus we have the following losses and profits:

$$18, \quad -7, \quad -5, \quad 11.$$

The total loss or profit is the sum of these numbers:

$$18 + (-7) + (-5) + 11 = 29 + (-12)$$
$$= 17.$$

The total profit was $17.

Example 14 In Churchill, Manitoba, Canada, the average daily low temperature in January is $-31°C$. The average daily low temperature in Key West, Florida, is $19°C$. How much higher is the average daily low temperature in Key West, Florida?

We first draw a picture of the situation. We can do this using a number line.

Churchill, Manitoba   Key West, Florida

$-31° \longleftarrow x \longrightarrow 19°$

To find the amount x, we subtract -31 from 19:

$$19 - (-31) = x$$
$$19 + 31 = x$$
$$50 = x.$$

The average daily low temperature in Key West is $50°C$ higher than the average daily low temperature in Churchill, Manitoba.

DO EXERCISES 19 AND 20.

Simplify.

18. $-6 - (-2) - (-4) - 12 + 3$

Solve.

19. A lab technician had 500 mL (milliliters) of acid in a beaker. After 16 mL had been poured out and 27 mL had been added, how much was in the beaker?

20. The elevation of Death Valley is -280 ft. The elevation of Dallas, Texas, is 512 ft. How much higher is Dallas than Death Valley?

ANSWERS ON PAGE A-3

CALCULATOR CORNER: NUMBER PATTERNS

In each of the following, do the first four calculations using your calculator. Look for a pattern. Then do the last calculation without using your calculator.

1. 1^2
    11^2
    111^2
    1111^2
    $11{,}111^2$

2. 6×4
    66×44
    666×444
    6666×4444
    $66{,}666 \times 44{,}444$
    $666{,}666 \times 444{,}444$

3. 8×6
    68×6
    668×6
    6668×6
    $66{,}668 \times 6$

4. 9^2
    99^2
    999^2
    9999^2
    $99{,}999^2$

• Subtract by adding the inverse of the subtrahend.

1. $3 - 7$

2. $4 - 9$

3. $0 - 7$

4. $0 - 10$

5. $-8 - (-2)$

6. $-6 - (-8)$

7. $-18 - (-18)$

8. $-8 - (-8)$

9. $12 - 16$

10. $14 - 19$

11. $20 - 27$

12. $30 - 40$

13. $-9 - (-3)$

14. $-7 - (-9)$

15. $-40 - (-40)$

16. $-9 - (-9)$

17. $7 - 7$

18. $9 - 9$

19. $7 - (-7)$

20. $4 - (-4)$

21. $8 - (-3)$

22. $-7 - 4$

23. $-6 - 8$

24. $6 - (-10)$

25. $-4 - (-9)$

26. $-14 - 2$

27. $2 - 9$

28. $2 - 8$

29. $-6 - (-5)$

30. $-4 - (-3)$

31. $8 - (-10)$

32. $5 - (-6)$

33. $0 - 5$

34. $0 - 6$

35. $-5 - (-2)$

36. $-3 - (-1)$

37. $-7 - 14$

38. $-9 - 16$

ANSWERS
1.
2.
3.
4.
5.
6.
7.
8.
9.
10.
11.
12.
13.
14.
15.
16.
17.
18.
19.
20.
21.
22.
23.
24.
25.
26.
27.
28.
29.
30.
31.
32.
33.
34.
35.
36.
37.
38.

39. $0 - (-5)$ **40.** $0 - (-1)$ **41.** $-8 - 0$ **42.** $-9 - 0$

43. $7 - (-5)$ **44.** $8 - (-3)$ **45.** $2 - 25$ **46.** $18 - 63$

47. $-42 - 26$ **48.** $-18 - 63$ **49.** $-71 - 2$ **50.** $-49 - 3$

51. $24 - (-92)$ **52.** $48 - (-73)$ **53.** $-50 - (-50)$ **54.** $-70 - (-70)$

Simplify.

55. $18 - (-15) - 3 - (-5) + 2$ **56.** $22 - (-18) + 7 + (-42) - 27$

57. $-31 + (-28) - (-14) - 17$ **58.** $-43 - (-19) - (-21) + 25$

59. $-34 - 28 + (-33) - 44$ **60.** $39 - (-88) - 29 - (-83)$

• • Solve.

61. The lowest temperature ever recorded in Minneapolis is $-34°F$. The lowest temperature ever recorded in Los Angeles is $28°F$. How much higher was the temperature in Los Angeles than in Minnesota?

62. A company experienced the following profits or losses over a five-day period: profit $320, loss $292, profit $814, loss $193, loss $138. What was the total profit or loss?

✔ **SKILL MAINTENANCE**

63. Find the area of a rectangle when the length is 36 ft and the width is 12 ft.

64. Find the prime factorization of 864.

☆ **EXTENSION**

Subtract.

65. $123,907 - 433,789$ **66.** 🖩 $23,011 - (-60,432)$

Tell whether each of the following is true or false for all integers a and b. If false, give a counterexample.

67. $a - 0 = 0 - a$ **68.** If $a + b = 0$, then $a = -b$.

Multiplication and Division of Integers

• **MULTIPLICATION**

Multiplication of integers is very much like multiplication of whole numbers. The only difference is that we must determine whether the answer is positive or negative. To see how this is done, consider the pattern in the following.

This number decreases by 1 each time.

$$4 \cdot 5 = 20$$
$$3 \cdot 5 = 15$$
$$2 \cdot 5 = 10$$
$$1 \cdot 5 = 5$$
$$0 \cdot 5 = 0$$
$$-1 \cdot 5 = -5$$
$$-2 \cdot 5 = -10$$
$$-3 \cdot 5 = -15$$

This number decreases by 5 each time.

DO EXERCISE 1.

According to this pattern, it looks as though the product of a negative integer and a positive integer is negative. This is the case.

To multiply a positive integer and a negative integer, multiply their absolute values. The answer is negative.

Examples Multiply.

1. $-4 \cdot 7 = -28$ *Think:* Multiply the absolute values: $4 \cdot 7 = 28$. Then the answer is negative.

2. $8 \cdot (-5) = -40$

3. $10 \cdot (-2) = -20$

4. $-7 \cdot 5 = -35$

DO EXERCISES 2–4.

How do we multiply two negative numbers? To see this we can again look for a pattern.

This number decreases by 1 each time.

$$4 \cdot (-5) = -20$$
$$3 \cdot (-5) = -15$$
$$2 \cdot (-5) = -10$$
$$1 \cdot (-5) = -5$$
$$0 \cdot (-5) = 0$$
$$-1 \cdot (-5) = 5$$
$$-2 \cdot (-5) = 10$$
$$-3 \cdot (-5) = 15$$

This number increases by 5 each time.

DO EXERCISE 5.

1. Complete, as in the example.

$$4 \cdot 10 = 40$$
$$3 \cdot 10 = 30$$
$$2 \cdot 10 =$$
$$1 \cdot 10 =$$
$$0 \cdot 10 =$$
$$-1 \cdot 10 =$$
$$-2 \cdot 10 =$$
$$-3 \cdot 10 =$$

Multiply.

2. $-3 \cdot 6$

3. $20 \cdot (-5)$

4. $4 \cdot (-20)$

5. Complete, as in the example.

$$3 \cdot (-10) = -30$$
$$2 \cdot (-10) = -20$$
$$1 \cdot (-10) =$$
$$0 \cdot (-10) =$$
$$-1 \cdot (-10) =$$
$$-2 \cdot (-10) =$$
$$-3 \cdot (-10) =$$

ANSWERS ON PAGE A-4

Multiply.

6. $-3 \cdot (-4)$

7. $-16 \cdot (-2)$

8. $-7 \cdot (-5)$

Divide.

9. $6 \div (-3)$

10. $\dfrac{-15}{-3}$

Divide.

11. $-24 \div 8$

12. $\dfrac{-32}{-4}$

13. $\dfrac{30}{-5}$

ANSWERS ON PAGE A-4

To multiply two negative integers, multiply their absolute values. The answer is positive.

DO EXERCISES 6–8.

• • DIVISION

To divide integers we use the definition of division: $a \div b$ or a/b is the number which when multiplied by b gives a. Consider the division

$$14 \div (-7).$$

We look for a number n such that $-7 \cdot n = 14$. We know from multiplication that -2 is the solution. That is,

$$-7 \cdot (-2) = 14 \quad \text{so} \quad 14 \div (-7) = -2$$

Examples Divide.

5. $-8 \div 2 = -4$ because $2 \cdot (-4) = -8$

6. $\dfrac{-10}{-2} = 5$ because $-2 \cdot 5 = -10$

DO EXERCISES 9 AND 10.

From these examples we see how to handle signs in division.

When we divide a positive integer by a negative, or a negative integer by a positive, the answer is negative. When we divide two negative integers, the answer is positive.

Examples Divide.

7. $-24 \div 6 = -4$ *Think:* $\dfrac{24}{6} = 4$. Then the answer is negative.

8. $\dfrac{-30}{-6} = 5$

9. $\dfrac{18}{-9} = -2$

There are some divisions of integers, such as $18/-5$, for which we do not get integers for answers. To get an answer, we need an expanded number system, which we will consider in the next section.

DO EXERCISES 11–13.

Chapter 2   Integers and Rational Numbers

• Multiply.

1. $-8 \cdot 2$

2. $-2 \cdot 5$

3. $-7 \cdot 6$

4. $-9 \cdot 2$

5. $8 \cdot (-3)$

6. $9 \cdot (-5)$

7. $-9 \cdot 8$

8. $-10 \cdot 3$

9. $-8 \cdot (-2)$

10. $-2 \cdot (-5)$

11. $-7 \cdot (-6)$

12. $-9 \cdot (-2)$

13. $-8 \cdot (-3)$

14. $-9 \cdot (-5)$

15. $-9 \cdot (-8)$

16. $-10 \cdot (-3)$

17. $15 \cdot (-8)$

18. $12 \cdot (-10)$

19. $-25 \cdot (-40)$

20. $-22 \cdot (-50)$

21. $-6 \cdot (-15)$

22. $-8 \cdot (-22)$

23. $-25 \cdot (-8)$

24. $-35 \cdot (-4)$

ANSWERS
1.
2.
3.
4.
5.
6.
7.
8.
9.
10.
11.
12.
13.
14.
15.
16.
17.
18.
19.
20.
21.
22.
23.
24.

Divide.

25. $36 \div (-6)$

26. $\dfrac{28}{-7}$

27. $\dfrac{26}{-2}$

28. $26 \div (-13)$

29. $\dfrac{-16}{8}$

30. $-22 \div (-2)$

31. $\dfrac{-48}{-12}$

32. $\dfrac{-63}{-9}$

33. $\dfrac{-72}{9}$

34. $\dfrac{-50}{25}$

35. $-100 \div (-50)$

36. $\dfrac{-200}{8}$

37. $-108 \div 9$

38. $\dfrac{-64}{-8}$

39. $\dfrac{200}{-25}$

40. $-300 \div (-75)$

✔ SKILL MAINTENANCE

41. Simplify: $\dfrac{54}{48}$.

42. Collect like terms: $27b + 9a + 5b + 3a$.

43. Solve: $3.1x = 387.5$.

44. 9 is what percent of 60?

☆ EXTENSION

Simplify.

45. $(-4)^2$

46. -4^2

47. $-4 \cdot 3 + 10 \cdot (-2)$

48. $\dfrac{|-24|}{3 - 15}$

49. $(-1)^{23}$

50. -1^{23}

Chapter 2   Integers and Rational Numbers

Rational Numbers

The coldest temperature on record is   126.9°F in Vostok, Antarctica, on August 24, 1960. −126.9 is a *rational* number.

The set of *rational numbers* consists of all the numbers of arithmetic and their additive inverses. The rational numbers include all the integers. There are positive and negative rational numbers, the same as for integers, but between any two integers there are many rational numbers.

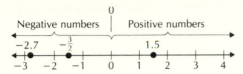

After finishing Section 2.5, you should be able to:

• Find the additive inverse and absolute value of any rational number and simplify additive inverse symbols.

•• Add rational numbers.

••• Subtract rational numbers.

• INVERSES AND ABSOLUTE VALUE

Everything we have said about integers so far holds true for rational numbers. For example, all rational numbers have additive inverses, and absolute value means the same as for integers.

Examples Find the additive inverse.

1. 4.2   The additive inverse is −4.2.
2. $-\frac{3}{2}$ The additive inverse is $\frac{3}{2}$.
3. −9.35   The additive inverse is 9.35.

DO EXERCISES 1–3.

Examples Simplify.

> "Simplify" means to rename with the least number of symbols.

4. $-(-\frac{4}{5}) = \frac{4}{5}$
5. $-(-9.9) = 9.9$
6. $-(\frac{3}{8}) = -\frac{3}{8}$

DO EXERCISES 4–6.

Example 7 Find $-a$ when a is −6.

$-a$ is the inverse of a. When a is −6, we have

$$-(-6), \quad \text{which is } 6.$$

Example 8 Find $-a$ when a is $-\frac{7}{8}$.

$$-a = -(-\tfrac{7}{8}) = \tfrac{7}{8}$$

> *Remember:* $-a$ should not be read "negative a." It should be read "the additive inverse of a." These examples illustrate that we can't know whether $-a$ is positive, negative, or 0 until we know what a represents.

DO EXERCISES 7–9.

Find the additive inverse.

1. 8.7

2. $-\dfrac{8}{9}$

3. −7.74

Simplify.

4. $-\left(-\dfrac{10}{3}\right)$

5. $-(-8.32)$

6. $-\left(\dfrac{5}{4}\right)$

Find $-x$ when x is each of the following.

7. −12

8. $-\dfrac{5}{6}$

9. 17.2

ANSWERS ON PAGE A-4

Simplify.

10. $|4.1|$

11. $\left|-\dfrac{8}{3}\right|$

12. $|-3.5|$

Add.

13. $-8.6 + 2.4$

14. $\dfrac{5}{9} + \left(-\dfrac{7}{9}\right)$

15. $-\dfrac{1}{5} + \left(-\dfrac{3}{4}\right)$

16. A lab technician had 500 mL of acid in a beaker. After 16.5 mL had been poured out and 27.3 mL had been added, how much was in the beaker?

Subtract.

17. $-3 - 6$

18. $\dfrac{7}{10} - \dfrac{9}{10}$

19. $-8.8 - (-1.3)$

20. $-\dfrac{2}{3} - \left(-\dfrac{5}{8}\right)$

ANSWERS ON PAGE A-4

Examples Simplify.

9. $|8.7| = 8.7$

10. $\left|-\dfrac{3}{4}\right| = \dfrac{3}{4}$

11. $|-17.3| = 17.3$

DO EXERCISES 10–12.

• • ADDITION

Addition is done as with integers.

Examples Add.

12. $-9.2 + 3.1 = -6.1$

13. $-\dfrac{3}{2} + \dfrac{9}{2} = \dfrac{6}{2} = 3$

14. $-\dfrac{2}{3} + \dfrac{5}{8} = -\dfrac{16}{24} + \dfrac{15}{24} = -\dfrac{1}{24}$

15. One day the value of a share of IBM stock was $253\frac{1}{4}$. That day it rose in value $\$\frac{5}{8}$. The next day it lost $\$\frac{3}{8}$. What was the value of the stock at the end of the two days?

$$253\tfrac{1}{4} + \tfrac{5}{8} + \left(-\tfrac{3}{8}\right) = 253\tfrac{2}{8} + \tfrac{5}{8} + \left(-\tfrac{3}{8}\right)$$
$$= 253\tfrac{4}{8}, \quad \text{or} \quad \$253\tfrac{1}{2}$$

DO EXERCISES 13–16.

• • • SUBTRACTION

Subtraction is done as with integers.

> **For any rational numbers a and b,**
>
> $$a - b = a + (-b).$$
>
> **(To subtract we add the inverse of the subtrahend.)**

Examples Subtract.

16. $3 - 5 = 3 + (-5) = -2$

17. $\dfrac{1}{8} - \dfrac{7}{8} = \dfrac{1}{8} + \left(-\dfrac{7}{8}\right) = -\dfrac{6}{8}, \quad \text{or} \quad -\dfrac{3}{4}$

18. $-4.6 - (-9.8) = -4.6 + 9.8 = 5.2$

19. $-\dfrac{3}{4} - \dfrac{7}{5} = -\dfrac{15}{20} + \left(-\dfrac{28}{20}\right) = -\dfrac{43}{20}$

DO EXERCISES 17–20.

Addition of rational numbers is associative and commutative. Each rational number has an additive inverse.

• Find the additive inverse.

1. -4.7

2. -5.2

3. $\dfrac{7}{2}$

4. $\dfrac{8}{3}$

5. -7

6. -9

7. -26.9

8. -17.83

Simplify.

9. $-\left(-\dfrac{1}{3}\right)$

10. $-\left(-\dfrac{1}{4}\right)$

11. $-\left(\dfrac{7}{6}\right)$

12. $-\left(\dfrac{9}{8}\right)$

13. $-(-9.3)$

14. $-(-8.6)$

15. $-(90.3)$

16. $-(78.8)$

Find $-x$ when x is each of the following.

17. 12.4

18. 23.8

19. $-\dfrac{9}{10}$

20. $-\dfrac{11}{12}$

21. -34.8

22. -44.1

23. 567

24. 890

Simplify.

25. $|19.2|$

26. $|16.8|$

27. $\left|-\dfrac{2}{3}\right|$

28. $\left|-\dfrac{5}{6}\right|$

29. $|-89.3|$

30. $|-17.4|$

31. $\left|\dfrac{14}{3}\right|$

32. $\left|\dfrac{16}{5}\right|$

•• Add.

33. $-6.5 + 4.7$

34. $-3.6 + 1.9$

35. $-2.8 + (-5.3)$

36. $-7.9 + (-6.5)$

ANSWERS
1.
2.
3.
4.
5.
6.
7.
8.
9.
10.
11.
12.
13.
14.
15.
16.
17.
18.
19.
20.
21.
22.
23.
24.
25.
26.
27.
28.
29.
30.
31.
32.
33.
34.
35.
36.

37. $-\dfrac{3}{5} + \dfrac{2}{5}$　　**38.** $-\dfrac{4}{3} + \dfrac{2}{3}$　　**39.** $-\dfrac{3}{7} + \left(-\dfrac{5}{7}\right)$　　**40.** $-\dfrac{4}{9} + \left(-\dfrac{6}{9}\right)$

41. $-\dfrac{5}{8} + \dfrac{1}{4}$　　**42.** $-\dfrac{5}{6} + \dfrac{2}{3}$　　**43.** $-\dfrac{3}{7} + \left(-\dfrac{2}{5}\right)$　　**44.** $-\dfrac{5}{8} + \left(-\dfrac{1}{3}\right)$

45. $-\dfrac{3}{5} + \left(-\dfrac{2}{15}\right)$　　　　　**46.** $-\dfrac{5}{9} + \left(-\dfrac{1}{18}\right)$

47. $-5.7 + (-7.2) + 6.6$　　　　　**48.** $-10.3 + (-7.5) + 3.1$

49. $-8.5 + 7.9 + (-3.7)$　　　　　**50.** $-9.6 + 8.4 + (-11.8)$

 Subtract.

51. $-9 - (-5)$　　**52.** $-8 - (-6)$　　**53.** $\dfrac{3}{8} - \dfrac{5}{8}$　　**54.** $\dfrac{3}{9} - \dfrac{9}{9}$

55. $\dfrac{3}{4} - \dfrac{2}{3}$　　**56.** $\dfrac{5}{8} - \dfrac{3}{4}$　　**57.** $-\dfrac{3}{4} - \dfrac{2}{3}$　　**58.** $-\dfrac{5}{8} - \dfrac{3}{4}$

59. $-\dfrac{5}{8} - \left(-\dfrac{3}{4}\right)$　　**60.** $-\dfrac{3}{4} - \left(-\dfrac{2}{3}\right)$　　**61.** $6.1 - (-13.8)$　　**62.** $1.5 - (-3.5)$

63. $-3.2 - 5.8$　　**64.** $-2.7 - 5.9$　　**65.** $0.99 - 1$　　**66.** $0.87 - 1$

✓ SKILL MAINTENANCE

67. Evaluate $\dfrac{3x}{y}$ when $x = 4$ and $y = 12$.

68. Translate to an algebraic expression: sixty-four percent of t.

69. Write decimal notation: $\dfrac{19}{25}$.

70. Solve: $x + \dfrac{2}{3} = \dfrac{5}{6}$.

☆ EXTENSION

71. Use the proper symbol $<$, $>$, or $=$.

　　a)　-8.6　　-9.0

　　b)　$-2\dfrac{1}{8}$　　$\dfrac{-35}{16}$

72. Arrange from smallest to largest: $-\dfrac{1}{3}$, $-\dfrac{1}{4}$, $-\dfrac{5}{6}$, $-\dfrac{1}{16}$, 0, $-\dfrac{5}{12}$.

Simplify.

73. $-(-3.8 + 7.2)$　　　**74.** $|-4.21 - 3.68|$　　　**75.** $-\left|-\dfrac{1}{3} + \dfrac{5}{6}\right|$

Multiplication and Division of Rational Numbers

• MULTIPLICATION

Multiplication of rational numbers is like multiplication of integers.

> To multiply a negative number and a positive number, multiply their absolute values. The answer is negative.

Examples Multiply.

1. $-9 \cdot 4 = -36$
2. $-\frac{1}{3} \cdot \frac{5}{7} = -\frac{5}{21}$
3. $\frac{5}{6} \cdot \left(-\frac{1}{5}\right) = -\frac{5}{30} = -\frac{1}{6}$
4. $-8.66 \times 4.3 = -37.238$

DO EXERCISES 1–4.

> To multiply two negative numbers, multiply their absolute values. The answer is positive.

Examples Multiply.

5. $-6(-8) = 48$
6. $-9.5(-7.4) = 70.3$
7. $-\frac{1}{2}\left(-\frac{1}{3}\right) = \frac{1}{6}$
8. $-\frac{3}{4}\left(-\frac{8}{9}\right) = \frac{24}{36} = \frac{2}{3}$

DO EXERCISES 5–8.

• • MULTIPLYING MORE THAN TWO NUMBERS

To multiply several numbers, we can multiply two at a time. We can do this because multiplication of rational numbers is also commutative and associative. This means we can group and order as we please.

Examples Multiply.

9. $3 \cdot (-2) \cdot 4 = (3 \cdot 4) \cdot (-2)$ Changing grouping and ordering
   $= 12 \cdot (-2) = -24$

10. $-5 \times (-3.8) \times 2.4 = 19 \times 2.4$ Multiplying the negatives
    $= 45.6$

11. $-\frac{1}{3} \cdot \left(-\frac{3}{5}\right) \cdot \left(-\frac{15}{7}\right) = \frac{1}{5} \cdot \left(-\frac{15}{7}\right)$ Multiplying the first two numbers
    $= -\frac{3}{7}$

12. $-5 \cdot (-2) \cdot (-3) \cdot (-6) = 10 \cdot 18$ Multiplying the first two numbers and the last two numbers
    $= 180$

After finishing Section 2.6, you should be able to:

• Multiply two rational numbers.

• • Multiply several positive and negative numbers.

• • • Find the reciprocal of a rational number, and divide by multiplying by a reciprocal.

Multiply.

1. $6(-5)$

2. $\frac{2}{3}\left(-\frac{5}{9}\right)$

3. $-\frac{4}{5} \cdot \frac{7}{8}$

4. -4.23×7.1

Multiply.

5. $-16(-4)$

6. $-\frac{4}{7}\left(-\frac{5}{9}\right)$

7. $-\frac{3}{2}\left(-\frac{4}{9}\right)$ $\frac{2}{3}$

8. $-3.25(-4.14)$

ANSWERS ON PAGE A-4

Multiply.

9. $5 \cdot (-3) \cdot 2$

10. $-3 \times (-4.1) \times (-2.5)$

11. $-\dfrac{1}{2} \cdot \left(-\dfrac{4}{3}\right) \cdot \left(-\dfrac{5}{2}\right)$

12. $-2 \cdot (-5) \cdot (-4) \cdot (-3)$

Find the reciprocal.

13. $\dfrac{2}{3}$

14. $-\dfrac{5}{4}$

15. -3

16. $-\dfrac{1}{5}$

17. Complete the following table.

Number	Additive inverse	Reciprocal
$\frac{2}{3}$		
$-\frac{5}{4}$		
0		
1		
-4.5		

ANSWERS ON PAGE A-4

The product of an odd number of negative numbers is negative. The product of an even number of negative numbers is positive.

DO EXERCISES 9–12.

••• RECIPROCALS AND DIVISION

Two numbers are reciprocals* of each other if their product is 1. (See Section 1.7.)

Examples

13. The reciprocal of $\frac{7}{8}$ is $\frac{8}{7}$ because $\frac{7}{8} \cdot \frac{8}{7} = \frac{56}{56} = 1$.

14. The reciprocal of $-\frac{2}{3}$ is $-\frac{3}{2}$ because $-\frac{2}{3}\left(-\frac{3}{2}\right) = \frac{6}{6} = 1$.

15. The reciprocal of a nonzero number a is $\dfrac{1}{a}$ because

$$a \cdot \frac{1}{a} = \frac{a}{a} = 1.$$

> Any nonzero number a has a reciprocal $1/a$. The reciprocal of a negative number is negative. The reciprocal of a positive number is positive. The reciprocal of a/b is b/a. The reciprocal of $-(b/a)$ is $-(a/b)$.

DO EXERCISES 13–16.

It is important *not* to confuse *additive inverse* with *reciprocal*. Keep in mind that the additive inverse of a number is what we add to it to get 0, whereas a reciprocal is what we multiply the number by to get 1. Compare the following.

Number	Additive inverse	Reciprocal
$-\frac{3}{8}$	$\frac{3}{8}$	$-\frac{8}{3}$
19	-19	$\frac{1}{19}$
$\frac{18}{7}$	$-\frac{18}{7}$	$\frac{7}{18}$
-7.9	7.9	$\dfrac{1}{-7.9}$ or $-\frac{10}{79}$
0	0	Does not exist

DO EXERCISE 17.

_____
*Also called *multiplicative inverses.*

Chapter 2   Integers and Rational Numbers

We can divide by multiplying by the reciprocal of the divisor.

Examples Divide.

16. $\dfrac{2}{3} \div \left(-\dfrac{5}{4} \right) = \dfrac{2}{3} \cdot \left(-\dfrac{4}{5} \right) = -\dfrac{8}{15}$

17. $-\dfrac{5}{6} \div \left(-\dfrac{3}{4} \right) = -\dfrac{5}{6} \cdot \left(-\dfrac{4}{3} \right) = \dfrac{20}{18} = \dfrac{10 \cdot 2}{9 \cdot 2} = \dfrac{10}{9} \cdot \dfrac{2}{2} = \dfrac{10}{9}$

> *Caution!* Do not change the sign of a number when taking its reciprocal.

18. $-\dfrac{3}{4} \div \dfrac{3}{10} = -\dfrac{3}{4} \cdot \left(\dfrac{10}{3} \right) = -\dfrac{30}{12} = -\dfrac{5}{2} \cdot \dfrac{6}{6} = -\dfrac{5}{2}$

19. $-6.3 \div 2.1 = -3$ Dividing 6.3 by 2.1 and attaching the appropriate sign. Multiplying by the reciprocal is not necessary here.

DO EXERCISES 18–21.

We can also describe the rational numbers as quotients of integers.

> **The set of rational numbers consists of all numbers that can be named with fractional notation a/b, where a and b are integers and b is not 0.**

The following are examples of rational numbers:

$$\dfrac{5}{4}, \quad \dfrac{-3}{8}, \quad \dfrac{10}{1}, \quad 8, \quad -467, \quad 0, \quad \dfrac{13}{-5}, \quad -\dfrac{2}{3}.$$

A negative number divided by a negative number is positive. Thus,

$$\dfrac{-2}{-3} = \dfrac{2}{3}.$$

We can verify this by multiplying by 1 using $\dfrac{-1}{-1}$:

$$\dfrac{-2}{-3} = 1 \cdot \dfrac{-2}{-3} = \dfrac{-1}{-1} \cdot \dfrac{-2}{-3} = \dfrac{(-1)(-2)}{(-1)(-3)} = \dfrac{2}{3}.$$

Example 20 Divide: $\dfrac{-7}{-8}$.

$$\dfrac{-7}{-8} = \dfrac{7}{8}$$

DO EXERCISES 22 AND 23.

Divide.

18. $\dfrac{4}{7} \div \left(-\dfrac{3}{5} \right)$

19. $-\dfrac{8}{5} \div \dfrac{2}{3}$

20. $-\dfrac{12}{7} \div \left(-\dfrac{3}{4} \right)$

21. $21.7 \div (-3.1)$

Divide.

22. $\dfrac{-19}{-20}$

23. $\dfrac{-8}{-5}$

ANSWERS ON PAGE A-4

Divide.

24. $\dfrac{-10}{3}$

25. $\dfrac{5}{-6}$

Find two equivalent expressions for the number with negative signs in different places.

26. $\dfrac{5}{-6}$

27. $-\dfrac{8}{7}$

28. $\dfrac{10}{-3}$

A negative number divided by a positive number is negative, and a positive number divided by a negative number is negative. Thus,

$$\dfrac{-2}{3} = \dfrac{2}{-3} = -\dfrac{2}{3}.$$

We can verify the preceding by working backwards from $-\frac{2}{3}$:

$$-\dfrac{2}{3} = -1 \cdot \dfrac{2}{3} = \dfrac{-1}{1} \cdot \dfrac{2}{3} = \dfrac{-1 \cdot 2}{1 \cdot 3} = \dfrac{-2}{3},$$

$$-\dfrac{2}{3} = -1 \cdot \dfrac{2}{3} = \dfrac{1}{-1} \cdot \dfrac{2}{3} = \dfrac{1 \cdot 2}{-1 \cdot 3} = \dfrac{2}{-3}.$$

Examples Divide.

21. $\dfrac{-4}{5} = -\dfrac{4}{5}$

22. $\dfrac{23}{-8} = -\dfrac{23}{8}$

Example 23 Find two different equivalent expressions for $\dfrac{-4}{5}$ with negative signs in different places.

Two different equivalent expressions are $\dfrac{4}{-5}$ and $-\dfrac{4}{5}$.

We can generalize the results of the preceding examples to a method for finding equivalent fractional notation.

For any numbers a and b, $b \neq 0$,

$$-\dfrac{a}{b} = \dfrac{-a}{b} = \dfrac{a}{-b}, \quad \dfrac{-a}{-b} = \dfrac{a}{b}, \quad \text{and} \quad -\dfrac{-a}{-b} = -\dfrac{a}{b}.$$

DO EXERCISES 24–28.

• Multiply.

1. $9 \cdot (-8)$

2. $7 \cdot (-9)$

3. $4 \cdot (-3.1)$

4. $3 \cdot (-2.2)$

5. $-6 \cdot (-4)$

6. $-5 \cdot (-6)$

7. $-7 \cdot (-3.1)$

8. $-4 \cdot (-3.2)$

9. $\frac{2}{3} \cdot \left(-\frac{3}{5}\right)$

10. $\frac{5}{7} \cdot \left(-\frac{2}{3}\right)$

11. $-\frac{3}{8} \cdot \left(-\frac{2}{9}\right)$

12. $-\frac{5}{8} \cdot \left(-\frac{2}{5}\right)$

13. -6.3×2.7

14. -4.1×9.5

15. $-\frac{5}{9} \cdot \frac{3}{4}$

16. $-\frac{8}{3} \cdot \frac{9}{4}$

•• Multiply.

17. $6 \cdot (-5) \cdot 3$

18. $8 \cdot (-7) \cdot 6$

19. $7 \cdot (-4) \cdot (-3) \cdot 5$

20. $9 \cdot (-2) \cdot (-6) \cdot 7$

21. $-\frac{2}{3} \cdot \frac{1}{2} \cdot \left(-\frac{6}{7}\right)$

22. $-\frac{1}{8} \cdot \left(-\frac{1}{4}\right) \cdot \left(-\frac{3}{5}\right)$

23. $-3 \cdot (-4) \cdot (-5)$

24. $-2 \cdot (-5) \cdot (-7)$

25. $-2 \cdot (-5) \cdot (-3) \cdot (-5)$

26. $-3 \cdot (-5) \cdot (-2) \cdot (-1)$

••• Find the reciprocal.

27. -5

28. -7

29. $\frac{1}{4}$

30. $\frac{1}{8}$

ANSWERS
1.
2.
3.
4.
5.
6.
7.
8.
9.
10.
11.
12.
13.
14.
15.
16.
17.
18.
19.
20.
21.
22.
23.
24.
25.
26.
27.
28.
29.
30.

31. $-\dfrac{7}{5}$ **32.** $-\dfrac{5}{3}$ **33.** $-\dfrac{4}{11}$ **34.** $-\dfrac{7}{12}$

Divide.

35. $\dfrac{3}{4} \div \left(-\dfrac{2}{3}\right)$ **36.** $\dfrac{7}{8} \div \left(-\dfrac{1}{2}\right)$ **37.** $-\dfrac{5}{4} \div \left(-\dfrac{3}{4}\right)$ **38.** $-\dfrac{5}{9} \div \left(-\dfrac{5}{6}\right)$

39. $-\dfrac{2}{7} \div \left(-\dfrac{4}{9}\right)$ **40.** $-\dfrac{3}{5} \div \left(-\dfrac{5}{8}\right)$ **41.** $-\dfrac{3}{8} \div \left(-\dfrac{8}{3}\right)$ **42.** $-\dfrac{5}{6} \div \left(-\dfrac{6}{5}\right)$

43. $\dfrac{5}{7} \div \left(-\dfrac{5}{7}\right)$ **44.** $\dfrac{7}{8} \div \left(-\dfrac{7}{8}\right)$ **45.** $-6.6 \div 3.3$ **46.** $-44.1 \div (-6.3)$

47. $\dfrac{-11}{-13}$ **48.** $\dfrac{-19}{20}$ **49.** $\dfrac{23}{-14}$ **50.** $\dfrac{-15}{-75}$

 SKILL MAINTENANCE

51. Find the area of a circle when $r =$ 15 cm. Use 3.14 for π.

52. Evaluate $m^3 + 2$ when $m = 5$.

53. Factor: $24x + 32y + 64$.

54. Collect like terms: $x + 12y + 11x + 14y + 9$.

⭐ **EXTENSION**

Evaluate.

55. $(-1)^4$ **56.** $(-1)^3$ **57.** $(-2)^3$ **58.** $(-2)^6$

Chapter 2   Integers and Rational Numbers

The Distributive Laws and Their Use

• THE DISTRIBUTIVE LAWS

For rational numbers the distributive law of multiplication over addition holds. There is also another distributive law. It is the distributive law of multiplication over subtraction. When we multiply a number by a difference, we can either subtract and then multiply or multiply and then subtract.

For any rational numbers a, b, and c,

$$a(b - c) = ab - ac$$

(the distributive law of multiplication over subtraction).

Examples Compute.

1. a) $3(4 - 2) = 3 \cdot 2 = 6$
   b) $3 \cdot 4 - 3 \cdot 2 = 12 - 6 = 6$

2. a) $6(3 - 8) = 6 \cdot (-5) = -30$
   b) $6 \cdot 3 - 6 \cdot 8 = 18 - 48$ $= -30$

DO EXERCISES 1–4.

In an expression like $(ab) - (ac)$, we can omit the parentheses. In other words, in $ab - ac$ we *agree* that the multiplications are to be done first. The distributive laws are used in algebra in several ways. The basic ones are factoring, multiplying, and collecting like terms.

•• TERMS

What do we mean by the *terms* of an expression? When they are all separated by plus signs, it is easy to tell. If there are subtraction signs, we can rewrite using addition signs.

Example 3 What are the terms of $3x - 4y + 2z$?

$3x - 4y + 2z = 3x + (-4y) + 2z$ Separating parts with + signs

The terms are $3x$, $-4y$, and $2z$.

DO EXERCISES 5 AND 6.

••• MULTIPLYING

The distributive laws are a basis for a procedure called *multiplying*. In an expression such as $8(a + 2b - 7)$, we multiply each term inside the parentheses by 8.

$$8 \cdot (a + 2b - 7) = 8 \cdot a + 8 \cdot 2b - 8 \cdot 7 = 8a + 16b - 56$$

After finishing Section 2.7, you should be able to:

- • Apply the distributive law of multiplication over subtraction in computing.
- •• Identify the terms of an expression.
- ••• Multiply, where one expression has several terms.
- :: Factor when terms have a common factor.
- ::• Collect like terms.

Compute.

1. a) $4(5 - 3)$

   b) $4 \cdot 5 - 4 \cdot 3$

2. a) $-2 \cdot (5 - 3)$

   b) $-2 \cdot 5 - (-2) \cdot 3$

3. a) $5(2 - 7)$

   b) $5 \cdot 2 - 5 \cdot 7$

4. a) $-4[3 - (-2)]$

   b) $-4 \cdot 3 - (-4)(-2)$

What are the terms of the expression?

5. $5x - 4y + 3$

6. $-4y - 2x + 3z$

ANSWERS ON PAGE A-4

Multiply.

7. $3(x - 5)$

8. $5(x - y + 4)$

9. $-2(x - 3)$

10. $b(x - 2y + 4z)$

Factor.

11. $6x - 12$

12. $3x - 6y + 9$

13. $bx + by - bz$

Collect like terms.

14. $6x - 3x$

15. $7x - x$

16. $x - 0.41x$

17. $5x + 4y - 2x - y$

18. $3x - 7x - 11 + 8y + 4 - 13y$

ANSWERS ON PAGE A-4

Examples Multiply.

4. $4(x - 2) = 4 \cdot x - 4 \cdot 2 = 4x - 8$

5. $b(s - 3t + 2w) = b \cdot s - b \cdot 3t + b \cdot 2w = bs - 3bt + 2bw$

6. $-4(x - 2y + 3z) = -4 \cdot x - (-4) \cdot 2y + (-4) \cdot 3z$
$$= -4x + 8y - 12z$$

DO EXERCISES 7–10.

∷∷ FACTORING

Factoring is the reverse of multiplying. Look at Example 4. To *factor* $4x - 8$ we would write the product $4(x - 2)$.

When all the terms of an expression have a factor in common, we can "factor it out" using the distributive laws. Note that.

$4x$ has the following factors: $4, x, 2, -2, 1, -1, -4$;

-8 has the following factors: $8, -1, 2, -2, 4, -4, 1, -8$.

We usually remove the largest common factor. In the case of $4x - 8$, that factor is 4.

Examples Factor.

7. $5x - 10 = 5 \cdot x - 5 \cdot 2 = 5(x - 2)$

8. $ax - ay + az = a(x - y + z)$

9. $9x + 27y - 9 = 9 \cdot x + 9 \cdot 3y - 9 \cdot 1 = 9(x + 3y - 1)$

> *Remember:* An expression is factored when it is written as a product.

> *Caution!* Note that $3(3x + 9y - 3)$ is also a factored form of $9x + 27y - 9$, but it is *not* the desired one. Factor out the largest common factor.

DO EXERCISES 11–13.

∷∷ COLLECTING LIKE TERMS

The process of collecting like terms is also based on the distributive laws.

Examples Collect like terms.

10. $4\,x - 2\,x = (4 - 2)\,x = 2x$ Factor out the x.

11. $2\,x + 3y - 5\,x - 2y = 2\,x - 5\,x + 3y - 2y$
$$= (2 - 5)\,x + (3 - 2)y = -3\,x + y$$

12. $3\,x - x = (3 - 1)\,x = 2x$

13. $x - 0.24x = 1 \cdot x - 0.24x$
$$= (1 - 0.24)x = 0.76x$$

14. $x - 6x = 1 \cdot x - 6 \cdot x$
$$= (1 - 6)x = -5x$$

15. $4x - 7y + 9x - 5 + 3y - 8 = 13x - 4y - 13$

DO EXERCISES 14–18.

Chapter 2   Integers and Rational Numbers

· Multiply.

1. $-3(3 - 7)$ 2. $-5(9 - 14)$ 3. $8(9 - 10)$ 4. $6(13 - 14)$

·· What are the terms of each expression?

5. $8x - 1.4y$ 6. $12x - 1.2y$

7. $-5x + 3y - 14z$ 8. $-8a - 10b + 18z$

⁛ Factor.

9. $8x - 24$ 10. $10x - 50$ 11. $32 - 4x$ 12. $24 - 6x$

13. $8x + 10y - 22$ 14. $9x + 6y - 15$ 15. $ax - 7a$ 16. $bx - 9b$

17. $ax + ay - az$ 18. $cx - cy - cz$

··· Multiply.

19. $7(x - 2)$ 20. $5(x - 8)$ 21. $-7(y - 2)$ 22. $-9(y - 7)$

23. $-3(7 - t)$ 24. $-5(14 - n)$ 25. $-4(x + 3y)$ 26. $-3(a + 2b)$

A N S W E R S
1.
2.
3.
4.
5.
6.
7.
8.
9.
10.
11.
12.
13.
14.
15.
16.
17.
18.
19.
20.
21.
22.
23.
24.
25.
26.

27. $7(-2x - 4y + 3)$

28. $9(-5x - 6y + 8)$

 Collect like terms.

29. $11x - 3x$

30. $17t - 9t$

31. $17y - y$

32. $6n - n$

33. $x - 12x$

34. $y - 14y$

35. $x - 0.83x$

36. $t - 0.02t$

37. $9x + 2y - 5x$

38. $8y + 3z - 4y$

39. $11x + 2y + 7 - 4x - 18 - y$

40. $13a + 9b - 14 + 10 - 2a - 4b$

41. $2.7x + 2.3y - 1.9x - 1.8y$

42. $6.7a + 4.3b - 4.1a - 2.9b$

43. $\frac{1}{5}x + \frac{4}{5}y + \frac{2}{5}x - \frac{1}{5}y$

44. $\frac{7}{8}x + \frac{5}{8}y + \frac{1}{8}x - \frac{3}{8}y$

✔ SKILL MAINTENANCE

45. Find the perimeter of a rectangle when l is 25 ft and w is 14 ft.

46. Evaluate y^0 for $y = 7$.

47. Solve: $\frac{3}{4}x = \frac{9}{2}$.

☆ EXTENSION

Factor.

48. $2\pi r + \pi rs$

49. $\frac{1}{2}ah + \frac{1}{2}bh$

50. A principal of P dollars was invested in a savings account at 8% simple interest. How much was in the account after one year?

51. Simplify: $\dfrac{5x - 15}{5} + \dfrac{2x + 6}{2}$.

O B J E C T I V E S

After finishing Section 2.8, you should be able to:

- Find an equivalent expression for an additive inverse without parentheses, where an expression has several terms.

- Simplify expressions by removing parentheses and collecting like terms.

- Simplify expressions with parentheses inside parentheses.

Multiplying by −1 and Simplifying

• MULTIPLYING BY −1

What happens when we multiply a rational number by −1?

Examples

1. $-1 \cdot 7 = -7$ 2. $-1 \cdot (-5) = 5$ 3. $-1 \cdot 0 = 0$

When we multiply a number by −1, the result is the additive inverse of that number.

> **For any rational number** a,
>
> $$-1 \cdot a = -a.$$
>
> **That is, negative one times** a **is the additive inverse of** a.

In other words, multiplying by −1 changes the sign. This fact enables us to find an equivalent expression for an additive inverse when an expression has one or more terms.

Example 4 Find an equivalent expression without parentheses.

$-(3x - 2y + 4) = -1 \cdot (3x - 2y + 4)$ Taking the additive inverse is the same as multiplying by −1. We replace − by −1.

$\quad\quad = -1 \cdot 3x - (-1) \cdot 2y + (-1) \cdot 4$

$\quad\quad = -3x + 2y - 4$ Multiplying

DO EXERCISES 1 AND 2.

Example 4 illustrates how we find the inverse when we have more than one term.

> **To find the additive inverse of an expression with more than one term, change the sign of** *every* **term.**

Examples Find an equivalent expression without parentheses.

5. $-(5 - y) = -5 + y$ 6. $-(2a - 7b + 6) = -2a + 7b - 6$

DO EXERCISES 3–6.

Find an equivalent expression without parentheses.

1. $-(x + 2)$

2. $-(5x - 2y - 8)$

Find an equivalent expression without parentheses. Try to do this in one step.

3. $-(6 - t)$

4. $-(x - y)$

5. $-(-4a + 3t - 10)$

6. $-(18 - m - 2n + 4z)$

ANSWERS ON PAGE A-4

Remove parentheses and simplify.

7. $5x - (3x + 9)$

8. $5y - 2 - (2y - 4)$

Remove parentheses and simplify.

9. $y - 9(x + y)$

10. $5a - 3(7a - 6)$

11. $4a - b - 6(5a - 7b + 8c)$

• • REMOVING CERTAIN PARENTHESES

When parentheses follow a minus sign, they can be removed by finding an equivalent expression, as above. Then we collect like terms.

Examples Remove parentheses and simplify.

7. $3x - (4x + 2) = 3x - 4x - 2$ Subtracting is adding an inverse: $3x - (4x + 2) = 3x + [-(4x + 2)]$. Change the sign of *every* term in the parentheses.

$= -x - 2$ Collecting like terms

8. $3y - 2 - (2y - 4) = 3y - 2 - 2y + 4$
$= y + 2$

Caution! Be sure to change the signs of all terms in parentheses.

When we remove parentheses that follow a subtraction or additive inverse sign, we change the sign of *every* term inside. Then we collect like terms.

DO EXERCISES 7 AND 8.

Next, consider subtracting an expression that consists of several terms preceded by a number other than 1 or -1. We use the property $a - b = a + (-b)$. That is, we subtract by adding the inverse of the subtrahend.

Examples Remove parentheses and simplify.

9. $x - 3(x + y) = x + [-3(x + y)]$ Adding the inverse of $3(x + y)$
$= x + (-3x - 3y)$ Multiplying $x + y$ by -3
$= x - 3x - 3y$ Removing parentheses. There are no sign changes since we are adding.
$= -2x - 3y$ Collecting like terms

10. $3y - 2(4y - 5) = 3y + [-2(4y - 5)]$ Adding the inverse of $2(4y - 5)$

$= 3y + (-8y + 10)$
$= 3y - 8y + 10$
$= -5y + 10$

11. $8x - 2y - 7(3x - 2y - 5z) = 8x - 2y + [-7(3x - 2y - 5z)]$
$= 8x - 2y + (-21x + 14y + 35z)$
$= 8x - 2y - 21x + 14y + 35z$
$= -13x + 12y + 35z$

DO EXERCISES 9–11.

••• PARENTHESES WITHIN PARENTHESES

Sometimes parentheses occur within parentheses. When this happens we may use parentheses of different shapes, such as [ ], called "brackets," or { }, called "braces."

> **When parentheses occur within parentheses, the computations in the inner ones are to be done first.**

Example 12 Simplify.

$$[(-4) \div (-\tfrac{1}{4})] \div \tfrac{1}{4} = [(-4) \cdot (-4)] \div \tfrac{1}{4} \qquad \text{Working from the inside out}$$
$$= 16 \div \tfrac{1}{4}$$
$$= 16 \cdot 4$$
$$= 64$$

DO EXERCISE 12.

Example 13 Simplify.

$$4(2 + 3) - \{6 - [3 - (7 + 3)]\}$$
$$= 4 \cdot 5 - \{6 - [3 - 10]\} \qquad \text{Working from the inside out}$$
$$= 20 - \{6 - [-7]\}$$
$$= 20 - \{13\}$$
$$= 7$$

DO EXERCISE 13.

Example 14 Simplify.

$$[5(x + 2) - 3x] - [3(y + 2) - 7(y + 2)]$$
$$= [5x + 10 - 3x] - [3y + 6 - 7y - 14] \qquad \text{Working from the inside out}$$
$$= [2x + 10] - [-4y - 8] \qquad \text{Collecting like terms inside}$$
$$= 2x + 10 + 4y + 8 \qquad \text{Removing parentheses by renaming}$$
$$= 2x + 4y + 18 \qquad \text{Collecting like terms. You need } not \text{ factor the final expresson.}$$

DO EXERCISE 14.

Simplify.

12. $[24 \div (-2)] \div (-2)$

Simplify.

13. $3(4 + 2) - \{7 - [4 - (6 + 5)]\}$

Simplify.

14. $[3(x + 2) + 2x]$
$-[4(y + 2) - 3(y - 2)]$

ANSWERS ON PAGE A-4

VOLUMES OF RECTANGULAR SOLIDS

The volume of a rectangular solid is the number of *unit* cubes it takes to fill it up. Unit cubes look like these.

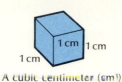

A cubic centimeter (cm³)

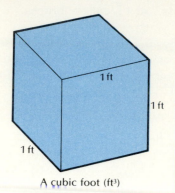

A cubic foot (ft³)

This rectangular solid has length 4 cm, width 3 cm, and height 2 cm. There are thus two layers of unit cubes, each having 3×4, or 12 cubes. The total volume is thus 24 cm³ (cubic centimeters).

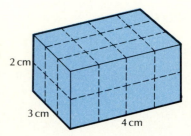

In any rectangular solid we can find the volume by multiplying the length by the width by the height.

> If a rectangular solid has length *l*, width *w*, and height *h*, then the volume is given by the formula $V = l \cdot w \cdot h$.

Example Find the volume of this solid.

$$V = lwh$$
$$= 10 \text{ m} \cdot 7 \text{ m} \cdot 8 \text{ m}$$
$$= 10 \cdot 56 \text{ m}^3 \text{ (cubic meters)}$$
$$= 560 \text{ m}^3$$

EXERCISES

Find the volume of each solid.

1.

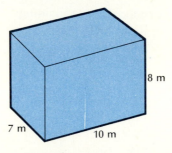

2.

Chapter 2   Integers and Rational Numbers

• Find an equivalent expression without parentheses.

1. $-(2x + 7)$

2. $-(3x + 5)$

3. $-(5x - 8)$

4. $-(6x - 7)$

5. $-(4a - 3b + 7c)$

6. $-(5x - 2y + 3z)$

7. $-(6x + 8y + 5)$

8. $-(8x + 3y + 9)$

9. $-(3x + 5y - 6)$

10. $-(6a + 4b - 7)$

11. $-(-8x - 6y - 43)$

12. $-(-2a - 9b - 5c)$

•• Remove parentheses and simplify. You need not factor the final expression.

13. $9x - (4x + 3)$

14. $7y - (2y + 9)$

15. $2a + (5a - 9)$

16. $11n + (3n - 7)$

17. $2x + 7x - (4x + 6)$

18. $3a + 2a - (4a + 7)$

19. $2x - 4y - (7x - 2y)$

20. $3a - 7b - (4a - 3b)$

21. $3x - y - (3x - 2y + 5z)$

22. $4a - b - (5a - 7b + 8c)$

23. $2x - 4y - 3(7x - 2y)$

24. $3a - 7b - 4(4a - 3b)$

ANSWERS
1.
2.
3.
4.
5.
6.
7.
8.
9.
10.
11.
12.
13.
14.
15.
16.
17.
18.
19.
20.
21.
22.
23.
24.

25. $4a - b - 5(5a - 7b + 8c)$

26. $15x - y - 6(3x - 2y + 5z)$

●●● Remove parentheses and simplify.

27. $[(-24) \div (-3)] \div (-\frac{1}{2})$

28. $[32 \div (-2)] \div (-2)$

29. $8 \cdot [9 - 2(5 - 3)]$

30. $10 \cdot [6 - 5(8 - 4)]$

31. $[4(9 - 6) + 11] - [14 - (6 + 4)]$

32. $[7(8 - 4) + 16] - [15 - (7 + 3)]$

33. $[3(8 - 4) + 12] - [10 - 2(3 + 5)]$

34. $[9(7 - 3) + 13] - [11 - 2(6 + 9)]$

35. $[10(x + 3) - 4] + [2(x - 17) + 6]$

36. $[9(x + 5) - 7] + [4(x - 12) + 9]$

37. $[4(2x - 5) + 7] + [3(x + 3) + 5x]$

38. $[8(3x - 2) + 9] + [7(x + 4) + 6x]$

39. $[7(x + 5) - 19] - [4(x - 6) + 10]$

40. $[6(x + 4) - 12] - [5(x - 8) + 11]$

41. $3\{[7(x - 2) + 4] - [2(2x - 5) + 6]\}$

42. $4\{[8(x - 3) + 9] - [4(3x - 7) + 2]\}$

43. $4\{[5(x - 3) + 2] - 3[2(x + 5) - 9]\}$

44. $3\{[6(x - 4) + 5] - 2[5(x + 8) - 10]\}$

⭐ **EXTENSION**

45. Find an equivalent expression for $6y + 2x - 3a + c$ by enclosing the last three terms in parentheses preceded by a minus sign.

Simplify.

46. $z - \{2z - [3z - (4z - 5z) - 6z] - 7z\} - 8z$

47. $\{x - [a - (a - x)] + [x - a]\} - 3x$

48. $x - \{x - 1 - [x - 2 - (x - 3 - \{x - 4 - [x - 5 - (x - 6)]\})]\}$

Order of Operations and Computer Language (Optional)

▪ · ORDER OF OPERATIONS

When several operations are to be done in a calculation or a problem, in what order should they be done? We have agreements about such operations, and we also use parentheses. In handwritten algebra, little, if any, difficulty arises in knowing in which order operations are to be performed. Consider, for example,

$$7 + 4^3.$$

It is rather natural to first find the power 4^3, and then add it to 7:

$$7 + 4^3 = 7 + 64 = 71.$$

With more complicated expressions and computers, however, there may be some problem determining in which order operations are to be performed. With computers the symbolism is different. The order in which a computer performs operations is the same as in ordinary algebra, but because of the different symbolism order of operations must be considered when using a computer language. The computer symbols we use to illustrate are from the computer language known as BASIC.

We now state the rules for order of operations.

BASIC Notation	Algebra Symbolism
$A + B$	$a + b$
$A - B$	$a - b$
$A * B$	ab, or $a \cdot b$
A / B	a/b, or $a \div b$
$A \wedge B$, or $A \uparrow B$	a^b
( )	( )

> **RULES FOR ORDER OF OPERATIONS**
> 1. Do all calculations within parentheses before operations outside.
> 2. Evaluate all exponential expressions.
> 3. Do all multiplications and divisions.
> 4. Do all additions and subtractions.

Often rules 3 and 4 contain the additional phrase "in order from left to right." Although this is how the computer operates, in actual practice it is not necessary. When computing by hand, you can treat each division as a multiplication by a reciprocal. Then by the commutative and associative laws you know that you can do the results in any order you prefer. A similar statement holds with respect to the order of addition and subtraction. Even though it is not necessary, we will do these operations in order from left to right as the computer would do them.

Let us apply the rules to sentences we see in arithmetic or algebra and then to sentences in BASIC language.

Example 1 Simplify: $-43 \cdot 68 - 17$.

There are no parentheses or powers, so we start with the third step.

$-43 \cdot 68 - 17 = -2924 - 17$ *Carrying out all multiplications and divisions in order from left to right*

$\qquad\qquad\qquad = -2941$ *Carrying out all additions and subtractions in order from left to right*

Simplify.

1. $23 - 42 \cdot 30$

2. $52 \cdot 5 + 5^3 - (4^2 - 48 \div 4)$

Simplify.

3. `5*6-(2+3)^3-4*10`

Write BASIC notation.

4. $\left(\dfrac{1}{4 \cdot 5}\right)^2$

5. $a^2 + b^2 - 2ab$

6. $\dfrac{x}{y} - \dfrac{t}{s}$

Write algebraic symbolism.

7. `2/(A+3)`

8. `A^2-2*A*B+B^2`

ANSWERS ON PAGE A-5

Example 2 Simplify: $2^4 + 51 \cdot 4 - 2(37 + 23 \cdot 2)$.

$2^4 + 51 \cdot 4 - 2(37 + 23 \cdot 2)$

$= 2^4 + 51 \cdot 4 - 2(37 + 46)$ **Carrying out operations inside parentheses first**

$= 2^4 + 51 \cdot 4 - 2(83)$ **Completing the addition inside parentheses**

$= 16 + 51 \cdot 4 - 2(83)$ **Evaluating exponential expressions**

$= 16 + 204 - 166$ **Doing all multiplications**

$= 220 - 166$ **Doing all additions and subtractions in order from left to right**

$= 54$

DO EXERCISES 1 AND 2.

Now let us apply the rules to BASIC language.

Example 3 Simplify: `3+4*7-(7+5)^2`

`3+4*7-(7+5)^2`

`= 3+4*7-12^2`   **Carrying out all operations inside parentheses first**

`= 3+4*7-144`   **Evaluating exponential expressions**

`= 3+28-144`   **Doing the multiplication**

`= 31-144`   **Doing all additions and subtractions in order from left to right**

`= -113`

DO EXERCISE 3.

•• CONVERTING BETWEEN BASIC NOTATION AND ALGEBRAIC SYMBOLISM

To program a computer to perform a task, it is important to be able to write BASIC notation for algebraic symbolism.

Examples Write BASIC notation for each of the following.

4. $\left(\dfrac{54 \cdot 49}{2}\right)^3$ `(54*49/2)^3`

5. $b^2 - 4ac$ `B^2-4*A*C`

6. $c + d - \dfrac{a^2}{b}$ `C+D-A^2/B`

DO EXERCISES 4–6.

It is also important to be able to convert from BASIC notation to algebraic symbolism.

Examples Write algebraic symbolism for each of the following.

7. `(5*6)^2-(7*8)^3` $(5 \cdot 6)^2 - (7 \cdot 8)^3$

8. `(A^2-B^2)/(A-B)` $\dfrac{a^2 - b^2}{a - b}$

DO EXERCISES 7 AND 8.

108   Chapter 2   Integers and Rational Numbers

• Simplify.

1. $[9 - 2(5 - 1)]$

2. $[6 - 5(8 - 4)]$

3. $8[7 - 6(4 - 2)]$

4. $10[7 - 4(7 - 5)]$

5. $[4(9 - 6) + 11] - [14 - (6 - 4)]$

6. $[7(8 - 4) + 16] - [15 - (7 + 3)]$

7. $[32 \div (-4)] \div (-2)$

8. $[24 \div (-3)] \div (-\frac{1}{4})$

9. $16(-24) + 50$

10. $10 \cdot 20 - 15 \cdot 24$

11. $2^4 + 2^3 - 10 \cdot 20$

12. $5 \cdot 8 - 3^2 - 2^3$

13. $5^3 + 26 \cdot 71 - (16 + 25 \cdot 3)$

14. $4^3 + 20 \cdot 30 + 8^2 - 23$

15. $(12 \cdot 3 - 7 \cdot 4)^2$

16. $(7 \cdot 8 + 2 \cdot 2)^2$

17. ▦ $3000 \cdot (1 + 0.16)^3$

18. ▦ $2000 \cdot (3 + 1.14)^2$

19. ▦ $(20 \cdot 4 + 12 \cdot 7)^2 - (34 \cdot 56)^3$

20. ▦ $(30 \cdot 5 - 13 \cdot 8)^3 + (23 \cdot 24)^2$

ANSWERS
1.
2.
3.
4.
5.
6.
7.
8.
9.
10.
11.
12.
13.
14.
15.
16.
17.
18.
19.
20.

●● Write BASIC notation for each of the following.

21. $a^2 + 2ab + b^2$

22. $a^3 - b^3$

23. $\dfrac{2(3 - b)}{c}$

24. $\dfrac{a + b}{c - d}$

25. $\dfrac{a}{b} - \dfrac{c}{d}$

26. $\left(\dfrac{a^2}{h^3}\right)^3$

Write algebraic symbolism for each of the following.

27. `2 * A + 7`

28. `A + 1 / B`

29. `3 * A ^ 2 - 5`

30. `( A + 4 ) / ( 2 * A )`

31. `( A + B ) ^ 2`

32. `A ^ 2 - 3 * A + 2`

✔ **SKILL MAINTENANCE**

33. An investment is made at 9% simple interest for one year. It grows to $5995. How much was originally invested?

34. Four-fifths of what number is 624?

35. Write an algebraic expression: two times y plus three times x.

36. Find the LCM: 16, 18, 24.

⭐ **EXTENSION**

37. Determine whether it is true that for any rational number x, $(-x)^2 = x^2$. Explain why or why not.

38. Determine whether it is true that for any rational numbers a and b, $ab = (-a)(-b)$. Explain why or why not.

Chapter 2   Integers and Rational Numbers

Summary and Review

The following contains a summary of what you should be able to do after completing this chapter. The review exercises are for practice. Answers are at the back of the book. If you miss an exercise, restudy the section and objective indicated alongside the answer.

Beginning with this chapter, material from certain sections of preceding chapters will be covered on the chapter tests. Accordingly, the summary and review sections will contain practice exercises. Review sections to be tested in addition to material in this chapter are 1.1, 1.2, 1.6, and 1.10. (See the discussion of skill maintenance exercises on p. 69.

You should be able to:

Tell which integer corresponds to a real-world situation.

1. Tell which integers correspond to this situation: A student has a debt of $45 and has $72 in a savings account.

Find the absolute value of any rational number.

Find the absolute value.

2. $|-38|$

3. $|7|$

4. $\left|\dfrac{5}{2}\right|$

5. $|-4.78|$

Write a true sentence using $<$ or $>$.

Write a true sentence using $<$ or $>$.

6. -3 10

7. -1 -6

8. 0 -23

Find the additive inverse and the reciprocal of a rational number.

Find the additive inverse of each number.

9. 3.8

10. $-\dfrac{3}{4}$

11. Find $-x$ when x is -34.

12. Find $-(-x)$ when x is 5.

Find the reciprocal.

13. $\dfrac{3}{8}$

14. -7

15. $-\dfrac{1}{4}$

16. 1.6

Add, subtract, multiply, and divide rational numbers.

Compute and simplify.

17. $4 + (-7)$

18. $-\dfrac{2}{3} + \dfrac{1}{12}$

19. $6 + (-9) + (-8) + 7$

20. $-3.8 + 5.1 + (-12)$

21. $-3 - (-7)$

22. $-\dfrac{9}{10} - \dfrac{1}{2}$

23. $-3.8 - 4.1$

24. $-9 \cdot (-6)$

25. $-2.7(3.4)$

26. $\dfrac{2}{3} \cdot \left(-\dfrac{3}{7}\right)$

27. $3 \cdot (-7) \cdot (-2) \cdot (-5)$

28. $35 \div (-5)$

29. $-5.1 \div 1.7$

30. $-\dfrac{3}{5} \div \left(-\dfrac{4}{5}\right)$

Use the distributive laws to multiply and factor algebraic expressions and to simplify expressions by collecting like terms.

Multiply.

31. $5(3x - 7)$

32. $-2(4x - 5)$

33. $10(0.4x + 1.5)$

34. $-8(3 - 6x)$

Factor.

35. $2x - 14$

36. $6x - 6$

37. $5x + 10$

38. $12 - 3x$

Collect like terms.

39. $11a + 2b - 4a - 5b$

40. $7x - 3y - 9x + 8y$

41. $6x + 3y - x - 4y$

42. $-3a + 9b + 2a - b$

43. $-5 - 3x + 8 - (-9x)$

44. $4y - 19 - (-7y) + 3$

Simplify expressions by removing parentheses and collecting like terms and using the rules for order of operations.

Remove parentheses and simplify.

45. $13 - 4[8 - (-2)]$

46. $20 - 3(5 + 2)$

47. $3[11 - 3(4 - 1)]$

48. $2^3 + 3^2 - (12 \cdot 4 + 35)$

49. $2a - 4(5a - 9)$

50. $3(b + 7) - 5b$

51. $2[6(y - 4) + 7]$

52. $[8(x + 4) - 10] - [3(x - 2) + 7]$

53. $5\{[6(x - 1) + 7] - [3(3x - 4) + 8]\}$

Convert between BASIC notation and algebraic symbolism.

Write BASIC notation for each of the following.

54. $a^2 - b^3$

55. $a - \dfrac{c}{d}$

Write algebraic symbolism for each of the following.

56. `3 − 4 * B`

57. `( A − B ) ^ 3`

SKILL MAINTENANCE

58. Find the area of a rectangle when l is 10.5 cm and w is 20 cm.

59. Find the LCM: 15, 27, and 30.

60. Find the prime factorization of 648.

61. Translate to an algebraic expression: the difference of two numbers.

62. Factor: $15x + 30y + 6$.

63. 2016 is what percent of 5600?

⭐ **EXTENSION**

64. Simplify: $-|\frac{7}{8} - (-\frac{1}{2}) - \frac{3}{4}|$.

65. Solve: $|y| = -1.237$.

66. Simplify: $-(-(-x))$.

67. For what numbers is $-x$ positive?

68. The population of a town is P. After a 6% increase, what is the new population?

Test: Chapter 2

Write a true sentence using < or >.

1. -4 0

2. -3 -8

3. -17 -8

Find the absolute value.

4. $|-9|$

5. $\left|\dfrac{9}{4}\right|$

Find the additive inverse.

6. $\dfrac{2}{3}$

7. -1.4

8. Find $-x$ when x is -8.

Find the reciprocal.

9. -2

10. $\dfrac{4}{7}$

Compute and simplify.

11. $3.1 + (-4.7)$

12. $-8 + 4 + (-7) + 3$

13. $2 - (-8)$

14. $3.2 - 5.7$

15. $\dfrac{1}{8} - \left(-\dfrac{3}{4}\right)$

16. $4 \cdot (-12)$

17. $-45 \div 5$

18. $-\dfrac{3}{7} \div \left(-\dfrac{4}{7}\right)$

19. $4.864 \div (-0.5)$

Multiply.

20. $3(6 - x)$

21. $-5(y - 1)$

Factor.

22. $12 - 22x$

23. $7x + 21 + 14y$

1. _____
2. _____
3. _____
4. _____
5. _____
6. _____
7. _____
8. _____
9. _____
10. _____
11. _____
12. _____
13. _____
14. _____
15. _____
16. _____
17. _____
18. _____
19. _____
20. _____
21. _____
22. _____
23. _____

Simplify.

24. $6 + 7 - 4 - (-3)$

25. $5x - (3x - 7)$

24. _____

25. _____

26. _____

26. $4(2a - 3b) + a - 7$

27. $4\{3[5(y - 3) + 2(y + 8)]\}$

27. _____

28. _____

28. $12 - 3(9 - 7)$

29. $23 \cdot 10 - (3 + 5)^2$

29. _____

30. _____

31. _____

30. Write BASIC notation for $2ab + b^3$.

31. Write algebraic symbolism for `( A − B ) / A ^ 2`.

32. _____

33. _____

34. _____

✓ **SKILL MAINTENANCE**

35. _____

32. Find the area of a rectangle when l is 32 yd and w is 15 yd.

33. 24 is what percent of 50?

36. _____

34. Find the prime factorization of 280.

35. Find the LCM: 16, 20, 30.

37. _____

⭐ **EXTENSION**

36. Solve: $|x| = -8$.

37. Simplify: $a - \{3a - [4a - (2a - 4a)]\}$.

Chapter 2   Integers and Rational Numbers

AN APPLICATION
You see a flash of lightning. It takes 10 seconds for the sound of the thunder to reach you. How far away is the storm?

THE MATHEMATICS
If it takes n seconds for the thunder to reach you, then the thunder is M miles away, where

$M = \frac{1}{5}n.$ } This is a formula

Substituting 10 for n, we get $M = \frac{1}{5}(10) = 2$ miles.

In this chapter we expand our equation-solving ability and consequently our ability to solve problems. Formulas like the one above will be considered in the latter part of the chapter. A study of the laws of exponents, useful in Chapter 4 and throughout the book, concludes the chapter.

The review sections to be tested in addition to the material in this chapter are 1.3, 1.7, 2.5, 2.6, and 2.8.

Solving Equations and Problems

3.1

The Addition Principle

• We now study the *addition principle*. We used this principle in Section 1.8, where we learned that we can subtract the same on both sides of an equation. We can actually subtract or add numbers on both sides of an equation.

Recall the following.

> The replacements that make an equation true are called *solutions*. To *solve* an equation means to find all its solutions.

There are various ways to solve equations. We develop one now, using an idea called the *addition principle.* An equation $a = b$ says that a and b stand for the same number. Suppose this is true and we then add a number c to the number a. We get the same answer if we add c to b, because a and b are the same number.

> THE ADDITION PRINCIPLE
>
> If an equation $a = b$ is true, then
>
> $$a + c = b + c$$
>
> is true for any number c.

The idea in using this principle is to obtain an equation for which you can see what the solution is. We do that by getting the variable alone on one side.

Example 1 Solve: $x + 5 = -7$.

$$x + 5 = -7$$
$$x + 5 + (-5) = -7 + (-5) \qquad \text{Using the addition principle;}$$
$$\text{adding } -5 \text{ on both sides}$$
$$x + 0 = -7 + (-5) \qquad \text{Simplifying}$$
$$x = -12$$

We can see that the solution of $x = -12$ is the number -12. To check the answer we substitute in the original equation.

Check:

$$\frac{x + 5 = -7}{\begin{array}{c|c} -12 + 5 & -7 \\ -7 & \end{array}}$$

The solution of the original equation is -12.

In Example 1, to get x alone, we added the *additive inverse* of 5. This "got rid of" the 5 on the left, giving us the *additive identity* 0, which when added to x is x. When using the addition principle, we sometimes say that we "add the same number on both sides of an equation." We started with $x + 5 = -7$ and, using the addition principle, we derived a simpler equation $x = -12$ for which it was easy to *see* what the solution is. Equations with the same solutions, such as $x + 5 = -7$ and $x = -12$, are called *equivalent equations*.

DO EXERCISE 1.

The addition principle also allows us to subtract on both sides of an equation. This is because subtracting is the same as adding an inverse. In Example 1, we could have subtracted 5 on both sides of the equation.

Now we "undo" a subtraction using the addition principle.

Example 2 Solve: $-6.5 = y - 8.4$.

$$-6.5 = y - 8.4$$
$$-6.5 + 8.4 = y - 8.4 + 8.4 \qquad \text{Adding 8.4 on both sides to get rid of } -8.4 \text{ on the right}$$

$$1.9 = y + 0$$
$$1.9 = y$$

Check:

$$\begin{array}{c|c} -6.5 = y - 8.4 \\ \hline -6.5 & 1.9 - 8.4 \\ & -6.5 \end{array}$$

The solution is 1.9.

Note that equations are reversible. That is, if $a = b$ is true, then $b = a$ is true. Thus when we solve $-6.5 = y - 8.4$, we can reverse it and solve $y - 8.4 = -6.5$ if we wish.

DO EXERCISES 2 AND 3.

Example 3 Solve: $x - \dfrac{2}{3} = 2\dfrac{1}{2}$.

$$x - \frac{2}{3} = 2\frac{1}{2}$$

$$x - \frac{2}{3} + \frac{2}{3} = 2\frac{1}{2} + \frac{2}{3} \qquad \text{Adding } \frac{2}{3}$$

$$x = 2\frac{1}{2} + \frac{2}{3} \qquad \text{Simplifying}$$

$$x = \frac{5}{2} + \frac{2}{3} \qquad \text{Converting to fractional notation to carry out the addition}$$

$$x = \frac{5}{2} \cdot \frac{3}{3} + \frac{2}{3} \cdot \frac{2}{2} \qquad \text{Multiplying by 1 to obtain the least common denominator}$$

$$x = \frac{15}{6} + \frac{4}{6}$$

$$x = \frac{19}{6}, \quad \text{or } 3\frac{1}{6}$$

The check is left to the student. The solution is $\dfrac{19}{6}$.

DO EXERCISES 4 AND 5.

Solve using the addition principle.

1. $x + 7 = 2$

Solve.

2. $8.7 = n - 4.5$

3. $y + 17.4 = 10.9$

Solve.

4. $x + \dfrac{1}{2} = -\dfrac{3}{2}$

5. $t - 3\dfrac{1}{4} = \dfrac{5}{8}$

ANSWERS ON PAGE A-5

CALCULATOR CORNER: NUMBER PATTERNS

There are many interesting number patterns in mathematics. Look for a pattern in the following. We can use a calculator for the computations.

$$6^2 = 36$$
$$66^2 = 4356$$
$$666^2 = 443556$$
$$6666^2 = 44435556$$

Do you see a pattern? If so, find 66666^2 without the use of your calculator.

EXERCISES

In each of the following, do the first four calculations using your calculator. Look for a pattern. Use the pattern to do the last calculation without the use of your calculator.

1.
$$3^2$$
$$33^2$$
$$333^2$$
$$3333^2$$
$$33333^2$$

2.
$$9 \cdot 6$$
$$99 \cdot 66$$
$$999 \cdot 666$$
$$9999 \cdot 6666$$
$$99999 \cdot 66666$$

3.
$37 \cdot 3$
$37 \cdot 33$
$37 \cdot 333$
$37 \cdot 3333$
$37 \cdot 33333$

4.
$37 \cdot 3$
$37 \cdot 6$
$37 \cdot 9$
$37 \cdot 12$
$37 \cdot 15$

Solve using the addition principle. Don't forget to check.

1. $x + 2 = 6$

2. $x + 5 = 8$

3. $x + 15 = -5$

Check: _____

Check: _____

Check: _____

4. $y + 25 = -6$

5. $r + \dfrac{2}{3} = 1$

6. $t + \dfrac{3}{4} = 1$

Check: _____

Check: _____

Check: _____

7. $x - \dfrac{5}{6} = \dfrac{7}{8}$

8. $x - \dfrac{2}{3} = \dfrac{7}{3}$

9. $8 + y = 12$

Check: _____

Check: _____

Check: _____

10. $5 + t = 7$

11. $\dfrac{1}{3} + a = \dfrac{5}{6}$

12. $-\dfrac{1}{5} + z = -\dfrac{1}{4}$

Check: _____

Check: _____

Check: _____

13. $x - 2.3 = -7.4$

14. $x - 3.7 = 8.4$

15. $-2.6 + x = 8.3$

Check: _____

Check: _____

Check: _____

1. _____

2. _____

3. _____

4. _____

5. _____

6. _____

7. _____

8. _____

9. _____

10. _____

11. _____

12. _____

13. _____

14. _____

15. _____

16. $-5.7 + t = 7.4$

17. $m + \dfrac{5}{6} = -\dfrac{11}{12}$

18. $x + \dfrac{2}{3} = -\dfrac{5}{6}$

19. $-6 = -2 + y$

20. $-8 = y - 2$

21. $10 = y - 6$

22. $20 = t - 7$

23. $-9.7 = -4.7 + y$

24. $-7.8 = 2.8 + x$

25. $5\dfrac{1}{6} + x = 7$

26. $4\dfrac{2}{3} + x = 5\dfrac{1}{4}$

27. $p + \dfrac{2}{3} = 7\dfrac{1}{3}$

28. $q + \dfrac{1}{3} = -\dfrac{1}{7}$

29. $22\dfrac{1}{7} = 30 + t$

30. $47\dfrac{1}{8} = -76 + z$

✔ **SKILL MAINTENANCE**

31. Add: $-3 + (-8)$.

32. Divide: $-\dfrac{3}{7} \div \left(-\dfrac{9}{7}\right)$.

33. Multiply: $-\dfrac{2}{3} \cdot \dfrac{5}{8}$.

34. Remove parentheses and simplify: $7w - 3 - (4w - 8)$.

☆ **EXTENSION**

Solve.

35. ▦ $-356.788 = -699.034 + t$

36. $x + 3 = 3 + x$

37. $x + 4 = 5 + x$

38. Explain why it is not necessary to prove a subtraction principle: If $a = b$, then $a - c = b - c$.

The Multiplication Principle

Suppose $a = b$ is true and we multiply a by some number c. We get the same answer if we multiply b by c because a and b are the same number.

> **THE MULTIPLICATION PRINCIPLE**
>
> If an equation $a = b$ is true, then
>
> $$a \cdot c = b \cdot c$$
>
> is true for any number c.

Example 1 Solve: $3x = 9$.

$$3x = 9$$
$$\tfrac{1}{3} \cdot 3x = \tfrac{1}{3} \cdot 9 \qquad \text{Using the multiplication principle; multiplying by } \tfrac{1}{3}$$
$$1 \cdot x = 3 \qquad \text{Simplifying}$$
$$x = 3$$

It is easy to see that the solution of $x = 3$ is 3.

Check:

$3x$	=	9
$3 \cdot 3$		9
9		

The solution of the original equation is 3.

In Example 1, to get x alone, we multiplied by the *multiplicative inverse*, or *reciprocal* of 3. When we multiplied we got the *multiplicative identity* 1 times x, or $1 \cdot x$, which simplified to x. This enabled us to "get rid of" the 3 on the left. The multiplication principle also allows us to divide on both sides of an equation by a nonzero number. This is because division by a number is the same as multiplying by a reciprocal.

DO EXERCISE 1.

Example 2 Solve: $\frac{3}{8} = -\frac{5}{4}x$.

$$\frac{3}{8} = -\frac{5}{4}x$$
$$-\frac{4}{5} \cdot \frac{3}{8} = -\frac{4}{5} \cdot \left(-\frac{5}{4}x\right) \qquad \text{Multiplying by } -\frac{4}{5} \text{ on both sides to get rid of } -\frac{5}{4} \text{ on the right (this is the same as dividing by } -\frac{5}{4})$$
$$-\frac{3}{10} = x \qquad \text{Simplifying}$$

Check:

$\frac{3}{8}$	=	$-\frac{5}{4}x$
$\frac{3}{8}$		$-\frac{5}{4}\left(-\frac{3}{10}\right)$
		$\frac{3}{8}$

The solution is $-\frac{3}{10}$.

DO EXERCISES 2 AND 3.

Solve.

1. $5x = 25$

Solve.

2. $4 = -\dfrac{1}{3}y$

3. $4x = -7$

ANSWERS ON PAGE A-5

Solve.

4. $1.12x = 8736$

5. $6.3 = -2.1y$

Solve.

6. $-x = -10$

Solve.

7. $-14 = -\dfrac{y}{2}$

Example 3 Solve: $1.16y = 9744$.

$$1.16y = 9744$$
$$\frac{1}{1.16} \cdot (1.16y) = \frac{1}{1.16} \cdot (9744) \qquad \text{Multiplying by 1/1.16 (or dividing by 1.16)}$$

$$y = \frac{9744}{1.16}$$
$$y = 8400$$

Check:

$$\begin{array}{c|c} 1.16y = 9744 \\ \hline 1.16(8400) & 9744 \\ 9744 & \end{array}$$

The solution is 8400.

DO EXERCISES 4 AND 5.

Example 4 Solve: $-x = 9$.

$$-x = 9$$
$$-1 \cdot x = 9 \qquad \text{Using the property } -1 \cdot x = -x$$
$$-1 \cdot (-1 \cdot x) = -1 \cdot 9 \qquad \text{Multiplying on both sides by } -1, \text{ the reciprocal of itself, or dividing by } -1$$

$$1 \cdot x = -9$$
$$x = -9$$

Check:

$$\begin{array}{c|c} -x = 9 \\ \hline -(-9) & 9 \\ 9 & \end{array}$$

The solution is -9.

DO EXERCISE 6.

Now we "undo" a division using the multiplication principle.

Example 5 Solve: $\dfrac{-y}{9} = 14$.

$$\frac{-y}{9} = 14$$
$$9 \cdot \frac{-y}{9} = 9 \cdot 14 \qquad \text{Thinking of } \frac{-y}{9} \text{ as } \frac{1}{9} \cdot (-y) \text{ and multiplying on both sides by 9}$$

$$-y = 126$$
$$y = -126$$

Check:

$$\begin{array}{c|c} \dfrac{-y}{9} = 14 \\ \hline \dfrac{-(-126)}{9} & 14 \\ \dfrac{126}{9} & \\ 14 & \end{array}$$

The solution is -126.

DO EXERCISE 7.

• Solve using the multiplication principle. Don't forget to check!

1. $6x = 36$

2. $3x = 39$

3. $9x = -36$

Check: _____|_____   Check: _____|_____   Check: _____|_____

4. $7x = -49$

5. $-12x = 72$

6. $-15x = -105$

Check: _____|_____   Check: _____|_____   Check: _____|_____

7. $\frac{1}{7}t = 9$

8. $\frac{1}{8}y = 11$

9. $\frac{1}{5} = \frac{1}{3}t$

Check: _____|_____   Check: _____|_____   Check: _____|_____

10. $\frac{1}{9} = \frac{1}{7}z$

11. $-2.7y = 54$

12. $-3.1y = 21.7$

Check: _____|_____   Check: _____|_____   Check: _____|_____

13. $\frac{3}{4}x = 27$

14. $\frac{4}{5}x = 16$

15. $-\frac{3}{5}r = -\frac{9}{10}$

Check: _____|_____   Check: _____|_____   Check: _____|_____

ANSWERS
1. _____
2. _____
3. _____
4. _____
5. _____
6. _____
7. _____
8. _____
9. _____
10. _____
11. _____
12. _____
13. _____
14. _____
15. _____

Copyright © 1987 Addison-Wesley Publishing Co., Inc.

ANSWERS
16. _____
17. _____
18. _____
19. _____
20. _____
21. _____
22. _____
23. _____
24. _____
25. _____
26. _____
27. _____
28. _____
29. _____
30. _____
31. _____
32. _____
33. _____
34. _____
35. _____
36. _____
37. _____
38. _____
39. _____

16. $-\dfrac{2}{5}y = -\dfrac{4}{15}$

17. $12 = \dfrac{6}{5}y$

18. $\dfrac{4}{5}y = 20$

19. $-3.3y = 6.6$

20. $-6.3x = 44.1$

21. $38.7m = 309.6$

22. $29.4x = 235.2$

23. $20.07 = \dfrac{3}{2}y$

24. $-\dfrac{9}{7}y = 12.06$

25. $-\dfrac{3}{2}r = -\dfrac{27}{4}$

26. $\dfrac{5}{7}x = -\dfrac{10}{14}$

27. $-\dfrac{2}{3}y = -10.6$

28. $-68 = -r$

29. $-x = 100$

30. $\dfrac{-t}{3} = 7$

31. $-9 = -\dfrac{x}{6}$

✔ SKILL MAINTENANCE

32. Subtract: $-7 - (-23)$.

33. Collect like terms: $x - 8x$.

34. Multiply: $3(x - 5)$.

35. Find the absolute value: $|-3.2|$.

☆ EXTENSION

Solve.

36. 🖩 $-0.2344m = 2028.732$

37. $0 \cdot x = 0$

38. $0 \cdot x = 9$

39. $4 \cdot |x| = 48$

3.3

Using the Principles Together

We now consider equation solving where we may need to use both the addition and multiplication principles. We also consider equations where collecting like terms is necessary.

After finishing Section 3.3, you should be able to:

• Solve simple equations using both the addition and multiplication principles.

•• Solve equations in which like terms are to be collected.

• APPLYING BOTH PRINCIPLES

Let's consider an equation in which we apply both principles. We usually apply the addition principle first. Then we apply the multiplication principle.

Example 1 Solve: $3x + 4 = 13$.

$$3x + 4 = 13$$
$$3x + 4 + (-4) = 13 + (-4) \quad \text{Using the addition principle; adding } -4 \text{ on both sides}$$
$$3x = 9 \quad \text{Simplifying}$$
$$\tfrac{1}{3} \cdot 3x = \tfrac{1}{3} \cdot 9 \quad \text{Using the multiplication principle; multiplying by } \tfrac{1}{3} \text{ on both sides}$$
$$x = 3 \quad \text{Simplifying}$$

Check:

$$\begin{array}{c|c} 3x + 4 = 13 & \\ \hline 3 \cdot 3 + 4 & 13 \\ 9 + 4 & \\ 13 & \end{array}$$

The solution is 3.

DO EXERCISE 1.

Example 2 Solve: $-5x - 6 = 16$.

$$-5x - 6 = 16$$
$$-5x - 6 + 6 = 16 + 6 \quad \text{Adding 6}$$
$$-5x = 22$$
$$-\tfrac{1}{5} \cdot (-5x) = -\tfrac{1}{5} \cdot 22 \quad \text{Multiplying by } -\tfrac{1}{5}$$
$$x = -\tfrac{22}{5}, \quad \text{or} \quad -4\tfrac{2}{5}$$

Check:

$$\begin{array}{c|c} -5x - 6 = 16 & \\ \hline -5\left(-\tfrac{22}{5}\right) - 6 & 16 \\ 22 - 6 & \\ 16 & \end{array}$$

The solution is $-\tfrac{22}{5}$.

DO EXERCISES 2 AND 3.

Solve.

1. $9x + 6 = 51$

Solve.

2. $8x - 4 = 28$

3. $-\dfrac{1}{2}x + 3 = 1$

ANSWERS ON PAGE A-5

Solve.

4. $-18 - x = -57$

Solve.

5. $-4 - 8x = 8$

6. $41.68 = 4.7 - 8.6y$

Solve.

7. $4x + 3x = -21$

8. $x - 0.09x = 728$

Example 3 Solve: $45 - x = 13$.

$$45 - x = 13$$
$$-45 + 45 - x = -45 + 13 \qquad \text{Adding } -45$$
$$-x = -32$$
$$-1 \cdot x = -32 \qquad \bigg\} \quad -x = -1 \cdot x$$
$$x = \frac{-32}{-1} \qquad \text{Dividing on both sides by } -1$$
$$x = 32$$

You could have multiplied on both sides by -1 instead. That would change the sign on both sides.

Check:

$45 - x$	$=$	13
$45 - 32$		13
13		

The solution is 32.

DO EXERCISE 4.

Example 4 Solve: $16.3 - 7.2y = -8.18$.

$$16.3 - 7.2y = -8.18$$
$$-7.2y = -16.3 + (-8.18) \qquad \text{Adding } -16.3$$
$$-7.2y = -24.48$$
$$y = \frac{-24.48}{-7.2} \qquad \text{Dividing by } -7.2$$
$$y = 3.4$$

Check:

$16.3 - 7.2y$	$=$	-8.18
$16.3 - 7.2(3.4)$		-8.18
$16.3 - 24.48$		
-8.18		

The solution is 3.4.

DO EXERCISES 5 AND 6.

• • COLLECTING LIKE TERMS

If there are like terms on one side of the equation, we collect them before using the principles.

Example 5 Solve: $3x + 4x = -14$.

$$3x + 4x = -14$$
$$7x = -14 \qquad \text{Collecting like terms}$$
$$\tfrac{1}{7} \cdot 7x = \tfrac{1}{7} \cdot (-14)$$
$$x = -2$$

The number -2 checks, so the solution is -2.

DO EXERCISES 7 AND 8.

Chapter 3 Solving Equations and Problems

If there are like terms on opposite sides of the equation, we get them on the same side by using the addition principle. Then we collect them.

Example 6 Solve: $2x - 2 = -3x + 3$.

$$2x - 2 = -3x + 3$$
$$2x - 2 + 2 = -3x + 3 + 2 \qquad \text{Adding 2}$$
$$2x = -3x + 5 \qquad \text{Simplifying}$$
$$2x + 3x = -3x + 3x + 5 \qquad \text{Adding } 3x$$
$$5x = 5 \qquad \text{Collecting like terms and simplifying}$$
$$\tfrac{1}{5} \cdot 5x = \tfrac{1}{5} \cdot 5 \qquad \text{Multiplying by } \tfrac{1}{5}$$
$$x = 1 \qquad \text{Simplifying}$$

Check:

$$\begin{array}{c|c} \multicolumn{2}{c}{2x - 2 \;=\; -3x + 3} \\ \hline 2 \cdot 1 - 2 & -3 \cdot 1 + 3 \\ 2 - 2 & -3 + 3 \\ 0 & 0 \end{array}$$

The solution is 1.

DO EXERCISE 9.

In Example 6, we used the addition principle to get all terms with a variable on one side and all numbers on the other side. Then we collected like terms and proceeded as before. If there are like terms on one side at the outset, they should be collected first.

Example 7 Solve: $6x + 5 - 7x = 10 - 4x + 3$.

$$6x + 5 - 7x = 10 - 4x + 3$$
$$-x + 5 = 13 - 4x \qquad \text{Collecting like terms}$$
$$-x + 4x = 13 - 5 \qquad \text{Adding } 4x \text{ and subtracting 5 to get all terms with variables on one side and all other terms on the other}$$
$$3x = 8 \qquad \text{Collecting like terms}$$
$$\tfrac{1}{3} \cdot 3x = \tfrac{1}{3} \cdot 8 \qquad \text{Multiplying by } \tfrac{1}{3}$$
$$x = \tfrac{8}{3} \qquad \text{Simplifying}$$

The number $\tfrac{8}{3}$ checks, so it is the solution.

DO EXERCISES 10–12.

If we multiply on both sides of $\tfrac{1}{2}x = \tfrac{3}{4}$ by 4, we get $2x = 3$, which has no fractions. We have "cleared the fractions." If we multiply on both sides of $2.3x = 5$ by 10, we get $23x = 50$, which has no decimals. We have "cleared the decimals." In what follows we first multiply on both sides to "clear" or "get rid of" fractions or decimals. For fractions, the number we multiply by is either the product of all the denominators or any number that contains all the denominators as factors.

Solve.

9. $7y + 5 = 2y + 10$

Solve.

10. $5 - 2y = 3y - 5$

11. $7x - 17 + 2x = 2 - 8x + 15$

12. $3x - 15 = 5x + 2 - 4x$

ANSWERS ON PAGE A-5

Solve.

13. $\frac{7}{8}x - \frac{1}{4} + \frac{1}{2}x = \frac{3}{4} + x$

Example 8 Solve: $\frac{2}{3}x - \frac{1}{6} + \frac{1}{2}x = \frac{7}{6} + 2x$.

The number 6 contains all the denominators as factors. In fact, it is the least common denominator. We multiply on both sides by 6.

$$6(\tfrac{2}{3}x - \tfrac{1}{6} + \tfrac{1}{2}x) = 6(\tfrac{7}{6} + 2x) \quad \text{Multiplying by 6 on both sides}$$

$$6 \cdot \tfrac{2}{3}x - 6 \cdot \tfrac{1}{6} + 6 \cdot \tfrac{1}{2}x = 6 \cdot \tfrac{7}{6} + 6 \cdot 2x \quad \text{Using the distributive laws}$$

Caution! Be sure to multiply *all* the terms by 6.

$$4x - 1 + 3x = 7 + 12x \quad \text{Simplifying. Note that the fractions are cleared.}$$

$$7x - 1 = 7 + 12x \quad \text{Collecting like terms}$$

$$7x - 12x = 7 + 1 \quad \text{Subtracting } 12x \text{ and adding 1 to get all the terms with variables on one side and all the other terms on the other}$$

$$-5x = 8 \quad \text{Collecting like terms}$$

$$-\tfrac{1}{5} \cdot (-5x) = -\tfrac{1}{5} \cdot 8 \quad \text{Multiplying by } -\tfrac{1}{5} \text{ or dividing by } -5$$

$$x = -\tfrac{8}{5}$$

The number $-\frac{8}{5}$ checks and is the solution.

DO EXERCISE 13.

Here is a procedure for solving the equations given in this section.

1. **Multiply on both sides to clear of fractions or decimals. (This is optional, but can ease computations.)**
2. **Collect like terms on each side, if possible.**
3. **Use the addition principle to get all terms with variables on one side and all the other terms on the other side.**
4. **Collect like terms again, if possible.**
5. **Use the multiplication principle to solve for the variable.**

Let's repeat Example 4, but clear of decimals first.

Example 9 Solve: $16.3 - 7.2y = -8.18$.

The greatest number of decimal places in any one number is two. Multiplying by 100, which has two 0's, will clear of decimals.

Solve. Clear of decimals first.

14. $41.68 = 4.7 - 8.6y$

$$100(16.3 - 7.2y) = 100(-8.18) \quad \text{Multiplying by 100}$$

$$100(16.3) - 100(7.2y) = 100(-8.18) \quad \text{Using a distributive law}$$

$$1630 - 720y = -818 \quad \text{Simplifying}$$

$$-720y = -818 - 1630 \quad \text{Subtracting 1630}$$

$$-720y = -2448 \quad \text{Collecting like terms}$$

$$y = \frac{-2448}{-720} = 3.4 \quad \text{Dividing by } -720$$

The number 3.4 checks and is the solution.

ANSWERS ON PAGE A-5

DO EXERCISE 14.

Chapter 3   Solving Equations and Problems

• Solve and check.

1. $5x + 6 = 31$

2. $3x + 6 = 30$

3. $4x - 6 = 34$

4. $6x - 3 = 15$

5. $7x + 2 = -54$

6. $5x + 4 = -41$

•• Solve and check.

7. $5x + 7x = 72$

8. $4x + 5x = 45$

9. $4y - 2y = 10$

10. $8y - 5y = 15$

11. $10.2y - 7.3y = -58$

12. $6.8y - 2.4y = -88$

13. $x + \frac{1}{3}x = 8$

14. $x + \frac{1}{4}x = 10$

15. $x + 0.08x = 9072$

16. $x + 0.06x = 7738$

17. $8y - 35 = 3y$

18. $4x - 6 = 6x$

19. $4x - 7 = 3x$

20. $y = 15 - 4y$

21. $x + 1 = 16 - 4x$

22. $6y + 8 - 4y = 18$

23. $2x - 1 = 4 + x$

24. $5y - 2 = 28 - y$

25. $5x + 2 = 3x + 6$

26. $6x + 3 = 2x + 11$

27. $5 - 2x = 3x - 7x + 25$

28. $10 - 3x = 2x - 8x + 40$

ANSWERS
1.
2.
3.
4.
5.
6.
7.
8.
9.
10.
11.
12.
13.
14.
15.
16.
17.
18.
19.
20.
21.
22.
23.
24.
25.
26.
27.
28.

29.

30.

31.

32.

00.

34.

35.

36.

37.

38.

39.

40.

41.

42.

43.

44.

45.

46.

29. $4 + 3x - 6 = 3x + 2 - x$

30. $5 + 4x - 7 = 4x + 3 - x$

31. $4y - 4 + y = 6y + 20 - 4y$

32. $5y - 7 + y = 7y + 21 - 5y$

33. $\dfrac{5}{2}x + \dfrac{1}{2}x = 3x + \dfrac{3}{2} + \dfrac{5}{2}x$

34. $\dfrac{7}{8}x - \dfrac{1}{4} + \dfrac{3}{4}x = \dfrac{1}{16} + x$

35. $2.1x + 45.2 = 3.2 - 8.4x$

36. $7\dfrac{1}{2}y - \dfrac{1}{2}y = \dfrac{15}{4}y + 39$

37. $\dfrac{1}{5}t - 0.4 + \dfrac{2}{5}t = 0.6 - \dfrac{1}{10}t$

38. $1.7t + 8 - 1.62t = 0.4t - 0.32 + 8$

✓ SKILL MAINTENANCE

39. Simplify: $\frac{80}{96}$.

40. Subtract: $-\frac{2}{3} - \frac{5}{8}$.

41. Find $-(-x)$ when $x = -14$.

42. Remove parentheses and simplify: $5(x - 3) - 4(5 - x)$.

☆ EXTENSION

Solve.

43. $0.008 + 9.62x - 42.8 = 0.944x + 0.0083 - x$

44. $\dfrac{y - 2}{3} = \dfrac{2 - y}{5}$

45. $0 = y - (-14) - (-3y)$

46. Solve the equation $4x - 8 = 32$ by first using the addition principle. Then solve it by first using the multiplication principle.

Equations Containing Parentheses

After finishing Section 3.4, you should be able to:

• Solve simple equations containing parentheses.

• Some equations containing parentheses can be solved by first multiplying to remove parentheses and then proceeding as before.

Example 1 Solve: $4x = 2(12 - 2x)$.

$$4x = 2(12 - 2x)$$
$$4x = 24 - 4x \qquad \text{Multiplying to remove parentheses}$$
$$4x + 4x = 24 \qquad \text{Adding } 4x \text{ to get all } x\text{-terms on one side}$$
$$8x = 24 \qquad \text{Collecting like terms}$$
$$x = 3 \qquad \text{Multiplying by } \tfrac{1}{8}$$

Check:

$$
\begin{array}{c|c}
4x & = 2(12 - 2x) \\
\hline
4 \cdot 3 & 2(12 - 2 \cdot 3) \\
12 & 2(12 - 6) \\
& 2 \cdot 6 \\
& 12
\end{array}
$$

The solution is 3.

DO EXERCISES 1 AND 2.

Example 2 Solve: $3(x - 2) - 1 = 2 - 5(x + 5)$.

$$3(x - 2) - 1 = 2 - 5(x + 5)$$
$$3x - 6 - 1 = 2 - 5x - 25 \qquad \text{Multiplying to remove parentheses}$$
$$3x - 7 = -5x - 23 \qquad \text{Simplifying}$$
$$3x + 5x = -23 + 7 \qquad \text{Adding } 5x \text{ and also 7, to get all } x\text{-terms on one side and all other terms on the other side}$$
$$8x = -16 \qquad \text{Simplifying}$$
$$x = -2 \qquad \text{Multiplying by } \tfrac{1}{8}$$

Check:

$$
\begin{array}{c|c}
3(x - 2) - 1 & = 2 - 5(x + 5) \\
\hline
3(-2 - 2) - 1 & 2 - 5(-2 + 5) \\
3 \cdot (-4) - 1 & 2 - 5(3) \\
-12 - 1 & 2 - 15 \\
-13 & -13
\end{array}
$$

The solution is -2.

DO EXERCISES 3 AND 4.

Solve.

1. $2(2y + 3) - 14$

2. $5(3x - 2) = 35$

Solve.

3. $3(7 + 2x) = 30 + 7(x - 1)$

4. $4(3 + 5x) - 4 = 3 + 2(x - 2)$

ANSWERS ON PAGE A-5

HANDLING DIMENSION SYMBOLS (PART 1)

Speed is often measured by measuring a distance and a time, and then dividing the distance by the time (this is *average* speed). If a distance is measured in kilometers (km) and the time required to travel that distance is measured in hours, the speed will be computed in *kilometers per hour* (km/hr). For example, if a car travels 100 km in 2 hr, the average speed is

$$\frac{100 \text{ km}}{2 \text{ hr}}, \quad \text{or} \quad 50 \frac{\text{km}}{\text{hr}}.$$

The standard notation for km/hr is km/h.

The symbol

$$\frac{100 \text{ km}}{2 \text{ hr}}$$

makes it look as though we are dividing 100 km by 2 hr. It may be argued that we cannot divide 100 km by 2 hr (we can only divide 100 by 2). Nevertheless, it is convenient to treat dimension symbols such as *kilometers, meters, hours, feet, seconds,* and *pounds* much like numerals or variables, since correct results can thus be obtained mechanically. Compare, for example,

$$\frac{100x}{2y} = \frac{100}{2} \cdot \frac{x}{y} = 50 \frac{x}{y} \quad \text{with} \quad \frac{100 \text{ km}}{2 \text{ hr}} = \frac{100}{2} \cdot \frac{\text{km}}{\text{hr}} = 50 \frac{\text{km}}{\text{hr}}.$$

The analogy holds in other situations.

Example 1 Compare

$$3 \text{ ft} + 2 \text{ ft} = (3 + 2) \text{ ft} = 5 \text{ ft}$$

with

$$3x + 2x = (3 + 2)x = 5x.$$

Compare 3 ft + 2 yd with $3x + 2y$. Note in this case that we cannot simplify further since the units are different. We could change units, but we will not do that here. (See pp. 402 and 512 for more on dimension symbols.)

EXERCISES

Distances and times are given. Use them to compute speed.

1. 45 mi, 9 hr

2. 680 km, 20 hr

3. 6.6 meters (m), 3 sec

4. 76 ft, 4 min

Add these measures.

5. 45 ft, 7 ft

6. 85 sec, 17 sec

7. 17 m, 14 m

8. 3 hr, 29 hr

9. $\frac{3}{4}$ lb, $\frac{2}{5}$ lb

10. 5 km, 7 km

11. 18 g, 4 g

12. 70 m/sec, 35 m/sec

• Solve the following equations. Check.

ANSWERS

1. $3(2y - 3) = 27$

2. $4(2y - 3) = 28$

3. $40 = 5(3x + 2)$

4. $9 = 3(5x - 2)$

5. $2(3 + 4m) - 9 = 45$

6. $3(5 + 3m) - 8 = 88$

7. $5r - (2r + 8) = 16$

8. $6b - (3b + 8) = 16$

9. $3g - 3 = 3(7 - g)$

10. $3d - 10 = 5(d - 4)$

11. $6 - 2(3x - 1) = 2$

12. $10 - 3(2x - 1) = 1$

1. _____

2. _____

3. _____

4. _____

5. _____

6. _____

7. _____

8. _____

9. _____

10. _____

11. _____

12. _____

13. 5(d + 4) = 7(d − 2)

14. 3(t − 2) = 9(t + 2)

13. _____

14. _____

15. 3(x − 2) = 5(x + 2)

16. 5(y + 4) = 3(y − 2)

15. _____

16. _____

17. 8(2t + 1) = 4(7t + 7)

18. 7(5x − 2) = 6(6x − 1)

17. _____

18. _____

19. 3(r − 6) + 2 = 4(r + 2) − 21

20. 5(t + 3) + 9 = 3(t − 2) + 6

19. _____

20. _____

21. 19 − (2x + 3) = 2(x + 3) + x

22. 13 − (2c + 2) = 2(c + 2) + 3c

21. _____

22. _____

23. _____

23. $\frac{1}{4}$(8y + 4) − 17 = −$\frac{1}{2}$(4y − 8)

24. $\frac{1}{3}$(6x + 24) − 20 = −$\frac{1}{4}$(12x − 72)

24. _____

25. _____

26. _____

 SKILL MAINTENANCE

25. The circumference of a circle is given by the expression 2πr, where r is the radius of the circle. Find the circumference when the radius is $\frac{7}{2}$ ft. Use $\frac{22}{7}$ for π.

26. Write a true sentence using < or >:

$$-15 \qquad -13.$$

27. _____

28. _____

27. Factor: 7x − 21 − 14y.

28. Compute and simplify: −22.1 ÷ 3.4.

29. _____

30. _____

⭐ **EXTENSION**

Solve.

29. 🖩 475(54x + 7856) + 9762 = 402(83x + 975)

31. _____

30. −2[3(x − 2) + 4] = 4(1 − x) + 8

31. 🖩 30,000 + 20,000x = 55,000(1 + 12,500x)

32. _____

32. 3(x + 4) = 3(4 + x)

Chapter 3   Solving Equations and Problems

3.5

Solving Problems

The first step in solving a problem is to translate it to mathematical language. Very often this means translating to an equation. Drawing a picture usually helps a great deal. Then we solve the equation and check to see if we have a solution to the problem.

Example 1 A 6-ft board is cut into two pieces, one twice as long as the other. How long are the pieces?

Drawing a picture

The picture can help in translating. Here is one way to do it.

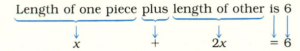

(We use x for the length of one piece and $2x$ for the length of the other.) Now we solve:

$$x + 2x = 6$$
$$3x = 6 \qquad \text{Collecting like terms}$$
$$x = 2 \qquad \text{Multiplying by } \tfrac{1}{3}$$

Do we have an answer to the *problem*? If one piece is 2 ft long, the other, to be twice as long, must be 4 ft long, and the lengths of the pieces add up to 6 ft. This checks.

Note that we did not have to check the solution in the equation. Rather, we checked in the problem itself.

When solving problems consider the following tips.

PROBLEM-SOLVING TIPS

1. *Familiarize* yourself with the situation. If the situation is described in words, as in a textbook, *read carefully.* In any case, think about the situation. Draw a picture whenever it makes sense to do so.
2. *Translate* the problem to an equation. Tell what the letters represent if this is not clear from the nature of the problem or the translation.
3. *Solve* the equation.
4. *Check* the answer in the original problem.
5. *State* the answer to the problem clearly.

DO EXERCISE 1.

---

1. An 8-ft board is cut into two pieces. One piece is 2 ft longer than the other. How long are the pieces?

ANSWER ON PAGE A-6

2. If 5 is subtracted from 3 times a certain number, the result is 10. What is the number?

3. Money is borrowed at 9% simple interest. After 1 year $6213 pays off the loan. How much was originally borrowed?

ANSWERS ON PAGE A-6

Example 2 Five plus three more than a number is nineteen. What is the number?

This time it does not make sense to draw a picture.

$$\underbrace{\text{Five}}_{5} \quad \underbrace{\text{plus}}_{+} \quad \underbrace{\text{three more than a number}}_{(x + 3)} \quad \underbrace{\text{is}}_{=} \quad \underbrace{\text{nineteen}}_{19}$$

We have used x to represent the unknown number.

Now solve: $5 + (x + 3) = 19$

$$x + 8 = 19 \qquad \text{Collecting like terms}$$
$$x = 11 \qquad \text{Adding } -8$$

Check: Three more than 11 is 14. Adding 5 to 14, we get 19. This checks and the answer is 11.

DO EXERCISE 2.

> Always read the problem carefully. Make a list of the information in the problem.

Example 3 Money is invested in a savings account at 8% simple interest. After one year there is $8208 in the account. How much was originally invested?

$$\underbrace{\text{Original investment}}_{x} \quad \underbrace{\text{plus}}_{+} \quad \underbrace{\text{interest}}_{8\%x} \quad \underbrace{\text{is}}_{=} \quad \underbrace{8208}_{8208}$$

We have used x to represent the original investment.

$$x + 8\%x = 8208$$
$$1x + 0.08x = 8208$$
$$1.08x = 8208$$
$$x = \frac{8208}{1.08} = 7600$$

Check: 8% of 7600 is 608. Adding this to 7600, we get 8208. This checks, so the original investment was $7600.

DO EXERCISE 3.

Example 4 The sum of two consecutive integers is 29. What are the integers? (*Consecutive integers* are next to each other, such as 3 and 4. The larger is 1 plus the smaller.)

$$\underbrace{\text{First integer}}_{x} \quad \underbrace{+}_{+} \quad \underbrace{\text{second integer}}_{(x + 1)} \quad \underbrace{= 29}_{= 29} \quad \text{Rewording}$$

We have let x represent the first integer. Then $x + 1$ represents the second.

> You should write down what your letters represent.

Chapter 3   Solving Equations and Problems

We solve:

$$x + (x + 1) = 29$$
$$2x + 1 = 29$$
$$2x = 28$$
$$x = 14.$$

Check: Our answers are 14 and 15. These are consecutive integers. Their sum is 29, so the answers check in the *problem.*

DO EXERCISE 4.

Example 5 Acme Rent-a-Car rents an intermediate-size car (such as a Chevrolet, Ford, or Plymouth) at a daily rate of $44.95 plus 29¢ per mile. A businessperson is not to exceed a daily car rental budget of $100. What mileage will allow the businessperson to stay within budget?

We translate to an equation, using $0.29 for 29¢ and m for mileage.

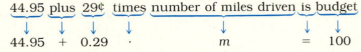

We solve:

$$44.95 + 0.29m = 100$$
$$0.29m = 55.05 \quad \text{Adding } -44.95$$
$$m = \frac{55.05}{0.29} \approx 189.8. \quad \text{Rounded to the nearest tenth}$$

This checks in the original problem. In this case it is an approximation. At least the businessperson now knows to stay under this mileage.

DO EXERCISE 5.

Example 6 The perimeter of a rectangle is 150 cm. The length is 15 cm greater than the width. Find the dimensions.

We first draw a picture.

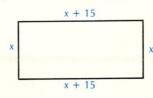

$$\underbrace{\text{Width}}_{} + \underbrace{\text{Width}}_{} + \underbrace{\text{Length}}_{} + \underbrace{\text{Length}}_{} = \underbrace{150.}_{} \quad \text{Rewording}$$
$$x + x + (x + 15) + (x + 15) = 150 \quad \text{Translating}$$

We have let x represent the width. Then $x + 15$ represents the length.

$$4x + 30 = 150$$
$$4x = 120$$
$$x = 30$$

The length is $x + 15$, or 45.

Check: The perimeter is $30 + 30 + 45 + 45$, which is 150. This checks, so the width is 30 cm and the length is 45 cm.

DO EXERCISE 6.

4. The sum of two consecutive even integers is 38. (Consecutive even integers are next to each other, such as 4 and 6. The larger is 2 plus the smaller.) What are the integers?

5. Acme also rents compact cars at a rate of $34.95 plus 27¢ per mile. What mileage will allow the businessperson to stay within a budget of $100?

6. The length of a rectangle is twice the width. The perimeter is 60 m. Find the dimensions.

ANSWERS ON PAGE A-6

7. The second angle of a triangle is 3 times as large as the first. The third angle measures 30° more than the first angle. Find the measures of the angles.

Example 7 The second angle of a triangle is twice as large as the first. The measure of the third angle is 20° greater than that of the first angle. How large are the angles?

We draw a picture. We use x for the measure of the first angle. The second is twice as large, so its measure will be $2x$. The third angle is 20° greater than the first angle so its measure will be $x + 20$.

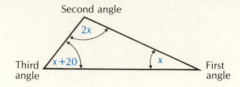

Now, to translate we need to recall a geometric fact. The measures of the angles of any triangle add up to 180°.

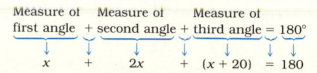

Now we solve:

$$x + 2x + (x + 20) = 180$$
$$4x + 20 = 180$$
$$4x = 160$$
$$x = 40.$$

The angles will have measures as follows:

First angle: $\qquad x = 40°$
Second angle: $\qquad 2x = 80°$
Third angle: $\qquad x + 20 = 60°$

These add up to 180° so they give the answer to the *problem*.

DO EXERCISE 7.

8. After a 30% reduction, an item is on sale for $8050. What was the marked price (the price before reduction)?

Example 8 After a 20% reduction, an item is on sale for $9600. What was the marked price (the price before reduction)?

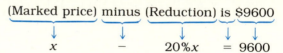

We have used x to represent the marked price:

$$x - 20\%x = 9600$$
$$1x - 0.2x = 9600$$
$$(1 - 0.2)x = 9600$$
$$0.8x = 9600$$
$$x = \frac{9600}{0.8} = 12{,}000$$

Check: 20% of $12,000 is $2400. Subtracting this from $12,000 we get $9600. This checks, so the marked price is $12,000.

DO EXERCISE 8.

ANSWERS ON PAGE A-6

• Solve.

1. The lowest temperature ever recorded in Minneapolis is $-34°$F. This is $62°$ less than the lowest temperature ever recorded in Los Angeles. What is the lowest temperature ever recorded in Los Angeles?

2. The elevation of the Dead Sea is -1286 ft. This is 1881 ft below the elevation of Chicago. What is the elevation of Chicago?

3. When 18 is subtracted from 6 times a certain number, the result is 96. What is the number?

4. When 28 is subtracted from 5 times a certain number, the result is 232. What is the number?

5. If you double a number and then add 16, you get $\frac{2}{5}$ of the original number. What is the original number?

6. If you double a number and then add 85, you get $\frac{3}{4}$ of the original number. What is the original number?

7. If you add $\frac{2}{5}$ of a number to the number itself, you get 56. What is the number?

8. If you add $\frac{1}{3}$ of a number to the number itself, you get 48. What is the number?

ANSWERS

1. _____

2. _____

3. _____

4. _____

5. _____

6. _____

7. _____

8. _____

9. A 180-m rope is cut into 3 pieces. The second piece is twice as long as the first. The third piece is 3 times as long as the second. How long is each piece of rope?

10. A 480-m wire is cut into 3 pieces. The second piece is 3 times as long as the first. The third piece is 4 times as long as the second. How long is each piece?

9. _____

10. _____

11. Consecutive odd integers are next to each other, such as 5 and 7. The larger is 2 plus the smaller. The sum of two consecutive odd integers is 76. What are the integers?

12. The sum of two consecutive odd integers is 84. What are the integers?

11. _____

12. _____

13. Consecutive even integers are next to each other, like 6 and 8. The larger is 2 plus the smaller. The sum of two consecutive even integers is 114. What are the integers?

14. The sum of two consecutive even integers is 106. What are the integers?

13. _____

14. _____

Chapter 3 Solving Equations and Problems

15. The sum of three consecutive integers is 108. What are the integers?

16. The sum of three consecutive integers is 126. What are the integers?

17. The sum of three consecutive odd integers is 189. What are the integers?

18. The sum of three consecutive odd integers is 255. What are the integers?

19. The perimeter of a rectangle is 310 m. The length is 25 m greater than the width. Find the width and the length of the rectangle.

20. The perimeter of a rectangle is 304 cm. The length is 40 cm greater than the width. Find the width and the length of the rectangle.

15. _____

16. _____

17. _____

18. _____

19. _____

20. _____

21.

22.

23.

24.

25.

26.

21. The perimeter of a rectangle is 152 m. The width is 22 m less than the length. Find the width and the length.

22. The perimeter of a rectangle is 280 m. The width is 26 m less than the length. Find the width and the length.

23. The second angle of a triangle is 4 times as large as the first. The third angle is 45° less than the sum of the other two angles. Find the measure of the first angle.

24. The second angle of a triangle is 3 times as large as the first. The third angle is 25° less than the sum of the other two angles. Find the measure of the first angle.

25. Money is invested in a savings account at 7% simple interest. After 1 year there is $4708 in the account. How much was originally invested?

26. Money is borrowed at 10% simple interest. After 1 year $7194 pays off the loan. How much was originally borrowed?

Chapter 3   Solving Equations and Problems

27. After a 40% reduction, a shirt is on sale for $9.60. What was the marked price (the price before reduction)?

28. After a 34% reduction, a blouse is on sale for $9.24. What was the marked price?

29. Badger Rent-a-Car rents an intermediate-size car at a daily rate of $34.95 plus 40¢ per mile. A businessperson is not to exceed a daily car rental budget of $200. What mileage will allow the businessperson to stay within budget?

30. Badger also rents compact cars at $43.95 plus 30¢ per mile. What mileage will allow the business-person to stay within the budget of $200?

31. The second angle of a triangle is 3 times as large as the first. The measure of the third angle is 40° greater than that of the first angle. How large are the angles?

32. The second angle of a triangle is 5 times as large as the first. The measure of the third angle is 10° greater than that of the first angle. How large are the angles?

33. The population of the world in 1984 was 4.8 billion. This was a 2% increase over the population one year earlier. What was the former population?

34. The population of the United States in 1985 was 234 million. This was a 0.8% increase over the population one year earlier. What was the former population?

33. _____

34. _____

35. The equation

$$R = -0.028t + 20.8$$

can be used to predict the world record in the 200-m dash, where R stands for the record in seconds, and t stands for the number of years since 1920. In what year should the record have been 19.0 sec? Check a sports record book to see if this happened.

36. The equation

$$F = \frac{1}{4}N + 40$$

can be used to determine temperatures given how many times a cricket chirps per minute, where F represents temperature in degrees Fahrenheit and N is the number of chirps per minute. Determine the chirps per minute necessary for the temperature to be 80°.

35. _____

36. _____

37. _____

⭐ **EXTENSION**

37. Abraham Lincoln's Gettysburg Address on November 19, 1863, refers to the year 1776 as "Four score and seven years ago." Write an equation and solve it to find what a *score* is.

38. ▦ The area of this triangle is 2.9047 in². Find x.

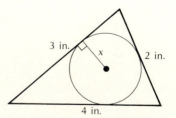

3 in. x 2 in. 4 in.

38. _____

39. In one city, a sales tax of 9% was added to the price of gasoline as registered on the pump. Suppose a driver asks for $10 worth of lead-free gas. The attendant fills the tank until the pump reads $9.10 and charges the driver $10. Something is wrong with this calculation. Find the error using algebra.

39. _____

Formulas

A formula is a kind of a "recipe" for doing a certain type of calculation. Formulas are often given as equations. Here is an example of a formula: $M = \frac{1}{5}n$. This formula has to do with weather. You see a flash of lightning. After a few seconds you hear the thunder associated with that flash. How far away was the lightning?

Your distance from the storm is M miles. You can find that distance by counting the number of seconds n it takes the sound of the thunder to reach you and then multiplying by $\frac{1}{5}$.

Example 1 Consider the formula $M = \frac{1}{5}n$. It takes 10 sec for the sound of the thunder to reach you. How far away is the storm?

We substitute 10 for n and calculate M:

$$M = \tfrac{1}{5}n = \tfrac{1}{5}(10) = 2.$$

The storm is 2 mi away.

DO EXERCISE 1.

Suppose we know how far we are from the storm and want to calculate the number of seconds it would take the sound of the thunder to reach us. We could substitute a number for M, say 2, and solve for n:

$$2 = \tfrac{1}{5}n$$
$$10 = n. \qquad \textbf{Multiplying by 5}$$

However, if we wanted to do this repeatedly, it might be easier to solve for n by getting it alone on one side. We "solve" the formula for n.

Example 2 Solve for n: $M = \frac{1}{5}n$.

$$M = \tfrac{1}{5}n \qquad \textbf{We want this letter alone.}$$
$$5 \cdot M = 5 \cdot \tfrac{1}{5}n \qquad \textbf{Multiplying on both sides by 5}$$
$$5M = n$$

In the above situation for $M = 2$, $n = 5(2)$, or 10.

DO EXERCISE 2.

To see how the principles apply to formulas, compare the following.

A. Solve.

$$5x + 2 = 12$$
$$5x = 12 - 2$$
$$5x = 10$$
$$x = \frac{10}{5} = 2$$

B. Solve.

$$5x + 2 = 12$$
$$5x = 12 - 2$$
$$x = \frac{12 - 2}{5}$$

C. Solve for x.

$$ax + b = c$$
$$ax = c - b$$
$$x = \frac{c - b}{a}$$

In (A) we solved as we did before. In (B) we did not carry out the calculations. In (C) we could not carry out the calculations because we had unknown numbers.

After finishing Section 3.6, you should be able to:

• Solve formulas for specified letters.

1. Suppose it takes the sound of thunder 14 sec to reach you. How far away is the storm?

2. Solve for I: $E = IR$. (This is a formula from electricity relating voltage E, current I, and resistance R.)

ANSWERS ON PAGE A-6

3. Solve for D: $C = \pi D$.
(This is a formula for the circumference C of a circle of diameter D.)

4. Solve for c: $A = \dfrac{a + b + c + d}{4}$.

5. Solve for I: $A = \dfrac{9R}{I}$.
(This is a formula for computing the earned run average A of a pitcher who has given up R earned runs in I innings of pitching.)

ANSWERS ON PAGE A-6

Example 3 Solve for r: $C = 2\pi r$.

This is a formula for the circumference C of a circle of radius r.

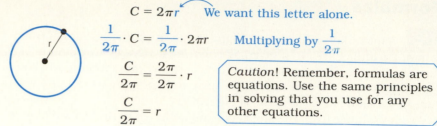

$$C = 2\pi r \quad \text{We want this letter alone.}$$

$$\frac{1}{2\pi} \cdot C = \frac{1}{2\pi} \cdot 2\pi r \quad \text{Multiplying by } \frac{1}{2\pi}$$

$$\frac{C}{2\pi} = \frac{2\pi}{2\pi} \cdot r$$

$$\frac{C}{2\pi} = r$$

Caution! Remember, formulas are equations. Use the same principles in solving that you use for any other equations.

DO EXERCISE 3.

With the formulas in this section we can use a procedure like that described in Section 3.3.

> **To solve a formula for a given letter, identify the letter, and:**
>
> 1. **Multiply on both sides to clear of fractions or decimals. (This is not always necessary, but can ease computations.)**
> 2. **Collect like terms on each side, if necessary.**
> 3. **Use the addition principle to get all terms with the letter to be solved for on one side of the equation and all other terms on the other side.**
> 4. **Collect like terms again, if necessary.**
> 5. **Use the multiplication principle to solve for the letter in question.**

Example 4 Solve for a: $A = \dfrac{a + b + c}{3}$.

This is a formula for the average A of three numbers a, b, and c.

$$A = \frac{a + b + c}{3} \quad \text{We want this letter alone.}$$

$$3A = a + b + c \quad \text{Multiplying by 3 to clear of the fraction}$$

$$3A - b - c = a$$

DO EXERCISE 4.

Example 5 Solve for C: $Q = \dfrac{100M}{C}$.

This is a formula used in psychology for intelligence quotient Q, where M is mental age and C is chronological, or actual, age.

$$Q = \frac{100M}{C} \quad \text{We want this letter alone.}$$

$$CQ = 100M \quad \text{Multiplying by } C \text{ to clear of the fraction}$$

$$C = \frac{100M}{Q} \quad \text{Multiplying by } \frac{1}{Q}$$

DO EXERCISE 5.

Chapter 3   Solving Equations and Problems

CLASS/SECTION: _____ DATE: _____

• Solve for the given letter.

1. $A = bh$, for b
   (Area of a parallelogram with base b and height h)

2. $A = bh$, for h

3. $d = rt$, for r
   (A distance formula, where d is distance, r is speed, and t is time)

4. $d = rt$, for t

5. $I = Prt$, for P
   (Simple interest formula, where I is interest, P is principal, r is interest rate, and t is time)

6. $I = Prt$, for t

7. $F = ma$, for a
   (A physics formula, where F is force, m is mass, and a is acceleration)

8. $F = ma$, for m

9. $P = 2l + 2w$, for w
   (Perimeter of a rectangle of length l and width w)

10. $P = 2l + 2w$, for l

11. $A = \pi r^2$, for r^2
    (Area of a circle with radius r)

12. $A = \pi r^2$, for π

13. $A = \frac{1}{2}bh$, for b
    (Area of a triangle with base b and height h)

14. $A = \frac{1}{2}bh$, for h

15. $E = mc^2$, for m
    (A relativity formula)

16. $E = mc^2$, for c^2

ANSWERS
1. _____
2. _____
3. _____
4. _____
5. _____
6. _____
7. _____
8. _____
9. _____
10. _____
11. _____
12. _____
13. _____
14. _____
15. _____
16. _____

17. $A = \dfrac{a + b + c}{3}$, for b

18. $A = \dfrac{a + b + c}{3}$, for c

19. $v = \dfrac{3k}{t}$, for t

20. $P = \dfrac{ab}{c}$, for c

21. $A = \dfrac{1}{2}ah + \dfrac{1}{2}bh$, for b

22. $A = \dfrac{1}{2}ah + \dfrac{1}{2}bh$, for a

23. The formula

$$H = \dfrac{D^2 N}{2.5}$$

is used to find the horsepower H of an N-cylinder engine. Solve for D^2.

24. Solve for N:

$$H = \dfrac{D^2 N}{2.5}.$$

25. The area of a sector of a circle is given by

$$A = \dfrac{\pi r^2 S}{360},$$

where r is the radius and S is the angle measure of the sector. Solve for S.

26. Solve for r^2:

$$A = \dfrac{\pi r^2 S}{360}.$$

27. The formula

$$R = -0.0075t + 3.85$$

can be used to estimate the world record in the 1500-m run t years after 1930. Solve for t.

28. The formula

$$F = \tfrac{9}{5}C + 32$$

can be used to convert from Celsius, or Centigrade, temperature C to Fahrenheit temperature F. Solve for C.

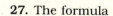 SKILL MAINTENANCE

29. What percent of 7500 is 2500?

30. Add: $-23 + (-67)$.

31. Subtract: $-45.8 - (-32.6)$.

32. Remove parentheses and simplify:

$$4a - 8b - 5(5a - 4b).$$

 EXTENSION

33. In $A = lw$, l and w both double. What is the effect on A?

34. In $P = 2a + 2b$, P doubles. Do a and b necessarily both double?

35. In $A = \tfrac{1}{2}bh$, b increases by 4 units and h does not change. What happens to A?

36. Solve for F:

$$D = \dfrac{1}{E + F}.$$

Integers as Exponents: Applications

After finishing Section 3.7, you should be able to:

· Rename a number with or without negative exponents.

· · Use exponents in multiplying (adding the exponents).

· · · Use exponents in dividing (subtracting the exponents).

: : Use exponents in raising a power to a power (multiplying the exponents).

:·: Solve problems involving interest compounded annually.

· NEGATIVE EXPONENTS

Negative numbers can be used as exponents. Look for a pattern in the following.

$$10^3 = 10 \cdot 10 \cdot 10 \qquad 8^3 = 8 \cdot 8 \cdot 8$$
$$10^2 = 10 \cdot 10 \qquad 8^2 = 8 \cdot 8$$
$$10^1 = 10 \qquad 8^1 = 8$$
$$10^0 = 1 \qquad 8^0 = 1$$
$$10^{-1} = ? \qquad 8^{-1} = ?$$
$$10^{-2} = ? \qquad 8^{-2} = ?$$

In the first case we divided by 10 each time. In the second we divided by 8. Continuing the pattern, we have

$$10^{-1} = \frac{1}{10} = \frac{1}{10^1} \quad \text{and} \quad 8^{-1} = \frac{1}{8} = \frac{1}{8^1};$$

$$10^{-2} = \frac{1}{10 \cdot 10} = \frac{1}{10^2} \quad \text{and} \quad 8^{-2} = \frac{1}{8 \cdot 8} = \frac{1}{8^2}.$$

We make the following agreement. It is a definition.

If n is any positive integer,

$$b^{-n} \text{ is given the meaning } \frac{1}{b^n}.$$

In other words, b^n and b^{-n} are reciprocals.

Example 1 Explain the meaning of 3^{-4} using positive exponents.

$$3^{-4} \text{ means } \frac{1}{3^4} \text{ or } \frac{1}{3 \cdot 3 \cdot 3 \cdot 3}, \text{ or } \frac{1}{81}$$

> 3^{-4} is *not* a negative number.

Example 2 Rename $1/5^2$ using a negative exponent.

$$\frac{1}{5^2} = 5^{-2}$$

DO EXERCISES 1–9.

· · MULTIPLYING USING EXPONENTS

Consider an expression with exponents, such as $a^3 \cdot a^2$. To simplify it, recall the definition of exponents:

$$a^3 \cdot a^2 \quad \text{means} \quad (a \cdot a \cdot a)(a \cdot a)$$

and $(a \cdot a \cdot a)(a \cdot a) = a^5$. The exponent in a^5 is the sum of those in $a^3 \cdot a^2$. Suppose one exponent is positive and one is negative.

$$a^5 \cdot a^{-2} \quad \text{means} \quad (a \cdot a \cdot a \cdot a \cdot a) \cdot \left(\frac{1}{a \cdot a}\right),$$

Explain the meaning of each of the following without using negative exponents.

1. 4^{-3}

2. 5^{-2}

3. 2^{-4}

Rename, using negative exponents.

4. $\frac{1}{3^2}$

5. $\frac{1}{5^4}$

6. $\frac{1}{7^3}$

Rename, using positive exponents.

7. 5^{-3}

8. 7^{-5}

9. 10^{-4}

ANSWERS ON PAGE A-6

Multiply and simplify.

10. $3^5 \cdot 3^3$

11. $5^{-2} \cdot 5^4$

12. $6^{-3} \cdot 6^{-4}$

13. $x \cdot x^{-5}$

14. $y^2 \cdot y^{-4}$

15. $x^{-2} \cdot x^{-6}$

ANSWERS ON PAGE A-6

which simplifies as follows:

$$a \cdot a \cdot a \cdot \left(\frac{a \cdot a}{1}\right)\left(\frac{1}{a \cdot a}\right) = a \cdot a \cdot a \cdot \frac{a \cdot a}{a \cdot a} = a \cdot a \cdot a = a^3.$$

If we add the exponents, we again get the correct result. Next suppose that both exponents are negative:

$$a^{-3} \cdot a^{-2} \quad \text{means} \quad \frac{1}{a \cdot a \cdot a} \cdot \frac{1}{a \cdot a}.$$

This is equal to

$$\frac{1}{a \cdot a \cdot a \cdot a \cdot a}, \quad \text{or} \quad \frac{1}{a^5}, \quad \text{or} \quad a^{-5}.$$

Again, adding the exponents gives the correct result. The same is true if one or both exponents are zero.

> **In multiplication with exponential notation, we can add exponents if the bases are the same:**
>
> $$a^m \cdot a^n = a^{m+n}.$$

Examples Multiply and simplify.

3. $8^4 \cdot 8^3 = 8^{4+3}$ Adding exponents

$\qquad\qquad = 8^7$

4. $7^{-3} \cdot 7^6 = 7^{-3+6}$

$\qquad\qquad\quad = 7^3$

5. $x \cdot x^8 = x^{1+8}$

$\qquad\quad = x^9$

6. $x^4 \cdot x^{-3} = x^{4+(-3)}$

$\qquad\qquad\quad = x^1$

$\qquad\qquad\quad = x$

DO EXERCISES 10–15.

••• DIVIDING USING EXPONENTS

Consider dividing with exponential notation.

$\dfrac{5^4}{5^2}$ means $\dfrac{5 \cdot 5 \cdot 5 \cdot 5}{5 \cdot 5}$, which is $5 \cdot 5 \cdot \dfrac{5 \cdot 5}{5 \cdot 5}$ and we have $5 \cdot 5$, or 5^2.

> **In division, we subtract exponents if the bases are the same:**
>
> $$\frac{a^m}{a^n} = a^{m-n}.$$

This is true whether the exponents are positive, negative, or zero.

Examples Divide and simplify.

7. $\dfrac{5^4}{5^{-2}} = 5^{4-(-2)}$ Subtracting exponents

$\qquad\quad = 5^6$

8. $\dfrac{x}{x^7} = x^{1-7}$

$\phantom{8. \dfrac{x}{x^7}} = x^{-6}$

9. $\dfrac{b^{-4}}{b^{-5}} = b^{-4-(-5)}$

$\phantom{9. \dfrac{b^{-4}}{b^{-5}}} = b^1$

$\phantom{9. \dfrac{b^{-4}}{b^{-5}}} = b$

In exercises such as Examples 7–9 above, it may help to think as follows: After writing the base, write the top exponent. Then write a subtraction sign. Then write the bottom exponent. Then do the subtraction. For example,

$$\dfrac{x^{-3}}{x^{-5}} = x^{-3-(-5)}$$

| Writing the base and the top exponent | Writing a subtraction sign | Writing the bottom exponent |

DO EXERCISES 16–21.

⦂⦂ RAISING A POWER TO A POWER

Consider raising a power to a power.

Example 10

$(3^2)^4$ means $3^2 \cdot 3^2 \cdot 3^2 \cdot 3^2$, or $3 \cdot 3 \cdot 3 \cdot 3 \cdot 3 \cdot 3 \cdot 3 \cdot 3$, or 3^8.

We could have multiplied the exponents in $(3^2)^4$.

Suppose the exponents are not positive.

Example 11

$(5^{-2})^3$ means $\dfrac{1}{5^2} \cdot \dfrac{1}{5^2} \cdot \dfrac{1}{5^2}$, or $\dfrac{1}{5 \cdot 5} \cdot \dfrac{1}{5 \cdot 5} \cdot \dfrac{1}{5 \cdot 5}$, or $\dfrac{1}{5^6}$, or 5^{-6}.

Again, we could have multiplied the exponents. This works for any integer exponents.

> To raise a power to a power we can multiply the exponents. For any exponents m and n,
> $$(a^m)^n = a^{mn}.$$

Examples Simplify.

12. $(3^5)^4 = 3^{5 \cdot 4}$ Multiply exponents.

$ = 3^{20}$

13. $(y^{-5})^7 = y^{-5 \cdot 7}$

$\phantom{13. (y^{-5})^7} = y^{-35}$

14. $(x^4)^{-2} = x^{4(-2)}$

$\phantom{14. (x^4)^{-2}} = x^{-8}$

15. $(a^{-4})^{-6} = a^{(-4)(-6)}$

$\phantom{15. (a^{-4})^{-6}} = a^{24}$

DO EXERCISES 22–25.

> When several factors are in parentheses, raise each to the given power:
> $$(a^m b^n)^t = (a^m)^t (b^n)^t = a^{mt} b^{nt}.$$

Divide and simplify.

16. $\dfrac{4^5}{4^2}$

17. $\dfrac{7^{-2}}{7^3}$

18. $\dfrac{a^2}{a^{-5}}$

19. $\dfrac{b^{-2}}{b^{-3}}$

20. $\dfrac{x}{x^{-3}}$

21. $\dfrac{x^8}{x}$

Simplify.

22. $(3^4)^5$

23. $(x^{-3})^4$

24. $(y^{-5})^{-3}$

25. $(x^{-4})^8$

ANSWERS ON PAGE A-6

26. $(2x^5y^{-3})^4$

27. $(5x^5y^{-6}z^{-3})^2$

28. $(3y^{-2}x^{-5}z^8)^3$

29. Suppose $2000 is invested at 16%, compounded annually. How much is in the account at the end of 3 years?

Examples Simplify.

> *Caution!* Be sure to raise *every* factor in parentheses to the power.

16. $(5x^2y^{-2})^3 = 5^3(x^2)^3(y^{-2})^3 = 125x^6y^{-6}$

17. $(3x^3y^{-5}z^2)^4 = 3^4(x^3)^4(y^{-5})^4(z^2)^4 = 81x^{12}y^{-20}z^8$

The following is a summary of the laws of exponents considered in this section.

$$a^m a^n = a^{m+n}, \qquad \frac{a^m}{a^n} = a^{m-n}, \qquad (a^m b^n)^t = a^{mt}b^{nt}$$

DO EXERCISES 26–28.

APPLICATION: INTEREST COMPOUNDED ANNUALLY

Suppose we invest P dollars at an interest rate of 8%, compounded annually. The amount to which this grows at the end of one year is given by

$$P + 8\%P = P + 0.08P \qquad \text{By definition of percent}$$
$$= (1 + 0.08)P \qquad \text{Factoring}$$
$$= 1.08P. \qquad \text{Simplifying}$$

Going into the second year, the new principal is $(1.08)P$ dollars since interest has been added to the account. By the end of the second year, the following amount will be in the account:

$$(1.08)[(1.08)P], \quad \text{or} \quad (1.08)^2P. \qquad \text{New principal}$$

Going into the third year, the principal will be $(1.08)^2P$ dollars. At the end of the third year, the following amount will be in the account:

$$(1.08)[(1.08)^2P], \quad \text{or} \quad (1.08)^3P. \qquad \text{New principal}$$

Note the pattern: At the end of years 1, 2, and 3 the amount is

$$(1.08)P, \quad (1.08)^2P, \quad (1.08)^3P, \quad \text{and so on.}$$

> If principal P is invested at interest rate r, compounded annually, in t years it will grow to the amount A given by
>
> $$A = P(1 + r)^t.$$

Compare the formula $A = P(1 + r)^t$ with the formula for simple interest, $A = P(1 + rt)$.

Example 18 Suppose $1000 is invested at 8%, compounded annually. How much is in the account at the end of 3 years?

Substituting 1000 for P, 0.08 for r, and 3 for t, we get

$$A = P(1 + r)^t$$
$$= 1000(1 + 0.08)^3 = 1000(1.08)^3 = 1000(1.259712) \approx \$1259.71.$$

DO EXERCISE 29.

• Explain the meaning of each without using negative exponents.

1. 3^{-2} 2. 2^{-3} 3. 10^{-4} 4. 5^{-6}

Rename using negative exponents.

5. $\dfrac{1}{4^3}$ 6. $\dfrac{1}{5^2}$ 7. $\dfrac{1}{x^3}$ 8. $\dfrac{1}{y^2}$

9. $\dfrac{1}{a^4}$ 10. $\dfrac{1}{t^5}$ 11. $\dfrac{1}{p^n}$ 12. $\dfrac{1}{m^n}$

Rename using positive exponents.

13. 7^{-3} 14. 5^{-2} 15. a^{-3} 16. x^{-2}

17. y^{-4} 18. t^{-7} 19. z^{-n} 20. h^{-m}

•• Multiply and simplify.

21. $2^4 \cdot 2^3$ 22. $3^5 \cdot 3^2$ 23. $3^{-5} \cdot 3^8$ 24. $5^{-8} \cdot 5^9$

25. $x^{-2} \cdot x$ 26. $x \cdot x$ 27. $x^4 \cdot x^3$ 28. $x^9 \cdot x^4$

29. $x^{-7} \cdot x^{-6}$ 30. $y^{-5} \cdot y^{-8}$ 31. $t^8 \cdot t^{-8}$ 32. $m^{10} \cdot m^{-10}$

ANSWERS
1.
2.
3.
4.
5.
6.
7.
8.
9.
10.
11.
12.
13.
14.
15.
16.
17.
18.
19.
20.
21.
22.
23.
24.
25.
26.
27.
28.
29.
30.
31.
32.

••• Divide and simplify.

33. $\dfrac{7^5}{7^2}$

34. $\dfrac{4^7}{4^3}$

35. $\dfrac{x}{x^{-1}}$

36. $\dfrac{x^6}{x}$

37. $\dfrac{x^7}{x^{-2}}$

38. $\dfrac{t^8}{t^{-3}}$

39. $\dfrac{z^{-6}}{z^{-2}}$

40. $\dfrac{y^{-7}}{y^{-3}}$

41. $\dfrac{x^{-5}}{x^{-8}}$

42. $\dfrac{y^{-4}}{y^{-9}}$

43. $\dfrac{m^{-9}}{m^{-9}}$

44. $\dfrac{x^{-8}}{x^{-8}}$

▫▫ Simplify.

45. $(2^3)^2$

46. $(3^4)^3$

47. $(5^2)^{-3}$

48. $(9^3)^{-4}$

49. $(x^{-3})^{-4}$

50. $(a^{-5})^{-6}$

51. $(x^4 y^5)^{-3}$

52. $(t^5 x^3)^{-4}$

53. $(x^{-6} y^{-2})^{-4}$

54. $(x^{-2} y^{-7})^{-5}$

55. $(3x^3 y^{-8} z^{-3})^2$

56. $(2a^2 y^{-4} z^{-5})^3$

▪▪ Solve.

57. Suppose $2000 is invested at 12%, compounded annually. How much is in the account at the end of 2 years?

58. Suppose $2000 is invested at 15%, compounded annually. How much is in the account at the end of 3 years?

59. Suppose $10,400 is invested at 16.5%, compounded annually. How much is in the account at the end of 5 years?

60. Suppose $20,800 is invested at 20.5%, compounded annually. How much is in the account at the end of 6 years?

✔ **SKILL MAINTENANCE**

61. Multiply: $-23.8(-5.5)$.

62. Simplify: $3^2 + 5^2 - (3+5)^2$.

☆ **EXTENSION**

If $x \neq 0$ and $y \neq 0$, tell whether each of the following is always true.

63. $x^m \cdot y^n = (xy)^{mn}$

64. $x^m \cdot x^m = x^{2m}$

65. $\left[\dfrac{x}{y}\right]^n = \dfrac{x^n}{y^n}$

Scientific Notation: Applications (Optional)

• SCIENTIFIC NOTATION

The following are examples of scientific notation, which is useful when calculations involve very large or very small numbers:

$$6.4 \times 10^{23}, \quad 4.6 \times 10^{-4}, \quad 10^{15}.$$

Scientific notation for a number consists of exponential notation for a power of 10 and, if needed, decimal notation for a number between 1 and 10, and a multiplication sign: $N \times 10^n$ or 10^n.

We can convert to scientific notation by multiplying by 1, choosing a name like $10^b \cdot 10^{-b}$ for the number 1.

Example 1 Light travels about 9,460,000,000,000 km in one year. Write scientific notation for the number.

We want to move the decimal point 12 places, between the 9 and the 4, so we choose $10^{-12} \times 10^{12}$ as a name for 1. Then we multiply.

$9,460,000,000,000 \times 10^{-12} \times 10^{12}$ Multiplying by 1
$= 9.46 \times 10^{12}$ The 10^{-12} moved the decimal point 12 places to the left and we have scientific notation.

DO EXERCISES 1 AND 2.

Example 2 Write scientific notation for 0.0000000000156.

We want to move the decimal point 11 places. We choose $10^{11} \times 10^{-11}$ as a name for 1, and then multiply.

$0.0000000000156 \times 10^{11} \times 10^{-11}$ Multiplying by 1
$= 1.56 \times 10^{-11}$ The 10^{11} moved the decimal point 11 places to the right and we have scientific notation.

DO EXERCISES 3 AND 4.

You should try to make conversions to scientific notation mentally as much as possible.

Examples Convert mentally to scientific notation.

3. $78,000 = 7.8 \times 10^4$ $7.8,000.$
   4 places
   Large number, so the exponent is positive.

4. $0.0000057 = 5.7 \times 10^{-6}$ $0.000005.7$
   6 places
   Small number, so the exponent is negative.

DO EXERCISES 5 AND 6.

1. Convert 460,000,000,000 to scientific notation.

2. The distance from the earth to the sun is about 93,000,000 mi. Write scientific notation for this number.

3. Convert 0.00000001235 to scientific notation.

4. The mass of a hydrogen atom is about 0.00000000000000000000000017 g. Write scientific notation for this number.

Convert mentally to scientific notation.

5. 0.000314

6. 218,000,000

ANSWERS ON PAGE A-6

Convert to decimal notation.

7. 7.893×10^{11}

8. 5.67×10^{-5}

Multiply and write scientific notation for the answer.

9. $(1.12 \times 10^{-8})(5 \times 10^{-7})$

10. $(9.1 \times 10^{-17})(8.2 \times 10^3)$

ANSWERS ON PAGE A-6

Examples Convert mentally to decimal notation.

5. $7.893 \times 10^5 = 789,300$ $7.89300.$

Positive exponent, so the answer is a large number.

6. $4.7 \times 10^{-8} = 0.000000047$ $0.00000004.7$

Negative exponent, so the answer is a small number.

DO EXERCISES 7 AND 8.

•• MULTIPLYING AND DIVIDING USING SCIENTIFIC NOTATION

Multiplying

Consider the product

$$400 \cdot 2000 = 800,000.$$

In scientific notation this would be

$$(4 \times 10^2) \cdot (2 \times 10^3) = 8 \times 10^5.$$

Note that we could find this product by multiplying $4 \cdot 2$ to get 8, and $10^2 \cdot 10^3$ to get 10^5 (we do this by adding the exponents).

Example 7 Multiply: $(1.8 \times 10^6) \cdot (2.3 \times 10^{-4})$.

a) Multiply 1.8 and 2.3:

$$1.8 \times 2.3 = 4.14.$$

b) Multiply 10^6 and 10^{-4}:

$$10^6 \cdot 10^{-4} = 10^{6+(-4)} \quad \text{Add the exponents.}$$
$$= 10^2.$$

c) The answer is

$$4.14 \times 10^2.$$

Example 8 Multiply: $(3.1 \times 10^5) \cdot (4.5 \times 10^{-3})$.

a) Multiply 3.1 and 4.5:

$$3.1 \times 4.5 = 13.95.$$

b) Multiply 10^5 and 10^{-3}:

$$10^5 \cdot 10^{-3} = 10^{5+(-3)} = 10^2.$$

c) The answer at this stage is

$$13.95 \times 10^2,$$

but this is *not* scientific notation, since 13.95 is not a number between 1 and 10. To find scientific notation we convert 13.95 to scientific notation and simplify:

$$13.95 \times 10^2 = (1.395 \times 10^1) \times 10^2$$
$$= 1.395 \times 10^3. \quad \text{Add the exponents.}$$

DO EXERCISES 9 AND 10.

Chapter 3   Solving Equations and Problems

Division

Consider the quotient

$$800{,}000 \div 400 = 2000.$$

In scientific notation this is

$$(8 \times 10^5) \div (4 \times 10^2) = 2 \times 10^3.$$

Note that we could find this product by dividing $8 \div 4$ to get 2, and $10^5 \div 10^2$ to get 10^3 (we do this by subtracting the exponents).

Example 9 Divide: $(3.41 \times 10^5) \div (1.1 \times 10^{-3})$.

a) Divide 3.41 by 1.1:

$$3.41 \div 1.1 = 3.1.$$

b) Divide 10^5 by 10^{-3}:

$$10^5 \div 10^{-3} = 10^{5-(-3)} \qquad \text{Subtract the exponents.}$$
$$= 10^8.$$

c) The answer is

$$3.1 \times 10^8$$

Example 10 Divide: $(6.4 \times 10^{-7}) \div (8.0 \times 10^6)$.

a) Divide 6.4 by 8.0:

$$6.4 \div 8.0 = 0.8.$$

b) Divide 10^{-7} by 10^6:

$$10^{-7} \div 10^6 = 10^{-7-6} \qquad \text{Subtract the exponents.}$$
$$= 10^{-13}.$$

c) The answer at this stage is

$$0.8 \times 10^{-13},$$

but this is *not* scientific notation, since 0.8 is not a number between 1 and 10. To find scientific notation we convert 0.8 to scientific notation and simplify:

$$0.8 \times 10^{-13} = (8.0 \times 10^{-1}) \times 10^{-13}$$
$$= 8.0 \times 10^{-14} \qquad \text{Add the exponents.}$$

DO EXERCISES 11 AND 12.

Divide and write scientific notation for the answer.

11. $\dfrac{4.2 \times 10^5}{2.1 \times 10^2}$

12. $\dfrac{1.1 \times 10^{-4}}{2.0 \times 10^{-7}}$

ANSWERS ON PAGE A-6

13. There are 300,000 words in the English language. The average person knows about 10,000 of them. What part of the total number of words does the average person know? Write scientific notation for the answer.

Example 11 There are 2864 members in the Professional Bowlers Association. There are 234 million people in the United States. What part of the population are members of the Professional Bowlers Association? Write scientific notation for the answer.

The part of the population that belongs to the Professional Bowlers Association is

$$\frac{2864}{234 \text{ million}}.$$

Now 1 million $= 1,000,000 = 10^6$, so 234 million $= 234 \times 10^6$, or 2.34×10^8. We also have $2864 = 2.864 \times 10^3$. Then we divide and write scientific notation for the answer:

$$\frac{2864}{234 \text{ million}} = \frac{2.864 \times 10^3}{2.34 \times 10^8}$$
$$\approx 1.2239 \times 10^{-5}.$$

DO EXERCISE 13.

Example 12 Americans drink 3 million gallons of orange juice in one day. How much orange juice is consumed in this country in one year? Write scientific notation for the answer.

There are 365 days in a year, so the amount of orange juice consumed is

$$(365 \text{ days}) \cdot (3 \text{ million}) = (3.65 \times 10^2)(3 \times 10^6)$$
$$= 10.95 \times 10^8$$
$$= (1.095 \times 10^1) \times 10^8$$
$$= 1.095 \times 10^9.$$

There are 1.095×10^9 gallons of orange juice consumed in this country in one year.

DO EXERCISE 14.

14. Americans eat 6.5 million gal of popcorn each day. How much popcorn do they eat in one year? Write scientific notation for the answer.

ANSWERS ON PAGE A-6

• Convert to scientific notation.

1. The average discharge of water at the mouth of the Amazon River is 4,200,000 ft^3/sec.

2. The elevation of Mt. McKinley in Alaska is 20,320 ft.

3. There are 4.8 billion people in the world.

4. The population of Canada is 24,882,000.

5. 78,000,000,000

6. 3,700,000,000,000

7. 907,000,000,000,000,000

8. 168,000,000,000,000

9. 0.00000374

10. 0.000000000275

11. 0.000000018

12. 0.0000000002

13. 10,000,000

14. 100,000,000,000

15. 0.000000001

16. 0.0000001

Convert to decimal notation.

17. 7.84×10^8

18. 1.35×10^7

19. 8.764×10^{-10}

20. 9.043×10^{-3}

21. 10^8

22. 10^4

23. 10^{-4}

24. 10^{-7}

•• Multiply or divide and write scientific notation for the result.

25. $(3 \times 10^4)(2 \times 10^5)$

26. $(1.9 \times 10^8)(3.4 \times 10^{-3})$

27. $(5.2 \times 10^5)(6.5 \times 10^{-2})$

28. $(7.1 \times 10^{-7})(8.6 \times 10^{-5})$

ANSWERS
1.
2.
3.
4.
5.
6.
7.
8.
9.
10.
11.
12.
13.
14.
15.
16.
17.
18.
19.
20.
21.
22.
23.
24.
25.
26.
27.
28.

ANSWERS

29. _____

30. _____

31. _____

32. _____

33. _____

34. _____

35. _____

36. _____

37. _____

38. _____

39. _____

40. _____

41. _____

42. _____

43. _____

44. _____

45. _____

29. $(9.9 \times 10^{-6})(8.23 \times 10^{-8})$

30. $(1.123 \times 10^4) \times 10^{-9}$

31. $\dfrac{8.5 \times 10^8}{3.4 \times 10^{-5}}$

32. $\dfrac{5.6 \times 10^{-2}}{2.5 \times 10^5}$

33. $(3.0 \times 10^6) \div (6.0 \times 10^9)$

34. $(1.5 \times 10^{-3}) \div (1.6 \times 10^{-6})$

35. $\dfrac{7.5 \times 10^{-9}}{2.5 \times 10^{12}}$

36. $\dfrac{4.0 \times 10^{-3}}{8.0 \times 10^{20}}$

••• Solve.

37. There are 300,000 words in the English language. The exceptional person knows 20,000 of them. What part of the total number of words does the exceptional person know? Write scientific notation for the answer.

38. Use the information in Exercise 1. How much water is discharged from the Amazon River in one hour?

✔ **SKILL MAINTENANCE**

39. Subtract: $\frac{2}{3} - \frac{3}{4}$.

40. Find the absolute value: $|-8.67|$.

41. Find $-x$ when x is -24.

42. Find the prime factorization: 2864.

☆ **EXTENSION**

43. Calculate. Write scientific notation for the answer.

$$\{2.1 \times 10^6[(2.5 \times 10^{-3}) \div (5.0 \times 10^{-5})]\} \div (3.0 \times 10^{17})$$

Find the reciprocal and write in scientific notation for the answer.

44. 6.25×10^{-3}

45. 4.0×10^{10}

Summary and Review

The following contains a summary of what you should be able to do after completing this chapter. The review exercises are for practice. Answers are at the back of the book. If you miss an exercise, restudy the section and objective indicated alongside the answer.

The review sections to be tested in addition to the material in this chapter are Sections 1.3, 1.7, 2.5, 2.6, and 2.8.

You should be able to:

Solve equations using the addition principle, the multiplication principle, the addition and multiplication principles together, and the distributive laws to collect like terms and to remove parentheses.

Solve.

1. $x + 5 = -17$

2. $-8x = -56$

3. $-\dfrac{x}{4} = 48$

4. $n - 7 = -6$

5. $15x = -35$

6. $x - 11 = 14$

7. $-\dfrac{2}{3} + x = -\dfrac{1}{6}$

8. $\dfrac{4}{5}y = -\dfrac{3}{16}$

9. $y - 0.9 = 9.09$

10. $5 - x = 13$

11. $5t + 9 = 3t - 1$

12. $7x - 6 = 25x$

13. $\dfrac{1}{4}x - \dfrac{5}{8} = \dfrac{3}{8}$

14. $14y = 23y - 17 - 10$

15. $0.22y - 0.6 = 0.12y + 3 - 0.8y$

16. $\dfrac{1}{4}x - \dfrac{1}{8}x = 3 - \dfrac{1}{16}x$

17. $4(x + 3) = 36$

18. $3(5x - 7) = -66$

19. $8(x - 2) = 5(x + 4)$

20. $-5x + 3(x + 8) = 16$

Solve a formula for a certain letter.

Solve.

21. $C = \pi D$ for D

22. $V = \dfrac{1}{3}Bh$ for B

23. $A = \dfrac{a + b}{2}$ for a

Solve problems involving equations of this chapter.

24. A 16-m board is cut into two pieces. One piece is 2 m longer than the other. How long are the pieces?

25. If 14 is added to 3 times a certain number, the result is 41. Find the number.

26. The sum of two consecutive odd integers is 116. Find the integers.

27. The perimeter of a rectangle is 56 cm. The width is 6 cm less than the length. Find the width and the length.

28. After a 30% reduction, an item is on sale for $154. What was the marked price (the price before reduction)?

29. A person's salary is $30,000. That is a 15% increase over the previous year's salary. What was the previous salary (to the nearest dollar)?

30. The measure of the second angle of a triangle is 50° more than that of the first. The measure of the third angle is 10° less than twice the first. Find the measures of the angles.

Rename a number with or without negative exponents, and use exponents in multiplying, dividing, and raising a power to a power.

31. Rename $\dfrac{1}{y^4}$ using negative exponents.

32. Rename 5^{-3} using positive exponents.

Simplify.

33. $x^{-6} \cdot x^4$

34. $\dfrac{t^{-2}}{t^{-11}}$

35. $7^{-5} \cdot 7^{-5}$

36. $\dfrac{4^{-7}}{4^8}$

37. $(8^3)^3$

38. $(3a^{-6})^4$

39. $(x^{-2}yz^7)^{-5}$

40. Suppose $4000 is invested at 9%, compounded annually. How much is in the account at the end of 3 years?

Convert between scientific notation and ordinary decimal notation, multiply and divide using scientific notation, and solve problems involving scientific notation.

Convert to scientific notation.

41. 0.0000278

42. 3,900,000,000

Convert to decimal notation.

43. 5×10^{-8}

44. 1.28×10^4

Multiply or divide and write scientific notation for the result.

45. $(3.8 \times 10^4)(5.5 \times 10^{-1})$

46. $\dfrac{1.28 \times 10^{-8}}{2.5 \times 10^{-4}}$

47. Each day Americans eat 170 million eggs. How many eggs are eaten in one year? Write scientific notation for the answer.

48. Each day Americans eat 170 million eggs. There are 234 million people in this country. How many eggs does each person eat in one year? Write scientific notation for the answer.

 SKILL MAINTENANCE

Compute and simplify.

49. $\dfrac{3}{4} + \dfrac{5}{8}$

50. $\dfrac{3}{4} \cdot \dfrac{5}{8}$

51. $\dfrac{3}{4} - \dfrac{5}{8}$

52. $\dfrac{3}{4} \div \dfrac{5}{8}$

53. Find the additive inverse of -45.78.

54. Add: $\dfrac{5}{8} + (-\dfrac{2}{3})$.

55. Subtract: $-2.3 - (-7.8)$.

56. Divide: $-12.42 \div (-5.4)$.

57. Remove parentheses and simplify: $5x - 8(6x - y)$.

 EXTENSION

Solve.

58. $2|n| + 4 = 50$

59. $0 \cdot x = 10$

60. $\dfrac{x - 4}{7} = \dfrac{4 - x}{8}$

61. $(4.5 \times 10^{-8})y = 9.0 \times 10^{-5}$

62. The total length of the Nile and Amazon rivers is 13,108 km. The Nile is 234 km longer than the Amazon. Find the length of each river.

63. Consumer experts advise us never to pay the sticker price for a car. A rule of thumb is to pay the sticker price minus 20% of the sticker price plus $200. A car is purchased for $11,520 using the rule. What was the sticker price?

Test: Chapter 3

Solve.

1. $x + 7 = 15$

2. $t - 9 = 17$

3. $3x = -18$

4. $-\dfrac{4}{7}x = -28$

5. $3t + 7 = 2t - 5$

6. $\dfrac{1}{2}x - \dfrac{3}{5} = \dfrac{2}{5}$

7. $8 - y = 16$

8. $-\dfrac{2}{5} + x = -\dfrac{3}{4}$

9. $3(x + 2) = 27$

10. $-3x + 6(x + 4) = 9$

11. $0.4p + 0.2 = 4.2p - 7.8 - 0.6p$

Solve.

12. The perimeter of a rectangle is 36 cm. The length is 4 cm greater than the width. Find the length and the width.

13. If you triple a number and then subtract 14, you get $\frac{2}{3}$ of the original number. What was the original number?

14. The sum of three consecutive odd integers is 249. Find the integers.

15. Money is invested in a savings account at 12% simple interest. After one year there is $840 in the account. How much was originally invested?

A N S W E R S
1. _____
2. _____
3. _____
4. _____
5. _____
6. _____
7. _____
8. _____
9. _____
10. _____
11. _____
12. _____
13. _____
14. _____
15. _____

16. _____

17. _____

18. _____

19. _____

20. _____

21. _____

22. _____

23. _____

24. _____

25. _____

26. _____

27. _____

28. _____

29. _____

30. _____

31. _____

32. _____

33. _____

34. _____

Solve.

16. $A = 2\pi rh$ for r

17. $w = \dfrac{P - 2l}{2}$ for l

18. Rename using a negative exponent: $\dfrac{1}{t^5}$.

19. Rename using a positive exponent: y^{-4}.

Simplify.

20. $\dfrac{x^{-3}}{x^9}$

21. $6^{-5} \cdot 6^{-8}$

22. $(2a^{-3})^4$

23. Convert to scientific notation: 0.0000328.

24. Convert to decimal notation: 8.3×10^6.

25. Suppose $1000 is invested at 13%, compounded annually. How much is in the account at the end of 3 years?

26. Divide $(1.242 \times 10^{11}) \div (5.4 \times 10^{15})$. Write scientific notation for the result.

 SKILL MAINTENANCE

27. Subtract: $\dfrac{2}{3} - \dfrac{8}{9}$.

28. Multiply and simplify: $\dfrac{11}{16} \cdot \dfrac{4}{5}$.

29. Add: $-56.8 + (-82.3)$.

30. Multiply: $(-3) \cdot (-4) \cdot (-20) \cdot (-6)$.

31. Remove parentheses and simplify: $2x - 3y - 5(4x - 8y)$.

⭐ **EXTENSION**

32. Solve $c = \dfrac{1}{a - d}$ for d.

33. Solve: $3|w| - 8 = 37$.

34. Solve: $8(5 + y) = 8(y + 5)$.

Cumulative Review

Evaluate.

1. $\dfrac{y - x}{4}$ for $y = 12$ and $x = 6$
2. $\dfrac{3x}{y}$ for $x = 5$ and $y = 4$
3. $x^3 - 3$ for $x = 3$

4. Find the perimeter of a rectangle when l is 12 cm and w is 14 cm.

Translate to an algebraic expression.

5. Four less than twice w
6. Thirty-four percent of some number

7. Find the prime factorization of 648.
8. Find the LCM: 8, 15, 24.

9. Simplify: $\dfrac{80}{144}$.
10. Find decimal notation: 2.6%.

11. Find fractional notation: 80%.

Find percent notation.

12. 1.9
13. $\dfrac{7}{8}$

Write a true sentence using $<$ or $>$.

14. $-4 \quad -6$
15. $0 \quad -5$
16. $-8 \quad 7$

17. Find the additive inverse, the absolute value, and the reciprocal of $\frac{2}{5}$.
18. Find $-(-x)$ when x is -10.

Compute and simplify.

19. $-6.7 + 2.3$
20. $-\dfrac{1}{6} - \dfrac{7}{3}$
21. $-\dfrac{5}{8}\left(-\dfrac{4}{3}\right)$
22. $(-7)(5)(-6)(-0.5)$

23. $81 \div (-9)$
24. $-10.8 \div 36$
25. $-\dfrac{4}{5} \div -\dfrac{25}{8}$

Multiply.

26. $5(3x + 5y + 2z)$
27. $4(-3x - 2)$
28. $-6(2y - 4x)$

Factor.

29. $64 + 18x + 24y$
30. $16y - 56$
31. $5a - 15b + 25$

Collect like terms.

32. $9b + 18y + 6b + 4y$
33. $3y + 4 + 6z + 6y$

34. $-4d - 6a + 3a - 5d + 1$
35. $3.2x + 2.9y - 5.8x - 8.1y$

Simplify.

36. $7 - 2x - (-5x) - 8$
37. $-3x - (-x + y)$

38. $-3(x - 2) - 4x$
39. $10 - 2(5 - 4x)$

40. $[3(x + 6) - 10] - [5 - 2(x - 8)]$

Solve.

41. $x + 1.75 = 6.25$

42. $\dfrac{5}{2}y = \dfrac{2}{5}$

43. $-2.6 + x = 8.3$

44. $4\dfrac{1}{2} + y = 8\dfrac{1}{3}$

45. $-\dfrac{3}{4}x = 36$

46. $-2.2y = -26.4$

47. $5.8x = -35.96$

48. $-4x + 3 = 15$

49. $-3x + 5 = -8x - 7$

50. $4y - 4 + y = 6y + 20 - 4y$

51. $-3(x - 2) = -15$

52. $\dfrac{1}{3}x - \dfrac{5}{6} = \dfrac{1}{2} + 2x$

53. $-3.7x + 6.2 = -7.3x - 5.8$

54. $A = \dfrac{1}{2}h(b + c)$ for h

Solve.

55. What percent of 60 is 18?

56. Two is four percent of what number?

57. If 25 is subtracted from a certain number, the result is 129. Find the number.

58. Jane and Becky purchased identical dresses for a total of $107. Jane paid $17 more for her dress than Becky did. What did Becky pay?

59. Money is invested in a savings account at 12% simple interest. After one year there is $1680 in the account. How much was originally invested?

60. A 143-m wire is cut into three pieces. The second is 3 m longer than the first. The third is $\frac{4}{5}$ as long as the first. How long is each piece?

Simplify.

61. $x^{-6} \cdot x^{12}$

62. $(2x^{-2}y^5)^4$

63. $\dfrac{x^{-14}}{x^{-7}}$

64. $(9 + 1)^2 - 23 \cdot 5$

65. $-5[2(x - 3) + 4] + 6(x - 5)$

 EXTENSION

66. It is known that males reach 96.1% of their final adult height by the time they are 15 years old. A male is 6 ft, 4 in. on his 15th birthday. What will be his final adult height?

Solve.

67. $4|x| - 13 = 3$

68. $4(x + 2) = 4(x - 2) + 16$

69. $0(x + 3) + 4 = 0$

70. $\dfrac{2 + 5x}{4} = \dfrac{11}{28} + \dfrac{8x + 3}{7}$

71. $5(7 + x) = (x + 7)5$

72. $P = \dfrac{2}{M + Q}$ for Q

AN APPLICATION
In a league of n teams in which each team plays every other team twice, the total number of games to be played is given by the algebraic expression

$$n^2 - n.$$

A softball league has 6 teams. What is the total number of games to be played?

THE MATHEMATICS
To solve the problem we substitute 6 for n and evaluate:

$$n^2 - n = 6^2 - 6 = 36 - 6 = 30.$$

— This is a polynomial.

One of the most important topics of introductory algebra is the study of polynomials. In this chapter we learn to add, subtract, and multiply polynomials. Particular emphasis is given to learning fast ways to multiply polynomials. Division of polynomials is discussed in Chapter 9.

The review sections to be tested in addition to the material in this chapter are 1.10, 2.7, 3.3, and 3.5.

Polynomials

Introduction to Polynomials

We have already learned to evaluate and manipulate certain kinds of algebraic expressions. We now consider algebraic expressions called *polynomials.*

Algebraic expressions like the following are called *monomials:*

$$5x^3, \qquad 7y^4, \qquad \tfrac{1}{4}t^2, \qquad x^1, \qquad x^0, \qquad 8, \qquad -2.7, \qquad 0.$$

Each expression is a number or a number times a variable to some whole-number power. More formally, a monomial is an expression of the type ax^n, where n is a whole number.

Algebraic expressions like the following are *polynomials:*

$$3x^2 + 2x - 5, \qquad -8a^3 + \tfrac{1}{4}a, \qquad 49p^{12}, \qquad x, \qquad 0, \qquad -2, \qquad 15y^6.$$

> A *polynomial* is a monomial or a combination of sums or differences of monomials.

1. Write three polynomials.

The following are examples of algebraic expressions that are *not* polynomials:

$$\frac{1}{x^2}, \qquad 5y^3 - 4y^2 + \frac{1}{y}, \quad \text{and} \quad \frac{x+7}{x-6}.$$

Each is *not* a polynomial because in some way it involves an expression with a negative exponent. Note the following:

$$\frac{1}{x^2} = x^{-2}, \qquad 5y^3 - 4y^2 + \frac{1}{y} = 5y^3 - 4y^2 + y^{-1},$$

$$\frac{x+7}{x-6} = (x+7)(x-6)^{-1}$$

DO EXERCISE 1.

Evaluate the polynomial for $x = 3$.

2. $-4x - 7$

• EVALUATING POLYNOMIALS

When we replace the variable in a polynomial by a number, the polynomial then represents a number. Finding that number is called *evaluating the polynomial.*

3. $-5x^3 + 7x + 10$

Example 1 Evaluate the polynomial $4x - 7$ for $x = 2$.

$$4x - 7 = 4(2) - 7 \qquad \text{Replacing } x \text{ by } 2$$
$$= 8 - 7 = 1$$

Evaluate the polynomial for $x = -4$.

4. $5x + 7$

Example 2 Evaluate the polynomial $3y^2 - 8y + 3$ for $y = -5$.

$$3y^2 - 8y + 3 = 3(-5)^2 - 8(-5) + 3 \qquad \text{Replacing } y \text{ by } -5$$
$$= 3(25) + 40 + 3$$
$$= 75 + 40 + 3 = 118$$

5. $2x^2 + 5x - 4$

DO EXERCISES 2–5.

In the following examples we see applications of polynomials in real-world situations.

Example 3 In a sports league of n teams in which each team plays every other team twice, the total number of games to be played is given by the polynomial $n^2 - n$. A slow-pitch softball league has 6 teams and each team plays every other team twice. What is the total number of games to be played?

We evaluate the polynomial for $n = 6$:

$$n^2 - n = 6^2 - 6 = 36 - 6 = 30.$$

The total number of games to be played is 30.

Example 4 The volume of a cube with side of length x is given by the polynomial x^3. It is known that if all the gold in the world could be gathered together it would form a cube 18 yd on a side. Find the volume of the world's gold.

We evaluate the polynomial x^3 for $x = 18$:

$$x^3 = 18^3 = 5832.$$

The volume of the world's gold is 5832 yd^3 (cubic yards).

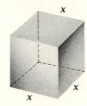

DO EXERCISES 6 AND 7.

• • IDENTIFYING TERMS

Subtractions can be rewritten as additions. We showed this in Sections 2.3 and 2.5. For any polynomial we can find an equivalent polynomial using only additions.

Examples Find an equivalent polynomial using only additions.

5. $-5x^2 - x = -5x^2 + (-x)$ **Adding the inverse of x instead of subtracting x**

6. $4y^5 - 2y^6 - 7y = 4y^5 + (-2y^6) + (-7y)$ **Adding the inverses of $2y^6$ and $7y$ instead of subtracting**

DO EXERCISES 8 AND 9.

When a polynomial has only additions, the parts being added are called *terms*.

Example 7 Identify the terms of the polynomial

$$4x^3 + 3x + 12 + 8x^4 + 5x.$$

Terms: $4x^3$, $3x$,   12, $8x^4$, $5x$

If there are subtractions you can think of them as additions without rewriting.

Example 8 Identify the terms of the polynomial $3t^4 - 5t^6 - 4t + 2$.

Terms: $3t^4$, $-5t^6$, $-4t$,   2

DO EXERCISES 10 AND 11.

6. In the situation of Example 3, what is the total number of games to be played in a league of 10 teams?

7. The perimeter of a square of side s is given by the polynomial $4s$.

A baseball diamond is a square 90 ft on a side. Find the perimeter of a baseball diamond.

Find an equivalent polynomial using only additions.

8. $-9x^3 - 4x^5$

9. $-2y^3 + 3y^7 - 7y$

Identify the terms of the polynomial.

10. $3x^2 + 6x + \frac{1}{2}$

11. $-4y^5 + 7y^2 - 3y - 2$

ANSWERS ON PAGE A-7

Identify the like terms in the polynomial.

12. $4x^3 - x^3 + 2$

13. $4t^4 - 9t^3 - 7t^4 + 10t^3$

Collect like terms.

14. $3x^2 + 5x^2$

15. $4x^3 - 2x^3 + 2 + 5$

16. $\frac{1}{2}x^5 - \frac{3}{4}x^5 + 4x^2 - 2x^2$

Collect like terms.

17. $24 - 4x^3 - 24$

18. $5x^3 - 8x^5 + 8x^5$

19. $-2x^4 + 16 + 2x^4 + 9 - 3x^5$

Collect like terms.

20. $7x - x$

21. $5x^3 - x^3 + 4$

22. $\frac{3}{4}x^3 + 4x^2 - x^3 + 7$

23. $8x^2 - x^2 + x^3 - 1 - 4x^2 + 10$

••• LIKE TERMS

Terms that have the same variable and the same exponent are called *like terms*, or *similar terms*.

Examples Identify the like terms in each polynomial.

9. $4x^3 + 5x - 4x^2 + 2x^3 + x^2$

Like terms: $4x^3$ and $2x^3$ Same exponent and variable
Like terms: $-4x^2$ and x^2 Same exponent and variable

10. $6 - 3a^2 + 8 - a - 5a$

Like terms: 6 and 8   No variable at all
Like terms: $-a$ and $-5a$

DO EXERCISES 12 AND 13.

∷ COLLECTING LIKE TERMS

We can often simplify polynomials by *collecting like terms*, or *combining similar terms*. To do this we use the distributive laws.

Examples Collect like terms.

11. $2x^3 - 6x^3 = (2 - 6)x^3$ Using a distributive law
$$= -4x^3$$

12. $5x^2 + 7 + 4x^4 + 2x^2 - 11 - 2x^4 = (5 + 2)x^2 + (4 - 2)x^4 + (7 - 11)$
$$= 7x^2 + 2x^4 - 4$$

DO EXERCISES 14–16.

In collecting like terms we may get zero.

Examples Collect like terms.

13. $5x^3 - 5x^3 = (5 - 5)x^3$ **14.** $3x^4 - 3x^4 + 2x^2 = (3 - 3)x^4 + 2x^2$
$$= 0x^3 = 0 \qquad\qquad = 0x^4 + 2x^2 = 2x^2$$

DO EXERCISES 17–19.

Multiplying a term of a polynomial by 1 does not change the polynomial, but it may make it easier to factor before collecting like terms.

Examples Collect like terms.

15. $5x^2 + x^2 = 5x^2 + 1x^2$ Replacing x^2 by $1x^2$
$$= (5 + 1)x^2 \quad \text{Using a distributive law}$$
$$= 6x^2$$

16. $5x^4 - 6x^3 - x^4 = 5x^4 - 6x^3 - 1x^4$ $x^4 = 1x^4$
$$= (5 - 1)x^4 - 6x^3$$
$$= 4x^4 - 6x^3$$

DO EXERCISES 20–23.

• Evaluate each polynomial for $x = 4$.

1. $-5x + 2$
2. $-3x + 1$
3. $2x^2 - 5x + 7$

4. $3x^2 + x + 7$
5. $x^3 - 5x^2 + x$
6. $7 - x + 3x^2$

The daily number of accidents involving a driver of age a in the United States is approximated by the polynomial
$$0.4a^2 - 40a + 1039.$$

7. Evaluate the polynomial for $a = 18$ to find the number of daily accidents involving an 18-year-old driver.

8. Evaluate the polynomial for $a = 20$ to find the number of daily accidents involving a 20-year-old driver.

Evaluate each polynomial for $x = -1$.

9. $3x + 5$
10. $6 - 2x$

11. $x^2 - 2x + 1$
12. $5x - 6 + x^2$

13. $-3x^3 + 7x^2 - 3x - 2$
14. $-2x^3 - 5x^2 + 4x + 3$

•• Identify the terms of each polynomial.

15. $2 - 3x + x^2$
16. $2x^2 + 3x - 4$

••• Identify the like terms in each polynomial.

17. $5x^3 + 6x^2 - 3x^2$
18. $3x^2 + 4x^3 - 2x^2$

19. $2x^4 + 5x - 7x - 3x^4$
20. $-3t + t^3 - 2t - 5t^3$

•• •• Collect like terms.

21. $2x - 5x$
22. $2x^2 + 8x^2$

23. $x - 9x$
24. $x - 5x$

ANSWERS
1.
2.
3.
4.
5.
6.
7.
8.
9.
10.
11.
12.
13.
14.
15.
16.
17.
18.
19.
20.
21.
22.
23.
24.

25. $5x^3 + 6x^3 + 4$

26. $6x^4 - 2x^4 + 5$

27. $5x^3 + 6x - 4x^3 - 7x$

28. $3a^4 - 2a + 2a + a^4$

29. $6b^5 + 3b^2 - 2b^5 - 3b^2$

30. $2x^2 - 6x + 3x + 4x^2$

31. $\dfrac{1}{4}x^5 - 5 + \dfrac{1}{2}x^5 - 2x - 37$

32. $\dfrac{1}{3}x^3 + 2x - \dfrac{1}{6}x^3 + 4 - 16$

33. $6x^2 + 2x^4 - 2x^2 - x^4 - 4x^2$

34. $8x^2 + 2x^3 - 3x^3 - 4x^2 - 4x^2$

35. $\dfrac{1}{4}x^3 - x^2 - \dfrac{1}{6}x^2 + \dfrac{3}{8}x^3 + \dfrac{5}{16}x^3$

36. $\dfrac{1}{5}x^4 + \dfrac{1}{5} - 2x^2 + \dfrac{1}{10} - \dfrac{3}{15}x^4 + 2x^2 - \dfrac{3}{10}$

✓ **SKILL MAINTENANCE**

37. Multiply: $3(s + t + 8)$.

38. Multiply: $-7(x + 4)$.

39. Collect like terms:
$9x + 2y - 4x - 2y$.

40. Add: $-2 + (-8)$.

☆ **EXTENSION**

Collect like terms.

41. $3x^2 + 2x - 2 + 3x^0$

42. $(3x^2)^3 + (4x^2)(4x^4) - x^4(2x)^2 + [(2x)^2]^3 - 100x^2(x^2)^2$

43. $\frac{9}{2}x^8 + \frac{1}{9}x^2 + \frac{1}{2}x^9 + \frac{9}{2}x^1 + \frac{9}{2}x^9 + \frac{8}{9}x^2 + \frac{1}{2}x - \frac{1}{2}x^8$

44. ▦ Evaluate $s^2 - 50s + 675$ and $-s^2 + 50s - 675$ for $s = 18$, $s = 25$, and $s = 32$.

Chapter 4   Polynomials

More on Polynomials

• DESCENDING ORDER

This polynomial is arranged in *descending order:*

$$8x^4 - 2x^3 + 5x^2 - x + 3.$$

The term with the largest exponent is first. The term with the next largest exponent is second, and so on. The associative and commutative laws allow us to arrange the terms of a polynomial in descending order.

Examples Arrange each polynomial in descending order.

1. $4x^5 + 4x^7 + x^2 + 2x^3 = 4x^7 + 4x^5 + 2x^3 + x^2$
2. $3 + 4x^5 - 4x^2 + 5x + 3x^3 = 4x^5 + 3x^3 - 4x^2 + 5x + 3$

We usually arrange polynomials in descending order. The opposite order is called *ascending.*

DO EXERCISES 1–3.

•• COLLECTING LIKE TERMS AND DESCENDING ORDER

Example 3 Collect like terms and then arrange in descending order.

$$2x^2 - 4x^3 + 3 - x^2 - 2x^3 = x^2 - 6x^3 + 3 \qquad \text{Collecting like terms}$$
$$= -6x^3 + x^2 + 3 \qquad \text{Descending order}$$

DO EXERCISES 4 AND 5.

••• DEGREES

The *degree* of a term is its exponent.

Example 4 Identify the degree of each term of $8x^4 + 3x + 7$.

The degree of $8x^4$ is 4.
The degree of $3x$ is 1.   Recall that $x = x^1$.
The degree of 7 is 0.   Think of 7 as $7x^0$. Recall that $x^0 = 1$.

The *degree of a polynomial* is its largest exponent, unless it is the polynomial 0. The polynomial 0 is a special case. Mathematicians agree that it has *no* degree either as a term or as a polynomial.

Example 5 Identify the degree of $3x^4 - 6x^3 + 7$.

$$3x^4 - 6x^3 + 7 \qquad \text{The largest exponent is 4.}$$

The degree of the polynomial is 4.

DO EXERCISE 6.

After finishing Section 4.2, you should be able to:

- • Arrange a polynomial in descending order.
- •• Collect the like terms of a polynomial and arrange in descending order.
- ••• Identify the degrees of terms of polynomials and degrees of polynomials.
- ∷ Identify the coefficients of the terms of a polynomial.
- ∷• Identify the missing terms of a polynomial.
- ⦂⦂ Tell whether a polynomial is a monomial, binomial, trinomial, or none of these.

Arrange each polynomial in descending order.

1. $x + 3x^5 + 4x^3 + 5x^2 + 6x^7 - 2x^4$

2. $4x^2 - 3 + 7x^5 + 2x^3 - 5x^4$

3. $-14 + 7t^2 - 10t^5 + 14t^7$

Collect like terms and then arrange in descending order.

4. $3x^2 - 2x + 3 - 5x^2 - 1 - x$

5. $-x + \frac{1}{2} + 14x^4 - 7x - 1 - 4x^4$

Identify the degree of each term and the degree of the polynomial.

6. $-6x^4 + 8x^2 - 2x + 9$

ANSWERS ON PAGE A-7

Identify the coefficient of each term.

7. $5x^9 + 6x^3 + x^2 - x + 4$

Identify the missing terms in the polynomial.

8. $2x^3 + 4x^2 - 2$

9. $-3x^4$

10. $x^3 + 1$

11. $x^4 - x^2 + 3x + 0.25$

Tell whether the polynomial is a monomial, binomial, trinomial, or none of these.

12. $5x^4$

13. $4x^3 - 3x^2 + 4x + 2$

14. $3x^2 + x$

15. $3x^2 + 2x - 4$

⠃ COEFFICIENTS

In this polynomial the color numbers are the *coefficients:*

$$3x^5 - 2x^3 + 5x + 4.$$

Example 6 Identify the coefficient of each term in the polynomial

$$3x^4 - 4x^3 + 7x^2 + x - 8.$$

The coefficient of the first term is 3.

The coefficient of the second term is -4.

The coefficient of the third term is 7.

The coefficient of the fourth term is 1.

The coefficient of the fifth term is -8.

DO EXERCISE 7.

⠣ MISSING TERMS

If a coefficient is 0, we usually do not write the term. We say that we have a *missing term.*

Example 7 In

$$8x^5 - 2x^3 + 5x^2 + 7x + 8,$$

there is no term with x^4. We say that the x^4-term (or the *fourth-degree term*) is missing.

We could write missing terms with zero coefficients or leave space. For example, we could write the polynomial $3x^2 + 9$ as

$$3x^2 + 0x + 9 \quad \text{or} \quad 3x^2 + \qquad 9,$$

but ordinarily we do not.

DO EXERCISES 8–11.

⠿ MONOMIALS, BINOMIALS, AND TRINOMIALS

Polynomials with just one term are called *monomials.* Polynomials with just two terms are called *binomials.* Those with just three terms are called *trinomials.*

Example 8

Monomials	*Binomials*	*Trinomials*
$4x^2$	$2x + 4$	$3x^3 + 4x + 7$
9	$3x^5 + 6x$	$6x^7 - 7x^2 + 4$
$-23x^{19}$	$-9x^7 - 6$	$4x^2 - 6x - \frac{1}{2}$

DO EXERCISES 12–15.

• Arrange each polynomial in descending order.

1. $x^5 + x + 6x^3 + 1 + 2x^2$

2. $3 + 2x^2 - 5x^6 - 2x^3 + 3x$

3. $5x^3 + 15x^9 + x - x^2 + 7x^8$

4. $9x - 5 + 6x^3 - 5x^4 + x^5$

5. $8y^3 - 7y^2 + 9y^6 - 5y^8 + y^7$

6. $p^8 - 4 + p + p^2 - 7p^4$

•• Collect like terms and then arrange in descending order.

7. $3x^4 - 5x^6 - 2x^4 + 6x^6$

8. $-1 + 5x^3 - 3 - 7x^3 + x^4 + 5$

9. $-2x + 4x^3 - 7x + 9x^3 + 8$

10. $-6x^2 + x - 5x + 7x^2 + 1$

11. $3x + 3x + 3x - x^2 - 4x^2$

12. $-2x - 2x - 2x + x^3 - 5x^3$

13. $-x + \dfrac{3}{4} + 15x^4 - x - \dfrac{1}{2} - 3x^4$

14. $2x - \dfrac{5}{6} + 4x^3 + x + \dfrac{1}{3} - 2x$

••• Identify the degree of each term of each polynomial and the degree of the polynomial.

15. $-7x^3 + 6x^2 + 3x + 7$

16. $5x^4 + x^2 - x + 2$

17. $x^2 - 3x + x^6 - 9x^4$

18. $8x - 3x^2 + 9 - 8x^3$

ANSWERS

1. _____

2. _____

3. _____

4. _____

5. _____

6. _____

7. _____

8. _____

9. _____

10. _____

11. _____

12. _____

13. _____

14. _____

15. _____

16. _____

17. _____

18. _____

Identify the coefficient of each term of each polynomial.

19. $-3x + 6$

20. $3x^2 - 5x + 2$

21. $6x^3 + 7x^2 - 8x - 2$

22. $-2 + 8x - 3x^2 + 6x^3 - 5x^4$

 Identify the missing terms in each polynomial.

23. $x^3 - 27$

24. $x^5 + x$

25. $x^4 - x$

26. $5x^4 - 7x + 2$

27. $2x^3 - 5x^2 + x - 3$

28. $-6x^3$

Tell whether each polynomial is a monomial, binomial, trinomial, or none of these.

29. $x^2 - 10x + 25$

30. $-6x^4$

31. $x^3 - 7x^2 + 2x - 4$

32. $x^2 - 9$

33. $4x^2 - 25$

34. $2x^4 - 7x^3 + x^2 + x - 6$

35. $40x$

36. $4x^2 + 12x + 9$

✔ SKILL MAINTENANCE

37. Find $-(-x)$ when x is 6.

38. Subtract: $7 - (-5)$.

39. Multiply: $(-6)(-5)$.

40. Find an equivalent expression for the additive inverse $-(4x - 7y + 2)$ that does not have parentheses.

★ EXTENSION

41. What is the degree of $(5m^5)^2$?

42. A polynomial in which x is the variable has degree 3. The coefficient of x^2 is 3 less than the coefficient of x^3. The coefficient of x is 3 times the coefficient of x^2. The remaining coefficient is 2 more than the coefficient of x^3. The sum of the coefficients is -4. Find the polynomial.

Addition of Polynomials

• ADDITION

> To *add* polynomials we can write a plus sign between them and collect like terms.

Depending on the situation, you may see polynomials written in descending order, ascending order, or neither. Generally, if an exercise is written in one kind of order, then we write the answer in that order.

Example 1 Add: $-3x^3 + 2x - 4$ and $4x^3 + 3x^2 + 2$.

$(-3x^3 + 2x - 4) + (4x^3 + 3x^2 + 2)$ Writing a plus sign between the polynomials

$= (-3 + 4)x^3 + 3x^2 + 2x + (-4 + 2)$ Collecting like terms. *No signs are changed.*

$= x^3 + 3x^2 + 2x - 2$

Example 2 Add: $\frac{2}{3}x^4 + 3x^2 - 2x + \frac{1}{2}$ and $-\frac{1}{3}x^4 + 5x^3 - 3x^2 + 3x - \frac{1}{2}$

$\left(\frac{2}{3}x^4 + 3x^2 - 2x + \frac{1}{2}\right) + \left(-\frac{1}{3}x^4 + 5x^3 - 3x^2 + 3x - \frac{1}{2}\right)$

$= \left(\frac{2}{3} - \frac{1}{3}\right)x^4 + 5x^3 + (3 - 3)x^2$

$+ (-2 + 3)x + \left(\frac{1}{2} - \frac{1}{2}\right)$ Collecting like terms

$= \frac{1}{3}x^4 + 5x^3 + x$

We can add polynomials as we do because they represent numbers.

DO EXERCISES 1–4.

After some practice you will be able to add mentally.

Example 3 Add: $3x^2 - 2x + 2$ and $5x^3 - 2x^2 + 3x - 4$.

$(3x^2 - 2x + 2) + (5x^3 - 2x^2 + 3x - 4)$
$= 5x^3 + (3 - 2)x^2 + (-2 + 3)x + (2 - 4)$ You might do this step mentally.

$= 5x^3 + x^2 + x - 2$ Then you would write only this.

DO EXERCISES 5 AND 6.

We can also add polynomials by writing like terms in columns.

Example 4 Add: $9x^5 - 2x^3 + 6x^2 + 3$ and $5x^4 - 7x^2 + 6$ and $3x^6 - 5x^5 + x^2 + 5$.

Add.

1. $3x^2 + 2x - 2$ and $-2x^2 + 5x + 5$

2. $-4x^5 + 3x^3 + 4$ and $7x^4 + 2x^2$

3. $31x^4 + x^2 + 2x - 1$ and $-7x^4 + 5x^3 - 2x + 2$

4. $17x^3 - x^2 + 3x + 4$ and $-15x^3 + x^2 - 3x - \frac{2}{3}$

Add mentally. Try to write just the answer.

5. $(4x^2 - 5x + 3) + (-2x^2 + 2x - 4)$

6. $(3x^3 - 4x^2 - 5x + 3) + (5x^3 + 2x^2 - 3x - \frac{1}{2})$

ANSWERS ON PAGE A-7

7. Add.

$$-2x^3 + 5x^2 - 2x + 4$$
$$x^4 \qquad + 6x^2 + 7x - 10$$
$$\underline{-9x^4 + 6x^3 + x^2 \qquad - 2}$$

8. Add $-3x^3 + 5x + 2$ and $x^3 + x^2 + 5$ and $x^3 - 2x - 4$.

9. Find the sum of the areas of the rectangles.

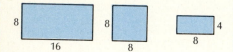

10. Find a polynomial for the sum of the areas of the rectangles.

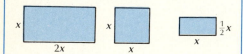

11. Find the sum of the areas in Exercise 9 by substituting 8 for x in the polynomial of Exercise 10.

ANSWERS ON PAGE A-7

Arrange the polynomials with like terms in columns.

$$9x^5 \qquad\quad - 2x^3 + 6x^2 + 3$$
$$\quad 5x^4 \qquad\quad - 7x^2 + 6$$
$$\underline{3x^6 - 5x^5 \qquad\qquad\quad + x^2 + 5}$$
$$3x^6 + 4x^5 + 5x^4 - 2x^3 \qquad\quad + 14$$

We leave spaces for missing terms.

DO EXERCISES 7 AND 8.

●● PROBLEMS

Suppose we want to find the sum of the areas of these rectangles.

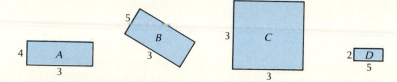

We can proceed as follows:

area of A	plus	area of B	plus	area of C	plus	area of D
$4 \cdot 3$	$+$	$5 \cdot 3$	$+$	$3 \cdot 3$	$+$	$2 \cdot 5$

$$= 12 + 15 + 9 + 10$$
$$= 46.$$

Now suppose certain sides were unknown, but represented by a variable x.

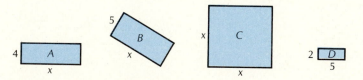

The sum of the areas is found as follows:

area of A	plus	area of B	plus	area of C	plus	area of D
$4x$	$+$	$5x$	$+$	$x \cdot x$	$+$	$2 \cdot 5$

$$= 4x + 5x + x^2 + 10$$
$$= x^2 + 9x + 10$$

Note that we could solve the first problem by replacing x by 3 in the preceding problem:

$$x^2 + 9x + 10 = 3^2 + 9 \cdot 3 + 10 = 9 + 27 + 10 = 46.$$

The polynomial gives us a formula for the sum of the areas of the rectangles with certain sides of length x. Thus we can substitute any length—say 8, 4, 78.6—for x and find the sum of the areas. This illustrates how we can manipulate unknown numbers in problems.

DO EXERCISES 9–11.

Chapter 4   Polynomials

• Add.

1. $3x + 2$ and $-4x + 3$

2. $5x^2 + 6x + 1$ and $-7x + 2$

3. $-6x + 2$ and $x^2 + x - 3$

4. $6x^4 + 3x^3 - 1$ and $4x^2 - 3x + 3$

5. $3y^5 + 6y^2 - 1$ and $7y^2 + 6y - 2$

6. $7t^3 + 3t^2 + 6t$ and $-3t^2 - 6$

7. $-4x^4 + 6x^2 - 3x - 5$ and
   $6x^3 + 5x + 9$

8. $5x^3 + 6x^2 - 3x + 1$ and
   $5x^4 - 6x^3 + 2x - 5$

9. $(7x^3 + 6x^2 + 4x + 1) +$
   $(-7x^3 + 6x^2 - 4x + 5)$

10. $(3x^4 - 5x^2 - 6x + 5) +$
    $(-4x^3 + 6x^2 + 7x - 1)$

11. $5x^4 - 6x^3 - 7x^2 + x - 1$ and
    $4x^3 - 6x + 1$

12. $8x^5 - 6x^3 + 6x + 5$ and
    $-4x^4 + 3x^3 - 7x$

13. $9x^8 - 7x^4 + 2x^2 + 5$ and
    $8x^7 + 4x^4 - 2x$

14. $4x^5 - 6x^3 - 9x + 1$ and
    $6x^3 + 9x^2 + 9x$

ANSWERS
1. _____
2. _____
3. _____
4. _____
5. _____
6. _____
7. _____
8. _____
9. _____
10. _____
11. _____
12. _____
13. _____
14. _____

15. $\dfrac{1}{4}x^4 + \dfrac{2}{3}x^3 + \dfrac{5}{8}x^2 + 7$ and

$-\dfrac{3}{4}x^4 + \dfrac{3}{8}x^2 - 7$

16. $\left(\dfrac{1}{3}x^9 + \dfrac{1}{5}x^5 - \dfrac{1}{2}x^2 + 7\right) +$

$\left(-\dfrac{1}{5}x^9 + \dfrac{1}{4}x^4 - \dfrac{3}{5}x^5 + \dfrac{3}{4}x^2 + \dfrac{1}{2}\right)$

17. $0.02x^5 - 0.2x^3 + x + 0.08$ and
$-0.01x^5 + x^4 - 0.8x - 0.02$

18. $(0.03x^6 + 0.05x^3 + 0.22x + 0.05) +$
$\left(\dfrac{7}{100}x^6 - \dfrac{3}{100}x^3 + 0.5\right)$

19. $\begin{array}{r} -3t^4 + 6t^2 + 2t - 1 \\ -\,3t^2 + 2t + 1 \\ \hline \end{array}$

20. $\begin{array}{r} -4y^3 + 8y^2 + 3y - 2 \\ -\,4y^2 + 3y + 2 \\ \hline \end{array}$

21. $\begin{array}{r} 3x^5 \quad\;\; - 6x^3 \quad\;\; + 3x \\ -\,3x^4 + 3x^3 + x^2 \\ \hline \end{array}$

22. $\begin{array}{r} 4x^5 \quad\;\; - 5x^3 \quad\;\; + 2x \\ -\,4x^4 + 2x^3 + 2x^2 \\ \hline \end{array}$

23. $\begin{array}{r} -3x^2 + x \\ 5x^3 - 6x^2 \quad\;\; + 1 \\ 3x - 8 \\ \hline \end{array}$

24. $\begin{array}{r} -4x^2 + 2x \\ 3x^3 - 5x^2 \quad\;\; + 3 \\ 5x - 5 \\ \hline \end{array}$

25. $\begin{array}{r} -\dfrac{1}{2}x^4 - \dfrac{3}{4}x^3 \qquad\;\; + 6x \\ \dfrac{1}{2}x^3 + \; x^2 + \dfrac{1}{4}x \\ \dfrac{3}{4}x^4 \qquad\;\; + \dfrac{1}{2}x^2 + \dfrac{1}{2}x + \dfrac{1}{4} \\ \hline \end{array}$

26. $\begin{array}{r} -\dfrac{1}{4}x^4 - \dfrac{1}{2}x^3 \qquad\;\; + 2x \\ \dfrac{3}{4}x^3 - \; x^2 + \dfrac{1}{2}x \\ \dfrac{1}{2}x^4 \qquad\;\; + \dfrac{1}{2}x^2 + \dfrac{1}{2}x + \dfrac{1}{2} \\ \hline \end{array}$

27. $\begin{array}{r} -4x^2 \\ 4x^4 - 3x^3 + 6x^2 + 5x \\ 6x^3 - 8x^2 \quad\;\; + 1 \\ -5x^4 \\ 6x^2 - 3x \\ \hline \end{array}$

28. $\begin{array}{r} 3x^2 \\ 5x^4 - 2x^3 + 4x^2 + 5x \\ 5x^3 - 5x^2 \quad\;\; + 2 \\ -7x^4 \\ 3x^2 - 2x \\ \hline \end{array}$

29. $\begin{array}{r} 3x^4 - 6x^2 + 7x \\ 3x^2 - 3x + 1 \\ -2x^4 + 7x^2 + 3x \\ 5x - 2 \\ \hline \end{array}$

30. $\begin{array}{r} 5x^4 - 8x^2 + 4x \\ 5x^2 - 2x + 3 \\ -3x^4 + 3x^2 + 5x \\ 3x - 5 \\ \hline \end{array}$

Chapter 4 Polynomials

31.
$$
\begin{aligned}
& 3x^5 - 6x^4 + 3x^3 \qquad\quad - 1 \\
& \qquad\;\; 6x^4 - 4x^3 + 6x^2 \\
& 3x^5 \qquad\qquad + 2x^3 \\
& \qquad\;\; - 6x^4 \qquad\quad - 7x^2 \\
& -5x^5 \qquad\qquad + 3x^3 \qquad\quad + 2 \\
\hline
\end{aligned}
$$

32.
$$
\begin{aligned}
& 4x^5 - 3x^4 + 2x^3 \qquad\quad - 2 \\
& \qquad\;\; 6x^4 + 5x^3 + 3x^2 \\
& 5x^5 \qquad\qquad + 4x^3 \\
& \qquad\;\; - 6x^4 \qquad\quad - 5x^2 \\
& -3x^5 \qquad\qquad + 2x^3 \qquad\quad + 5 \\
\hline
\end{aligned}
$$

33.
$$
\begin{aligned}
& \qquad - p^3 + 6p^2 + 3p + 5 \\
& p^4 \qquad\;\; - 3p^2 \qquad\;\; + 2 \\
& \qquad\qquad\qquad\;\; - 5p + 3 \\
& 6p^4 \qquad\;\; + 4p^2 \qquad\;\; - 1 \\
& \qquad - p^3 \qquad\qquad + 6p \\
\hline
\end{aligned}
$$

34.
$$
\begin{aligned}
& \qquad - 2q^3 + 3q^2 + 5q + 3 \\
& q^4 \qquad\;\; - 5q^2 \qquad\;\; + 1 \\
& \qquad\qquad\qquad\;\; - 7q + 4 \\
& 4q^4 \qquad\;\; + 6q^2 \qquad\;\; - 2 \\
& \qquad - q^3 \qquad\qquad + 5q \\
\hline
\end{aligned}
$$

35.
$$
\begin{aligned}
& \qquad - 3x^4 + 6x^3 - 6x^2 + 5x + 1 \\
& 5x^5 \qquad\;\; - 3x^3 \qquad\qquad - 5x \\
& \qquad\;\; 4x^4 + 7x^3 \qquad\quad + 3x + 1 \\
& -2x^5 \qquad\qquad\qquad + 7x^2 \qquad\;\; - 8 \\
\hline
\end{aligned}
$$

36.
$$
\begin{aligned}
& \qquad - 5x^4 + 4x^3 - 7x^2 + 3x + 2 \\
& 3x^5 \qquad\;\; - 7x^3 \qquad\qquad - 6x \\
& \qquad\;\; 3x^4 + 5x^3 \qquad\quad + 5x - 3 \\
& -5x^5 \qquad\qquad\qquad + 10x^2 \qquad + 4 \\
\hline
\end{aligned}
$$

37.
$$
\begin{aligned}
& 0.15x^4 + 0.10x^3 - 0.9x^2 \\
& \qquad\qquad - 0.01x^3 + 0.01x^2 + x \\
& 1.25x^4 \qquad\qquad + 0.11x^2 \qquad + 0.01 \\
& \qquad\;\; 0.27x^3 \qquad\qquad\quad + 0.99 \\
& -0.35x^4 \qquad\qquad + 15x^2 \qquad - 0.03 \\
\hline
\end{aligned}
$$

38.
$$
\begin{aligned}
& 0.05x^4 + 0.12x^3 - 0.5x^2 \\
& \qquad\qquad - 0.02x^3 + 0.02x^2 + 2x \\
& 1.5x^4 \qquad\qquad + 0.01x^2 \qquad + 0.15 \\
& \qquad\;\; 0.25x^3 \qquad\qquad\quad + 0.85 \\
& -0.25x^4 \qquad\qquad + 10x^2 \qquad - 0.04 \\
\hline
\end{aligned}
$$

A N S W E R S

31.

32.

33.

34.

35.

36.

37.

38.

39. a) _____

b) _____

40. a) _____

b) _____

41. _____

42. _____

43. _____

44. _____

45. _____

 Solve.

39. a) Find a polynomial for the sum of the areas of these rectangles.

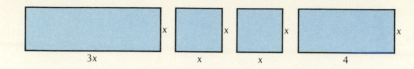

b) Find the sum of the areas when $x = 3$ and $x = 8$.

40. a) Find a polynomial for the sum of the areas of these circles. Leave answer in terms of π.

b) Find the sum of the areas when $r = 5$ and $r = 11.3$.

 SKILL MAINTENANCE

41. Multiply: $x^7 \cdot x^3$.

42. Remove parentheses and simplify:

$$5y - 8 - (9y - 6).$$

43. 3478 is thirty-seven percent of what number?

⭐ **EXTENSION**

44. 🔢 Add: $(-20.344x^6 - 70.789x^5 + 890x) + (68.888x^6 + 69.994x^5)$.

45. Addition of rational numbers is commutative. That is, $a + b = b + a$ for any rational numbers.

   a) Show that addition of binomials $ax + b$ and $cx + d$ is commutative.

   b) Show that addition of trinomials $ax^2 + bx + c$ and $dx^2 + ex + f$ is commutative.

Chapter 4   Polynomials

Subtraction of Polynomials

● ADDITIVE INVERSE OF POLYNOMIALS

We know that two numbers are additive inverses of each other if their sum is zero. The same definition holds for polynomials.

After finishing Section 4.4, you should be able to:

● Find two equivalent polynomials for the additive inverse of a polynomial, and simplify an expression such as $-(3x^2 - 2x + 5)$ by replacing each term by its additive inverse.

●● Subtract polynomials.

> Two polynomials are *additive inverses* of each other if their sum is zero.

To find a way to determine an additive inverse, look for a pattern in the following examples.

a) $3x + (-3x) = 0$

b) $-8y^2 + 8y^2 = 0$

c) $(5t^3 - 2) + (-5t^3 + 2) = 0$

d) $(7x^3 - 6x^2 - x + 4) + (-7x^3 + 6x^2 + x - 4) = 0$

Since $(5t^3 - 2) + (-5t^3 + 2) = 0$, we know that the additive inverse of $5t^3 - 2$ is $-5t^3 + 2$. To say this another way with algebraic symbolism we have

The additive inverse of $(5t^3 - 2)$ is $-5t^3 + 2$.

$$-(5t^3 - 2) = -5t^3 + 2.$$

> We can find an equivalent polynomial for the additive inverse of a polynomial by replacing each term by its additive inverse (changing the sign of every term).

Find two equivalent expressions for the additive inverse of the polynomial.

1. $12x^4 - 3x^2 + 4x$

2. $-4x^4 + 3x^2 - 4x$

3. $-13x^6 + 2x^4 - 3x^2 + x - \dfrac{5}{13}$

4. $-7y^3 + 2y^2 - y + 3$

Example 1 Find two equivalent expressions for the additive inverse of $4x^5 - 7x^3 - 8x + \frac{1}{4}$.

a) $-\left(4x^5 - 7x^3 - 8x + \dfrac{1}{4}\right)$

b) $-4x^5 + 7x^3 + 8x - \dfrac{1}{4}$ **Changing the sign of every term**

DO EXERCISES 1–4.

Example 2 Simplify: $-(-7y^4 - \frac{4}{9}y^3 + 8y^2 - y + 98)$.

$$-\left(-7y^4 - \dfrac{4}{9}y^3 + 8y^2 - y + 98\right) = 7y^4 + \dfrac{4}{9}y^3 - 8y^2 + y - 98.$$

DO EXERCISES 5–7.

Simplify.

5. $-(4x^3 - 6x + 3)$

6. $-(5x^4 + 3x^2 + 7x - 5)$

7. $-\left(14x^{10} - \dfrac{1}{2}x^5 + 5x^3 - x^2 + 3x\right)$

ANSWERS ON PAGE A-8

Subtract.

8. $(7x^3 + 2x + 4) - (5x^3 - 4)$

9. $(-3x^2 + 5x - 4) - (-4x^2 + 11x - 2)$

Subtract mentally. Try to write just the answer.

10. $(-6x^4 + 3x^2 + 6) - (2x^4 + 5x^3 - 5x^2 + 7)$

11. $(\frac{3}{2}x^3 - \frac{1}{2}x^2 + 0.3) - (\frac{1}{2}x^3 + \frac{1}{2}x^2 + \frac{4}{3}x + 1.2)$

Subtract the second polynomial from the first. Use columns.

12. $4x^3 + 2x^2 - 2x - 3$,
    $2x^3 - 3x^2 + 2$

Subtract.

13. $\quad\quad 2x^3 + x^2 - 6x + 2$
    $\underline{x^5 + 4x^3 - 2x^2 - 4x}$

•• SUBTRACTION OF POLYNOMIALS

Recall that we can subtract a rational number by adding its additive inverse: $a - b = a + (-b)$. This allows us to find an equivalent expression for the difference of two polynomials.

Example 3 Subtract: $(9x^5 + x^3 - 2x^2 + 4) - (2x^5 + x^4 - 4x^3 - 3x^2)$.

$(9x^5 + x^3 - 2x^2 + 4) - (2x^5 + x^4 - 4x^3 - 3x^2)$

$= (9x^5 + x^3 - 2x^2 + 4)$
  $+ [-(2x^5 + x^4 - 4x^3 - 3x^2)]$ Adding an inverse

$= (9x^5 + x^3 - 2x^2 + 4)$
  $+ (-2x^5 - x^4 + 4x^3 + 3x^2)$ Finding the inverse by changing the sign of *every* term

$= 7x^5 - x^4 + 5x^3 + x^2 + 4$ Collecting like terms

DO EXERCISES 8 AND 9.

After some practice you will be able to subtract mentally.

Example 4 Subtract: $(9x^5 + x^3 - 2x) - (-2x^5 + 5x^3 + 6)$.

$(9x^5 + x^3 - 2x) - (-2x^5 + 5x^3 + 6)$

$= (9x^5 + 2x^5) + (x^3 - 5x^3) - 2x - 6$ Subtract the like terms mentally.

$= 11x^5 - 4x^3 - 2x - 6$ Write only this.

Caution! If you make errors by trying to subtract mentally, then don't do it!

DO EXERCISES 10 AND 11.

We can use columns to subtract. We replace coefficients by their inverses, as shown in Example 2. You can also do it mentally.

Example 5 Subtract: $(5x^2 - 3x + 6) - (9x^2 - 5x - 3)$.

a) $\quad 5x^2 - 3x + 6$ Writing similar
   $\quad \underline{9x^2 - 5x - 3}$ terms in columns

b) $\quad 5x^2 - 3x + 6$
   $\quad \underline{-9x^2 \pm 5x \pm 3}$ Changing signs

c) $\quad\quad 5x^2 - 3x + 6$
   $\quad \underline{-9x^2 + 5x + 3}$ Adding
   $\quad\quad -4x^2 + 2x + 9$

If you can do so without error, you should skip steps (b) and (c). Write just the answer.

Example 6 Subtract: $(x^3 + x^2 + 2x - 12) - (2x^3 + x^2 - 3x)$.

$\quad\quad x^3 + x^2 + 2x - 12$
$\quad \underline{2x^3 + x^2 - 3x}$
$\quad -x^3 \quad\quad\quad + 5x - 12$

DO EXERCISES 12 AND 13.

• Find two equivalent expressions for the additive inverse of each polynomial.

1. $-5x$

2. $x^2 - 3x$

3. $-x^2 + 10x - 2$

4. $-4x^3 - x^2 - x$

5. $12x^4 - 3x^3 + 3$

6. $4x^3 - 6x^2 - 8x + 1$

Simplify.

7. $-(3x - 7)$

8. $-(-2x + 4)$

9. $-(4x^2 - 3x + 2)$

10. $-(-6a^3 + 2a^2 - 9a + 1)$

11. $-\left(-4x^4 - 6x^2 + \frac{3}{4}x - 8\right)$

12. $-(-5x^4 + 4x^3 - x^2 + 0.9)$

•• Subtract.

13. $(5x^2 + 6) - (3x^2 - 8)$

14. $(7x^3 - 2x^2 + 6) - (7x^2 + 2x - 4)$

15. $(6x^5 - 3x^4 + x + 1) - (8x^5 + 3x^4 - 1)$

16. $\left(\frac{1}{2}x^2 - \frac{3}{2}x + 2\right) - \left(\frac{3}{2}x^2 + \frac{1}{2}x - 2\right)$

17. $(6x^2 + 2x) - (-3x^2 - 7x + 8)$

18. $7x^3 - (-3x^2 - 2x + 1)$

19. $\left(\frac{5}{8}x^3 - \frac{1}{4}x - \frac{1}{3}\right) - \left(-\frac{1}{8}x^3 + \frac{1}{4}x - \frac{1}{3}\right)$

20. $\left(\frac{1}{5}x^3 + 2x^2 - 0.1\right) - \left(-\frac{2}{5}x^3 + 2x^2 + 0.01\right)$

21. $(0.08x^3 - 0.02x^2 + 0.01x) - (0.02x^3 + 0.03x^2 - 1)$

22. $(0.8x^4 + 0.2x - 1) - \left(\frac{7}{10}x^4 + \frac{1}{5}x - 0.1\right)$

ANSWERS	
1.	
2.	
3.	
4.	
5.	
6.	
7.	
8.	
9.	
10.	
11.	
12.	
13.	
14.	
15.	
16.	
17.	
18.	
19.	
20.	
21.	
22.	

ANSWERS

23. _____

24. _____

25. _____

26. _____

27. _____

28. _____

29. _____

30. _____

31. _____

32. _____

33. _____

34. _____

35. _____

36. _____

37. _____

38. _____

39. _____

40. _____

Subtract.

23. $x^2 + 5x + 6$
$\underline{x^2 + 2x}$

24. $x^3 + 1$
$\underline{x^3 + x^2}$

25. $x^4 - 3x^2 + x + 1$
$\underline{x^4 - 4x^3 }$

26. $3x^2 - 6x + 1$
$\underline{6x^2 + 8x - 3}$

27. $5x^4 + 6x^3 - 9x^2$
$\underline{-6x^4 -6x^3 + 8x + 9}$

28. $5x^4 + 6x^2 - 3x + 6$
$\underline{ 6x^3 + 7x^2 - 8x - 9}$

29. $3x^4 + 6x^2 + 8x - 1$
$\underline{4x^5 - 6x^4 - 8x - 7}$

30. $6x^5 + 3x^2 - 7x + 2$
$\underline{10x^5 + 6x^3 - 5x^2 - 2x + 4}$

31. $x^5 - 1$
$\underline{x^5 - x^4 + x^3 - x^2 + x - 1}$

32. $x^5 + x^4 - x^3 + x^2 - x + 2$
$\underline{x^5 - x^4 + x^3 - x^2 - x + 2}$

✓ SKILL MAINTENANCE

33. Multiply: $8(x - 2)$.

34. Factor: $10x - 8y - 24$.

35. Solve: $3x - 3 = -4x + 4$.

36. Solve: $4(x - 5) = 7(x + 8)$.

★ EXTENSION

37. 🖩 Subtract: $(345.099x^3 - 6.178x) - (-224.508x^3 + 8.99x)$.

Simplify.

38. $(5x^3 - 4x^2 + 6) - (2x^3 + x^2 - x) + (x^3 - x)$

39. $(-y^4 - 7y^3 + y^2) + (-2y^4 + 5y - 2) - (-6y^3 + y^2)$

40. A 4-ft by 4-ft sandbox is placed on a square lawn x ft on a side. Express the area left over as a polynomial.

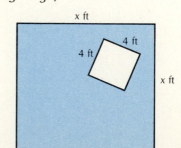

Chapter 4   Polynomials

Multiplication of Polynomials

• MULTIPLYING MONOMIALS

To find an equivalent expression for the product of two monomials, we multiply the coefficients and then use properties of exponents. We use parentheses to show multiplication: $(3x)(4x)$ means $(3x) \cdot (4x)$.

Examples Multiply.

1. $(3x)(4x) = (3 \cdot 4)(x \cdot x)$ Multiplying the coefficients
    $= 12x^2$ Simplifying

2. $(3x)(-x) = (3x)(-1x)$
    $= (3)(-1)(x \cdot x)$
    $= -3x^2$

3. $(-7x^5)(4x^3) = (-7 \cdot 4)(x^5 \cdot x^3)$
    $= -28x^{5+3}$
    $= -28x^8$ Adding exponents and simplifying

After some practice you can do this mentally. Multiply the coefficients and add the exponents. Write only the answer.

DO EXERCISES 1–8.

•• MULTIPLYING A MONOMIAL AND A BINOMIAL

Multiplications are based on the distributive laws.

Example 4 Multiply: $2x$ and $5x + 3$.

$(2x)(5x + 3) = (2x)(5x) + (2x)(3)$ Using a distributive law
    $= 10x^2 + 6x$ Multiplying the monomials

DO EXERCISES 9 AND 10.

••• MULTIPLYING TWO BINOMIALS

To find an equivalent expression for the product of two binomials, we use the distributive laws more than once.

Example 5 Multiply: $x + 5$ and $x + 4$.

$(x + 5)(x + 4) = (x + 5)x + (x + 5)4$ Using a distributive
                      (a)   (b)   law

We now use a distributive law with parts (a) and (b).

After finishing Section 4.5, you should be able to:

• Multiply two monomials.

•• Multiply a monomial and a binomial.

••• Multiply two binomials.

∷ Multiply a binomial and a trinomial.

⦂⦂ Multiply any polynomials.

Multiply.

1. $3x$ and -5

2. $-x$ and x

3. $-x$ and $-x$

4. $-x^2$ and x^3

5. $3x^5$ and $4x^2$

6. $4y^5$ and $-2y^6$

7. $-7y^4$ and $-y$

8. $7x^5$ and 0

Multiply.

9. $4x$ and $2x + 4$

10. $3t^2$ and $-5t + 2$

ANSWERS ON PAGE A-8

Multiply.

11. $x + 8$ and $x + 5$

12. $(x + 5)(x - 4)$

Multiply.

13. $5x + 3$ and $x - 4$

14. $(2x - 3)(3x - 5)$

Multiply.

15. $(x^2 + 3x - 4)(x^2 + 5)$

16. $(2y^3 - 2y + 5)(3y^2 - 7)$

a) $(x + 5)x = x \cdot x + 5 \cdot x$ Distributive law
$= x^2 + 5x$ Multiplying the monomials

b) $(x + 5)4 = x \cdot 4 + 5 \cdot 4$ Distributive law
$= 4x + 20$ Multiplying the monomials

Now we replace parts (a) and (b) in the original expression with their answers and collect like terms:

$$(x + 5)(x + 4) = (x^2 + 5x) + (4x + 20)$$
$$= x^2 + 9x + 20.$$

DO EXERCISES 11 AND 12.

Example 6 Multiply: $4x + 3$ and $x - 2$.

$$(4x + 3)(\ x - 2\ ) = 4x(\ x - 2\ ) + 3(\ x - 2\ )\quad \text{Using a}$$
$$\text{distributive law}$$
$$\text{(a)}\qquad\qquad \text{(b)}$$

Now we consider the two parts (a) and (b) separately.

a) $4x\ (x - 2) = 4x\ (x) - 4x\ (2)$ Using a distributive law
$= 4x^2 - 8x$ Multiplying the monomials

b) $3\ (x - 2) = 3\ x - 3\ \cdot 2$ Using a distributive law
$= 3x - 6$ Multiplying the monomials

We replace parts (a) and (b) in the original expression with their answers and collect like terms:

$$(4x + 3)(x - 2) = (4x^2 - 8x) + (3x - 6)$$
$$= 4x^2 - 5x - 6.$$

DO EXERCISES 13 AND 14.

⠒ MULTIPLYING A BINOMIAL AND A TRINOMIAL

Example 7 Multiply: $(x^2 + 2x - 3)(x^2 + 4)$.

$$(x^2 + 2x - 3)(x^2 + 4) = (x^2 + 2x - 3)(x^2) + (x^2 + 2x - 3)(4)$$
$$\text{(a)}\qquad\qquad\qquad\qquad \text{(b)}$$

Consider parts (a) and (b).

a) $(x^2 + 2x - 3)x^2 = (x^2)(x^2) + 2x(x^2) - 3(x^2)$
$= x^4 + 2x^3 - 3x^2$

b) $(x^2 + 2x - 3)4 = (x^2)4 + (2x)4 + (-3)4$
$= 4x^2 + 8x - 12$

Now we replace parts (a) and (b) in the original expression and collect like terms:

$$(x^2 + 2x - 3)(x^2 + 4) = (x^4 + 2x^3 - 3x^2) + (4x^2 + 8x - 12)$$
$$= x^4 + 2x^3 + x^2 + 8x - 12.$$

DO EXERCISES 15 AND 16.

Chapter 4   Polynomials

:: MULTIPLYING ANY POLYNOMIALS

Perhaps you have discovered the following.

> **To multiply two polynomials, multiply each term of one by every term of the other. Then add the results.**

We can use columns for long multiplications. We multiply each term at the top by every term at the bottom. Then we add.

Example 8 Multiply: $(4x^2 - 2x + 3)(x + 2)$.

$$
\begin{array}{l}
4x^2 - 2x + 3 \\
x + 2 \\
\hline
4x^3 - 2x^2 + 3x \qquad \text{\color{blue}Multiplying the top row by } x \\
 8x^2 - 4x + 6 \qquad \text{\color{blue}Multiplying the top row by } 2 \\
\hline
4x^3 + 6x^2 - x + 6 \qquad \text{\color{blue}Adding}
\end{array}
$$

Example 9 Multiply: $(5x^3 - 3x + 4)(-2x^2 - 3)$.

$$
\begin{array}{l}
5x^3 - 3x + 4 \\
{-2x^2 - 3} \\
\hline
-10x^5 + 6x^3 - 8x^2 \qquad \text{\color{blue}Multiplying by } -2x^2 \\
 - 15x^3 + 9x - 12 \qquad \text{\color{blue}Multiplying by } -3 \\
\hline
-10x^5 - 9x^3 - 8x^2 + 9x - 12 \qquad \text{\color{blue}Adding}
\end{array}
$$

> When we multiplied $-2x^2$ by $-3x$, the power dropped from x^5 to x^3 so we left a space for the missing x^4-term.

In addition, we leave space for "missing terms." (Recall, from Section 4.2, that a *missing term* is a term with a 0 coefficient.)

DO EXERCISES 17 AND 18.

Example 10 Multiply: $(2x^2 + 3x - 4)(2x^2 - x + 3)$.

$$
\begin{array}{l}
2x^2 + 3x - 4 \\
2x^2 - x + 3 \\
\hline
4x^4 + 6x^3 - 8x^2 \qquad \text{\color{blue}Multiplying by } 2x^2 \\
 - 2x^3 - 3x^2 + 4x \qquad \text{\color{blue}Multiplying by } -x \\
 6x^2 + 9x - 12 \qquad \text{\color{blue}Multiplying by } 3 \\
\hline
4x^4 + 4x^3 - 5x^2 + 13x - 12 \qquad \text{\color{blue}Adding}
\end{array}
$$

DO EXERCISE 19.

Multiply.

17. $3x^2 - 2x + 4$
    $\underline{x + 5}$

18. $-5x^2 + 4x + 2$
    $\underline{-4x^2 - 8}$

Multiply.

19. $3x^2 - 2x - 5$
    $\underline{2x^2 + x - 2}$

ANSWERS ON PAGE A-8

EXPANDED NOTATION AND POLYNOMIALS

The number 6345 can be named with *expanded notation* as follows:

$$6345 = 6000 + 300 + 40 + 5$$
$$= 6 \cdot 1000 + 3 \cdot 100 + 4 \cdot 10 + 5$$
$$= 6 \cdot 10^3 + 3 \cdot 10^2 + 4 \cdot 10 + 5. \qquad \text{Expanded notation}$$

Note that this is the result of replacing x in the related polynomial

$$6x^3 + 3x^2 + 4x + 5 \qquad \text{Related polynomial}$$

by the number 10:

$$6 \cdot 10^3 + 3 \cdot 10^2 + 4 \cdot 10 + 5.$$

For each of the following, write expanded notation and the related polynomial.

1. 8762　　　　2. 786　　　　3. 16,432　　　　4. 7063

• Multiply.

1. $3x$ and -4

2. $4x$ and 5

3. $6x^2$ and 7

4. $-4x$ and -3

5. $-5x$ and -6

6. $3x^2$ and -7

7. y^2 and $-2y$

8. $-t^3$ and $-t$

9. x^4 and x^2

10. $-x^5$ and x^3

11. $3x^4$ and $2x^2$

12. $-\frac{1}{5}x^3$ and $-\frac{1}{3}x$

13. $-4x^4$ and 0

14. $7x^5$ and x^5

15. $-0.1x^6$ and $0.2x^4$

•• Multiply.

16. $3x$ and $-x + 5$

17. $2x$ and $4x - 6$

18. $4x^2$ and $3x + 6$

19. $-6x^2$ and $x^2 + x$

20. $3x^2$ and $6x^4 + 8x^3$

21. $4x^4$ and $x^2 - 6x$

••• Multiply.

22. $(x + 6)(x + 3)$

23. $(x + 6)(-x + 2)$

24. $(x + 3)(x - 3)$

25. $(y - 5)(2y - 5)$

26. $(2t + 5)(2t + 5)$

27. $(3x - 5)(3x + 5)$

ANSWERS
1.
2.
3.
4.
5.
6.
7.
8.
9.
10.
11.
12.
13.
14.
15.
16.
17.
18.
19.
20.
21.
22.
23.
24.
25.
26.
27.

A N S W E R S

28.

29.

30.

31.

32.

33.

34.

35.

36.

37.

38.

39.

40.

41.

42.

43.

44.

45.

46.

47.

28. $(3x + 1)(3x + 1)$ 29. $\left(2x - \frac{1}{2}\right)\left(x + \frac{3}{2}\right)$ 30. $(2x + 0.1)(3x - 0.1)$

⋅⋅ Multiply.

31. $(x^2 + 6x + 1)(x + 1)$ 32. $(2x^3 + 6x + 1)(2x + 1)$

33. $(-5q^2 - 7q + 3)(2q + 1)$ 34. $(2z^3 - 5z + 6)(2z + 2)$

⋅⋅⋅ Multiply.

35. $(3x^2 - 6x + 2)(x^2 - 3)$ 36. $(x^2 + 6x - 1)(-3x^2 + 2)$

37. $(2t^2 - t - 4)(3t^2 + 2t - 1)$ 38. $(3a^2 - 5a + 2)(2a^2 - 3a + 4)$

39. $(x^3 + x^2 + x + 1)(x - 1)$ 40. $(x + 2)(x^2 - 2x + 4)$

✓ SKILL MAINTENANCE

41. Solve $y = mx + b$ for x.   42. Divide and simplify: $\dfrac{x^{-4}}{x^{-9}}$.

43. Simplify: $(t^{-4})^5$.   44. Convert to decimal notation: 97.3%.

⭐ EXTENSION

Compute.

45. $(x + 3)(x + 6) + (x + 3)(x + 6)$ 46. $(x - 6)(x - 2) - (x - 6)(x - 2)$

47. A box with a square bottom is to be made from a 12-inch square piece of cardboard. Squares with side x are cut out of the corners and the sides folded up. Find polynomials for the volume and the outside surface area of the box.

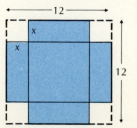

Special Products of Polynomials

We now consider special products of polynomials. These are products that we encounter so often that it is helpful to have ways to compute them that are faster than multiplying each term of one by each term of the other.

• Multiply a monomial and a polynomial mentally.

•• Multiply two binomials mentally.

• PRODUCT OF A MONOMIAL AND ANY POLYNOMIAL

There is a quick way to multiply a monomial and any polynomial. Use the distributive law mentally, multiplying every term inside the parentheses by the monomial. Just write the answer.

Example 1 Multiply: $5x(2x^2 - 3x + 4)$.

$$5x(2x^2 - 3x + 4) = 10x^3 - 15x^2 + 20x$$

DO EXERCISES 1 AND 2.

•• PRODUCTS OF TWO BINOMIALS

To multiply two binomials, we multiply each term of one by every term of the other. We can do it like this.

$$(A + B)(C + D) = AC + AD + BC + BD$$

1. Multiply First terms: AC.
2. Multiply Outside terms: AD.
3. Multiply Inside terms: BC.
4. Multiply Last terms: BD.

FOIL   This will help you remember the rule.

Example 2 Multiply: $(x + 8)(x^2 + 5)$.

$$(x + 8)(x^2 + 5) = x^3 + 5x + 8x^2 + 40$$

Often we can collect like terms after we multiply.

Examples Multiply.

3. $(x + 6)(x - 6) = x^2 - 6x + 6x - 36$ Using FOIL
   $= x^2 - 36$ Collecting like terms
4. $(y + 3)(y - 2) = y^2 - 2y + 3y - 6$
   $= y^2 + y - 6$
5. $(x^3 + 5)(x^3 - 5) = x^6 - 5x^3 + 5x^3 - 25$
   $= x^6 - 25$
6. $(4t^3 + 5)(3t^2 - 2) = 12t^5 - 8t^3 + 15t^2 - 10$

DO EXERCISES 3–9.

Multiply. Write just the answer.

1. $4x(2x^2 - 3x + 4)$

2. $2y^3(5y^3 + 4y^2 - 5y)$

Multiply mentally. Write just the answer.

3. $(x + 3)(x + 4)$

4. $(x + 3)(x - 5)$

5. $(2x + 1)(x + 4)$

6. $(2x^2 - 3)(x - 2)$

7. $(6x^2 + 5)(2x^3 + 1)$

8. $(y^3 + 7)(y^3 - 7)$

9. $(2x^4 + x^2)(-x^3 + x)$

ANSWERS ON PAGE A-8

Multiply.

10. $(x + \frac{4}{5})(x - \frac{4}{5})$

11. $(x^3 - 0.5)(x^2 + 0.5)$

12. $(2 + 3x^2)(4 - 5x^2)$

13. $(6x^3 - 3x^2)(5x^2 + 2x)$

ANSWERS ON PAGE A-8

Examples Multiply.

7. $\left(x - \frac{2}{3}\right)\left(x + \frac{2}{3}\right) = x^2 + \frac{2}{3}x - \frac{2}{3}x - \frac{4}{9}$

$$= x^2 - \frac{4}{9}$$

8. $(x^2 - 0.3)(x^2 - 0.3) = x^4 - 0.3x^2 - 0.3x^2 + 0.09$

$$= x^4 - 0.6x^2 + 0.09$$

9. $(3 - 4x)(7 - 5x^3) = 21 - 15x^3 - 28x + 20x^4$

$$= 21 - 28x - 15x^3 + 20x^4$$

> If the original polynomials are in ascending order, it is natural to write the product in ascending order, but this is not a "must."

10. $(5x^4 + 2x^3)(3x^2 - 7x) = 15x^6 - 35x^5 + 6x^5 - 14x^4$

$$= 15x^6 - 29x^5 - 14x^4$$

DO EXERCISES 10–13.

Chapter 4   Polynomials

Multiply. Write only the answer.

1. $4x(x + 1)$

2. $3x(x + 2)$

3. $-3x(x - 1)$

4. $-5x(-x - 1)$

5. $x^2(x^3 + 1)$

6. $-2x^3(x^2 - 1)$

7. $3x(2x^2 - 6x + 1)$

8. $-4x(2x^3 - 6x^2 - 5x + 1)$

Multiply. Write only the answer.

9. $(x + 1)(x^2 + 3)$

10. $(x^2 - 3)(x - 1)$

11. $(x^3 + 2)(x + 1)$

12. $(x^4 + 2)(x + 12)$

13. $(x + 2)(x - 3)$

14. $(x + 2)(x + 2)$

15. $(3x + 2)(3x + 3)$

16. $(4x + 1)(2x + 2)$

17. $(5x - 6)(x + 2)$

18. $(x - 8)(x + 8)$

19. $(3x - 1)(3x + 1)$

20. $(2x + 3)(2x + 3)$

21. $(4x - 2)(x - 1)$

22. $(2x - 1)(3x + 1)$

23. $(x - \frac{1}{4})(x + \frac{1}{4})$

24. $(x + \frac{3}{4})(x + \frac{3}{4})$

25. $(t - 0.1)(t + 0.1)$

26. $(3y^2 + 1)(y + 1)$

27. $(2x^2 + 6)(x + 1)$

28. $(2x^2 + 3)(2x - 1)$

ANSWERS
1.
2.
3.
4.
5.
6.
7.
8.
9.
10.
11.
12.
13.
14.
15.
16.
17.
18.
19.
20.
21.
22.
23.
24.
25.
26.
27.
28.

29. $(-2x + 1)(x + 6)$

30. $(3x + 4)(2x - 4)$

31. $(a + 7)(a + 7)$

32. $(2b + 5)(2b + 5)$

33. $(1 + 2x)(1 - 3x)$

34. $(-3x - 2)(x + 1)$

35. $(x^2 + 3)(x^3 - 1)$

36. $(x^4 - 3)(2x + 1)$

37. $(x^2 - 2)(x - 1)$

38. $(x^3 + 2)(x - 3)$

39. $(3q^2 - 2)(q^4 - 2)$

40. $(p^{10} + 3)(p^{10} - 3)$

41. $(3x^5 + 2)(2x^2 + 6)$

42. $(1 - 2x)(1 + 3x^2)$

43. $(8x^3 + 1)(x^3 + 8)$

44. $(4 - 2x)(5 - 2x^2)$

45. $(4x^2 + 3)(x - 3)$

46. $(7x - 2)(2x - 7)$

47. $(4x^4 + x^2)(x^2 + x)$

48. $(5x^6 + 3x^3)(2x^6 + 2x^3)$

✓ **SKILL MAINTENANCE**

49. Convert to decimal notation: 4.7×10^{-5}.

50. Apollo 10 reached a speed of 24,790 miles per hour. That was 37 times the speed of the first supersonic flight in 1947. What was the speed of the first supersonic flight?

⭐ **EXTENSION**

Multiply.

51. $4y(y + 5)(2y + 8)$

52. $[(x + 1) - x^2][(x - 2) + 2x^2]$

53. Solve: $(x + 2)(x - 5) = (x + 1)(x - 3)$.

54. Find two equivalent expressions for the shaded region.

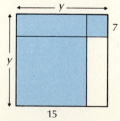

Chapter 4   Polynomials

More Special Products

• MULTIPLYING SUMS AND DIFFERENCES OF TWO EXPRESSIONS

Look for a pattern.

a) $(x + 2)(x - 2) = x^2 - 2x + 2x - 4$

$$= x^2 - 4$$

b) $(3x - 5)(3x + 5) = 9x^2 + 15x - 15x - 25$

$$= 9x^2 - 25$$

DO EXERCISES 1 AND 2.

Perhaps you discovered the following.

> The product of the sum and difference of two expressions is the square of the first expression minus the square of the second:
>
> $$(A + B)(A - B) = A^2 - B^2.$$

> *Caution!* You must memorize the rule. Otherwise, you will have trouble later on.

Examples Multiply.

$$(A + B)(A - B) = A^2 - B^2$$

> Carry out the rule. Say the words as you go.

1. $(x + 4)(x - 4) = x^2 - 4^2$

"The square of the first expression, x^2, minus the square of the second, 4^2."

$$= x^2 - 16 \quad \text{Simplifying}$$

2. $(5 + 2w)(5 - 2w) = 5^2 - (2w)^2$

$$= 25 - 4w^2$$

3. $(3x^2 - 7)(3x^2 + 7) = (3x^2)^2 - 7^2$

$$= 9x^4 - 49$$

4. $(-4x - 10)(-4x + 10) = (-4x)^2 - 10^2 = 16x^2 - 100$

DO EXERCISES 3–6.

• • SQUARING BINOMIALS

In this special product we multiply a binomial by itself. This is also called "squaring a binomial." Look for a pattern.

a) $(x + 3)^2 = (x + 3)(x + 3) = x^2 + 3x + 3x + 9 = x^2 + 6x + 9$

b) $(5 + 3y)^2 = (5 + 3y)(5 + 3y) = 25 + 15y + 15y + 9y^2$

$$= 25 + 30y + 9y^2$$

After finishing Section 4.7, you should be able to:

• Multiply the sum and difference of two expressions mentally.

• • Square a binomial mentally.

• • • Find special products, such as those above and those in Section 4.6, mentally, when they are mixed together.

Multiply.

1. $(x + 5)(x - 5)$

2. $(2x - 3)(2x + 3)$

Multiply.

3. $(x + 2)(x - 2)$

4. $(x - 7)(x + 7)$

5. $(6 - 4y)(6 + 4y)$

6. $(2x^3 - 1)(2x^3 + 1)$

ANSWERS ON PAGE A-8

Multiply.

7. $(x + 8)(x + 8)$

8. $(x - 5)(x - 5)$

Multiply.

9. $(x + 2)^2$

10. $(a - 4)^2$

11. $(2x + 5)^2$

12. $(4x^2 - 3x)^2$

13. $(7 + y)(7 + y)$

14. $(3x^2 - 5)(3x^2 - 5)$

ANSWERS ON PAGE A-8

c) $(x - 3)^2 = (x - 3)(x - 3) = x^2 - 3x - 3x + 9 = x^2 - 6x + 9$

d) $(3x - 5)^2 = (3x - 5)(3x - 5) = 9x^2 - 15x - 15x + 25$
$$= 9x^2 - 30x + 25$$

DO EXERCISES 7 AND 8.

Perhaps you discovered a quick way to square a binomial.

> The square of a binomial is the square of the first term, plus or minus twice the product of the two terms, plus the square of the last:
> $$(A + B)^2 = A^2 + 2AB + B^2;$$
> $$(A - B)^2 = A^2 - 2AB + B^2.$$

Examples Multiply.

$$(A + B)^2 = A^2 + 2 \quad A \quad B + B^2$$

5. $(x + 3)^2 = x^2 + 2 \cdot x \cdot 3 + 3^2$
$$= x^2 + 6x + 9$$

> Carry out the rule. Say the words as you go.

6. $(t - 5)^2 = t^2 - 2 \cdot t \cdot 5 + 5^2$
$$= t^2 - 10t + 25$$

7. $(2x + 7)^2 = (2x)^2 + 2 \cdot 2x \cdot 7 + 7^2$
$$= 4x^2 + 28x + 49$$

8. $(5x - 3x^2)^2 = (5x)^2 - 2 \cdot 5x \cdot 3x^2 + (3x^2)^2$
$$= 25x^2 - 30x^3 + 9x^4$$

> *Caution!* Note carefully in these examples that the square of a sum is *not* the sum of squares:
> $$(A + B)^2 \neq A^2 + B^2.$$
> To see this, note that
> $$(20 + 5)^2 = 25^2 = 625,$$
> but
> $$20^2 + 5^2 = 400 + 25 = 425 \neq 625.$$

DO EXERCISES 9–14.

••• MULTIPLICATIONS OF VARIOUS TYPES

Now that we have considered how to multiply quickly certain kinds of polynomials, let us try several kinds mixed together so we can learn to sort them out. When you multiply, first see what kind of multiplication you have. Then use the best method. The methods and the questions you should ask yourself are as follows.

Chapter 4   Polynomials

To multiply two polynomials:

1. Is the product the square of a binomial? If so, use the following:

$$(A + B)(A + B) = (A + B)^2 = A^2 + 2AB + B^2, \text{ or}$$
$$(A - B)(A - B) = (A - B)^2 = A^2 - 2AB + B^2.$$

   The square of a binomial is the square of the first term, plus or minus *twice* the product of the two terms, plus the square of the last term. [The answer has 3 terms.]

2. Is it the product of the sum and difference of the *same* two terms? If so, use the following:

$$(A + B)(A - B) = A^2 - B^2.$$

   The product of the sum and difference of the same two terms is the difference of the squares. [The answer has 2 terms.]

3. Is it the product of two binomials other than those above? If so, use FOIL. [The answer will have 3 or 4 terms.]

4. Is it the product of two polynomials other than those above? If so, multiply each term of one by every term of the other. Use columns if you wish. [The answer will have 2 or more terms, usually more than 2 terms.]

Note that although FOIL will actually work instead of either of the first two rules, those rules will make your work go faster.

Example 9 Multiply: $(x + 3)(x - 3)$.

$(x + 3)(x - 3) = x^2 - 9$ Using method 2 (the product of the sum and difference of two expressions)

Example 10 Multiply: $(t + 7)(t - 5)$.

$(t + 7)(t - 5) = t^2 + 2t - 35$ Using method 3 (the product of two binomials, but neither the square of a binomial nor the product of the sum and difference of two expressions)

Example 11 Multiply: $(x + 7)(x + 7)$.

$(x + 7)(x + 7) = x^2 + 14x + 49$ Using method 1 (the square of a binomial sum)

Example 12 Multiply: $2x^3(9x^2 + x - 7)$.

$2x^3(9x^2 + x - 7) = 18x^5 + 2x^4 - 14x^3$ Using method 4 (the product of a monomial and a trinomial: multiply each term of the trinomial by the monomial)

DO EXERCISES 15–17.

Multiply.

15. $(x + 5)(x + 6)$

16. $(t - 4)(t + 4)$

17. $4x^2(-2x^3 + 5x^2 + 10)$

ANSWERS ON PAGE A-8

Multiply.

18. $(9x^2 + 1)^2$

19. $(2a - 5)(2a + 8)$

20. $\left(5x + \dfrac{1}{2}\right)^2$

21. $\left(2x - \dfrac{1}{2}\right)^2$

ANSWERS ON PAGE A-8

Example 13 Multiply: $(3x^2 - 7x)^2$.

$$(3x^2 - 7x)^2 = 9x^4 - 42x^3 + 49x^2$$

Using method 1 (the square of a binomial difference)

Example 14 Multiply: $\left(3x + \dfrac{1}{4}\right)^2$.

$$\left(3x + \frac{1}{4}\right)^2 = 9x^2 + 2(3x)\frac{1}{4} + \frac{1}{16}$$
$$= 9x^2 + \frac{3}{2}x + \frac{1}{16}$$

Using method 1 (the square of a binomial sum. To get the middle term, we multiply $3x$ by $\frac{1}{4}$ and double.)

Example 15 Multiply: $\left(4x - \dfrac{3}{4}\right)^2$.

$$\left(4x - \frac{3}{4}\right)^2 = 16x^2 - 2(4x)\frac{3}{4} + \frac{9}{16}$$

Using method 1

$$= 16x^2 - 6x + \frac{9}{16}$$

DO EXERCISES 18–21.

S O M E T H I N G   E X T R A

FACTORS AND SUMS

In the table, the top number has been factored in such a way that the sum of the factors is the bottom number. For example, in the first column 56 has been factored as $(-7)(-8)$ and $-7 + (-8) = -15$, the bottom number.

Product	56	63	-36	-72	140	-96		168	-110			
Factor	-7									-9	-24	-3
Factor	-8					-8	-8			-10	18	
Sum	-15	-16	-16	34	-24	-4	-14		-1			-24

EXERCISE

Find the missing numbers in the table.

200

Chapter 4   Polynomials

• Multiply mentally.

1. $(x + 4)(x - 4)$

2. $(x + 1)(x - 1)$

3. $(2x + 1)(2x - 1)$

4. $(x^2 + 1)(x^2 - 1)$

5. $(5m - 2)(5m + 2)$

6. $(3x^4 + 2)(3x^4 - 2)$

7. $(2x^2 + 3)(2x^2 - 3)$

8. $(6x^5 - 5)(6x^5 + 5)$

9. $(3x^4 - 4)(3x^4 + 4)$

10. $(t^2 - 0.2)(t^2 + 0.2)$

11. $(x^6 - x^2)(x^6 + x^2)$

12. $(2x^3 - 0.3)(2x^3 + 0.3)$

13. $(a^4 + 3a)(a^4 - 3a)$

14. $\left(\dfrac{3}{4} + 2t^3\right)\left(\dfrac{3}{4} - 2t^3\right)$

15. $(x^{12} - 3)(x^{12} + 3)$

16. $(12 - 3x^2)(12 + 3x^2)$

17. $(2x^8 + 3)(2x^8 - 3)$

18. $\left(x - \dfrac{2}{3}\right)\left(x + \dfrac{2}{3}\right)$

•• Multiply mentally.

19. $(x + 2)^2$

20. $(2x - 1)^2$

21. $(3x^2 + 1)^2$

22. $\left(3x + \dfrac{3}{4}\right)^2$

23. $\left(x - \dfrac{1}{2}\right)^2$

24. $\left(2x - \dfrac{1}{5}\right)^2$

25. $(3 + x)^2$

26. $(x^3 - 1)^2$

27. $(y^2 + 1)^2$

28. $(8p - p^2)^2$

29. $(2 - 3x^4)^2$

30. $(6x^3 - 2)^2$

31. $(5 + 6t^2)^2$

32. $(3p^2 - p)^2$

ANSWERS
1.
2.
3.
4.
5.
6.
7.
8.
9.
10.
11.
12.
13.
14.
15.
16.
17.
18.
19.
20.
21.
22.
23.
24.
25.
26.
27.
28.
29.
30.
31.
32.

••• Multiply mentally.

33. $(3 - 2x^3)^2$

34. $(x - 4x^3)^2$

35. $4x(x^2 + 6x - 3)$

36. $8x(-x^5 + 6x^2 + 9)$

37. $\left(2x^2 - \dfrac{1}{2}\right)\left(2x^2 - \dfrac{1}{2}\right)$

38. $(-x^2 + 1)^2$

39. $(-1 + 3p)(1 + 3p)$

40. $(-3x + 2)(3x + 2)$

41. $3t^2(5t^3 - t^2 + t)$

42. $-6x^2(x^3 + 8x - 9)$

43. $(6x^4 + 4)^2$

44. $(8a + 5)^2$

45. $(3x + 2)(4x^2 + 5)$

46. $(2x^2 - 7)(3x^2 + 9)$

47. $(8 - 6x^4)^2$

48. $\left(\dfrac{1}{5}x^2 + 9\right)\left(\dfrac{3}{5}x^2 - 7\right)$

✔ SKILL MAINTENANCE

49. In an apartment, lamps, an air conditioner, and a television set are all operating at the same time. The lamps take 10 times as many watts as the television set, and the air conditioner takes 40 times as many watts as the television set. The total wattage used in the apartment is 2550 watts. How many watts are used by each appliance?

50. Solve: $3x - 8x = 4(7 - 8x)$.

☆ EXTENSION

51. ▨ Multiply: $(67.58x + 3.225)^2$.

Multiply.

52. $[(2x - 1)(2x + 1)](4x^2 + 1)$

53. $(5t^2 - 3)^2(5t^2 + 3)^2$

54. Solve: $(x + 4)^2 = (x + 8)(x - 8)$.

Chapter 4   Polynomials

Summary and Review

The following contains a summary of what you should be able to do after completing this chapter. The review exercises are for practice. Answers are at the back of the book. If you miss an exercise, restudy the section and objective indicated alongside the answer.

The review sections to be tested in addition to the material in this chapter are 1.10, 2.7, 3.3, and 3.5.

You should be able to:

Evaluate a polynomial for a given value of a variable.

Evaluate each polynomial for $x = -1$.

1. $7x - 10$

2. $x^2 - 3x + 6$

For a polynomial identify the terms, the missing terms, the coefficients of each term, the degree of each term, and the degree of the polynomial, and classify the polynomial as a monomial, binomial, trinomial, or none of these.

Identify the terms of each polynomial.

3. $3x^2 + 6x + \dfrac{1}{2}$

4. $-4y^5 + 7y^2 - 3y - 2$

5. Identify the missing terms in $x^3 + x$.

Identify the coefficient of each term of each polynomial.

6. $6x^2 + 17$

7. $4x^3 + 6x^2 - 5x + 0.43$

Identify the degree of each term and the degree of the polynomial.

8. $x^3 + 4x - 6$

9. $3 - 2x^4 + 3x^9 + x^6 - \dfrac{3}{4}x^3$

Tell whether each polynomial is a monomial, binomial, trinomial, or none of these.

10. $4x^3 - 1$

11. $4 - 9t^3 - 7t^4 + 10t^2$

12. $7y^2$

Collect the like terms of a polynomial and arrange them in descending order.

Collect like terms.

13. $5x - x^2 + 4x$

14. $-t^3 + 4t^2 - t^3 + 7$

15. $-2x^4 + 16 + 2x^4 + 9 - 3x^5$

Collect like terms and then arrange them in descending order.

16. $3y^2 - 2y + 3 - 5y^2 - 1 - y$

17. $-x + \dfrac{1}{2} + 14x^4 - 7x^2 - 1 - 4x^4$

Add and subtract polynomials.

Add.

18. $(3x^4 - x^3 + x - 4) + (x^5 + 7x^3 - 3x^2 - 5) + (-5x^4 + 6x^2 - x)$

19. $(3t^5 - 4t^4 + t^3 - 3) + (3t^4 - 5t^3 + 3t^2) + (4t^5 + 4t^3) + (-5t^5 - 5t^2) + (-5t^4 + 2t^3 + 5)$

20.
$$-\frac{3}{4}x^4 + \frac{1}{2}x^3 \qquad\qquad +\frac{7}{8}$$
$$\qquad -\frac{1}{4}x^3 - \quad x^2 - \frac{7}{4}x$$
$$\overline{\quad \frac{3}{4}x^4 \qquad\quad +\frac{2}{3}x^2 \qquad -\frac{1}{2}\quad}$$

Subtract.

21. $(5x^2 - 4x + 1) - (3x^2 + 7)$

22. $(3x^5 - 4x^4 + 2x^2 + 3) - (2x^5 - 4x^4 + 3x^3 + 4x^2 - 5)$

23.
$$2y^5 \qquad - \quad y^3 \qquad\quad + y + 3$$
$$\overline{3y^5 - y^4 + 4y^3 + 2y^2 - y + 3}$$

Multiply two monomials, a monomial and a binomial, two binomials, a binomial and a trinomial, and any two polynomials. Find special products such as the sum and difference of two expressions and the square of a binomial.

Multiply.

24. $3x(-4x^2)$

25. $(7x + 1)^2$

26. $(y + \frac{2}{3})(y + \frac{1}{2})$

27. $(1.5t - 6.5)(0.2t + 1.3)$

28. $(4x^2 - 5x + 1)(3x - 2)$

29. $(x - 9)^2$

30. $5x^4(3x^3 - 8x^2 + 10x + 2)$

31. $(x + 4)(x - 7)$

32. $(x - 0.3)(x - 0.75)$

33. $(x^4 - 2x + 3)(x^3 + x - 1)$

34. $(3y^2 - 2y)^2$

35. $(2t^2 + 3)(t^2 - 7)$

36. $(x^3 - 2x + 3)(4x^2 - 5x)$

37. $(3x^2 + 4)(3x^2 - 4)$

38. $(2 - m)(2 + m)$

39. $(13x - 3)(x - 13)$

 SKILL MAINTENANCE

40. What is 78% of 95?

41. 128.8 is what percent of 560?

42. Factor: $64t - 32m + 16$.

43. Multiply: $8(-3x + 5y - 4)$.

44. Collect like terms: $12x + 3y + 8 - 5x - 19 - 2y$.

Solve.

45. $7x - 4x - 2 = 37$

46. $\frac{1}{2}x - 3x = \frac{1}{4}x - 6x + 43$

47. The first angle of a triangle is four times as large as the second. The measure of the third angle is 30° greater than that of the second. How large are the angles?

⭐ EXTENSION

48. Solve: $(x - 7)(x + 10) = (x - 4)(x - 6)$.

49. Multiply: $[(1 - 8t)(1 + 8t)](1 - 64t^2)$.

50. Collect like terms:
$$-3x^5(3x^3) - x^6(2x)^2 + (3x^4)^2 + (2x^2)^4 - 40x^2(x^3)^2.$$

51. Simplify:
$$(-4 + x^2 + 2x^3) - (-6 - x + 3x^3) - (-x^2 - 5x^3).$$

52. Compute: $(x + 2)(x - 5) - (x - 2)(x + 5)$.

Test: Chapter 4

1. Evaluate the polynomial $x^2 + 5x - 1$ for $x = -2$.

2. Identify the coefficient of each term of the polynomial $\frac{1}{3}x^5 - x + 7$.

3. Identify the degree of each term and the degree of the polynomial.

$$2x^4 - 4 + 5x + 3x^6$$

4. Tell whether the polynomial is a monomial, binomial, trinomial, or none of these.

$$x^2 + x - 7$$

Collect like terms.

5. $4a^2 - 6 + a^2$

6. $y^2 - 3y - y + \frac{3}{4}y^2$

7. Collect like terms and then arrange them in descending order.

$$3 - x^2 + 2x^3 + 5x^2 - 6x - 2x + x^5$$

Add.

8. $(3x^5 + 5x^3 - 5x^2 - 3) + (x^5 + x^4 - 3x^3 - 3x^2 + 2x - 4)$

9. $\left(y^4 + \frac{2}{3}y + 5\right) + \left(y^4 + 5y^2 + \frac{1}{3}y\right)$

Subtract.

10. $(2x^4 + x^3 - 8x^2 - 6x - 3) - (6x^4 - 8x^2 + 2x)$

11. $(t^3 - 0.4t^2 - 12) - (t^5 + 0.3t^3 + 0.4t^2 + 9)$

A N S W E R S
1.
2.
3.
4.
5.
6.
7.
8.
9.
10.
11.

Multiply.

12. $-3x^2(4x^2 - 3x - 5)$

13. $\left(x - \dfrac{1}{3}\right)^2$

14. $(10 - 3y)(10 + 3y)$

15. $(3b + 5)(b - 3)$

16. $(x^6 - 4)(x^8 + 4)$

17. $(8 - y)(6 + 5y)$

18. $(2x + 1)(3x^2 - 5x - 3)$

19. $(5t + 2)^2$

12. _____

13. _____

14. _____

15. _____

16. _____

17. _____

18. _____

19. _____

20. _____

21. _____

22. _____

23. _____

24. _____

25. _____

✔ SKILL MAINTENANCE

20. 16 is what percent of 50?

22. Factor: $25t - 50 + 100m$.

21. Solve: $7x + 6 - 8x = 11 - 5x + 4$.

23. The perimeter of a rectangle is 540 m. The width is 19 m less than the length. Find the width and the length.

⭐ EXTENSION

24. Solve: $(x - 5)(x + 5) = (x + 6)^2$.

25. Find a polynomial for the shaded region in the figure.

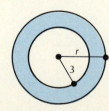

AN APPLICATION
A 12-oz beverage can has a height of 4.7 in. and a radius of 1.2 in. Find the area of the can.

THE MATHEMATICS
The area is given by the polynomial
$$2\pi rh + 2\pi r^2$$
where h is the height and r is the radius. Evaluating the polynomial for $h = 4.7$ and $r = 1.2$, we get
$$2\pi(1.2)(4.7) + 2\pi(1.2)^2$$
as the area.

In this chapter we learn to factor polynomials. Factoring is the reverse of multiplication. To factor a polynomial is to find an equivalent expression that is a product. By learning to factor, we are able to solve certain second-degree equations, the solution of which will allow us to solve certain kinds of problems.

The review sections to be tested in addition to the material in this chapter are 2.6, 3.4, 3.6, and 4.7.

Polynomials and Factoring

Factoring Polynomials

To *factor* an expression is to find an equivalent expression that is a product. In this chapter we factor polynomials. To factor polynomials quickly, we study the quick methods of multiplication that we learned in Chapter 4.

• FACTORING MONOMIALS

To factor a monomial we find two monomials whose product is that monomial. Compare. Note that factoring is the reverse of multiplication.

	Multiplying	*Factoring*
a)	$(4x)(5x) = 20x^2$	$20x^2 = (4x)(5x)$
b)	$(2x)(10x) = 20x^2$	$20x^2 = (2x)(10x)$
c)	$(-4x)(-5x) = 20x^2$	$20x^2 = (-4x)(-5x)$
d)	$(x)(20x) = 20x^2$	$20x^2 = (x)(20x)$

The monomial $20x^2$ thus has many factorizations. There are still other ways to factor $20x^2$.

DO EXERCISES 1 AND 2

To factor a monomial, factor the coefficient first. Then shift some of the letters to one factor and some to the other. We can do this because of the commutative and associative laws.

Example 1 Find three factorizations of $15x^3$.

a) $15x^3 = (3 \cdot 5)x^3$
   $= (3x)(5x^2)$

b) $15x^3 = (3 \cdot 5)x^3$
   $= (3x^2)(5x)$

c) $15x^3 = (-1) \cdot (-15)x^3$
   $= (-x)(-15x^2)$

Later we will be more selective about the way we factor a monomial depending on the uses of such factoring.

DO EXERCISES 3–5.

•• FACTORING WHEN TERMS HAVE A COMMON FACTOR

To multiply a monomial and a polynomial with more than one term, we multiply each term by the monomial. To factor, we do the reverse.

---

1. a) Multiply: $(3x)(4x)$.

   b) Factor: $12x^2$.

2. a) Multiply: $(2x)(8x^2)$.

   b) Factor: $16x^3$.

Find three factorizations of the monomial.

3. $8x^4$

4. $21x^2$

5. $6x^5$

ANSWERS ON PAGE A-9

Compare.

Multiply:

$$3x(x^2 + 2x - 4) = 3x \cdot x^2 + 3x \cdot 2x + 3x(-4)$$
$$= 3x^3 + 6x^2 - 12x$$

Factor:

$$3x^3 + 6x^2 - 12x = 3x \cdot x^2 + 3x \cdot 2x + 3x(-4)$$
$$= 3x(x^2 + 2x - 4)$$

A common error: $3x^3 + 6x^2 - 12x = 3 \cdot x \cdot x \cdot x + 6 \cdot x \cdot x - 4 \cdot 3x$. The parts, or terms, of the expression have been factored but the expression itself has not been factored.

DO EXERCISES 6 AND 7.

We are finding a factor in common with all the terms. There may not always be one other than 1. When there is, we generally use the factor with the largest possible coefficient and the largest exponent. In this way we "factor completely."

Example 2 Factor: $3x^2 + 6$.

$$3x^2 + 6 = 3 \cdot x^2 + 3 \cdot 2$$
$$= 3(x^2 + 2) \quad \text{Factoring out the common factor, 3}$$

Example 3 Factor: $16y^3 + 20y^2$.

$$16y^3 + 20y^2 = (4y^2)(4y) + 4y^2 \cdot 5$$
$$= 4y^2(4y + 5) \quad \text{Factoring out } 4y^2$$

Example 4 Factor: $15x^5 - 12x^4 + 27x^3 - 3x^2$.

$$15x^5 - 12x^4 + 27x^3 - 3x^2$$
$$= (3x^2)(5x^3) - (3x^2)(4x^2) + (3x^2)(9x) - (3x^2)(1)$$
$$= 3x^2(5x^3 - 4x^2 + 9x - 1) \quad \text{Factoring out } 3x^2$$

If you can spot the common factor without factoring each term, you should write just the answer.

Example 5 Factor: $\dfrac{4}{5}x^2 + \dfrac{1}{5}x + \dfrac{2}{5}$.

$$\frac{4}{5}x^2 + \frac{1}{5}x + \frac{2}{5} = \frac{1}{5}(4x^2 + x + 2)$$

DO EXERCISES 8–12.

6. a) Multiply: $3(x + 2)$.

 b) Factor: $3x + 6$.

7. a) Multiply: $2x(x^2 + 5x + 4)$.

 b) Factor: $2x^3 + 10x^2 + 8x$.

Factor.

8. $x^2 + 3x$

9. $3y^6 - 5y^3 + 2y^2$

10. $9x^4 - 15x^3 + 3x^2$.

11. $\dfrac{3}{4}t^3 + \dfrac{5}{4}t^2 + \dfrac{7}{4}t + \dfrac{1}{4}$

12. $35x^7 - 49x^6 + 14x^5 - 63x^3$

ANSWERS ON PAGE A-9

Factor.

13. $x^2 + 5x + 2x + 10$

14. $y^2 - 4y + 3y - 12$

15. $4x^3 - 6x^2 - 6x + 9$

16. $8t^3 + 2t^2 + 12t + 3$

17. $3m^5 - 15m^3 + 2m^2 - 10$

ANSWERS ON PAGE A-9

●●● **FACTORING BY GROUPING**

The method we are about to consider is called *factoring by grouping*. Certain polynomials with four terms can be factored using this method. Consider

$$x^2 + 3x + 4x + 12.$$

There is no factor common to all the terms other than 1. But we can factor $x^2 + 3x$ and $4x + 12$:

$$x^2 + 3x = x(x + 3); \qquad \text{Factoring } x^2 + 3x$$
$$4x + 12 = 4(x + 3). \qquad \text{Factoring } 4x + 12$$

Then

$$x^2 + 3x + 4x + 12 = x(x + 3) + 4(x + 3).$$

Note the common *binomial* factor $x + 3$. We can use the distributive law again like this:

$$x(x + 3) + 4(x + 3) = (x + 4)(x + 3). \qquad \text{Factoring out the common factor, } x + 3$$

Because of the commutative laws the factorization $(x + 3)(x + 4)$ is also correct.

Examples Factor. For purposes of learning this method, do not collect like terms.

6. $x^2 + 7x + 2x + 14$
   $= (x^2 + 7x) + (2x + 14)$ Separating into two binomials
   $= x(x + 7) + 2(x + 7)$ Factoring each binomial
   $= (x + 2)(x + 7)$ Factoring out the common factor, $x + 7$

7. $5x^2 - 10x + 2x - 4 = (5x^2 - 10x) + (2x - 4)$ Separating into two binomials
   $= 5x(x - 2) + 2(x - 2)$ Factoring each binomial
   $= (5x + 2)(x - 2)$ Factoring out the common factor, $x - 2$

8. $x^2 + 3x - x - 3 = (x^2 + 3x) + (-x - 3)$
   $= x(x + 3) - 1(x + 3)$ Factoring -1 out of the second binomial
   $= (x - 1)(x + 3)$

9. $2x^4 - 12x^3 - 3x + 18 = (2x^4 - 12x^3) + (-3x + 18)$
   $= 2x^3(x - 6) - 3(x - 6)$
   $= (2x^3 - 3)(x - 6)$

10. $12x^5 + 20x^2 - 21x^3 - 35 = (12x^5 + 20x^2) + (-21x^3 - 35)$
    $= 4x^2(3x^3 + 5) - 7(3x^3 + 5)$
    $= (4x^2 - 7)(3x^3 + 5)$

Not all expressions with four terms can be factored by grouping. An example is $x^3 - 7x^2 + 12x - 10$.

DO EXERCISES 13–17.

• Find three factorizations of each monomial.

1. $6x^3$ **2.** $9x^4$ **3.** $-9x^5$

4. $-12x^6$ **5.** $24x^4$ **6.** $15x^5$

•• Factor.

7. $x^2 - 4x$ **8.** $2x^2 + 6x$

9. $x^3 + 6x^2$ **10.** $2x^2 + 2x - 8$

11. $8x^4 - 24x^2$ **12.** $x^5 + x^4 + x^3 - x^2$

13. $17x^5 + 34x^3 + 51x$ **14.** $6x^4 - 10x^3 + 3x^2$

15. $10x^3 + 25x^2 + 15x - 20$ **16.** $16x^4 - 24x^3 + 32x^2 + 64x$

17. $\dfrac{5}{3}x^6 + \dfrac{4}{3}x^5 + \dfrac{1}{3}x^4 + \dfrac{1}{3}x^3$ **18.** $\dfrac{5}{7}x^7 + \dfrac{3}{7}x^5 - \dfrac{6}{7}x^3 - \dfrac{1}{7}x$

ANSWERS
1.
2.
3.
4.
5.
6.
7.
8.
9.
10.
11.
12.
13.
14.
15.
16.
17.
18.

••• Factor.

19. $y^2 + 4y + y + 4$

20. $x^2 + 5x + 2x + 10$

21. $x^2 + 5x - 2x - 10$

22. $a^2 - 4a - a + 4$

23. $16 - 12x - 4x + 3x^2$

24. $24 - 18y - 20y + 15y^2$

25. $2x^3 + 6x^2 + x + 3$

26. $3x^3 + 2x^2 + 3x + 2$

27. $x^3 + 8x^2 - 3x - 24$

28. $2x^3 + 12x^2 - 5x - 30$

29. $12t^3 - 16t^2 + 3t - 4$

30. $18a^3 - 21a^2 + 30a - 35$

31. $4x^5 + 6x^3 + 6x^2 + 9$

32. $4x^5 + 6x^4 + 6x^3 + 9x^2$

 SKILL MAINTENANCE

Multiply.

33. $(x - 4)(x + 4)$

34. $(3 - 5x^2)(3 + 5x^2)$

35. Solve: $2 - 5(x + 5) = 3(x - 2) - 1$.

36. Solve $A = \dfrac{p + q}{3}$ for p.

⭐ EXTENSION

Factor.

37. $x^6 + x^4 + x^2 + 1$

38. $x^{13} + x^7 + x^6 + 1$

Two polynomials are *relatively prime* if they have no common factors other than constants. Determine whether each pair is relatively prime.

39. $5x, 6x^2$ **40.** $x + x^2, 3x^3$ **41.** $3x, 9x - 1$ **42.** $x^2 + 3, x$

Chapter 5 **Polynomials and Factoring**

Differences of Squares

The following are differences of squares:

$$x^2 - 4, \qquad 9 - 25t^4.$$

We now learn how to factor differences of squares.

• RECOGNIZING DIFFERENCES OF SQUARES

A difference of squares is an expression like the following:

$$A^2 - B^2.$$

How can we recognize such expressions? Look at $A^2 - B^2$. In order for a binomial to be a difference of squares:

a) There must be two expressions, both squares, such as

$$4x^2, \qquad 9, \qquad 25t^4, \qquad 1, \qquad x^6.$$

b) There must be a minus sign between them.

Example 1 Determine whether $9x^2 - 64$ is a difference of squares.

a) The first expression is a square: $9x^2 = (3x)^2$. The second expression is a square: $64 = 8^2$.

b) There is a minus sign between them.

Therefore, we have a difference of squares.

Example 2 Determine whether $t^2 + 25$ is a difference of squares.

a) The first expression is a square: t^2. The second expression is a square: $25 = 5^2$.

b) There is *no* minus sign between them.

Thus we do not have a difference of squares.

DO EXERCISES 1–7.

•• FACTORING DIFFERENCES OF SQUARES

To factor a difference of squares we can use the following equation:

$$A^2 - B^2 = (A - B)(A + B).$$

Where does this equation come from? If you multiply out the two expressions on the right, you do get the expression on the left. So the two expressions are equivalent. In effect, we are going to use the equation $(A - B)(A + B) = A^2 - B^2$ in reverse so we can factor.

Determine whether each is a difference of squares. Write "yes" or "no."

1. $x^2 - 25$

2. $t^2 - 24$

3. $y^2 + 36$

4. $4x^2 - 15$

5. $16x^4 - 49$

6. $9w^6 - 1$

7. $-49 + 25t^2$

ANSWERS ON PAGE A-9

Factor.

8. $x^2 - 9$

9. $t^2 - 64$

10. $32y^2 - 8y^6$

Factor.

11. $64x^4 - 25x^6$

12. $5 - 20t^6$
[Hint: $1 = 1^2$, $t^6 = (t^3)^2$.]

Factor.

13. $81x^4 - 1$

14. $49m^4 - 25m^{10}$

ANSWERS ON PAGE A-9

To factor a difference of squares, $A^2 - B^2$, we find the numbers we square to get A^2 and B^2. These are called *square roots.* We write a plus sign between the square roots in one of the factors and a minus sign between them in the other factor.

Example 3 Factor: $x^2 - 4$.

$$x^2 - 4 = x^2 - 2^2 = (x - 2)(x + 2)$$
$$A^2 - B^2 = (A - B)(A + B)$$

Always remember in any kind of factoring to look first for a common factor. Not only will it lead to the correct answer, but it will also ease your work.

Example 4 Factor: $18x^2 - 50x^6$.

Always look first for a factor common to all terms. This time there is one.

$$18x^2 - 50x^6 = 2x^2(9 - 25x^4) \quad \text{Factoring out the common factor first}$$

$$= 2x^2[3^2 - (5x^2)^2]$$
$$= 2x^2(3 - 5x^2)(3 + 5x^2)$$

DO EXERCISES 8–10.

Example 5 Factor: $49t^4 - 9t^6$.

$$49t^4 - 9t^6 = t^4(49 - 9t^2) \quad \text{Factoring out the common factor first}$$

$$= t^4(7 - 3t)(7 + 3t)$$

DO EXERCISES 11 AND 12.

••• FACTORING COMPLETELY

When a factor with more than one term can still be factored, you should do so. When no factor can be factored into polynomials of smaller degree and there are no common factors, you have *factored completely.* Always factor completely even when directions do not say to do so.

Example 6 Factor: $1 - 16x^{12}$.

$$1 - 16x^{12} = (1 + 4x^6)(1 - 4x^6) \quad \text{Factoring a difference of squares}$$
$$= (1 + 4x^6)(1 - 2x^3)(1 + 2x^3) \quad \text{Factoring further. The factor } 1 - 4x^6 \text{ is a difference of squares.}$$

FACTORING HINTS

1. Always look first for a common factor.
2. Always factor completely.
3. Check by multiplying.
4. Never try to factor a sum of squares, $A^2 + B^2$.

DO EXERCISES 13 AND 14.

Chapter 5   Polynomials and Factoring

• Determine whether each of the following is a difference of squares.

1. $x^2 - 36$ 2. $y^2 - 16$ 3. $x^2 - 35$ 4. $x^2 + 36$

5. $16x^2 - 25$ 6. $36t^2 - 1$ 7. $49 + 16t^4$ 8. $x^2 + 6x + 9$

•• Factor.

9. $x^2 - 4$ 10. $x^2 - 36$ 11. $t^2 - 9$ 12. $m^2 - 1$

13. $16a^2 - 9$ 14. $25x^2 - 4$ 15. $4x^2 - 25$ 16. $9a^2 - 16$

17. $8x^2 - 98$ 18. $24x^2 - 54$ 19. $36x - 49x^3$ 20. $16x - 81x^3$

21. $16y^2 - 25y^4$ 22. $x^{16} - 9x^2$ 23. $49a^4 - 81$ 24. $25a^4 - 9$

25. $a^{12} - 4a^2$ 26. $121a^8 - 100$ 27. $81y^6 - 25$ 28. $100y^6 - 49$

ANSWERS
1.
2.
3.
4.
5.
6.
7.
8.
9.
10.
11.
12.
13.
14.
15.
16.
17.
18.
19.
20.
21.
22.
23.
24.
25.
26.
27.
28.

••• Factor.

29. $x^4 - 1$ **30.** $x^4 - 16$ **31.** $4x^4 - 64$ **32.** $5x^4 - 80$

33. $1 - y^8$ **34.** $x^8 - 1$ **35.** $x^{12} - 16$ **36.** $y^8 - 81$

37. $\dfrac{1}{16} - y^2$ **38.** $\dfrac{1}{25} - x^2$ **39.** $25 - \dfrac{1}{49}y^2$ **40.** $4 - \dfrac{1}{9}y^2$

41. $16 - t^4$ **42.** $1 - a^4$

 SKILL MAINTENANCE

Multiply.

43. $(t - 9)^2$ **44.** $(5x + 3)^2$

45. Multiply: $(-16)(-5)$.   **46.** Divide: $\dfrac{-124}{31}$.

⭐ **EXTENSION**

Factor.

47. $3.24x^2 - 0.81$ **48.** $3x^2 - \frac{1}{3}$ **49.** $1.28t^2 - 2$

50. $y^8 - 256$ **51.** $x^6 - x^4 - x^2 + 1$

Trinomial Squares

Recall that a trinomial is a polynomial with just three terms. Some trinomials are squares of binomials. For example, the trinomial $x^2 - 8x + 16$ is the square of $(x - 4)$. To see this we can calculate $(x - 4)^2$. It is $x^2 - 2 \cdot 4 \cdot x + 4^2$, or $x^2 - 8x + 16$.

A trinomial that is the square of a binomial is called a *trinomial square*.

• RECOGNIZING TRINOMIAL SQUARES

We can use the equations for squaring in reverse in order to factor trinomial squares:

$$A^2 + 2AB + B^2 = (A + B)^2;$$
$$A^2 - 2AB + B^2 = (A - B)^2.$$

How can we recognize when an expression to be factored is a trinomial square? Look at $A^2 + 2AB + B^2$ and $A^2 - 2AB + B^2$. In order for an expression to be a trinomial square:

a) Two of the terms, A^2 and B^2, must be squares, such as

$$4, \quad x^2, \quad 25y^4, \quad 16t^2.$$

b) There must be no minus sign before A^2 or B^2.

c) If we multiply A and B (the square roots of these expressions) and double the result, we get the remaining term $2 \cdot A \cdot B$, or its additive inverse, $-2 \cdot A \cdot B$.

Example 1 Determine whether $x^2 + 6x + 9$ is a trinomial square.

a) x^2 and 9 are squares.

b) There is no minus sign before x^2 or 9.

c) If we multiply the square roots x and 3 and double the product, we get the remaining term: $2 \cdot 3 \cdot x = 6x$.

Thus $x^2 + 6x + 9$ is the square of a binomial.

Example 2 Determine whether $x^2 + 6x + 11$ is a trinomial square.

The answer is *no*, because only one term is a square.

Example 3 Determine whether $16x^2 + 49 - 56x$ is a trinomial square.

a) $16x^2$ and 49 are squares.

b) There is no minus sign before $16x^2$ or 49.

c) If we multiply the square roots $4x$ and 7, and double the product, we get the additive inverse of the remaining term: $2 \cdot 4x \cdot 7 = 56x$. And $56x$ is the additive inverse of $-56x$.

Thus $16x^2 + 49 - 56x$ is a trinomial square.

Determine whether each of the following is a trinomial square. Write "yes" or "no."

1. $x^2 + 8x + 16$

2. $25 - x^2 + 10x$

3. $t^2 - 12t + 4$

4. $25 + 20y + 4y^2$

5. $5x^2 + 16 - 14x$

6. $16x^2 + 40x + 25$

7. $m^2 + 6m - 9$

8. $25x^2 + 9 - 30x$

ANSWERS ON PAGE A-10

Factor.

9. $x^2 + 2x + 1$

10. $1 - 2x + x^2$

11. $4 + t^2 + 4t$

12. $25x^2 - 70x + 49$

13. $49 - 56y + 16y^2$

14. $48m^2 + 75 + 120m$

DO EXERCISES 1–8 ON THE PRECEDING PAGE.

● ● FACTORING TRINOMIAL SQUARES

To factor trinomial squares we use the following equations:

$$A^2 + 2AB + B^2 = (A + B)^2;$$
$$A^2 - 2AB + B^2 = (A - B)^2.$$

We use the square roots of the squared terms and the sign of the remaining term.

Example 4 Factor: $x^2 + 6x + 9$.

$$x^2 + 6x + 9 = x^2 + 2 \cdot x \cdot 3 + 3^2 = (x + 3)^2$$
$$A^2 + 2 \cdot A \cdot B + B^2 = (A + B)^2$$

The sign of the middle term is positive.

Example 5 Factor: $x^2 + 49 - 14x$.

We first change the order.

$$x^2 + 49 - 14x = x^2 - 14x + 49 \quad \text{Changing the order}$$
$$= x^2 - 2 \cdot x \cdot 7 + 7^2$$
$$= (x - 7)^2 \quad \text{The sign of the middle term is negative.}$$

Example 6 Factor: $16y^2 - 40y + 25$.

$$16y^2 - 40y + 25 = (4y)^2 - 2 \cdot 4y \cdot 5 + 5^2 = (4y - 5)^2$$
$$A^2 - 2 \cdot A \cdot B + B^2 = (A - B)^2$$

DO EXERCISES 9–14.

• Determine whether each of the following is a trinomial square.

1. $x^2 - 14x + 49$

2. $x^2 - 16x + 64$

3. $x^2 - 64 + 16x$

4. $x^2 - 49 - 14x$

5. $y^2 - 3y + 9$

6. $t^2 + 2t + 4$

7. $25 + 40x + 8x^2$

8. $24 - 36x + 9x^2$

•• Factor. Remember to look first for a common factor.

9. $x^2 - 14x + 49$

10. $x^2 - 16x + 64$

11. $y^2 + 16y + 64$

12. $t^2 + 14t + 49$

13. $m^2 + 1 - 2m$

14. $1 + 2p + p^2$

15. $4 + 4x + x^2$

16. $4 + x^2 - 4x$

17. $2x^2 - 4x + 2$

18. $2x^2 - 40x + 200$

19. $x^3 - 18x^2 + 81x$

20. $x^3 + 24x^2 + 144x$

ANSWERS
1.
2.
3.
4.
5.
6.
7.
8.
9.
10.
11.
12.
13.
14.
15.
16.
17.
18.
19.
20.

21. $20t^2 + 100t + 125$

22. $12q^2 + 36q + 27$

23. $49 - 42x + 9x^2$

24. $64 - 112x + 49x^2$

25. $5y^4 + 10y^2 + 5$

26. $a^4 + 14a^2 + 49$

27. $y^6 + 26y^3 + 169$

28. $y^6 - 18y^3 + 81$

29. $16x^{10} - 8x^5 + 1$

30. $9x^{10} + 12y^5 + 4$

 SKILL MAINTENANCE

Multiply.

31. $(x + 6)(x - 4)$

32. $(-6x^8)(2x^5)$

33. About 5 L of oxygen can be dissolved in 100 L of water at 0°C. This is 1.6 times the amount that can be dissolved in the same volume of water at 20°C. How much oxygen can be dissolved at 20°C?

EXTENSION

Factor.

34. $\dfrac{1}{81}x^6 + \dfrac{8}{27}x^3 + \dfrac{16}{9}$

35. $8.1x^2 - 6.4$

36. $(y + 3)^2 + 2(y + 3) + 1$

37. $4(a + 5)^2 + 20(a + 5) + 25$

38. Determine whether $(x + 2)^2(x - 2)^2$ is a factorization of $x^4 - 8x^2 + 16$. Prove your answer.

Factoring Trinomials of the Type $x^2 + bx + c$

After finishing Section 5.4, you should be able to:

• Factor trinomials of the type $x^2 + bx + c$ by examining the constant term c.

• There are trinomials that are not trinomial squares—for example,

$$x^2 + 5x + 6 \quad \text{and} \quad x^2 + 3x - 10.$$

To try to factor such trinomials we use a trial-and-error procedure.

CONSTANT TERM POSITIVE

Recall the FOIL method of multiplying two binomials.

$$(x + 2)(x + 5) = x^2 + \underbrace{5x + 2x}_{} + 10$$
$$= x^2 + \quad 7x \quad + 10$$

The product is a trinomial. In this example, the term of highest degree, called the leading term, has a coefficient of 1. The constant term is positive. To factor $x^2 + 7x + 10$, we think of FOIL in reverse. We multiplied x times x to get the first term of the trinomial. Thus the first term of each binomial factor is x.

$$(x + \underline{\quad})(x + \underline{\quad})$$

To get the middle term and the last term of the trinomial, we look for two numbers whose product is 10 and whose sum is 7. Those numbers are 2 and 5. Thus the factorization is

$$(x + 2)(x + 5), \quad \text{or} \quad (x + 5)(x + 2)$$

by the commutative law of multiplication.

Example 1 Factor: $x^2 + 5x + 6$.

Think of FOIL in reverse. The first term of each factor is x.

$$(x + \underline{\quad})(x + \underline{\quad})$$

Then we look for two numbers whose product is 6 and whose sum is 5. Since both 5 and 6 are positive, we need consider only positive factors.

Pairs of Factors	Sums of Factors
1, 6	7
2, 3	5

The numbers we want are 2 and 3. The factorization is $(x + 2)(x + 3)$. We can check by multiplying to see whether we get the original trinomial.

DO EXERCISES 1 AND 2.

Factor.

1. $x^2 + 7x + 12$

2. $x^2 + 13x + 36$

ANSWERS ON PAGE A-10

3. $x^2 - 8x + 15$

Consider this multiplication.

$$
\begin{array}{ccc}
 & \text{F} \quad \text{O} \quad \text{I} \quad \text{L} & \\
(x - 3)(x - 4) = x^2 - \underline{4x - 3x} + 12 & \\
\downarrow \qquad \downarrow \qquad \downarrow & \\
= x^2 - \quad 7x \quad + 12 &
\end{array}
$$

When the constant term of a trinomial is positive, we look for two numbers with the same sign. The sign is that of the middle term.

$$x^2 - 7x + 12 = (x - 3)(x - 4)$$

Example 2 Factor: $y^2 - 8y + 12$.

Since the constant term is positive and the coefficient of the middle term is negative, we look for a factorization of 12 in which both factors are negative. Their sum must be -8.

Pairs of Factors	Sums of Factors
$-1, -12$	-13
$-2, \ -6$	-8
$-3, \ -4$	-7

The numbers we want are -2 and -6. Thus the factorization is $(x - 2)(x - 6)$.

DO EXERCISES 3 AND 4.

CONSTANT TERM NEGATIVE

Sometimes when we use FOIL, the product has a negative constant term. Consider these multiplications.

$$
\begin{array}{ccc}
 & \text{F} \quad \text{O} \quad \text{I} \quad \text{L} & \\
(x - 5)(x + 2) = x^2 + \underline{2x - 5x} - 10 & \\
\downarrow \qquad \downarrow \qquad \downarrow & \\
= x^2 - \quad 3x \quad - 10 &
\end{array}
$$

$$
\begin{array}{ccc}
 & \text{F} \quad \text{O} \quad \text{I} \quad \text{L} & \\
(x + 5)(x - 2) = x^2 - \underline{2x + 5x} - 10 & \\
\downarrow \qquad \downarrow \qquad \downarrow & \\
= x^2 + \quad 3x \quad - 10 &
\end{array}
$$

4. $t^2 - 9t + 20$

When the constant term is negative, the middle term may be positive or negative. In these cases, we still look for two factors whose product is -10. One of them must be positive and the other negative. Their sum must still be the coefficient of the middle term.

Example 3 Factor: $x^2 - 8x - 20$.

Since the constant term is negative, we look for a factorization of -20 in which one factor is positive and one factor is negative. Their sum must be -8.

Pairs of Factors	Sums of Factors
-1, 20	19
1, -20	-19
-2, 10	8
2, -10	-8

The numbers we want are 2 and -10. Thus the factorization is $(x + 2)(x - 10)$.

Example 4 Factor: $t^2 + 5t - 24$.

We look for a factorization of -24 in which one factor is positive and the other is negative. Their sum must be 5.

Pairs of Factors	Sums of Factors
1, -24	-23
-1, 24	23
2, -12	-10
-2, 12	10
-8, 3	-5
8, -3	5
4, -6	-2
-4, 6	2

The numbers we want are 8 and -3. The factorization is $(t + 8)(t - 3)$.

Example 5 Factor: $x^2 - x - 110$.

Since the constant term is negative, we look for a factorization of -110 in which one factor is positive and one factor is negative. Their sum must be -1.

Pairs of Factors	Sums of Factors
-2, 55	53
2, -55	-53
-5, 22	17
5, -22	-17
-10, 11	1
10, -11	-1

The numbers we want are 10 and -11. Thus the factorization is $(x + 10)(x - 11)$.

There are polynomials that are not factorable.

Example 6 Factor: $x^2 - x + 5$.

Since 5 has very few factors we can easily check all possibilities.

Pairs of Factors	Sums of Factors
5, 1	6
-5, -1	-6

There are no factors whose sum is -1. The polynomial is *not* factorable (at least not into polynomials with integer or rational coefficients).

DO EXERCISES 5–9.

Factor.

5. $x^2 + 4x - 12$

6. $y^2 - 4y - 12$

7. $t^2 + 5t - 14$

8. $x^2 - 30 - x$

9. $x^2 + 2x + 7$

ANSWERS ON PAGE A-10

Factor.

10. $y^y + 8y + 16$

Can we factor a trinomial that is a perfect square using this method? The answer is "yes."

Example 7 Factor: $x^2 - 10x + 25$.

Since the constant term is positive and the coefficient of the middle term is negative, we look for a factorization of 25 in which both factors are negative. Their sum must be -10.

Pairs of Factors	Sums of Factors
25, 1	26
$-25, -1$	-26
5, 5	10
$-5, -5$	-10

The numbers we want are -5 and -5. Thus the factorization is $(x - 5)(x - 5)$.

Note that $(x - 5)(x - 5) = (x - 5)^2$. Thus the trinomial is a square and could be factored by another method. In practice, when you encounter a trinomial you should check to see if it is a square *before* considering trial-and-error.

DO EXERCISE 10.

SOMETHING EXTRA

CALCULATOR CORNER: NESTED EVALUATION

To evaluate polynomials with a calculator, it helps to first use factoring to change the polynomial to *nested form*, as shown in the example below.

Example Write nested form and evaluate for $x = 5.3$:

$$2x^4 + 4x^3 - x^2 + 7x - 9$$
$$= x\{2x^3 + 4x^2 - x + 7\} - 9 \quad \text{Factoring out } x$$
$$= x\{x[2x^2 + 4x - 1] + 7\} - 9 \quad \text{Factoring out } x \text{ again}$$
$$= x\{x[x(2x + 4) - 1] + 7\} - 9. \quad \text{Factoring out } x \text{ again}$$

To evaluate $x\{x[x(2x + 4) - 1] + 7\} - 9$, start from the inside and work out. For $x = 5.3$, you enter 5.3, multiply by 2, add 4, multiply by 5.3, subtract 1, multiply by 5.3, add 7, multiply by 5.3, and subtract 9. The answer is 2173.6142.

EXERCISES

Write nested form for each polynomial. Then evaluate for $x = 5.3$, $x = 10.6$, and $x = -23$. Round to four decimal places.

1. $x^5 + 3x^4 - x^3 + x^2 - x + 9$

2. $5x^4 - 17x^3 + 2x^2 - x + 11$

3. $2x^4 - 3x^3 + 5x^2 - 2x + 18$

4. $-2x^5 + 4x^4 + 8x^3 - 4x^2 - 3x + 24$

Chapter 5   Polynomials and Factoring

● Factor.

1. $x^2 + 8x + 15$

2. $x^2 + 7x + 6$

3. $x^2 - x - 2$

4. $x^2 - 10x + 25$

5. $x^2 + 7x + 12$

6. $x^2 + x - 42$

7. $x^2 + 2x - 15$

8. $x^2 + 8x + 12$

9. $y^2 + 9y + 8$

10. $x^2 - 6x + 9$

11. $x^4 + 5x^2 + 6$

12. $y^2 + 16y + 64$

13. $x^2 + 3x - 28$

14. $x^2 - 11x + 10$

15. $16 + 8x + x^2$

16. $6 + 5x + x^2$

17. $a^2 - 12a + 11$

18. $c^2 - 10c + 21$

19. $x^2 - \dfrac{2}{5}x + \dfrac{1}{25}$

20. $x^2 + \dfrac{2}{3}x + \dfrac{1}{9}$

21. $y^2 - 0.2y - 0.08$

22. $t^2 - 0.3t - 0.10$

ANSWERS
1.
2.
3.
4.
5.
6.
7.
8.
9.
10.
11.
12.
13.
14.
15.
16.
17.
18.
19.
20.
21.
22.

23. $y^2 + 11y + 28$

24. $m^2 + 9m + 14$

25. $30 + 11a + a^2$

26. $4 + 5b + b^2$

27. $x^2 - x - 42$

28. $x^2 + x - 2$

29. $x^2 - 2x - 99$

30. $x^2 - 7x - 60$

31. $6x - 72 + x^2$

32. $c - 56 + c^2$

33. $x^2 + 20x + 100$

34. $y^2 - 21y + 108$

 SKILL MAINTENANCE

Multiply.

35. $8x(2x^2 - 6x + 1)$

36. $(7w + 6)(4w - 1)$

37. Solve: $4x + 9 = 17$.

38. Simplify: $(3x^4)^3$.

⭐ **EXTENSION**

Factor.

39. $x^2 - \dfrac{1}{2}x - \dfrac{3}{16}$

40. $x^2 + \dfrac{30}{7}x - \dfrac{25}{7}$

41. Find all integers m for which $y^2 + my + 50$ can be factored.

42. Find a polynomial in factored form for the area of the shaded region. Leave the answer in terms of π.

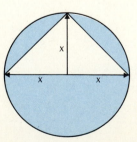

O B J E C T I V E S

After finishing Section 5.5, you should be able to:

• Factor trinomials of the type $ax^2 + bx + c$, $a \neq 1$.

Factoring Trinomials of the Type $ax^2 + bx + c$, $a \neq 1$

• In Section 5.4 we learned to factor trinomials of the type $x^2 + bx + c$. Now we consider trinomials where the leading, or x^2, coefficient is not 1. We consider two methods. You should use the one that works best for you, or you may use the one that your instructor chooses for you. Both methods involve trial and error, but the first requires trial and error in only one step.

METHOD 1

We know how to factor the trinomial $x^2 + 5x + 6$. We look for factors of the constant term, 6, whose sum is the coefficient, 5, of the middle term:

$$x^2 + 5x + 6$$

(1) Factor: $6 = 2 \cdot 3$.
(2) Sum of factors: $2 + 3 = 5$.

What happens when the coefficient of the first, or x^2, term is not 1? Consider the trinomial $6x^2 + 23x + 20$. The method we use is similar to what we used for the preceding trinomial, but we need two more steps. We first multiply the leading coefficient 6 and the constant 20, and get 120. Then we look for a factorization of 120 in which the sum of the factors is the coefficient, 23, of the middle term. Next we split the middle term as a sum or difference using these factors.

$$6x^2 + 23x + 20$$

(1) Multiply 6 and 20: $6 \cdot 20 = 120$.
(2) Factor 120: $120 = 8 \cdot 15$ and $8 + 15 = 23$.
(3) Split the middle term: $23x = 8x + 15x$.
(4) Factor by grouping.

We factor by grouping as follows:

$$6x^2 + 23x + 20 = 6x^2 + 8x + 15x + 20$$
$$= 2x(3x + 4) + 5(3x + 4) \qquad \text{Factoring by grouping;}$$
$$\text{see Section 5.1}$$
$$= (2x + 5)(3x + 4).$$

It does not matter which way we split the middle term, so long as we split it correctly. We still get the same factorization, although the factors may be in a different order. Note the following:

$$6x^2 + 23x + 20 = 6x^2 + 15x + 8x + 20$$
$$= 3x(2x + 5) + 4(2x + 5)$$
$$= (3x + 4)(2x + 5).$$

The method we used to factor trinomials of the type $x^2 + bx + c$ was based on FOIL. The method we have just introduced is also based on FOIL. Before we show why, we state the method more formally, and consider more examples.

Factor.

1. $2x^2 - x - 15$

2. $12x^2 - 17x - 5$

To factor $ax^2 + bx + c$:

a) **First look for a common factor.**

b) **Multiply the leading coefficient** a **and the constant** c**.**

c) **Try to factor the product** ac **so that the sum of the factors is** b**.**

d) **Split the middle term. That is, write it as a sum using the factors found in (c).**

e) **Then factor by grouping.**

Example 1 Factor: $3x^2 - 10x - 8$.

a) First look for a common factor. There is none (other than 1).

b) Multiply the leading coefficient and the constant, 3 and -8:

$$3(-8) = -24.$$

c) Try to factor -24 so that the sum of the factors is -10.

Pairs of Factors	Sums of Factors
4, −6	−2
−4, 6	2
12, −2	10
−12, 2	−10

d) Split $-10x$ using the results of part (c). That is, split the middle term as follows:

$$-10x = -12x + 2x.$$

e) Factor by grouping:

$3x^2 - 10x - 8 = 3x^2 - 12x + 2x - 8$ Substituting $-12x + 2x$ for $-10x$

$$= 3x(x - 4) + 2(x - 4)$$
$$= (3x + 2)(x - 4).$$

DO EXERCISES 1 AND 2.

Example 2 Factor: $8x^2 + 8x - 6$.

a) First look for a common factor. The number 2 is common to all three terms, so we factor it out:

$$2(4x^2 + 4x - 3).$$

b) Now we factor the trinomial $4x^2 + 4x - 3$. Multiply the leading coefficient and the constant, 4 and -3:

$$4(-3) = -12.$$

c) Try to factor -12 so that the sum of the factors is 4.

Pairs of Factors	Sums of Factors
−3, 4	1
3, −4	−1
12, −1	11
−12, 1	−11
−6, 2	−4
6, −2	4

d) We split the middle term $4x$, as follows:

$$4x = 6x - 2x.$$

e) Factor by grouping:

$$4x^2 + 4x - 3 = 4x^2 + 6x - 2x - 3$$
$$= 2x(2x + 3) - (2x + 3)$$
$$= 2x(2x + 3) - 1(2x + 3)$$
$$= (2x - 1)(2x + 3)$$

The factorization of $4x^2 + 4x - 3$ is $(2x - 1)(2x + 3)$. But, don't forget the common factor! We must include it to get a factorization of the original trinomial:

$$8x^2 + 8x - 6 = 2(2x - 1)(2x + 3).$$

This method of factoring is based on FOIL as shown here:

$$(ax + b)(cx + d) = acx^2 + adx + bcx + bd$$
$$= acx^2 + (ad + bc)x + bd$$
$$(ac)(bd) = (ad)(bc).$$

We multiply the outside coefficients and factor the result in such a way that we can split the middle term.

DO EXERCISES 3 AND 4.

METHOD 2

We now consider an alternative method for factoring trinomials of the type $ax^2 + bx + c$. Consider the following multiplication.

$$\overset{\text{F} \quad \text{O} \quad \text{I} \quad \text{L}}{(2x + 5)(3x + 4) = 6x^2 + 8x + 15x + 20}$$
$$= 6x^2 + 23x + 20$$

F	O + I	L
$2 \cdot 3$	$2 \cdot 4 + 5 \cdot 3$	$5 \cdot 4$

Now to factor $6x^2 + 23x + 20$, we do the reverse of what we just did:

$$\text{F} \quad \text{O} + \text{I} \quad \text{L}$$
$$6x^2 + 23x + 20$$
$$= (2x + 5)(3x + 4)$$

Factor.

3. $3x^2 - 19x + 20$

4. $20x^2 - 46x + 24$

ANSWERS ON PAGE A-10

5. $6x^2 + 7x + 2$

We look for numbers (x-coefficients) whose product is 6—in this case, 2 and 3—and numbers whose product is 20—in this case, 4 and 5. The product of the outside terms plus the product of the inside terms must, of course, be 23.

In order to factor $ax^2 + bx + c$, we look for two binomials like this:

$$(\underline{\quad}x + \underline{\quad})(\underline{\quad}x + \underline{\quad}),$$

where products of numbers in the blanks are as follows.

> 1. The numbers in the *first* blanks of each binomial have product a.
> 2. The numbers in the *last* blanks of each binomial have product c.
> 3. The *outside* product and the *inside* product add up to b.

Example 3 Factor: $3x^2 + 5x + 2$.

We look for two numbers whose product is 3. These are

$$1, 3 \quad \text{and} \quad -1, -3.$$

We have these possibilities:

$$(x + \underline{\quad})(3x + \underline{\quad}), \quad \text{or} \quad (-x + \underline{\quad})(-3x + \underline{\quad}).$$

Now we look for numbers whose product is 2. These are

$$1, 2 \quad \text{and} \quad -1, -2.$$

Here are some possibilities for factorizations. There are eight possibilities, but we have not listed all of them here.

$$(x + 1)(3x + 2) \qquad (-x + 1)(-3x + 2) \qquad (x + 2)(3x + 1)$$
$$(x - 1)(3x - 2) \qquad (-x - 1)(-3x - 2) \qquad (x - 2)(3x - 1)$$

When we multiply, we must get $3x^2 + 5x + 2$. When we multiply, we find that both of the color expressions are factorizations. We usually choose the one in which the first coefficients are positive. Thus the factorization is

$$(x + 1)(3x + 2).$$

DO EXERCISE 5.

We always look first for a common factor. If there is one, we remove that common factor before proceeding.

Example 4 Factor: $8x^2 + 8x - 6$.

First look for a factor common to all three terms. The number 2 is a common factor, so we factor it out:

$$2(4x^2 + 4x - 3).$$

Now we factor the trinomial $4x^2 + 4x - 3$. We look for pairs of numbers whose product is 4. These are

$$4, 1 \quad \text{and} \quad 2, 2. \qquad \text{Both positive}$$

We then have these possibilities:

$$(4x + \underline{\qquad})(x + \underline{\qquad}) \quad \text{and} \quad (2x + \underline{\qquad})(2x + \underline{\qquad}).$$

Next we look for pairs of numbers whose product is -3. They are

$$3, -1 \quad \text{and} \quad -3, 1.$$

Then we have these possibilities for factorizations:

$$(4x + 3)(x - 1), \qquad (2x + 3)(2x - 1),$$
$$(4x - 1)(x + 3), \qquad (2x - 3)(2x + 1).$$
$$(4x + 1)(x - 3),$$
$$(4x - 3)(x + 1),$$

We usually do not write all of these. We multiply until we find the factors that give the product $4x^2 + 4x - 3$. We find that the factorization is

$$(2x + 3)(2x - 1).$$

But don't forget the common factor. We must include it to get a factorization of the original polynomial:

$$8x^2 + 8x - 6 = 2(2x + 3)(2x - 1).$$

Keep in mind that no matter which of the two methods you use to factor trinomials of the type $ax^2 + bx + c$, you will use trial and error. This is the way such factoring is done. As you practice you will find that you can make better and better guesses. Don't forget: When factoring any polynomials, always look first for a common factor. Failing to do so is a common error.

DO EXERCISES 6–9.

Factor.

6. $6x^2 + 15x + 9$

7. $2y^2 + 4y - 6$

8. $4t^2 + 2t - 6$

9. $6x^2 - 5x + 1$

ANSWERS ON PAGE A-10

CALCULATOR CORNER: A NUMBER PATTERN

1. Calculate each of the following. Look for a pattern.

$$(20 \cdot 20) - (21 \cdot 19)$$
$$(34 \cdot 34) - (35 \cdot 33)$$
$$(1999 \cdot 1999) - (2000 \cdot 1998)$$
$$(5.8 \times 5.8) - (6.8 \times 4.8)$$
$$[0.7 \times 0.7] - [(1.7)(-0.3)]$$
$$(999 \cdot 999) - (1000)(998)$$

2. Use the pattern to find the following without using a calculator.

$$(3778 \cdot 3778) - (3779 \cdot 3777)$$
$$(14.78 \times 14.78) - (15.78 \times 13.78)$$
$$[0.375 \times 0.375] - [(1.375) \times (-0.625)]$$

3. Find an equation that describes the pattern above. Verify it.

• Factor.

1. $3x^2 + 4x + 1$

2. $6x^2 + 13x + 6$

3. $12x^2 + 28x - 24$

4. $6x^2 + 33x + 15$

5. $2x^2 - x - 1$

6. $15x^2 - 19x + 6$

7. $9x^2 + 18x - 16$

8. $14x^2 + 35x + 14$

9. $15x^2 - 25x - 10$

10. $18x^2 - 3x - 10$

11. $12x^2 + 31x + 20$

12. $15x^2 + 19x - 10$

13. $14x^2 + 19x - 3$

14. $35x^2 + 34x + 8$

15. $9x^4 + 18x^2 + 8$

16. $6 - 13x + 6x^2$

17. $9x^2 - 42x + 49$

18. $15x^4 - 19x^2 + 6$

19. $6x^3 + 4x^2 - 10x$

20. $18x^3 - 21x^2 - 9x$

21. $12a - 5 + 9a^2$

22. $15 + t - 2t^2$

ANSWERS
1.
2.
3.
4.
5.
6.
7.
8.
9.
10.
11.
12.
13.
14.
15.
16.
17.
18.
19.
20.
21.
22.

23. $25x^2 + 40x + 16$

24. $49 - 70x + 25x^2$

25. $7 + 23m + 6m^2$

26. $3w^2 - 15 - 4w$

27. $4x + 6x^2 - 10$

28. $30x^2 - 54 - 24x$

29. $12x^3 + 31x^2 + 20x$

30. $15x^3 + 19x^2 - 10x$

31. $14x^4 + 19x^3 - 3x^2$

32. $70x^4 + 68x^3 + 16x^2$

33. $x^2 + 3x - 7$

34. $x^2 + 11x + 12$

 SKILL MAINTENANCE

35. Divide and simplify: $\dfrac{y^{-12}}{y^{-8}}$.

36. Convert to scientific notation: 967,500,000.

37. Solve: $3(2y + 3) = 21$.

38. The elevation of Death Valley is -280 ft. This is 792 ft less than the elevation of Dallas, Texas. What is the elevation of Dallas?

⭐ **EXTENSION**

Factor.

39. $27x^3 - 63x^2 - 147x + 343$

40. $20x^{2n} + 16x^n + 3$

41. $3x^{6a} - 2x^{3a} - 1$

42. $x^{2n+1} - 2x^{n+1} + x$

Factoring: A General Strategy

After finishing Section 5.6, you should be able to:

• Factor polynomials completely using any of the methods considered thus far in this chapter.

• We now try to put all our factoring strategies together and consider a general strategy for factoring polynomials. Here we will encounter all types of polynomials so that you will have to determine which method to use.

> To factor a polynomial:
>
> (A) Always look first for a common factor.
>
> (B) Then look at the number of terms.
>
> *Two terms:* Determine whether you have a difference of squares. Do not try to factor a sum of squares: $A^2 + B^2$!
>
> *Three terms:* Determine whether the trinomial is a square. If so, you know how to factor. If not, use trial and error.
>
> *Four terms:* Try factoring by grouping.
>
> (C) Always *factor completely.* If a factor with more than one term can still be factored further into polynomials of smaller degree, you should do so. When no factor can be factored further, you have finished.

Example 1 Factor: $10x^3 - 40x$.

A. We look first for a common factor:

$$10x^3 - 40x = 10x(x^2 - 4).$$ Factoring out the largest common factor

B. The factor $x^2 - 4$ has only two terms. It is a difference of squares. We factor it, being careful to include the common factor:

$$10x(x - 2)(x + 2).$$

> We did not forget to include the common factor.

C. Have we factored completely? Yes, because none of the factors with more than one term can be factored further into a polynomial of smaller degree.

Example 2 Factor: $t^4 - 16$.

A. We look for a common factor. There isn't one.

B. There are only two terms. It is a difference of squares: $(t^2)^2 - 4^2$. We factor it:

$$(t^2 + 4)(t^2 - 4).$$

Factor completely.

1. $3m^4 - 3$

2. $x^6 + 8x^3 + 16$

3. $2x^4 + 8x^3 + 6x^2$

4. $3x^3 + 12x^2 - 2x - 8$

5. $8x^3 - 200x$

ANSWERS ON PAGE A-10

We see that one of the factors is still a difference of squares. We factor it:

$$(t^2 + 4)(t - 2)(t + 2).$$

This is a sum of squares. It cannot be factored!

C. We have factored completely because none of the factors with more than one term can be factored further into a polynomial of smaller degree.

Example 3 Factor: $2x^3 + 10x^2 + x + 5$.

A. We look for a common factor. There isn't one.

B. There are four terms. We try factoring by grouping.

$$2x^3 + 10x^2 + x + 5$$
$$= (2x^3 + 10x^2) + (x + 5) \qquad \text{Separating into two binomials}$$
$$= 2x^2(x + 5) + 1(x + 5) \qquad \text{Factoring each binomial}$$
$$= (2x^2 + 1)(x + 5) \qquad \text{Factoring out the common factor, } x + 5$$

C. None of the factors with more than one term can be factored further into a polynomial of smaller degree, so we have factored completely.

Example 4 Factor: $x^5 - 2x^4 - 35x^3$.

A. We look first for a common factor. This time there is one.

$$x^5 - 2x^4 - 35x^3 = x^3(x^2 - 2x - 35)$$

Don't forget to look for a common factor!

B. The factor $x^2 - 2x - 35$ has three terms, but it is not a trinomial square. We factor it using trial and error:

$$x^5 - 2x^4 - 35x^3 = x^3(x^2 - 2x - 35) = x^3(x - 7)(x + 5).$$

Don't forget to include the common factor in your final answer!

C. None of the factors with more than one term can be factored further into a polynomial of smaller degree, so we have factored completely.

Example 5 Factor: $x^4 - 10x^2 + 25$.

A. We look first for a common factor. There isn't one.

B. There are three terms. We see that this is a trinomial square. We factor it:

$$x^4 - 10x^2 + 25 = (x^2)^2 - 2 \cdot 5 \cdot x^2 + 5^2 = (x^2 - 5)^2.$$

C. None of the factors with more than one term can be factored further into a polynomial of smaller degree, so we have factored completely.

DO EXERCISES 1–5.

Factor completely.

1. $2x^2 - 128$

2. $3t^2 - 27$

3. $a^2 + 25 - 10a$

4. $y^2 + 49 + 14y$

5. $2x^2 - 11x + 12$

6. $8y^2 - 18y - 5$

7. $x^3 + 24x^2 + 144x$

8. $x^3 - 18x^2 + 81x$

9. $x^2 + 3x + 2x + 6$

10. $x^2 - 8x - 2x + 25$

11. $24x^2 - 54$

12. $8x^2 - 98$

13. $20x^3 - 4x^2 - 72x$

14. $9x^3 + 12x^2 - 45x$

15. $x^2 + 4$

16. $t^2 + 25$

17. $x^4 + 7x^2 - 3x^2 - 21$

18. $m^4 + 8m^2 + 8m^2 + 64$

19. $x^5 - 14x^4 + 49x^3$

20. $2x^6 + 8x^5 + 8x^4$

ANSWERS
1.
2.
3.
4.
5.
6.
7.
8.
9.
10.
11.
12.
13.
14.
15.
16.
17.
18.
19.
20.

21. $20 - 6x - 2x^2$

22. $45 - 3x - 6x^2$

23. $x^2 + 3x + 1$

24. $x^2 + 5x + 2$

25. $4x^4 - 64$

26. $5x^4 - 80$

27. $1 - y^8$

28. $t^8 - 1$

29. $x^5 - 4x^4 - 3x^3$

30. $x^6 - 2x^5 + 7x^4$

31. $36a^2 - 15a + \dfrac{25}{16}$

32. $\dfrac{1}{81}x^6 - \dfrac{8}{27}x^3 + \dfrac{16}{9}$

33. $a^4 - 2a^2 + 1$

34. $x^4 + 9$

 SKILL MAINTENANCE

35. The population of London, England, is about 7,028,000. This is about 95% of the population of New York City. What is the population of New York City?

36. Add: $\dfrac{7}{5} + \left[-\dfrac{11}{10}\right]$.

37. Subtract: $\dfrac{11}{6} - \left[-\dfrac{11}{18}\right]$.

38. Find the area and the perimeter of a rectangle when l is 12.5 cm and w is 16.4 cm.

⭐ **EXTENSION**

Factor.

39. $12.25x^2 - 7x + 1$

40. $\dfrac{1}{5}x^2 - x + \dfrac{4}{5}$

41. $5x^2 + 13x + 7.2$

42. $x^3 - (x - 3x^2) - 3$

43. $y^2(y - 1) - 2y(y - 1) + (y - 1)$

44. $acx^{m+n} + adx^n + bcx^m + bd$, where a, b, c, and d are constants

O B J E C T I V E S

After finishing Section 5.7, you should be able to:

- Solve equations (already factored) using the principle of zero products.

- Solve certain equations by factoring and then using the principle of zero products.

Solving Equations by Factoring

In this section we introduce a new equation-solving method and use it and factoring to solve certain equations.

· THE PRINCIPLE OF ZERO PRODUCTS

The product of two numbers is 0 if one of the numbers is 0. Furthermore, *if any product is 0, then a factor must be 0.* For example, if $7x = 0$, then we know that $x = 0$. If $x(2x - 9) = 0$, then we know that $x = 0$ or $2x - 9 = 0$. If $(x + 3)(x - 2) = 0$, then we know that $x + 3 = 0$ or $x - 2 = 0$.

Example 1 Solve: $(x + 3)(x - 2) = 0$.

We have a product of 0. This equation will be true when either factor is 0. Hence it is true when

$$x + 3 = 0 \quad \text{or} \quad x - 2 = 0.$$

Here we have two simple equations, which we know how to solve:

$$x = -3 \quad \text{or} \quad x = 2.$$

There are two solutions, -3 and 2.

We have another principle to help in solving equations.

> **THE PRINCIPLE OF ZERO PRODUCTS**
> An equation with 0 on one side and with factors on the other can be solved by finding those numbers that make the factors 0.

Example 2 Solve: $(5x + 1)(x - 7) = 0$. *Caution!* Don't multiply these.

$5x + 1 = 0 \quad \text{or} \quad x - 7 = 0$	Using the principle of zero products
$5x = -1 \quad \text{or} \qquad x = 7$	
$x = -\frac{1}{5} \quad \text{or} \qquad x = 7$	Solving the two equations separately

Check: For $-\frac{1}{5}$:

$$\begin{array}{c|c} (5x + 1)(x - 7) = 0 & \\ \hline (5(-\frac{1}{5}) + 1)(-\frac{1}{5} - 7) & 0 \\ (-1 + 1)(-7\frac{1}{5}) & \\ 0(-7\frac{1}{5}) & \\ \hline 0 & \end{array}$$

For 7:

$$\begin{array}{c|c} (5x + 1)(x - 7) = 0 & \\ \hline (5 \cdot 7 + 1)(7 - 7) & 0 \\ (35 + 1) \cdot 0 & \\ \hline 0 & \end{array}$$

The solutions are $-\frac{1}{5}$ and 7.

Solve using the principle of zero products.

1. $(x - 3)(x + 4) = 0$

2. $(x - 7)(x - 3) = 0$

3. $(4t + 1)(3t - 2) = 0$

Solve.

4. $y(3y - 17) = 0$

Solve.

5. $x^2 - x - 6 = 0$

The "possible solutions" we get by using the principle of zero products are actually always solutions, unless we have made an error in solving. Thus, when we use this principle, a check is not necessary, except to detect errors.

Caution! Do not make the mistake of using this principle when there is not a zero on one side.

DO EXERCISES 1–3.

Example 3 Solve: $x(2x - 9) = 0$

$$x = 0 \quad \text{or} \quad 2x - 9 = 0 \qquad \text{Using the principle of zero products}$$
$$x = 0 \quad \text{or} \qquad 2x = 9$$
$$x = 0 \quad \text{or} \qquad x = \frac{9}{2}$$

When some factors have only one term, you can still use the principle of zero products in the same way.

The solutions are 0 and $\frac{9}{2}$.

DO EXERCISE 4.

•• USING FACTORING TO SOLVE EQUATIONS

For certain equations we can factor and then use the principle of zero products.

Example 4 Solve: $x^2 + 5x + 6 = 0$.

We first factor the polynomial. Then we use the principle of zero products.

$$x^2 + 5x + 6 = 0$$
$$(x + 2)(x + 3) = 0 \qquad \text{Factoring}$$
$$x + 2 = 0 \quad \text{or} \quad x + 3 = 0 \qquad \text{Using the principle of zero products}$$
$$x = -2 \quad \text{or} \qquad x = -3$$

Check:

$x^2 + 5x + 6 = 0$		$x^2 + 5x + 6 = 0$	
$(-2)^2 + 5(-2) + 6$	0	$(-3)^2 + 5(-3) + 6$	0
$4 - 10 + 6$		$9 - 15 + 6$	
$-6 + 6$		$-6 + 6$	
0		0	

The solutions are -2 and -3.

DO EXERCISE 5.

Chapter 5   Polynomials and Factoring

Example 5 Solve: $x^2 - 8x = -16$.

We first add 16 to get 0 on one side.

> You *must* have 0 on one side before you can use the principle of zero products. That is why it is called the principle of *zero* products.

$$x^2 - 8x + 16 = 0 \qquad \text{Adding 16}$$
$$(x - 4)(x - 4) = 0 \qquad \text{Factoring}$$
$$x - 4 = 0 \quad \text{or} \quad x - 4 = 0 \qquad \text{Using the principle of zero products}$$
$$x = 4 \quad \text{or} \qquad x = 4$$

There is only one solution, 4. The check is left to the student.

DO EXERCISES 6 AND 7.

Example 6 Solve: $x^2 + 5x = 0$.

> When some factors have only one term, you can still use the principle of zero products in the same way.

$$x(x + 5) = 0 \qquad \text{Factoring out a common factor}$$
$$x = 0 \quad \text{or} \quad x + 5 = 0 \qquad \text{Using the principle of zero products}$$
$$x = 0 \quad \text{or} \qquad x = -5$$

The solutions are 0 and −5. The check is left to the student.

Example 7 Solve: $4x^2 = 25$.

$$4x^2 = 25$$
$$4x^2 - 25 = 0 \qquad \text{Adding } -25 \text{ to both sides to get 0 on one side}$$
$$(2x - 5)(2x + 5) = 0 \qquad \text{Factoring}$$
$$2x - 5 = 0 \quad \text{or} \quad 2x + 5 = 0$$
$$2x = 5 \quad \text{or} \qquad 2x = -5$$
$$x = \frac{5}{2} \quad \text{or} \qquad x = -\frac{5}{2}$$

The solutions are $\frac{5}{2}$ and $-\frac{5}{2}$. The check is left to the student.

DO EXERCISES 8 AND 9.

Solve.

6. $x^2 - 3x = 28$

7. $x^2 + 9 = 6x$

Solve.

8. $x^2 - 4x = 0$

9. $x^2 = 16$

ANSWERS ON PAGE A-10

CALCULATOR CORNER: NUMBER PATTERNS

1. Calculate each of the following using your calculator. Look for a pattern.

$$1$$
$$1 + 3$$
$$1 + 3 + 5$$
$$1 + 3 + 5 + 7$$
$$1 + 3 + 5 + 7 + 9$$
$$1 + 3 + 5 + 7 + 9 + 11$$

Use the pattern to find each of the following without using your calculator.

$$1 + 3 + 5 + 7 + 9 + 11 + 13$$
$$1 + 3 + 5 + 7 + 9 + 11 + 13 + 15$$

2. Verify each of the following using your calculator.

$$1 = \frac{1 \cdot 2}{2}$$

$$1 + 2 = \frac{2 \cdot 3}{2}$$

$$1 + 2 + 3 = \frac{3 \cdot 4}{2}$$

$$1 + 2 + 3 + 4 = \frac{4 \cdot 5}{2}$$

$$1 + 2 + 3 + 4 + 5 = \frac{5 \cdot 6}{2}$$

Use the pattern to find each of the following without using your calculator.

$$1 + 2 + 3 + 4 + 5 + 6 + 7 + 8$$
$$1 + 2 + 3 + 4 + 5 + 6 + 7 + 8 + 9$$

See if you can find a formula for

$$1 + 2 + 3 + 4 + \cdots + n.$$

• Solve.

1. $(x + 8)(x + 6) = 0$

2. $(x + 3)(x + 2) = 0$

3. $(x - 3)(x + 5) = 0$

4. $(x + 9)(x - 3) = 0$

5. $(x - 12)(x - 11) = 0$

6. $(x - 13)(x - 53) = 0$

7. $y(y - 13) = 0$

8. $x(x - 4) = 0$

9. $0 = x(x + 21)$

10. $0 = y(y + 10)$

11. $(2x + 5)(x + 4) = 0$

12. $(2x + 9)(x + 8) = 0$

13. $(3x - 1)(x + 2) = 0$

14. $(5x + 1)(x - 3) = 0$

15. $2x(3x - 2) = 0$

16. $5x(8x - 9) = 0$

17. $\left(\frac{1}{3}y - \frac{2}{3}\right)\left(\frac{1}{4}y - \frac{3}{2}\right) = 0$

18. $\left(\frac{7}{4}x - \frac{1}{12}\right)\left(\frac{2}{3}x - \frac{12}{11}\right) = 0$

19. $(0.03x - 0.01)(0.05x - 1) = 0$

20. $(0.01x - 0.03)(0.04x - 2) = 0$

1. _____
2. _____
3. _____
4. _____
5. _____
6. _____
7. _____
8. _____
9. _____
10. _____
11. _____
12. _____
13. _____
14. _____
15. _____
16. _____
17. _____
18. _____
19. _____
20. _____

Solve.

21. $x^2 + 6x + 5 = 0$ **22.** $x^2 + 7x + 6 = 0$

23. $x^2 - 5x = 0$ **24.** $4x^2 - 9 = 0$

25. $x^2 + 6x + 9 = 0$ **26.** $3t^2 + t = 2$

27. $6y^2 - 4y - 10 = 0$ **28.** $5x^2 = 6x$

29. $3x^2 = 7x + 20$ **30.** $0 = 2y^2 + 12y + 10$

31. $-5x = -12x^2 + 2$ **32.** $14 = x^2 - 5x$

33. $0 = -3x + 5x^2$ **34.** $x^2 - 5x = 18 + 2x$

35. $x(x - 5) = 14$ **36.** $t(3t + 1) = 0$

37. $64m^2 = 81$ **38.** $100t^2 = 49$

✓ SKILL MAINTENANCE

39. Multiply: $(-9)(16)$.   **40.** Divide: $-24.3 \div 5.4$.

☆ EXTENSION

Solve.

41. $b(b + 9) = 4(5 + 2b)$ **42.** $(t - 3)^2 = 36$

43. $(m - 5)^2 = 2(5 - m)$ **44.** $x^2 - \dfrac{1}{64} = 0$

45. $(0.00005x + 0.1)(0.0097x + 0.5) = 0$ **46.** $\dfrac{27}{25}x^2 = \dfrac{1}{3}$

47. $(x + 3)(4x - 5)(x - 7) = 0$

Solving Problems

- Recall that to solve a problem we can consider the following tips.

After finishing Section 5.8, you should be able to:

- Solve applied problems involving equations that can be solved by factoring.

PROBLEM-SOLVING TIPS

1. *Familiarize* yourself with the situation. If the situation is described in words, as in a textbook, *read carefully*. In any case, think about the situation. Draw a picture whenever it makes sense to do so.
2. *Translate* the problem to an equation. Tell what the letters represent, if it is not clear from the nature of the problem or the translation.
3. *Solve* the equation.
4. *Check* the answer in the original problem.
5. *State* the answer to the problem clearly.

Translate to an equation. Then solve and check.

1. One more than a number times one less than the number is 24.

Example 1 Solve this problem.

One more than a number times one less than that number is 8.

$(x + 1)$ $\cdot$ $(x - 1)$ $= 8$

Translating

Solve:

$$(x + 1)(x - 1) = 8$$
$$x^2 - 1 = 8 \qquad \text{Multiplying}$$
$$x^2 - 1 - 8 = 0 \qquad \text{Adding } -8 \text{ to get 0 on one side}$$
$$x^2 - 9 = 0$$
$$(x - 3)(x + 3) = 0 \qquad \text{Factoring}$$
$$x - 3 = 0 \quad \text{or} \quad x + 3 = 0 \qquad \text{Using the principle of zero products}$$
$$x = 3 \quad \text{or} \qquad x = -3$$

2. Seven less than a number times eight less than the number is 0.

Check for 3: One more than 3 (this is 4) times one less than 3 (this is 2) is 8. Thus, 3 checks.

Check for −3: Left to the student.

There are two such numbers, 3 and −3.

DO EXERCISES 1 AND 2.

ANSWERS ON PAGE A-11

Translate to an equation. Then solve and check.

3. The square of a number minus the number is 20.

4. The width of a rectangle is 2 cm less than the length. The area is 15 cm². Find the length and the width.

Example 2 Solve this problem.

The square of a number minus twice the number is 48.

$$x^2 \qquad - \qquad 2x \qquad = 48 \qquad \text{Translating}$$

$$x^2 - 2x = 48$$
$$x^2 - 2x - 48 = 0 \qquad \text{Adding } -48 \text{ to get 0 on one side}$$
$$(x - 8)(x + 6) = 0$$
$$x - 8 = 0 \quad \text{or} \quad x + 6 = 0 \qquad \text{Using the principle of zero products}$$
$$x = 8 \quad \text{or} \qquad x = -6$$

There are two such numbers, 8 and −6. They both check.

DO EXERCISE 3.

Sometimes it helps to reword a problem before translating.

Example 3 The height of a triangular sail is 7 ft more than the base. The area of the sail is 30 ft². Find the height and the base. (Area is $\frac{1}{2} \cdot$ base $\cdot$ height.)

We first make a drawing.

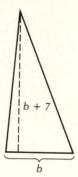

$\frac{1}{2}$ times the base times the base plus 7 is 30   **Rewording**

$$\frac{1}{2} \quad \cdot \quad b \quad \cdot \quad (b + 7) \quad = 30 \qquad \text{Translating}$$

$$\frac{1}{2} \cdot b \cdot (b + 7) = 30$$

$$\frac{1}{2}(b^2 + 7b) = 30 \qquad \text{Multiplying}$$

$$b^2 + 7b = 60 \qquad \text{Multiplying by 2}$$
$$b^2 + 7b - 60 = 0 \qquad \text{Adding } -60 \text{ to get } 0 \text{ on one side}$$
$$(b + 12)(b - 5) = 0 \qquad \text{Factoring}$$
$$b + 12 = 0 \quad \text{or} \quad b - 5 = 0 \qquad \text{Using principle of zero products}$$
$$b = -12 \quad \text{or} \qquad b = 5$$

The solutions of the equation are −12 and 5. The base of a triangle cannot have a negative length. Thus the base is 5 ft. The height is 7 ft more than the base, so the height is 12 ft. These numbers check.

DO EXERCISE 4.

Chapter 5   **Polynomials and Factoring**

```
       SLOW-PITCH STANDINGS
  TEAM                       W   L
  Exponents . . . . . . . . . . . . . . . . . . .  7   0
  The Hogs . . . . . . . . . . . . . . . . . . .  6   1
  Hazardous Waist . . . . . . . . . . . . .  5   2
  No Big Deal . . . . . . . . . . . . . . . . .  2   5
  The King's Kids . . . . . . . . . . . . . .  1   6
  We're the Best . . . . . . . . . . . . . .  0   7
```

Example 4 In a sports league of n teams in which each team plays every other team twice, the total number N of games to be played is given by

$$N = n^2 - n.$$

(a) A slow-pitch softball league has 17 teams and each team plays every other team twice. What is the total number of games to be played? (b) A basketball league plays a total of 90 games and each team plays every other team twice. How many teams are in the league?

a) We substitute 17 for n:

$$N = n^2 - n = 17^2 - 17 = 289 - 17 = 272.$$

b) We substitute 90 for N and solve for n:

$$n^2 - n = 90 \qquad \text{Substituting 90 for } N$$
$$n^2 - n - 90 = 0 \qquad \text{Adding } -90 \text{ to get 0 on one side}$$
$$(n - 10)(n + 9) = 0 \qquad \text{Factoring}$$
$$n - 10 = 0 \quad \text{or} \quad n + 9 = 0 \qquad \text{Principle of zero products}$$
$$n = 10 \quad \text{or} \qquad n = -9$$

Since the number of teams cannot be negative, -9 cannot be a solution. But 10 checks, so there are 10 teams in the league.

DO EXERCISE 5.

5. Use $N = n^2 - n$ for the following.

a) A volleyball league has 19 teams. What is the total number of games to be played?

b) A slow-pitch softball league plays a total of 72 games. How many teams are in the league?

ANSWERS ON PAGE A-11

6. The product of two consecutive integers is 462. Find the integers.

Example 5 The product of two consecutive integers is 156. Find the integers.

(*Consecutive* integers are next to each other, such as 49 and 50 or −6 and −5. The larger is 1 plus the smaller.)

(First integer) times (Second integer) is 156   Rewording

$$x \quad \cdot \quad (x + 1) \quad = 156 \quad \text{Translating}$$

We have let x represent the first integer. Then $x + 1$ represents the second.

We solve:

$$x(x + 1) = 156$$
$$x^2 + x = 156 \qquad \text{Multiplying}$$
$$x^2 + x - 156 = 0 \qquad \text{Adding } -156 \text{ to get 0 on one side}$$
$$(x - 12)(x + 13) = 0 \qquad \text{Factoring}$$
$$x - 12 = 0 \quad \text{or} \quad x + 13 = 0 \qquad \text{Using the principle of zero products}$$
$$x = 12 \quad \text{or} \qquad x = -13$$

The solutions of the equation are 12 and −13. When x is 12, then $x + 1$ is 13, and $12 \cdot 13 = 156$. So 12 and 13 are solutions to the problem. When x is −13, then $x + 1$ is −12, and $(-13)(-12) = 156$. Thus in this problem we have two pairs of solutions: 12 and 13, and −13 and −12. Both are pairs of consecutive integers whose product is 156.

DO EXERCISE 6.

ANSWERS ON PAGE A-11

Chapter 5   Polynomials and Factoring

Solve.

1. If you subtract a number from four times its square, the result is three. Find the number.

2. Eight more than the square of a number is six times the number. Find the number.

3. The product of two consecutive integers is 182. Find the integers.

4. The product of two consecutive even integers is 168. Find the integers.

5. The product of two consecutive odd integers is 195. Find the integers.

6. The length of a rectangle is 4 m greater than the width. The area of the rectangle is 96 m². Find the length and the width.

7. _____

8. _____

9. _____

10. _____

11. _____

12. _____

7. The area of a square is 5 more than the perimeter. Find the length of a side.

8. The base of a triangle is 10 cm greater than the height. The area is 28 cm^2. Find the height and the base.

9. If the sides of a square are lengthened by 3 m, the area becomes 81 m^2. Find the length of a side of the original square.

10. The sum of the squares of two consecutive odd positive integers is 74. Find the integers.

Use $N = n^2 - n$ for Exercises 11–14.

11. A slow-pitch softball league has 23 teams. What is the total number of games to be played?

12. A basketball league has 14 teams. What is the total number of games to be played?

13. A slow-pitch softball league plays a total of 132 games. How many teams are in the league?

14. A basketball league plays a total of 240 games. How many teams are in the league?

The number N of possible doubles teams that can be chosen from a group of n people at a tournament is given by

$$N = \frac{1}{2}n(n - 1).$$

15. At a tournament there are 16 people. How many doubles teams are possible?

16. At a tournament there are 100 people. How many doubles teams are possible?

17. At a tournament there were 190 possible doubles teams. How many people were at the tournament?

18. At a tournament there were 300 possible doubles teams. How many people were at the tournament?

13. _____

14. _____

15. _____

16. _____

17. _____

18. _____

SKILL MAINTENANCE

19. Solve for h: $S = 2\pi rh$.

20. Solve: $4(5 - t) = 6(7 + 3t)$.

21. Add: $-67.3 + (-32.8)$.

22. Find the absolute value: $|-19.4|$.

⭐ **EXTENSION**

23. A number is less than 100. The ones digit of the number is four greater than the tens digit. The sum of the number and the product of the digits is 58. Find the number.

24. A rectangular piece of cardboard is twice as long as it is wide. A 4-cm square is cut out of each corner, and the sides are turned up to make a box. The volume of the box is 616 cm^3. Find the original dimensions of the cardboard.

25. An open rectangular gutter is made by turning up the sides of a piece of metal 20 in. wide. The area of the cross-section of the gutter is 50 in^2. Find the depth of the gutter.

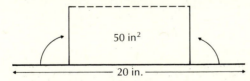

26. The length of each side of a square is increased by 5 cm to form a new square. The area of the new square is $2\frac{1}{4}$ times the area of the original square. Find the area of each square.

5.9

Polynomials in Several Variables

Most of the polynomials you have studied so far have had only one variable. A *polynomial in several variables* is an expression like those you have already seen, but we allow that there can be more than one variable. Here are some examples:

$$3x + xy^2 + 5y + 4, \qquad 8xy^2z - 2x^3z - 13x^4y^2 + 5.$$

• EVALUATING POLYNOMIALS

Example 1 Evaluate the polynomial

$$4 + 3x + xy^2 + 8x^3y^3$$

for $x = -2$ and $y = 5$.

We replace x by -2 and y by 5:

$$4 + 3(-2) + (-2) \cdot 5^2 + 8(-2)^3 \cdot 5^3 = 4 - 6 - 50 - 8000 = -8052.$$

DO EXERCISES 1 AND 2.

Example 2 (*The magic number*). The Boston Red Sox are leading the New York Yankees for the Eastern Division championship of the American League. The magic number is 8. This means that any combination of Red Sox wins and Yankee losses that totals 8 will ensure the championship for the Red Sox. The magic number is given by the polynomial

$$G - P - L + 1,$$

where G is the number of games in the season, P is the number of games the leading team has played, and L is the number of games ahead in the loss column.

Given the situation shown in the table and assuming a 162-game season, what is the magic number for the Philadelphia Phillies?

EASTERN DIVISION			
	W	L	Pct.	GB
Philadelphia	77	40	.658	—
Pittsburgh	65	53	.551	$12\frac{1}{2}$
New York	61	60	.504	18
Chicago	55	67	.451	$24\frac{1}{2}$
St. Louis	51	65	.440	$25\frac{1}{2}$
Montreal	41	73	.360	$34\frac{1}{2}$

We evaluate the polynomial for $G = 162$, $P = 77 + 40$, or 117, and $L = 53 - 40$, or 13:

$$162 - 117 - 13 + 1 = 33.$$

DO EXERCISE 3.

OBJECTIVES

After finishing Section 5.9, you should be able to:

• Evaluate a polynomial in several variables for given values of the variables.

•• Identify the coefficients and degrees of the terms of a polynomial and degrees of polynomials.

••• Collect like terms of a polynomial.

⦂⦂ Add, subtract, and multiply polynomials.

⦂⦂⦂ Factor polynomials.

1. Evaluate the polynomial

$$4 + 3x + xy^2 + 8x^3y^3$$

for $x = 2$ and $y = -5$.

2. Evaluate the polynomial

$$8xy^2 - 2x^3z - 13x^4y^2 + 5$$

for $x = -1$, $y = 3$, and $z = 4$.

3. Given the situation below, what is the magic number for the Cincinnati Reds? Assume $G = 162$.

WESTERN DIVISION			
	W	L	Pct.	GB
Cincinnati	77	44	.636	—
Los Angeles	65	54	.546	11
San Diego	60	64	.484	$18\frac{1}{2}$
Houston	59	64	.480	19
Atlanta	56	65	.463	21
San Francisco	52	70	.426	$25\frac{1}{2}$

ANSWERS ON PAGE A-11

Identify the coefficient of each term.

4. $-3xy^2 + 3x^2y - 2y^3 + xy + 2$

Identify the degree of each term and the degree of the polynomial.

5. $4xy^2 + 7x^2y^3z^2 - 5x + 2y + 4$

Collect like terms.

6. $4x^2y + 3xy - 2x^2y$

7. $-3pq - 5pqr^3 + 8pq + 5pqr^3 + 4$

⠆ COEFFICIENTS AND DEGREES

The *degree* of a term is the sum of the exponents of the variables. The *degree of a polynomial* is the degree of the term of highest degree.

Example 3 Identify the coefficient and degree of each term of

$$9x^2y^3 - 14xy^2z^3 + xy + 4y + 5x^2 + 7.$$

Term	Coefficient	Degree	
$9x^2y^3$	9	5	
$-14xy^2z^3$	-14	6	
xy	1	2	
$4y$	4	1	Think: $4y = 4y^1$
$5x^2$	5	2	
7	7	0	Think: $7 = 7x^0$

Example 4 What is the degree of $5x^3y + 9xy^4 - 8x^3y^3$?

The term of highest degree is $-8x^3y^3$. Its degree is 6. The degree of the polynomial is 6.

DO EXERCISES 4 AND 5.

⠒ COLLECTING LIKE TERMS

Like terms (or *similar terms*) have exactly the same variables with exactly the same exponents. For example,

$3x^2y^3$ and $-7x^2y^3$ are like terms;

$9x^4z^7$ and $12x^4z^7$ are like terms.

But

$13xy^2$ and $-2x^2y$ are *not* like terms;

$3xyz^2$ and $4xy$ are *not* like terms.

Collecting like terms is based on the distributive law.

Examples Collect like terms.

5. $5x^2y + 3xy^2 - 5x^2y - xy^2 = (5 - 5)x^2y + (3 - 1)xy^2 = 2xy^2$

6. $3xy - 5xy^2 + 3xy^2 + 9xy = -2xy^2 + 12xy$

7. $4ab^2 - 7a^2b^2 + 9a^2b^2 - 4a^2b = 4ab^2 + 2a^2b^2 - 4a^2b$

8. $3pq + 5pqr^3 - 8pq - 5pqr^3 - 4 = -5pq - 4$

DO EXERCISES 6 AND 7.

⠿ ADDITION, SUBTRACTION, AND MULTIPLICATION

Addition

To add polynomials in several variables, we collect like terms.

Example 9 Add: $(5xy^2 - 4x^2y + 5x^3 + 2) + (3xy^2 - 2x^2y + 3x^3y - 5)$.

We look for like terms. The like terms are $5xy^2$ and $3xy^2$, $-4x^2y$ and $-2x^2y$, and 2 and -5. We collect these. There are no more like terms. The answer is

$$8xy^2 - 6x^2y + 5x^3 + 3x^3y - 3.$$

DO EXERCISES 8–10.

Subtraction

We subtract a polynomial by adding its additive inverse. An equivalent expression for the additive inverse of a polynomial is found by replacing each coefficient by its additive inverse, or by changing the sign of every term. For example, the additive inverse of the polynomial

$$4x^2y - 6x^3y^2 + x^2y^2 - 5y$$

can be represented by

$$-(4x^2y - 6x^3y^2 + x^2y^2 - 5y).$$

An equivalent expression can be found by replacing each coefficient by its additive inverse. Thus,

$$-(4x^2y - 6x^3y^2 + x^2y^2 - 5y) = -4x^2y + 6x^3y^2 - x^2y^2 + 5y.$$

Example 10 Subtract.

$(4x^2y + x^3y^2 + 3x^2y^3 + 6y) - (4x^2y - 6x^3y^2 + x^2y^2 - 5y)$

$= 4x^2y + x^3y^2 + 3x^2y^3 + 6y - 4x^2y + 6x^3y^2 - x^2y^2 + 5y$

                                        Adding the inverse

$= 7x^3y^2 + 3x^2y^3 - x^2y^2 + 11y$ **Collecting like terms (Try to write just the answer!)**

DO EXERCISES 11 AND 12.

Multiplication

To multiply polynomials in several variables we can multiply each term of one by every term of the other. Where appropriate, we use special products.

Example 11 Multiply: $(3x^2y - 2xy + 3y)(xy + 2y)$.

$$
\begin{array}{l}
3x^2y - 2xy \ + 3y \\
\underline{xy + 2y} \\
3x^3y^2 - 2x^2y^2 + 3xy^2 \qquad \text{Multiplying by } xy \\
\underline{\qquad\quad 6x^2y^2 - 4xy^2 + 6y^2} \quad \text{Multiplying by } 2y \\
3x^3y^2 + 4x^2y^2 - \ xy^2 + 6y^2 \qquad \text{Adding}
\end{array}
$$

DO EXERCISES 13 AND 14.

Add.

8. $4x^3 + 4x^2 - 8x - 3$ and
$-8x^3 - 2x^2 + 4x + 5$

9. $(13x^3y + 3x^2y - 5y)$
$+ (x^3y + 4x^2y - 3xy + 3y)$

10. $(-5p^2q^4 + 2p^2q^2 + 3q)$
$+ (6pq^2 + 3p^2q + 5)$

Subtract.

11. $(-4s^4t + s^3t^2 + 2s^2t^3)$
$- (4s^4t - 5s^3t^2 + s^2t^2)$

12. $(-5p^4q + 5p^3q^2 - 3p^2q^3 - 7q^4)$
$- (4p^4q - 5p^3q^2 + p^2q^3 + 2q^4)$

Multiply.

13. $(x^2y^3 + 2x)(x^3y^2 + 3x)$

14. $(p^4q - 2p^3q^2 + 3q^3)(p + 2q)$

ANSWERS ON PAGE A-11

Multiply.

15. $(3xy + 2x)(x^2 + 2xy^2)$

16. $(x - 3y)(2x - 5y)$

17. $(4x + 5y)^2$

18. $(3x^2 - 2xy^2)^2$

19. $(2xy^2 + 3x)(2xy^2 - 3x)$

20. $(3xy^2 + 4y)(-3xy^2 + 4y)$

21. $(3y + 4 - 3x)(3y + 4 + 3x)$

22. $(2a + 5b + c)(2a - 5b - c)$

Examples Multiply.

$$\qquad\qquad\qquad\quad \text{F} \qquad \text{O} \qquad \text{I} \qquad \text{L}$$

12. $(x^2y + 2x)(xy^2 + y^2) = x^3y^3 + x^2y^3 + 2x^2y^2 + 2xy^2$

13. $(p + 5q)(2p - 3q) = 2p^2 - 3pq + 10pq - 15q^2$
$$= 2p^2 + 7pq - 15q^2$$

$$(A + B)^2 = A^2 + 2\ A\ B + B^2$$

14. $(3x + 2y)^2 = (3x)^2 + 2(3x)(2y) + (2y)^2$
$$= 9x^2 + 12xy + 4y^2$$

15. $(2y^2 - 5x^2y)^2 = (2y^2)^2 - 2(2y^2)(5x^2y) + (5x^2y)^2$
$$= 4y^4 - 20x^2y^3 + 25x^4y^2$$

$$(A + B)\ (A - B) = A^2 - B^2$$

16. $(3x^2y + 2y)(3x^2y - 2y) = (3x^2y)^2 - (2y)^2$
$$= 9x^4y^2 - 4y^2$$

17. $(-2x^3y^2 + 5t)(2x^3y^2 + 5t) = (5t - 2x^3y^2)(5t + 2x^3y^2)$
$$= (5t)^2 - (2x^3y^2)^2 = 25t^2 - 4x^6y^4$$

$$(A - B)\ (A + B) = A^2 - B^2$$

18. $(2x + 3 - 2y)(2x + 3 + 2y) = (2x + 3)^2 - (2y)^2$
$$= 4x^2 + 12x + 9 - 4y^2$$

DO EXERCISES 15–22.

:•: FACTORING

To factor polynomials in several variables, we can use the same general strategy that we considered in Section 5.6. You might review that before studying the following examples.

Example 19 Factor: $20x^3y + 12x^2y$.

A. We look first for a common factor:

$$20x^3y + 12x^2y = (4x^2y)(5x) + (4x^2y) \cdot 3$$
$$= 4x^2y(5x + 3). \qquad \text{Factoring out the greatest common factor}$$

B. Then look at the number of terms. There are only two terms, but the binomial $5x + 3$ is not a difference of squares. It cannot be factored further.

C. We have factored completely because no factors with more than one term can be factored further.

Example 20 Factor: $6x^2y - 21x^3y^2 + 3x^2y^3$.

A. We look first for a common factor:

$$6x^2y - 21x^3y^2 + 3x^2y^3 = 3x^2y(2 - 7xy + y^2).$$

B. There are three terms in $2 - 7xy + y^2$. Determine whether the trinomial is a square. Since only y^2 is a square, we do not have a trinomial square. Can the trinomial be factored by trial? The fact that x is only in the term $-7xy$ is a key to the answer. The polynomial might be in a form like $(1 - y)(2 + y)$, but then there would be no x in the middle term.

C. Have we factored completely? Yes, because no factor with more than one term can be factored further.

DO EXERCISES 23 AND 24.

Example 21 Factor: $(p + q)(x + 2) + (p + q)(x + y)$.

A. We look first for a common factor:

$$(p + q)(x + 2) + (p + q)(x + y) = (p + q)[(x + 2) + (x + y)]$$
$$= (p + q)(2x + y + 2).$$

B. There are three terms in $2x + y + 2$, but this trinomial cannot be factored further.

C. No factor with more than one term can be factored further, so we have factored completely.

Example 22 Factor: $px + py + qx + qy$.

A. We look first for a common factor. There isn't one.

B. There are four terms. We try factoring by grouping:

$$px + py + qx + qy = p(x + y) + q(x + y)$$
$$= (p + q)(x + y).$$

C. Have we factored completely? Since no factor with more than one term can be factored further, we have factored completely.

DO EXERCISES 25 AND 26.

Example 23 Factor: $25x^2 + 20xy + 4y^2$.

A. We look first for a common factor. There isn't one.

B. There are three terms. Determine whether the trinomial is a square. The first term and the last term are squares:

$$25x^2 = (5x)^2 \quad \text{and} \quad 4y^2 = (2y)^2.$$

Twice the product of $5x$ and $2y$ should be the other term:

$$2 \cdot 5x \cdot 2y = 20xy.$$

Factor.

23. $x^4y^2 + 2x^3y + 3x^2y$

24. $10p^6q^2 - 4p^5q^3 + 2p^4q^4$

Factor.

25. $(a - b)(x + 5) + (a - b)(x + y^2)$

26. $ax^2 + ay + bx^2 + by$

ANSWERS ON PAGE A-11

Factor.

27. $x^4 + 2x^2y^2 + y^4$

28. $-4x^2 + 12xy - 9y^2$
(*Hint:* First factor out -1.)

Factor.

29. $x^2y^2 + 5xy + 4$

30. $2x^4y^6 + 6x^2y^3 - 20$

31. $25x^2y^4 - 4a^2$

Thus the trinomial is a perfect square. We factor by writing the square roots of the square terms and the sign of the other term:

$$25x^2 + 20xy + 4y^2 = (5x + 2y)^2.$$

We can check by squaring $5x + 2y$.

C. No factor with more than one term can be factored further, so we have factored completely.

DO EXERCISES 27 AND 28.

Example 24 Factor: $p^2q^2 + 7pq + 12$.

A. We look first for a common factor. There isn't one.

B. There are three terms. Determine whether the trinomial is a square. The first term is a square, but since neither of the other terms is a square, we do not have a trinomial square. We use trial and error, treating pq as though it were a single variable:

$$p^2q^2 + 7pq + 12 = (pq)^2 + (3 + 4)pq + 3 \cdot 4$$
$$= (pq + 3)(pq + 4).$$

C. No factor with more than one term can be factored further, so we have factored completely.

Example 25 Factor: $8x^4 - 20x^2y - 12y^2$.

A. We look first for a common factor:

$$8x^4 - 20x^2y - 12y^2 = 4(2x^4 - 5x^2y - 3y^2).$$

B. There are three terms in $2x^4 - 5x^2y - 3y^2$. Determine whether the trinomial is a square. Since none of the terms is a square, we do not have a trinomial square. We use trial and error:

$$8x^4 - 20x^2y - 12y^2 = 4(2x^4 - 5x^2y - 3y^2)$$
$$= 4[(2x^2)(x^2) + (-6 + 1)x^2y + (-3y)y]$$
$$= 4(2x^2 + y)(x^2 - 3y).$$

C. No factor with more than one term can be factored further, so we have factored completely.

Example 26 Factor: $a^4 - 16b^4$.

$$a^4 - 16b^4 = (a^2 - 4b^2)(a^2 + 4b^2)$$
$$= (a - 2b)(a + 2b)(a^2 + 4b^2)$$

DO EXERCISES 29–31.

• Evaluate each polynomial for $x = 3$ and $y = -2$.

1. $x^2 - y^2 + xy$ **2.** $x^2 + y^2 - xy$

Evaluate each polynomial for $x = 2$, $y = -3$, and $z = -1$.

3. $xyz^2 + z$ **4.** $xy - xz + yz$

An amount of money P is invested at interest rate r. In 3 years it will grow to an amount given by the polynomial

$$P + 3rP + 3r^2P + r^3P.$$

5. Evaluate the polynomial for $P = 10,000$ and $r = 0.08$ to find the amount to which $10,000 will grow at 8% interest for 3 years.

6. Evaluate the polynomial for $P = 10,000$ and $r = 0.07$ to find the amount to which $10,000 will grow at 7% interest for 3 years.

The area of a right circular cylinder is given by the polynomial

$$2\pi rh + 2\pi r^2,$$

where h is the height and r is the radius of the base.

7. A 12-oz beverage can has height 4.7 in. and radius 1.2 in. Evaluate the polynomial for $h = 4.7$ and $r = 1.2$ to find the area of the can. Use 3.14 for π.

8. A 16-oz beverage can has height 6.3 in. and radius 1.2 in. Evaluate the polynomial for $h = 6.3$ and $r = 1.2$ to find the area of the can. Use 3.14 for π.

•• Identify the coefficient and degree of each term of the following polynomials. Then find the degree of the polynomial.

9. $x^3y - 2xy + 3x^2 - 5$ **10.** $5y^3 - y^2 + 15y + 1$

11. $17x^2y^3 - 3x^3yz - 7$ **12.** $6 - xy + 8x^2y^2 - y^5$

••• Collect like terms.

13. $a + b - 2a - 3b$ **14.** $y^2 - 1 + y - 6 - y^2$

15. $3x^2y - 2xy^2 + x^2$ **16.** $m^3 + 2m^2n - 3m^2 + 3mn^2$

ANSWERS

1. _____

2. _____

3. _____

4. _____

5. _____

6. _____

7. _____

8. _____

9. _____

10. _____

11. _____

12. _____

13. _____

14. _____

15. _____

16. _____

17. $2u^2v - 3uv^2 + 6u^2v - 2uv^2$

18. $3x^2 + 6xy + 3y^2 - 5x^2 - 10xy - 5y^2$

19. $6au + 3av - 14au + 7av$

20. $3x^2y - 2z^2y + 3xy^2 + 5z^2y$

⁘ Perform the indicated operations.

Add.

21. $(2x^2 - xy + y^2) + (-x^2 - 3xy + 2y^2)$

22. $(2z - z^2 + 5) + (z^2 - 3z + 1)$

23. $(r - 2s + 3) + (2r + s) + (s + 4)$

24. $(b^3a^2 - 2b^2a^3 + 3ba + 4) + (b^2a^3 - 4b^3a^2 + 2ba - 1)$

25. $(2x^2 - 3xy + y^2) + (-4x^2 - 6xy - y^2) + (x^2 + xy - y^2)$

Subtract.

26. $(x^3 - y^3) - (-2x^3 + x^2y - xy^2 + 2y^3)$

27. $(xy - ab) - (xy - 3ab)$

28. $(3y^4x^2 + 2y^3x - 3y) - (2y^4x^2 + 2y^3x - 4y - 2x)$

29. $(-2a + 7b - c) - (-3b + 4c - 8d)$

30. Find the sum of $2a + b$ and $3a - 4b$. Then subtract $5a + 2b$.

Multiply.

31. $(3z - u)(2z + 3u)$

32. $(a - b)(a^2 + b^2 + 2ab)$

33. $(a^2b - 2)(a^2b - 5)$

34. $(xy + 7)(xy - 4)$

35. $(a^2 + a - 1)(a^2 - y + 1)$

36. $(tx + r)(vx + s)$

37. $(a^3 + bc)(a^3 - bc)$

38. $(m^2 + n^2 - mn)(m^2 + mn + n^2)$

39. $(y^4x + y^2 + 1)(y^2 + 1)$

40. $(a - b)(a^2 + ab + b^2)$

41. $(3xy - 1)(4xy + 2)$

42. $(m^3n + 8)(m^3n - 6)$

43. $(3 - c^2d^2)(4 + c^2d^2)$

44. $(6x - 2y)(5x - 3y)$

45. $(m^2 - n^2)(m + n)$

46. $(pq + 0.2)(0.4pq - 0.1)$

47. $(x^5y^5 + xy)(x^4y^4 - xy)$

48. $(x - y^3)(x + 2y^3)$

49. $(x + h)^2$

50. $(3a + 2b)^2$

51. $(r^3t^2 - 4)^2$

52. $(3a^2b - b^2)^2$

53. $(p^4 + m^2n^2)^2$

54. $(ab + cd)^2$

55. $(2a^3 - \frac{1}{2}b^3)^2$

56. $-5x(x + 3y)^2$

57. $3a(a - 2b)^2$

58. $(a^2 + b + 2)^2$

59. $(2a - b)(2a + b)$

60. $(x - y)(x + y)$

61. $(c^2 - d)(c^2 + d)$

62. $(p^3 - 5q)(p^3 + 5q)$

63. $(ab + cd^2)(ab - cd^2)$

64. $(xy + pq)(xy - pq)$

Factor.

65. $2\pi rh + 2\pi r^2$

66. $10p^4q^4 + 35p^3q^3 + 10p^2q^2$

67. $(a + b)(x - 3) + (a + b)(x + 4)$

68. $5c(a^3 + b) - (a^3 + b)$

69. $(x - 1)(x + 1) - y(x + 1)$

70. $x^2 + x + xy + y$

71. $n^2 + 2n + np + 2p$

72. $a^2 - 3a + ay - 3y$

73. $2x^2 - 4x + xz - 2z$

74. $6y^2 - 3y + 2py - p$

75. $x^2 - 2xy + y^2$

76. $a^2 - 4ab + 4b^2$

77. $9c^2 + 6cd + d^2$

78. $16x^2 + 24xy + 9y^2$

79. $49m^4 - 112m^2n + 64n^2$

80. $4x^2y^2 + 12xyz + 9z^2$

81. $y^4 + 10y^2z^2 + 25z^4$

82. $0.01x^4 - 0.1x^2y^2 + 0.25y^4$

83. $\frac{1}{4}a^2 + \frac{1}{3}ab + \frac{1}{9}b^2$

84. $4p^2q + pq^2 + 4p^3$

85. $a^2 - ab - 2b^2$

86. $3b^2 - 17ab - 6a^2$

Chapter 5   Polynomials and Factoring

87. $m^2 + 2mn - 360n^2$

88. $x^2y^2 + 8xy + 15$

89. $m^2n^2 - 4mn - 32$

90. $p^2q^2 + 7pq + 6$

91. $a^5b^2 + 3a^4b - 10a^3$

92. $m^2n^6 + 4mn^5 - 32n^4$

93. $a^5 + 4a^4b - 5a^3b^2$

94. $2s^6t^2 + 10s^3t^3 + 12t^4$

95. $x^6 + x^3y - 2y^2$

96. $a^4 + a^2bc - 2b^2c^2$

97. $x^2 - y^2$

98. $a^2 - h^2$

99. $a^2b^2 - 9$

100. $p^2q^2 - r^2$

101. $9x^4y^2 - b^2$

102. $36t^2 - 49p^2q^2$

103. $3x^2 - 48y^2$

104. $9s^4 - 9s^2$

105. $64z^2 - 25c^2d^2$

106. $5t^2 - 20m^2$

107. $7p^4 - 7q^4$

108. $a^4b^4 - 16$

ANSWERS
87.
88.
89.
90.
91.
92.
93.
94.
95.
96.
97.
98.
99.
100.
101.
102.
103.
104.
105.
106.
107.
108.

109. $81a^4 - b^4$

110. $1 - 16x^{12}y^{12}$

111. $18m^4 + 12m^3 + 2m^2$

112. $75a^3 + 60a^2b + 12ab^2$

113. $xy^2 + 3y^2 - 4x - 12$

114. $ay^2 - a - y^2 + 1$

115. $p^3 - p^2t - 2pt^2$

116. $15x^2y - 20xy - 35y$

117. $-a^2 - ab + 6b^2$
(*Hint:* Factor out -1.)

118. $r^2 + 6rs + 9s^2$

119. $ab^3 - ab^2 - ab$

120. $x^4 + x^3 + x^2$

⭐ **EXTENSION**

Find a polynomial for the area of each shaded region. (Leave answers in terms of π where appropriate). Find an equivalent factored expression.

121.

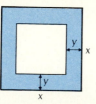

122.

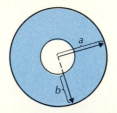

123.

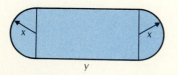

124.

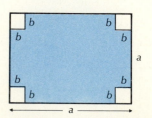

125. Find a formula for $(A + B)^3$.

Summary and Review

The following contains a summary of what you should be able to do after completing this chapter. The review exercises are for practice. Answers are at the back of the book. If you miss an exercise, restudy the section and objective indicated alongside the answer.

The review sections to be tested in addition to the material in this chapter are 2.6, 3.4, 3.6, and 4.7.

You should be able to:

Factor monomials.

Find three factorizations of each monomial.

1. $-10x^2$

2. $36x^5$

Factor polynomials when the terms have a common factor; factor differences of two squares, trinomial squares, trinomials of the type $x^2 + bx + c$, trinomials of the type $ax^2 + bx + c$, $a \neq 1$, and polynomials with four terms by grouping. Apply the general strategy for factoring.

Factor.

3. $5 - 20x^6$

4. $x^2 - 3x$

5. $9x^2 - 4$

6. $x^2 + 4x - 12$

7. $x^2 + 14x + 49$

8. $6x^3 + 12x^2 + 3x$

9. $x^3 + x^2 + 3x + 3$

10. $6x^2 - 5x + 1$

11. $x^4 - 81$

12. $9x^3 + 12x^2 - 45x$

13. $2x^2 - 50$

14. $x^4 + 4x^3 - 2x - 8$

15. $16x^4 - 1$

16. $8x^6 - 32x^5 + 4x^4$

17. $75 + 12x^2 + 60x$

18. $x^2 + 9$

19. $x^3 - x^2 - 30x$

20. $4x^2 - 25$

21. $9x^2 + 25 - 30x$

22. $6x^2 - 28x - 48$

23. $x^2 - 6x + 9$

24. $2x^2 - 7x - 4$

25. $18x^2 - 12x + 2$

26. $3x^2 - 27$

27. $15 - 8x + x^2$

28. $25x^2 - 20x + 4$

Solve equations by factoring and then using the principle of zero products.

Solve.

29. $(x - 1)(x + 3) = 0$

30. $x^2 + 2x - 35 = 0$

31. $x^2 + x - 12 = 0$

32. $3x^2 + 2 = 5x$

33. $2x^2 + 5x = 12$

34. $16 = x(x - 6)$

Solve problems involving equations that can be factored.

Solve.

35. The square of a number is six more than the number. Find the number.

36. The product of two consecutive even integers is 288. Find the integers.

37. The product of two consecutive odd integers is 323. Find the integers.

38. Twice the square of a number is 10 more than the number. Find the number.

Evaluate a polynomial in several variables for given values of the variables, and identify the coefficients and the degrees of the terms and the degree of the polynomial. Also collect like terms of a polynomial in several variables.

39. Evaluate the polynomial $2 - 5xy + y^2 - 4xy^3 + x^6$ for $x = -1$ and $y = 2$.

Identify the coefficient and degree of each term of the following polynomials. Then find the degree of the polynomial.

40. $x^5y - 7xy + 9x^2 - 8$

41. $x^2y^5z^9 - y^{40} + x^{13}z^{10}$

Collect like terms.

42. $y + w - 2y + 8w - 5$

43. $m^6 - 2m^2n + m^2n^2 + n^2m - 6m^3 + m^2n^2 + 7n^2m$

Add, subtract, multiply, and factor polynomials in several variables.

44. Add: $(5x^2 - 7xy + y^2) + (-6x^2 - 3xy - y^2) + (x^2 + xy - 2y^2)$.

45. Subtract: $(6x^3y^2 - 4x^2y - 6x) - (-5x^3y^2 + 4x^2y + 6x^2 - 6)$.

Multiply.

46. $(p - q)(p^2 + pq + q^2)$

47. $\left(3a^4 - \dfrac{1}{3}b^3\right)^2$

Factor.

48. $x^2y^2 + xy - 12$

49. $12a^2 + 84ab + 147b^2$

50. $m^2 + 5m + mt + 5t$

51. $32x^4 - 128y^4z^4$

SKILL MAINTENANCE

Multiply.

52. -5.9×4.7

53. $-\dfrac{5}{2} \cdot \left[-\dfrac{4}{7}\right]$

Divide.

54. $-\dfrac{12}{25} \div \left[-\dfrac{21}{10}\right]$

55. $-\dfrac{144}{12}$

Solve.

56. $60 = 5(5x + 2)$

57. $20 - (3x + 2) = 2(x + 5) + x$

58. $A = a + 2b$ for b

59. $S = \dfrac{1}{2}gt^2$ for g

Multiply.

60. $(2a + 3)^2$

61. $(2a - 3)(2a + 3)$

62. $(2a - 3)(5a + 7)$

EXTENSION

Solve.

63. $(x - 2)(x + 3)(2x - 5) = 0$

64. $(y + 4)^2 = 49$

Factor.

65. $y^2(y + 1) - 4y(y + 1) + 4(y + 1)$

66. $x^2 - 2.25$

67. $x^2 - \dfrac{1}{4}x - \dfrac{1}{8}$

68. $x^5 + 2x^4 + 2x + 1$

69. $x^{2n} - y^{2m}$

70. The cube of a number is the same as twice the square of a number. Find the number.

1. Find three factorizations of $4x^3$.

Factor.

2. $x^2 - 7x + 10$

3. $x^2 + 25 - 10x$

4. $6y^2 - 8y^3 + 4y^4$

5. $x^3 + x^2 + 2x + 2$

6. $x^2 - 5x$

7. $x^3 + 2x^2 - 3x$

8. $28x - 48 + 10x^2$

9. $4x^2 - 9$

10. $x^2 - x - 12$

11. $6m^3 + 9m^2 + 3m$

12. $3w^2 - 75$

13. $60x + 45x^2 + 20$

14. $3x^4 - 48$

15. $49x^2 - 84x + 36$

16. $5x^2 - 26x + 5$

17. $x^4 + 2x^3 - 3x - 6$

18. $80 - 5x^4$

19. $4x^2 - 4x - 15$

20. $6t^3 + 9t^2 - 15t$

Solve.

21. $x^2 - x - 20 = 0$

22. $2x^2 + 7x = 15$

23. $x(x - 3) = 28$

ANSWERS
1.
2.
3.
4.
5.
6.
7.
8.
9.
10.
11.
12.
13.
14.
15.
16.
17.
18.
19.
20.
21.
22.
23.

24. _____

25. _____

26. _____

27. _____

28. _____

29. _____

30. _____

31. _____

32. _____

33. _____

34. _____

35. _____

24. The square of a number is 24 more than five times the number. Find the number.

25. The length of a rectangle is 6 m more than the width. The area of the rectangle is 40 m². Find the length and the width.

26. Collect like terms:

$$x^3y - y^3 + xy^3 + 8 - 6x^3y - x^2y^2 + 11.$$

27. Subtract:

$$(8a^2b^2 - ab + b^3) - (-6ab^2 - 7ab - ab^3 + 5b^3).$$

28. Multiply: $(3x^5 - 4y^5)(3x^5 + 4y^5)$.

29. Factor: $3m^2 - 9mn - 30n^2$.

 SKILL MAINTENANCE

30. Divide: $\dfrac{5}{8} \div \left(-\dfrac{11}{16}\right)$.

31. Solve: $10(x - 3) = 4(x + 2)$.

32. Solve for T: $I = PRT$.

33. Multiply: $(5t^2 - 7)^2$.

EXTENSION

34. The length of a rectangle is 5 times its width. When the length is decreased by 3 and the width is increased by 2, the area of the new rectangle is 60. Find the original length and width.

35. Factor: $(a + 3)^2 - 2(a + 3) - 35$.

AN APPLICATION
The truck in the photo has been slanted in order to dump grain for storage in the silo. The measure of the slant is called the slope. The vertical "rise" of the truck is 5.5 ft vertically for every 8.1 ft of horizontal "run." Find the slope.

THE MATHEMATICS
The slope is given by

$$\frac{\text{Rise}}{\text{Run}} = \frac{5.5}{8.1} \approx 0.68, \text{ or } 68\%.$$

└── This is slope.

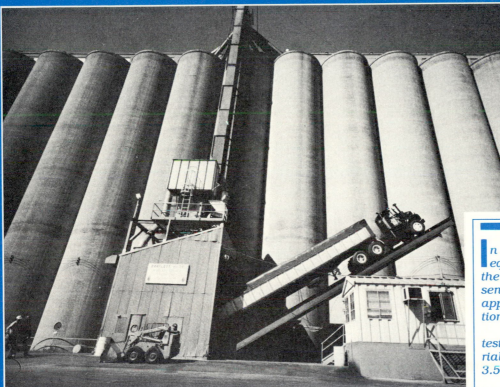

In this chapter graphs of equations that are lines and the concept of slope are presented. Direct variation, an application of linear equations, is also studied.

The review sections to be tested in addition to the material in this chapter are 2.9, 3.5. 4.4, and 5.6.

Graphs, Linear Equations, and Slope

After finishing Section 6.1, you should be able to:

- **·** Plot points associated with ordered pairs of numbers.
- **··** Determine the quadrant in which a point lies.
- **···** Find coordinates of a point on a graph.

Graphs and Equations

Graphs are often used in newspapers and magazines to convey information. Typically, certain units, in this case *years*, are shown horizontally. With each horizontal number there is associated a vertical number, or unit. In this case the vertical unit is *dollars*. Note the arrow pointing to $46.5 billion. That location on the graph can be thought of as a pair of numbers (1984, $46.5).

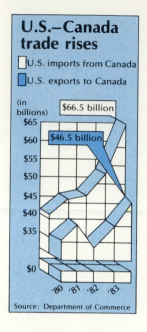

U.S.–Canada trade rises

☐ U.S. imports from Canada
☐ U.S. exports to Canada

(in billions)
$66.5 billion
$46.5 billion

Source: Department of Commerce

Plot these points on the graph below.

1. (4, 5)   2. (5, 4)

3. (−2, 5)   4. (−3, −4)

5. (5, −3)   6. (−2, −1)

7. (0, −3)   8. (2, 0)

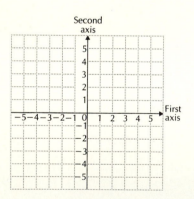

POINTS AND ORDERED PAIRS

On a number line each point is the graph of a number. On a plane each point is the graph of a number pair. We use two perpendicular number lines called *axes*. They cross at a point called the *origin*. It has coordinates (0, 0) but is usually labeled 0. The arrows show the positive directions.

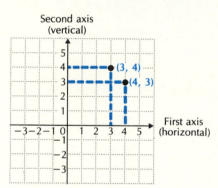

· **PLOTTING POINTS**

Note that (4, 3) and (3, 4) give different points. They are called *ordered pairs* of numbers because it makes a difference which number comes first.

Example 1 Plot the point (−3, 4).

The first number, −3, is negative. We go −3 units in the first direction (3 units to the left). The second number, 4, is positive. We go 4 units in the second direction (up).

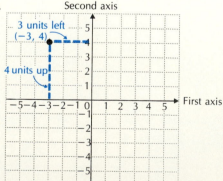

DO EXERCISES 1–8.

ANSWERS ON PAGE A-12

The numbers in an ordered pair are called *coordinates*. In (−3, 4), the *first coordinate* is −3 and the *second coordinate* is 4.

▪▪ QUADRANTS

This drawing shows some points and their coordinates. In region I (the first *quadrant*) both coordinates of any point are positive. In region II (the second quadrant) the first coordinate is negative and the second positive, and so on.

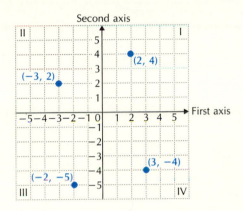

DO EXERCISES 9–14.

▪▪▪ FINDING COORDINATES

To find the coordinates of a point, we see how far to the right or left of zero it is located and how far up or down.

Example 2 Find the coordinates of point *A*.

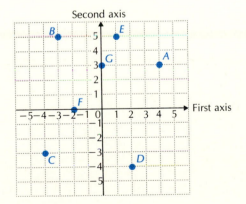

Point *A* is 4 units to the right (first direction) and 3 units up (second direction). Its coordinates are (4, 3).

DO EXERCISE 15.

9. What can you say about the coordinates of a point in the third quadrant?

10. What can you say about the coordinates of a point in the fourth quadrant?

In which quadrant is the point located?

11. (5, 3)

12. (−6, −4)

13. (10, −14)

14. (−13, 9)

15. Find the coordinates of points *B, C, D, E, F,* and *G* in the drawing of Example 2.

ANSWERS ON PAGE A-12

AN APPLICATION: COORDINATES

Three-dimensional objects can also be coordinatized. 0° latitude is the equator. 0° longitude is a line from the North Pole to the South Pole through France and Spain. In the drawing below, the hurricane Clara is at a point about 260 miles northwest of Bermuda near latitude 36.0 North, longitude 69.0 West.

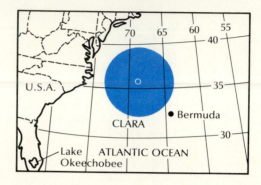

EXERCISES

1. Approximate the latitude and longitude of Bermuda.
2. Approximate the latitude and longitude of Lake Okeechobee.

•

1. Plot these points.

   (2, 5)   (−1, 3)   (3, −2)   (−2, −4)

   (0, 4)   (0, −5)   (5, 0)   (−5, 0)

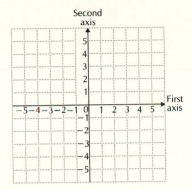

2. Plot these points.

   (4, 4)   (−2, 4)   (5, −3)   (−5, −5)

   (0, 4)   (0, −4)   (3, 0)   (−4, 0)

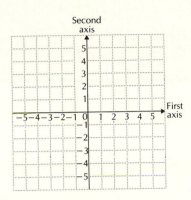

• • In which quadrant is each point located?

3. (−5, 3)   **4.** (−12, 1)   **5.** (100, −1)   **6.** (35.6, −2.5)

7. (−6, −29)   **8.** (−3.6, −105.9)   **9.** (3.8, 9.2)   **10.** (1895, 1492)

11. In quadrant III, first coordinates are always _____ and second coordinates are always _____ .

12. In quadrant II, _____ coordinates are always positive and _____ coordinates are always negative.

ANSWERS

1. _____

2. _____

3. _____

4. _____

5. _____

6. _____

7. _____

8. _____

9. _____

10. _____

11. _____

12. _____

•••

13. Find the coordinates of points A, B, C, D, and E.

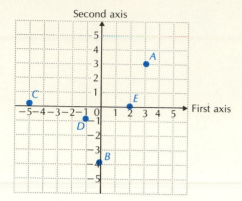

14. Find the coordinates of points A, B, C, D, and E.

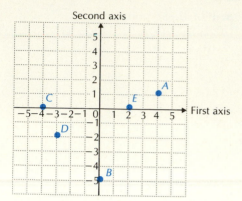

Plot these points.

15. $(0, -3)$, $(-1, -5)$, $(1, -1)$, $(2, 1)$

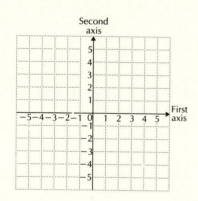

16. $(0, 1)$, $(1, 4)$, $(-1, -2)$, $(-2, -5)$

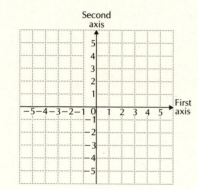

 SKILL MAINTENANCE

Solve.

17. $\dfrac{5}{2} + y = \dfrac{1}{3}$

18. $-2x + 9 = -11$

19. $-8x + 3x = 25$

20. Evaluate $9 - 5x$ for $x = 4$.

⭐ **EXTENSION**

In Exercises 21–24, tell in which quadrant(s) the points could be located.

21. The first coordinate is positive.

22. The second coordinate is negative.

23. The first and second coordinates are the same.

24. The first coordinate is the additive inverse of the second coordinate.

Chapter 6   Graphs, Linear Equations, and Slope

Graphing Equations

• SOLUTIONS OF EQUATIONS

An equation with two variables has *pairs* of numbers for solutions. We usually take the variables in alphabetical order. Then we get *ordered pairs* for solutions.

Example 1 Determine whether (3, 7) is a solution of $y = 2x + 1$.

$$y = 2x + 1$$

7	$2 \cdot 3 + 1$	We substitute 3 for x and 7 for y (alphabetical order of variables).
	$6 + 1$	
	7	The equation becomes true: (3, 7) is a solution.

Example 2 Determine whether (−2, 3) is a solution of $2t = 4s − 8$.

$$2t = 4s − 8$$

$2 \cdot 3$	$4(-2) - 8$	We substitute −2 for s and 3 for t.
6	$-8 - 8$	
	-16	The equation becomes false: (−2, 3) is not a solution.

DO EXERCISES 1 AND 2.

• • GRAPHING EQUATIONS $y = mx$ AND $y = mx + b$

> To *graph* an equation means to make a drawing of its solutions. Such a drawing is called the *graph* of the equation.

If an equation has a graph that is a line, we can graph it by plotting a few points and then drawing a line through them.

Example 3 Graph $y = x$.

We will use alphabetical order. Thus the first axis will be the x-axis and the second axis will be the y-axis. Next, we find some solutions of the equation. In this case it is easy. Here are a few:

(0, 0), (1, 1), (5, 5),
(−2, −2), (−4, −4).

Now we plot these points. We can see that if we were to plot a million solutions, the dots we draw would resemble a solid line. We see the pattern, so we can draw the line with a ruler. The line is the graph of the equation $y = x$. We label the line $y = x$ on the graph paper.

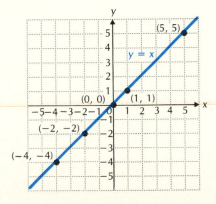

1. Determine whether (2, 3) is a solution of $y = 2x + 3$.

2. Determine whether (−2, 4) is a solution of $4q − 3p = 22$.

Graph.

3. $y = 3x$

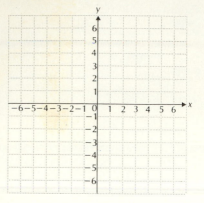

4. $y = \frac{1}{2}x$

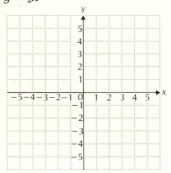

A graph of an equation is a picture of its solution set. Each point of the picture gives an ordered pair (a, b) that is a solution. No other points give solutions.

Example 4 Graph $y = 2x$.

We find some ordered pairs that are solutions, keeping the results in a table. We choose *any* number for x and then determine y by substitution. Suppose we choose 0 for x. Then $y = 2x = 2 \cdot 0 = 0$.

We get a solution: the ordered pair $(0, 0)$. Suppose we choose 3 for x. Then $y = 2x = 2 \cdot 3 = 6$.

We get a solution: the ordered pair $(3, 6)$. We make some negative choices for x, as well as some positive ones. If a number takes us off the graph paper, we generally do not use it. Continuing in this manner we get a table like the one shown below. In this case, since $y = 2x$, we get y by doubling x.

Now we plot these points. If we had enough of them, they would make a line. We draw it with a ruler and label it $y = 2x$.

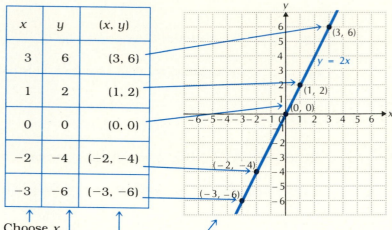

x	y	(x, y)
3	6	$(3, 6)$
1	2	$(1, 2)$
0	0	$(0, 0)$
-2	-4	$(-2, -4)$
-3	-6	$(-3, -6)$

1. Choose x.
2. Compute y.
3. Form the pair (x, y).
4. Plot the points.

DO EXERCISES 3 AND 4.

ANSWERS ON PAGE A-12

Chapter 6   Graphs, Linear Equations, and Slope

Example 5 Graph $y = -3x$.

We make a table of solutions. Then we plot the points. If we had enough of them, they would make a line. We draw it with a ruler and label it $y = -3x$.

x	y	(x, y)
0	0	$(0, 0)$
1	-3	$(1, -3)$
-1	3	$(-1, 3)$
2	-6	$(2, -6)$
-2	6	$(-2, 6)$

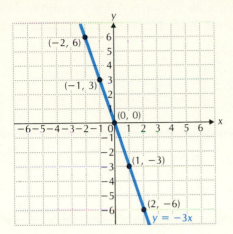

DO EXERCISES 5 AND 6.

Example 6 Graph $y = -\frac{5}{3}x$.

We make a table of solutions.

When $x = 0$, $y = -\frac{5}{3} \cdot 0 = 0$.

When $x = 3$, $y = -\frac{5}{3} \cdot 3 = -5$.

When $x = -3$, $y = -\frac{5}{3}(-3) = 5$.

When $x = 1$, $y = -\frac{5}{3} \cdot 1 = -\frac{5}{3}$.

Note that if we substitute multiples of 3 we can avoid fractions.

Next we plot the points. If we had enough of them, they would make a line.

x	y
0	0
3	-5
-3	5
1	$-\dfrac{5}{3}$

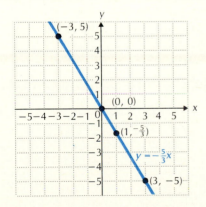

Every equation $y = mx$ has a graph that is a straight line. It contains the origin. The number m, called the *slope*, tells us how the line slants. For a positive slope a line slants up from left to right, as in Examples 3 and 4. For a negative slope a line slants down from left to right, as in Examples 5 and 6.

DO EXERCISES 7 AND 8.

Graph.

5. $y = -x$ (or $-1 \cdot x$)

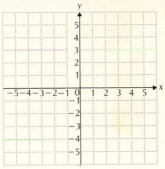

6. $y = -2x$

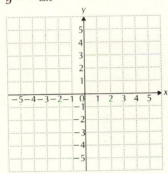

Graph.

7. $y = \frac{3}{4}x$

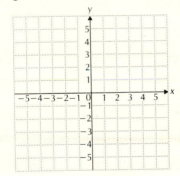

8. $y = -\frac{4}{5}x$

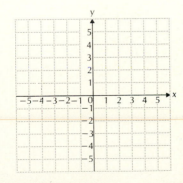

ANSWERS ON PAGE A-12

9. Graph $y = x + 3$ and compare it with $y = x$.

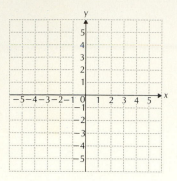

10. Graph $y = x - 1$ and compare it with $y = x$.

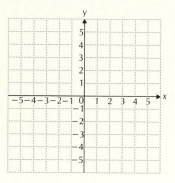

11. Graph $y = 2x + 3$ and compare it with $y = 2x$.

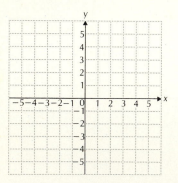

We know that the graph of any equation $y = mx$ is a straight line through the origin, with slope m. What will happen if we add a number b on the right side to get an equation $y = mx + b$?

Example 7 Graph $y = x + 2$ and compare it with $y = x$.

We first make a table of values.

x	y (or $x + 2$)
0	2
1	3
-1	1
2	4
-2	0
3	5

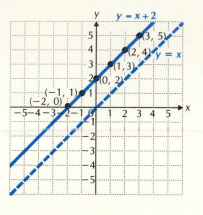

We plot these points. If we had enough of them they would make a line. We draw this line with a ruler and label it $y = x + 2$. The graph of $y = x$ is drawn for comparison. Note that the graph of $y = x + 2$ looks just like the graph of $y = x$, but it is moved up 2 units.

DO EXERCISES 9 AND 10.

Example 8 Graph $y = 2x - 3$ and compare it with $y = 2x$.

We first make a table of values.

x	y (or $2x - 3$)
0	-3
1	-1
2	1
-1	-5

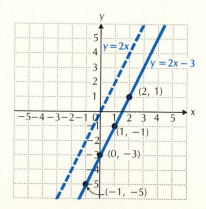

We draw the graph of $y = 2x - 3$. It looks just like the graph of $y = 2x$, but it is moved down 3 units.

DO EXERCISE 11.

The graph of $y = mx$ goes through the origin $(0, 0)$. The graph of any equation $y = mx + b$ is also a line. It is parallel to $y = mx$, but moved up or down. It goes through the point $(0, b)$. That point is called the *y-intercept*. We can also refer to the number b as the y-intercept. The number m is still called the *slope*. It tells us how steeply the line slants. We will study slope in more detail in Section 6.5.

Example 9 Graph $y = \frac{2}{5}x + 4$.

We first make a table of values. Using multiples of 5 avoids fractions.

When $x = 0$, $y = \frac{2}{5} \cdot 0 + 4 = 0 + 4 = 4$.

When $x = 5$, $y = \frac{2}{5} \cdot 5 + 4 = 2 + 4 = 6$.

When $x = -5$, $y = \frac{2}{5}(-5) + 4 = -2 + 4 = 2$.

Since two points determine a line, that is all you really need to graph a line, but you should always plot a third point as a check. We draw the graph of $y = \frac{2}{5}x + 4$.

x	y
0	4
5	6
-5	2

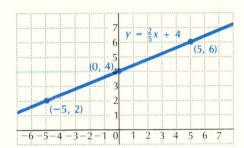

Example 10 Graph $y = -\frac{3}{4}x - 2$.

We first make a table of values.

When $x = 0$, $y = -\frac{3}{4} \cdot 0 - 2 = 0 - 2 = -2$.

When $x = 4$, $y = -\frac{3}{4} \cdot 4 - 2 = -3 - 2 = -5$.

When $x = -4$, $y = -\frac{3}{4}(-4) - 2 = 3 - 2 = 1$.

x	y
0	-2
4	-5
-4	1

We plot these points and draw a line through them.

We plot this point for a check to see whether it is on the line.

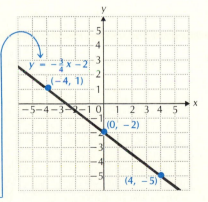

We draw the graph of $y = -\frac{3}{4}x - 2$.

Don't forget to label the graph.

DO EXERCISES 12–15.

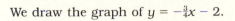

Any equation $y = mx + b$ has a graph that is a straight line. It goes through the point $(0, b)$, the y-intercept, and has slope m.

Graph.

12. $y = \frac{3}{5}x + 2$

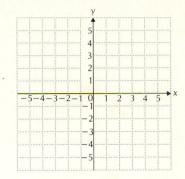

13. $y = \frac{3}{5}x - 2$

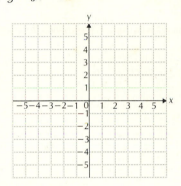

14. $y = -\frac{3}{5}x - 1$

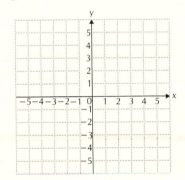

15. $y = -\frac{3}{5}x + 4$

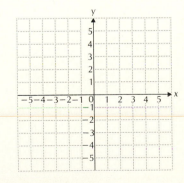

ANSWERS ON PAGE A-13

APPLICATIONS

The following are some other applications of polynomials in several variables.

The area A of a rectangle of length l and width w:

$$A = lw.$$

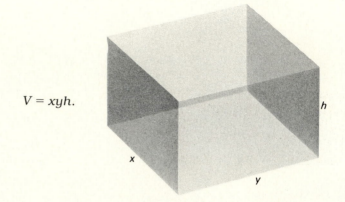

The volume V of a rectangular solid of length x, width y, and height h:

$$V = xyh.$$

The total revenue R from the sale of x units of one product at \$4 each and y units of another product at \$7 each:

$$R = 4x + 7y.$$

The amount A that P dollars will grow to at interest rate r, compounded annually for 3 years:

$$A = P + 3rP + 3r^2P + r^3P.$$

The height S, in feet, of an object t seconds after being given an upward velocity of v feet per second from an altitude h:

$$S = -16t^2 + vt + h.$$

The approximate length L of a pulley belt around pulleys whose centers are D units apart and whose circumferences are C_1 and C_2:

$$L = \frac{1}{2}C_1 + \frac{1}{2}C_2 + 2D.$$

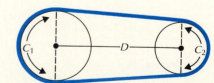

EXERCISE

Solve the pulley formula for D.

• Determine whether the given point is a solution of the equation.

1. $(2, 5)$; $y = 3x - 1$

2. $(1, 7)$; $y = 2x + 5$

3. $(2, -3)$; $3x - y = 4$

4. $(-1, 4)$; $2x + y = 6$

5. $(-2, -1)$; $2a + 2b = -7$

6. $(0, -4)$; $4m + 2n = -9$

A N S W E R S
1. _____
2. _____
3. _____
4. _____
5. _____
6. _____

•• Graph.

7. $y = 4x$

8. $y = 2x$

9. $y = -2x$

10. $y = -4x$

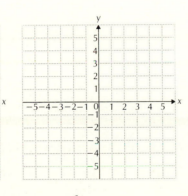

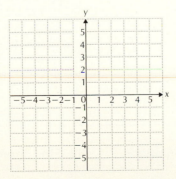

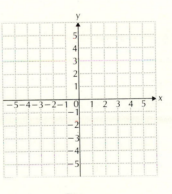

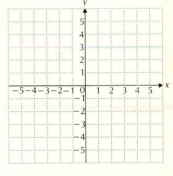

11. $y = \frac{1}{3}x$

12. $y = \frac{1}{4}x$

13. $y = -\frac{3}{2}x$

14. $y = -\frac{5}{4}x$

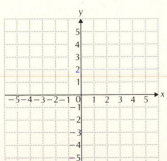

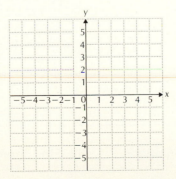

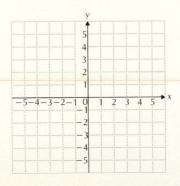

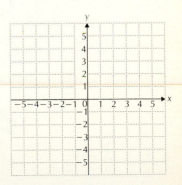

15. $y = x + 1$

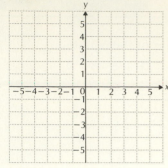

16. $y = -x + 1$

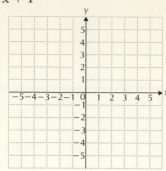

27.

17. $y = 2x + 2$

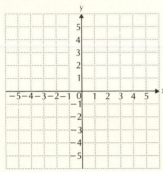

18. $y = 3x - 2$

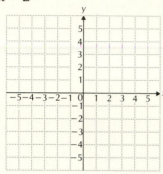

28.

19. $y = \frac{1}{3}x - 1$

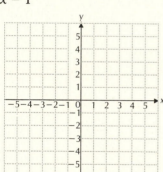

20. $y = \frac{1}{2}x + 1$

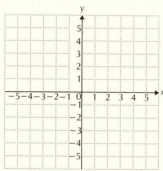

29.

Use graph paper. Draw and label x- and y-axes. Graph these equations.

21. $y = -x - 3$

22. $y = -x - 2$

23. $y = \frac{5}{2}x + 3$

30.

24. $y = \frac{5}{3}x - 2$

25. $y = -\frac{5}{2}x - 2$

26. $y = -\frac{5}{3}x + 3$

 SKILL MAINTENANCE

Factor.

27. $16 - t^4$

28. $2y^3 - 10y^2 + y - 5$

31.

29. $x^5 - 2x^4 - 35x^3$

30. $m^4 - 10m^2 + 25$

 EXTENSION

32.

31. Find all the whole-number solutions of $x + y = 6$.

32. Find three solutions of $y = |x|$.

Chapter 6   Graphs, Linear Equations, and Slope

6.3

Linear Equations

• GRAPHING USING INTERCEPTS

The fastest method for graphing equations whose graphs are straight lines involves the use of intercepts. Look at the graph of $y - 2x = 4$ shown below. We could graph this equation by solving for y to get $y = 2x + 4$ and proceed as before, but we want to develop a faster method.

After finishing Section 6.3, you should be able to:

• Graph using intercepts.

•• Graph equations of the type $x = a$ or $y = b$.

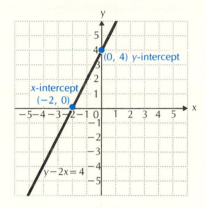

The y-intercept is (0, 4). It occurs where the line crosses the y-axis and always has 0 as the first coordinate. The x-intercept is (−2, 0). It occurs where the line crosses the x-axis and always has 0 as the second coordinate.

DO EXERCISE 1.

> The x-intercept is $(a, 0)$. To find a, let $y = 0$.
> The y-intercept is $(0, b)$. To find b, let $x = 0$.

Now let us draw a graph using intercepts.

Example 1 Graph $4x + 3y = 12$.

To find the x-intercept, let $y = 0$. Then

$$4x + 3 \cdot 0 = 12$$
$$4x = 12$$
$$x = 3.$$

Thus (3, 0) is the x-intercept.

Note that this amounts to covering up the y-term and looking at the rest of the equation.

To find the y-intercept, let $x = 0$. Then

$$4 \cdot 0 + 3y = 12$$
$$3y = 12$$
$$y = 4.$$

Thus (0, 4) is the y-intercept.

1. Look at the graph shown below.

a) Find the coordinates of the x-intercept.

b) Find the coordinates of the y-intercept.

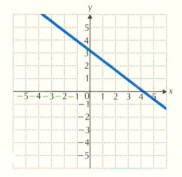

ANSWERS ON PAGE A-13

Graph using intercepts.

2. $2x + 3y = 6$

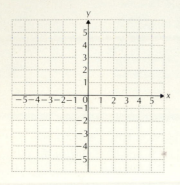

3. $3y - 4x = 12$

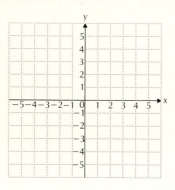

We plot these points and draw the line.

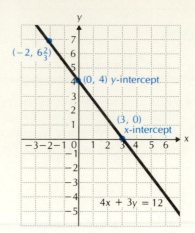

A third point should be used as a check. We substitute any arbitrary value for x and solve for y. We let $x = -2$. Then

$$4(-2) + 3y = 12 \qquad \text{Substituting } -2 \text{ for } x$$
$$-8 + 3y = 12$$
$$3y = 12 + 8$$
$$3y = 20$$
$$y = \frac{20}{3}, \quad \text{or} \quad 6\frac{2}{3} \qquad \text{Solving for } y$$

We see that the point $(-2, 6\frac{2}{3})$ is on the graph, so the graph is probably correct.

DO EXERCISES 2 AND 3.

• • EQUATIONS WITH A MISSING VARIABLE

Consider the equation $y = 3$. We can think of it as $y = 0 \cdot x + 3$. No matter what number we choose for x, we find that $y = 3$.

Example 2 Graph $y = 3$.

Any ordered pair $(x, 3)$ is a solution. So the line is parallel to the x-axis with y-intercept $(0, 3)$.

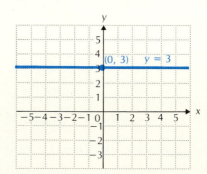

Chapter 6   Graphs, Linear Equations, and Slope

Example 3 Graph $x = -4$.

Any ordered pair $(-4, y)$ is a solution. So the line is parallel to the y-axis with x-intercept $(-4, 0)$.

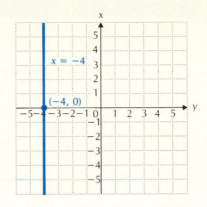

The graph of $y = b$ is a horizontal line. The graph of $x = a$ is a vertical line.

DO EXERCISES 4–7.

Equations whose graphs are straight lines are *linear equations*. We summarize the best procedure for graphing linear equations.

TO GRAPH LINEAR EQUATIONS

1. Is the equation of the type $x = a$ or $y = b$? If so, the graph will be a line parallel to an axis.
2. If the line is not of the type $x = a$ or $y = b$, find the intercepts. Graph using the intercepts if this is feasible.
3. If the intercepts are too close together, choose another point farther from the origin.
4. In any case, use a third point as a check.

If you have trouble remembering whether a graph such as $y = 3$ or $x = -4$ is horizontal or vertical, the following may help. Consider $y = 3$. Make up a table with all 3's in the y-column.

x	y
	3
	3 (y must be 3)
	3

Choose any numbers for x.

x	y
-2	3
0	3
4	3

Now when you plot the ordered pairs $(-2, 3)$, $(0, 3)$, and $(4, 3)$ and connect them, you will obtain a horizontal line. Similarly, consider $x = -4$. Make up a table with all -4's in the x-column.

x	y
-4	
-4	
-4	

Choose any numbers for y.

x	y
-4	-5
-4	1
-4	3

Now when you plot the ordered pairs $(-4, -5)$, $(-4, 1)$, and $(-4, 3)$ and connect them, you will obtain a vertical line.

Graph.

4. $x = 5$

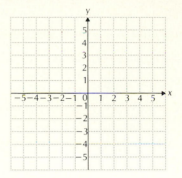

5. $y = -2$

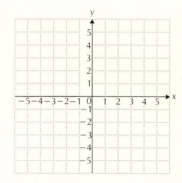

6. $x = 0$

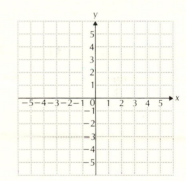

7. $x = -3$

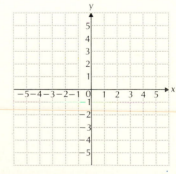

ANSWERS ON PAGE A-13

AN APPLICATION

Depreciation: The Straight-line Method

A company buys a machine for $5200. The machine is expected to last for 8 years at which time its trade-in, or scrap, value will be $1300. Over its lifetime it depreciates $5200 − $1300, or $3900. If the company figures the decline in value to be the same each year— that is, $\frac{1}{8}$, or 12.5% of $3900, which is $487.50—then they are using what is called *straight-line depreciation.* We see this in the graph below. We see that the book value after one year is $5200 − $487.50, or $4712.50. After two years it is $4712.50 − $487.50, or $4225. After 3 years it is $4225 − $487.50, or $3737.50, etc.

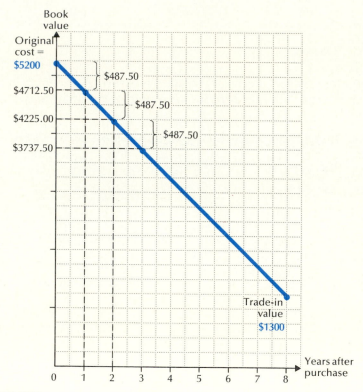

EXERCISE

Find the book values of the machine after each of the remaining years.

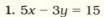
Find the intercepts of each equation. Then graph.

1. $5x - 3y = 15$

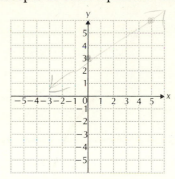

2. $2x - 4y = 8$

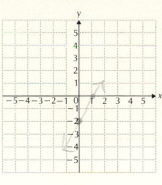

3. $4x + 2y = 8$

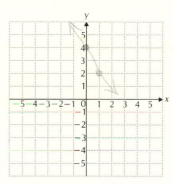

4. $3x + 5y = 15$

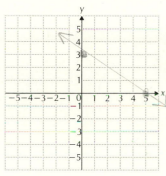

5. $x - 1 = y$

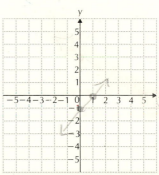

6. $x - 3 = y$

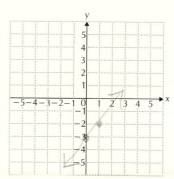

7. $2x - 1 = y$

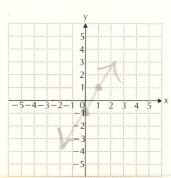

8. $3x - 2 = y$

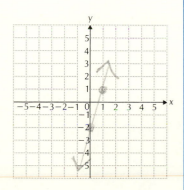

Use graph paper. Find the intercepts of each equation. Then graph. Attach graphs to this page.

9. $4x - 3y = 12$

10. $2x - 5y = 10$

11. $7x + 2y = 6$

12. $3x + 3y = 5$

13. $y = -4 - 4x$

14. $y = -3 - 3x$

ANSWERS

1. _____

2. _____

3. _____

4. _____

5. _____

6. _____

7. _____

8. _____

9. _____

10. _____

11. _____

12. _____

13. _____

14. _____

•• Graph.

15. $x = -2$

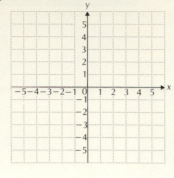

16. $x = -1$

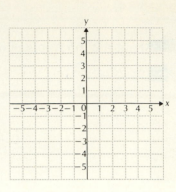

17. $y = 2$

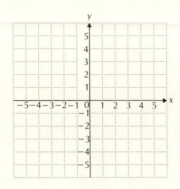

18. $y = 4$

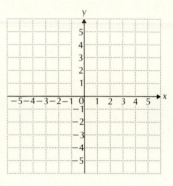

Use graph paper. Graph each of the following. Attach graphs to this page.

19. $x = 2$ **20.** $x = 3$ **21.** $y = 0$

22. $y = -1$ **23.** $x = \dfrac{3}{2}$ **24.** $x = -\dfrac{5}{2}$

SKILL MAINTENANCE

Subtract.

25. $(6x^2 + 7) - (4x^2 - 9)$ **26.** $(8x^3 - 3x^2 + 7) - (8x^2 + 3x - 5)$

27. $\left(\dfrac{1}{4}x^2 - \dfrac{3}{4}x + 3\right) - \left(\dfrac{3}{4}x^2 + \dfrac{1}{4}x - 3\right)$

28. $(0.08y^3 - 0.04y^2 + 0.01y) - (0.02y^3 + 0.05y^2 + 1)$

EXTENSION

29. Write an equation for the x-axis. **30.** Write an equation for the y-axis.

31. Find the coordinates of the point of intersection of the graphs of the equations $x = -3$ and $y = 6$.

32. Find the value of m in $y = mx + 3$ so that the x-intercept of its graph will be $(2, 0)$.

Direct Variation

• EQUATIONS OF DIRECT VARIATION

• Find an equation of direct variation given a pair of values of the variables.

•• Solve problems involving direct variation.

A bicycle is traveling at 10 km/h. In 1 hr it goes 10 km; in 2 hr it goes 20 km; in 3 hr it goes 30 km; and so on. We will use the number of hours as the first coordinate and the number of kilometers traveled as the second coordinate. This creates a set of ordered pairs,

$$(1, 10), \quad (2, 20), \quad (3, 30), \quad (4, 40), \quad \text{and so on.}$$

Note that as the first number gets larger, so does the second. Note too that the ratio of distance to time for each of these ordered pairs is $\frac{10}{1}$, or 10. The *distance varies directly as the time*:

$$\frac{d}{t} = 10 \text{ (a constant)}, \quad \text{or} \quad d = 10t.$$

Find an equation of variation where y varies directly as x, and the following is true.

1. $y = 84$ when $x = 12$

If a situation translates to a function described by $y = kx$, where k is a positive constant, then $y = kx$ is called an *equation of direct variation* and k is called the *variation constant*.

The graph of $y = kx$, $k > 0$, always goes through the origin and rises from left to right. Note that as x increases, y increases.

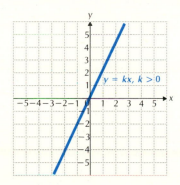

2. $y = 50$ when $x = 80$

When there is direct variation $y = kx$, the variation constant can be found if one pair of values of x and y is known. Then other values can be found.

Example 1 Find an equation of variation where y varies directly as x and $y = 7$ when $x = 25$.

We substitute to find k:

$$y = kx$$
$$7 = k \cdot 25 \qquad \text{Substituting 7 for } y \text{ and 25 for } x$$
$$\tfrac{7}{25} = k, \quad \text{or} \quad k = 0.28.$$

Thus the equation is $y = 0.28x$. Note that the answer is an equation.

DO EXERCISES 1 AND 2.

ANSWERS ON PAGE A-14

3. The cost C of operating a TV varies directly as the number n of hours it is in operation. It costs $14.00 to operate a standard-size color TV continuously for 30 days. At this rate, how much would it cost to operate the TV for 1 day? for 1 hour?

4. The weight V of an object on Venus varies directly as its weight E on earth. A person weighing 75 kg on earth would weigh 66 kg on Venus. How much would a person weighing 90 kg on earth weigh on Venus?

●● SOLVING PROBLEMS INVOLVING DIRECT VARIATION

Example 2 It is known that the karat rating K of a gold object varies directly as the actual percentage of gold in the object. A 14-karat gold object is 58.25% gold. What is the percentage of gold in a 24-karat object?

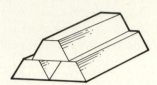

The problem states that we have direct variation between the variables K and P. Thus an equation $K = kP$, $k > 0$, applies. As the percentage of gold increases, so does the karat rating.

a) First, find an equation of variation.

$$K = kP$$
$$14 = k \cdot 0.5825 \qquad \text{Substituting 14 for } K \text{ and 0.5825 for } P$$
$$\frac{14}{0.5825} = k$$
$$24.03 \approx k \qquad \text{Dividing and rounding to the nearest hundredth}$$

The equation of variation is $K = 24.03P$.

b) Use the equation to find the percentage of gold in a 24-karat object.

$$K = 24.03P$$
$$24 = 24.03P$$
$$\frac{24}{24.03} = P$$
$$0.999 = P$$
$$99.9\% = P$$

A 24-karat object is 99.9% gold.

DO EXERCISES 3 AND 4.

Chapter 6   Graphs, Linear Equations, and Slope

Find an equation of variation where y varies directly as x, and the following are true.

1. $y = 28$ when $x = 7$

2. $y = 30$ when $x = 8$

3. $y = 0.7$ when $x = 0.4$

4. $y = 0.8$ when $x = 0.5$

5. $y = 400$ when $x = 125$

6. $y = 630$ when $x = 175$

7. $y = 200$ when $x = 300$

8. $y = 500$ when $x = 60$

Solve.

9. A person's paycheck P varies directly as the number H of hours worked. For working 15 hr the pay is $78.75. Find the pay for working 35 hr.

10. The number B of bolts a machine can make varies directly as the time T it operates. It can make 6578 bolts in 2 hr. How many can it make in 5 hr?

11. The number S of servings of meat that can be obtained from a turkey varies directly as its weight W. From a turkey weighing 14 kg one can get 40 servings of meat. How many servings can be obtained from an 8-kg turkey?

12. The number S of servings of meat that can be obtained from round steak varies directly as the weight W. From 9 kg of round steak one can get 70 servings of meat. How many servings can one get from 12 kg of round steak?

ANSWERS
1.
2.
3.
4.
5.
6.
7.
8.
9.
10.
11.
12.

13. The weight M of an object on the moon varies directly as its weight E on earth. A person who weighs 78 kg on earth weighs 13 kg on the moon. How much would a 100-kg person weigh on the moon?

14. The weight M of an object on Mars varies directly as its weight E on earth. A person who weighs 95 kg on earth weighs 36.1 kg on Mars. How much would an 80-kg person weigh on Mars?

15. The number W of kilograms of water in a human body varies directly as the total body weight B. A person weighing 75 kg contains 54 kg of water. How many kilograms of water are in a person weighing 95 kg?

16. The amount C that a family spends on car expenses varies directly as its income I. A family making $21,760 a year will spend $3264 a year for car expenses. How much will a family making $30,000 a year spend for car expenses?

 SKILL MAINTENANCE

17. Evaluate $x^2 - 5x + 7$ for $x = -3$.

Solve.

18. $x^2 - 5x = 0$

19. $121 = x^2$

20. $x^2 = x + 2$

⭐ EXTENSION

Write an equation of variation for each situation in Exercises 21 and 22. Give a value for k and graph the equation.

21. The circumference C of a circle varies directly as the radius r.

22. The perimeter P of a square varies directly as the length s of a side.

23. The area of a circle varies directly as the square of the length of the radius. What is the variation constant?

24. Show that if p varies directly as q, then q varies directly as p.

Equations of Lines and Slope

• SLOPE

As we have seen, graphs of some linear equations slant upward from left to right. Others slant downward. Some are vertical and some are horizontal. Some slant more steeply than others. We are looking for a way to describe such possibilities with numbers.

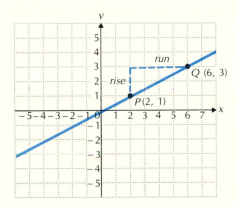

Consider a line with two points marked P and Q. As we move from P to Q, the x-coordinate changes from 2 to 6 and the y-coordinate changes from 1 to 3. The change in x is $6 - 2$, or 4. The change in y is $3 - 1$, or 2.

We call the change in y the *rise* and the change in x the *run*. The ratio rise/run is the same for any two points on a line. We call this ratio the *slope*. Slope describes the slant of a line. The slope of the line in the graph is given by

$$\frac{\text{rise}}{\text{run}}, \quad \text{or} \quad \frac{2}{4}, \quad \text{or} \quad \frac{1}{2}.$$

The *slope m* of a line containing points (x_1, y_1) and (x_2, y_2) is given by

$$m = \frac{\text{rise}}{\text{run}} = \frac{\text{the change in } y}{\text{the change in } x} = \frac{y_2 - y_1}{x_2 - x_1}.$$

OBJECTIVES

After finishing Section 6.5, you should be able to:

• Given the coordinates of two points, find the slope of the line containing them.

•• Determine the slope, if it exists, of a horizontal or vertical line.

••• Find the slope of a line from an equation of the line.

:: Given an equation of a line, find the slope–intercept equation, the slope, and the y-intercept of the line.

:•: Find an equation of a line given a point on the line and the slope, or given two points on the line.

Graph the line containing the points and find the slope.

1. $(-2, 3)$ and $(3, 5)$

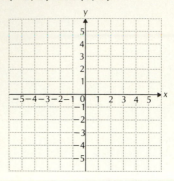

2. $(0, -3)$ and $(-3, 2)$

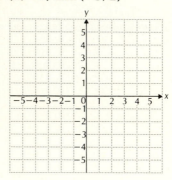

3. Find the slope in Exercise 1, but subtract in a different order.

4. Find the slope in Exercise 2, but subtract in a different order.

ANSWERS ON PAGE A-14

Example 1 Graph the line containing the points $(-4, 3)$ and $(2, -6)$ and find the slope.

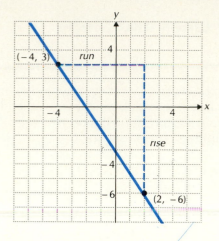

The graph is shown here. From $(-4, 3)$ to $(2, -6)$ the change in y, or rise, is $3 - (-6)$, or 9. The change in x, or run, is $-4 - 2$, or -6.

$$\text{Slope} = \frac{\text{rise}}{\text{run}} = \frac{\text{change in } y}{\text{change in } x} = \frac{3 - (-6)}{-4 - 2}$$

$$= \frac{9}{-6} = -\frac{9}{6}, \quad \text{or} \quad -\frac{3}{2}.$$

DO EXERCISES 1 AND 2.

When we use the formula

$$m = \frac{y_2 - y_1}{x_2 - x_1}$$

we can subtract in two ways. We just have to remember to subtract the y-coordinates in the same order in which we subtract the x-coordinates. Let's do Example 1 again.

$$\text{Slope} = \frac{\text{change } y}{\text{change in } x} = \frac{-6 - 3}{2 - (-4)} = \frac{-9}{6} = -\frac{3}{2}.$$

DO EXERCISES 3 AND 4.

The slope of a line tells how it slants. A line with a positive slope slants up from left to right. The larger the positive slope, the steeper the slant. A line with negative slope slants downward from left to right.

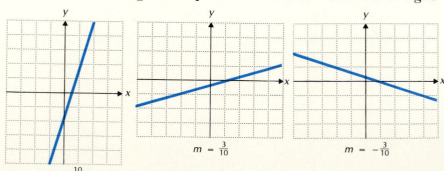

·· HORIZONTAL AND VERTICAL LINES

What about the slope of a horizontal or a vertical line?

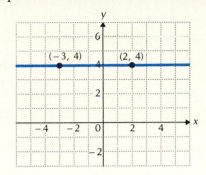

Example 2 Find the slope of the line $y = 4$.

Consider the points $(-3, 4)$ and $(2, 4)$, which are on the line.

The change in $y = 4 - 4$, or 0.

The change in $x = -3 - 2$, or -5.

$$m = \frac{4 - 4}{-3 - 2} = \frac{0}{-5} = 0$$

Any two points on a horizontal line have the same y-coordinate. Thus the change in y is 0, so the slope is 0.

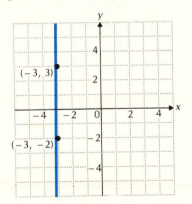

Example 3 Find the slope of the line $x = -3$.

Consider the points $(-3, 3)$ and $(-3, -2)$, which are on the line.

The change in $y = 3 - (-2)$, or 5.

The change in $x = -3 - (-3)$, or 0.

$$m = \frac{3 - (-2)}{-3 - (-3)} = \frac{5}{0}$$

Since division by 0 is not defined, this line has no slope.

> **A horizontal line has slope 0. A vertical line has *no* slope.**

DO EXERCISES 5–8.

Find the slope, if it exists, of the line containing these points.

5. $(9, 7)$ and $(3, 7)$

6. $(4, -6)$ and $(4, 0)$

7. Find the slope of the line $x = 7$.

8. Find the slope of the line $y = -5$.

ANSWERS ON PAGE A-14

Find the slope of the line.

9. $y = -8x - 14$

10. $4x + 5y = 7$

11. $\frac{1}{4}x = 7 + y$

12. $5x - 4y = 8$

●●● **FINDING SLOPE FROM AN EQUATION**

It is possible to find the slope of a line from its equation. Let us consider the equation

$$y = 2x + 3.$$

Two points on the line are (0, 3) and (1, 5). We can find such points by picking arbitrary values for x, say 0 and 1, and substituting to find corresponding y-values. The slope of the line is found as follows:

$$m = \frac{\text{change in } y}{\text{change in } x}$$

$$= \frac{5 - 3}{1 - 0}$$

$$= \frac{2}{1} = 2.$$

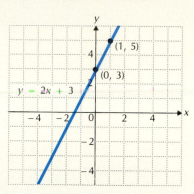

The slope is 2. This is also the coefficient of the x-term in the equation $y = 2x + 3$.

> The slope of the line $y = mx + b$ is m. To find the slope of a nonvertical line given by a linear equation in x and y, first solve the equation for y, and get the resulting equation in the form $y = mx + b$. The coefficient of the x-term, m, is the slope of the line.

Example 4 Find the slope of the line $2x + 3y = 7$.

We solve for y:

$$2x + 3y = 7$$
$$3y = -2x + 7$$
$$y = \tfrac{1}{3}(-2x + 7) = \tfrac{1}{3}(-2x) + \tfrac{1}{3}(7)$$
$$y = -\tfrac{2}{3}x + \tfrac{7}{3}$$

The slope is $-\tfrac{2}{3}$.

DO EXERCISES 9–12.

⁛⁛ **THE SLOPE–INTERCEPT EQUATION OF A LINE**

In the equation $y = mx + b$, we know that m is the slope. What is the y-intercept? To find out we let $x = 0$ and solve for y:

$$y = mx + b$$
$$y = m(0) + b$$
$$y = b.$$

Thus the y-intercept is (0, b).

> The equation $y = mx + b$ is called the *slope–intercept* equation. The slope is m and the y-intercept is (0, b).

Example 5 Find the slope and y-intercept of $y = 3x - 4$.

Since the equation is already in the form $y = mx + b$, we simply interpret the slope and y-intercept from the equation.

$$y = 3x - 4 \qquad \text{The } y\text{-intercept is } (0, -4).$$

The slope is 3.

Example 6 Find the slope and y-intercept of $2x + 3y = 8$.

We first solve for y:

$$2x + 3y = 8$$
$$3y = -2x + 8$$
$$y = -\tfrac{2}{3}x + \tfrac{8}{3}.$$

The slope is $-\tfrac{2}{3}$ and the y-intercept is $(0, \tfrac{8}{3})$.

DO EXERCISES 13–17.

▓▓ THE POINT–SLOPE EQUATION OF A LINE

Suppose we know the slope of a line and that it contains a certain point. We can use the slope–intercept equation to find an equation of the line.

Example 7 Find an equation of the line with slope 3 that contains the point (4, 1).

Step 1. Use the point (4, 1) and substitute 4 for x and 1 for y in $y = mx + b$. We also substitute 3 for m, the slope. Then we solve for b.

$$y = mx + b$$
$$1 = 3 \cdot 4 + b \qquad \text{Substituting}$$
$$-11 = b \qquad \text{Solving for } b, \text{ the } y\text{-intercept}$$

Step 2. Write the equation $y = mx + b$ by substituting 3 for m and -11 for b in $y = mx + b$, we get $y = 3x - 11$.

DO EXERCISES 18 AND 19.

Now consider a line with slope 2 and containing the point (1, 3) as shown. Suppose (x, y) is any point on this line. Using the definition of slope and the two points (1, 3) and (x, y), we get

$$m = \frac{\text{change in } y}{\text{change in } x} = \frac{y - 3}{x - 1}.$$

We know that the slope is 2, so

$$2 = \frac{y - 3}{x - 1}, \quad \text{or} \quad \frac{y - 3}{x - 1} = 2.$$

Multiplying on both sides by $x - 1$, we get

$$y - 3 = 2(x - 1).$$

Solving for y we obtain $y - 3 = 2x - 2$, or $y = 2x + 1$.

Find the slope and y-intercept.

13. $y = 5x$

14. $y = -\dfrac{3}{2}x - 6$

15. $2y = 4x - 17$

16. $3x + 4y = 15$

17. $-7x - 5y = 22$

Find an equation of the line that contains the given point and has the given slope.

18. $(4, 2)$, $m = 5$

19. $(-2, 1)$, $m = -3$

ANSWERS ON PAGE A-14

Find an equation of the line containing the given point and having the given slope.

20. $(3, 5)$, $m = 6$

21. $(1, 4)$, $m = -\dfrac{2}{3}$

Find an equation of the line containing the given points.

22. $(2, 4)$ and $(3, 5)$

23. $(-1, 2)$ and $(-3, -2)$

ANSWERS ON PAGE A-14

Example 8 Find an equation of the line with slope 5 that contains the point $(-2, -3)$.

We use the point–slope equation. We substitute 5 for m, -2 for x_1, and -3 for y_1:

$$y - y_1 = m(x - x_1)$$
$$y - (-3) = 5[x - (-2)]$$
$$y + 3 = 5(x + 2) = 5x + 10$$
$$y = 5x + 7.$$

DO EXERCISES 20 AND 21.

We can also use the point–slope equation to find an equation of a line that contains two given points.

Example 9 Find an equation of the line containing $(2, 3)$ and $(-6, 1)$.

First, we find the slope:

$$m = \frac{3 - 1}{2 - (-6)} = \frac{1}{4}.$$

Then we use the point–slope equation:

$$y - y_1 = m(x - x_1)$$
$$y - 3 = \tfrac{1}{4}(x - 2) \qquad \text{Substituting 2 for } x_1,$$
$$\qquad\qquad\qquad 3 \text{ for } y_1, \text{ and } \tfrac{1}{4} \text{ for } m$$
$$y - 3 = \tfrac{1}{4}x - \tfrac{1}{2}$$
$$y = \tfrac{1}{4}x + \tfrac{5}{2}.$$

DO EXERCISES 22 AND 23.

APPLICATIONS OF SLOPE

Slope has many real-world applications. For example, numbers like 2%, 3%, and 6% are often used to represent the *grade* of a road. Such a number is meant to tell how steep a road up a hill or mountain is. For example, a 3% grade means that for every horizontal distance of 100 ft, the road rises 3 ft. The concept of grade also occurs in cardiology when a person runs on a treadmill. A physician may change the steepness of the treadmill to measure its effect on heartbeat.

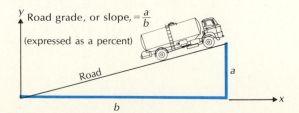

Road grade, or slope, $= \dfrac{a}{b}$ (expressed as a percent)

Chapter 6   Graphs, Linear Equations, and Slope

Find the slope, if it exists, of the line containing each pair of points.

1. $(3, 2)$ and $(-1, 2)$

2. $(4, 1)$ and $(-2, 3)$

3. $(-2, 4)$ and $(3, 0)$

4. $(-4, 2)$ and $(2, -3)$

5. $(4, 0)$ and $(5, 7)$

6. $(3, 0)$ and $(6, 2)$

7. $(-3, -2)$ and $(-5, -6)$

8. $(-2, -4)$ and $(-6, -7)$

9. $(-2, \frac{1}{2})$ and $(-5, \frac{1}{2})$

10. $(8, -3)$ and $(10, -3)$

11. $(9, -4)$ and $(9, -7)$

12. $(-10, 3)$ and $(-10, 4)$

Find the slope, if it exists, of each line.

13. $x = -8$

14. $x = -4$

15. $y = 2$

16. $y = 17$

17. $x = 9$

18. $y = -9$

19. $y = -4$

20. $x = 6$

Find the slope of each line.

21. $3x + 2y = 6$

22. $4x - y = 5$

23. $x + 4y = 8$

24. $x + 3y = 6$

25. $-2x + y = 4$

26. $-5x + y = 5$

Find the slope and y-intercept of each line.

27. $y = -4x - 9$

28. $y = -3x - 5$

29. $y = 1.8x$

30. $y = -27.4x$

31. $-8x - 7y = 21$

32. $-2x - 9y = 13$

33. $9x = 3y + 5$

34. $4x = 9y + 7$

35. $-6x = 4y + 2$

36. $5x + 4y = 12$

37. $y = -17$

38. $y = 23$

ANSWERS
1.
2.
3.
4.
5.
6.
7.
8.
9.
10.
11.
12.
13.
14.
15.
16.
17.
18.
19.
20.
21.
22.
23.
24.
25.
26.
27.
28.
29.
30.
31.
32.
33.
34.
35.
36.
37.
38.

Find an equation of the line containing the given point and having the given slope.

39. $(2, 5)$, $m = 5$ **40.** $(-3, 0)$, $m = -2$

41. $(2, 4)$, $m = \frac{3}{4}$ **42.** $(\frac{1}{2}, 2)$, $m = -1$

43. $(2, -6)$, $m = 1$ **44.** $(4, -2)$, $m = 6$

45. $(-3, 0)$, $m = -3$ **46.** $(0, 3)$, $m = -3$

Find an equation of the line that contains the given pair of points.

47. $(-6, 1)$ and $(2, 3)$ **48.** $(12, 16)$ and $(1, 5)$

49. $(0, 4)$ and $(4, 2)$ **50.** $(0, 0)$ and $(4, 2)$

51. $(3, 2)$ and $(1, 5)$ **52.** $(-4, 1)$ and $(-1, 4)$

53. $(-2, -4)$ and $(2, -1)$ **54.** $(-3, 5)$ and $(-1, -3)$

✓ SKILL MAINTENANCE

55. A post is placed through some water into the mud at the bottom of a lake. Half of the post is in the mud and $\frac{1}{3}$ is in the water, and the part above the water is $5\frac{1}{2}$ ft long. How long is the post?

56. The sum of two consecutive even integers is 130. Find the product of the integers.

Simplify.

57. $[10 - 3(7 - 2)]$

58. $5^3 - 4^2 + 6(5 \cdot 7 + 4 \cdot 3)$

☆ EXTENSION

59. Find an equation of the line that contains the point $(2, -3)$ and has the same slope as the line $3x - y + 4 = 0$.

60. Find an equation of the line that has the same y-intercept as the line $x - 3y = 6$ and contains the point $(5, -1)$.

61. Find an equation of the line with the same slope as $3x - 2y = 8$ and the same y-intercept as $2y + 3x = -4$.

62. Graph several equations that have the same slope. How are they related?

Summary and Review

The following contains a summary of what you should be able to do after completing this chapter. The review exercises are for practice. Answers are at the back of the book. If you miss an exercise, restudy the section and objective indicated alongside the answer.

The review sections to be tested in addition to the material in this chapter are 2.9, 3.5, 4.4, and 5.6.

You should be able to:

Plot points associated with ordered pairs of numbers, determine the quadrant in which a point lies, and find coordinates of a point on a graph.

Use graph paper. Plot these points.

1. $(2, 5)$

2. $(0, -3)$

3. $(-4, -2)$

In which quadrant is each point located?

4. $(3, -8)$

5. $(-20, -14)$

6. $(4.9, 1.3)$

Find the coordinates of each point.

7. A

8. B

9. C

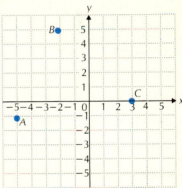

Determine whether an ordered pair is a solution of an equation in two variables.

Determine whether the given point is a solution of the equation $2y - x = 10$.

10. $(2, -6)$

11. $(0, 5)$

Graph linear equations.

Graph.

12. $y = 2x - 5$

13. $y = -\dfrac{3}{4}x$

14. $y = -x + 4$

15. $y = 3 - 4x$

16. $5x - 2y = 10$

17. $y = 3$

18. $x = -\dfrac{3}{4}$

19. $x - 2y = 6$

Find an equation of variation given a pair of values of the variables, and solve problems involving direct variation.

Find an equation of variation where y varies directly as x, and the following are true.

20. $y = 12$ when $x = 4$

21. $y = 0.4$ when $x = 0.5$

22. A person's paycheck P varies directly as the number H of hours worked. The pay is \$165.00 for working 20 hr. Find the pay for 30 hr of work.

Find the slope, if it exists, of the line containing a given pair of points. Determine the slope of a horizontal or a vertical line. Find the slope of a line given an equation of the line.

Find the slope, if it exists, of the line containing each pair of points.

23. $(6, 8)$ and $(-2, -4)$

24. $(5, 1)$ and $(-1, 1)$

25. $(-3, 0)$ and $(-3, 5)$

26. $(-8.3, 4.6)$ and $(-9.9, 1.4)$

Find the slope, if it exists, of each line.

27. $y = -6$

28. $x = 76$

29. $4x + 3y = -12$

Given an equation of a line, find the slope–intercept equation, the slope, and the y-intercept of the line.

Find the slope and y-intercept of each line.

30. $y = -9x + 46$

31. $x + y = 9$

32. $3x - 5y = 4$

Find an equation of a line given a point on the line and the slope, or given two points on the line.

Find an equation of the line containing the given point and having the given slope.

33. $(1, 2)$, $m = 3$

34. $(-2, -5)$, $m = \dfrac{2}{3}$

35. $(0, 4)$, $m = -2$

Find an equation of the line that contains the given pair of points.

36. $(5, 7)$ and $(-1, 1)$

37. $(2, 0)$ and $(-4, -3)$

SKILL MAINTENANCE

38. A salesperson gets a weekly salary of $235 plus a $2 commission for each tire that is sold. How much did the salesperson make in a four-week period in which he sold 84 tires?

39. The perimeter of a rectangle is 326 m. The length is 87 m greater than the width. Find the length and the width.

Subtract.

40. $(9r^4 - 7r^3 + 6r^2 - 8) - (-8r^4 - 10r^2 + 5r - 9)$

41. $(2.13x^2 - 5.6x + 19) - (4.56x^2 + 7.8x - 37)$

Factor.

42. $200y - 8y^3$

43. $6x^3 + 24x^2 - 4x - 16$

44. $10x^4 + 40x^3 + 30x^2$

45. $x^7 + 8x^4 + 16x$

Simplify.

46. $4000(2 + 4.56)^2$

47. $20 \cdot 30 - 2^3 + 2^4$

EXTENSION

48. Find m in $y = mx + 3$ so that $(-2, 5)$ will be on the graph.

49. Find three solutions of $y = 4 - |x|$.

50. Write an equation of a line parallel to the x-axis and 5 units below it.

51. Find an equation of the line that has the same y-intercept as the line $5x - 3y = 15$ and the same slope as $y = -2x + 7$.

Test: Chapter 6

In which quadrant is each point located?

Find the coordinates of each point.

1. $\left(-\dfrac{1}{2}, 7\right)$ **2.** $(-5, -6)$

3. A **4.** B

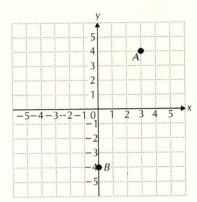

5. Determine whether the ordered pair $(2, -4)$ is a solution of the equation $y - 3x = -10$.

Graph.

6. $y = 2x - 1$

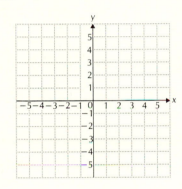

7. $2x - 4y = -8$

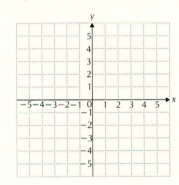

8. $y = 5$

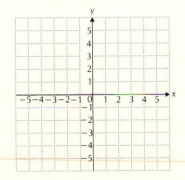

9. $y = -\dfrac{3}{2}x$

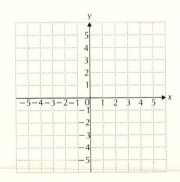

Find an equation of variation where y varies directly as x, and the following are true.

10. $y = 6$ when $x = 3$ **11.** $y = 1.5$ when $x = 3$

12. The distance d traveled by a train varies directly as the time t it travels. The train travels 60 km in $\frac{1}{2}$ hr. How far will it travel in 2 hr?

Find the slope, if it exists, of the line containing each pair of points.

13. $(4, 7)$ and $(4, -1)$ **14.** $(9, 2)$ and $(-3, -5)$

Find the slope, if it exists, of each line.

15. $y = -7$ **16.** $x = 6$

Find the slope and y-intercept.

17. $y = 2x - \dfrac{1}{4}$ **18.** $-4x + 3y = -6$

Find an equation of the line containing the given point and having the given slope.

19. $(3, 5)$, $m = 1$ **20.** $(-2, 0)$, $m = -3$

Find an equation of the line containing the given pair of points.

21. $(1, 1)$ and $(2, -2)$ **22.** $(4, -1)$ and $(-4, -3)$

 SKILL MAINTENANCE

23. Subtract: **24.** Simplify: $12 \cdot 15 - 6^2 + 5^3$.

$\left(5x^3 - 4x^2 + 6x - \dfrac{1}{2}\right) - \left(8x^3 - 7x^2 + 9x - \dfrac{1}{4}\right)$.

25. Factor: $x^7 - 3x^5 - 28x^3$.   **26.** The sum of two consecutive odd integers is 280. Find the integers.

⭐ **EXTENSION**

27. Find three solutions of $y = |x| + 1$.   **28.** Find the slope–intercept equation of the line that contains the point $(-4, 1)$ and has the same slope as the line $2x - 3y = -6$.

Chapter 6   Graphs, Linear Equations, and Slope

Cumulative Review

Evaluate each polynomial for $x = -2$.

1. $3x - 7$

2. $x^3 - x^2 + x - 1$

3. Identify the coefficient of each term of the polynomial

$$x^3 - 2x^2 + x - 1.$$

4. Identify the degree of each term and the degree of the polynomial

$$x^3 - 2x^2 + x - 1.$$

5. Tell whether the polynomial is a monomial, binomial, trinomial, or none:

$$x^3 - 2x^2 + x - 1.$$

Solve.

6. $x(x - 4) = 0$

7. $x^2 + x - 20 = 0$

8. $x^2 - 10x = 0$

9. $2x^2 + 7x = 4$

Collect like terms.

10. $x^2 - 3x^3 - 4x^2 + 5x^3 - 2$

11. $2x^3 - 7 + \frac{3}{7}x^2 - 6x^3 - \frac{4}{7}x^2 + 5$

12. Add: $(4x^4 + 6x^3 - 6x^2 - 4) + (2x^5 + 2x^4 - 4x^3 - 4x^2 + 3x - 5)$.

13. Subtract: $(-8y^2 - y + 2) - (y^3 - 6y^2 + y - 5)$

Multiply.

14. $4(3x^3 + 4x^2 + x)$

15. $(2.5a + 7.5)(0.4a - 1.2)$

16. $(2x^2 - 1)(x^3 + x - 3)$

17. $(1 - 3x^2)(2 - 4x^2)$

18. $(2x^5 + 3)(3x^2 - 6)$

19. $(2x^3 + 1)(2x^3 - 1)$

20. $(6x - 5)^2$

21. $(8 - \frac{1}{3}x)(8 + \frac{1}{3}x)$

Factor.

22. $36 - 81y$

23. $-6 - 2x - 12y$

24. $x^2 - 10x + 24$

25. $8x^2 + 10x + 3$

26. $6x^5 - 36x^3 + 9x^2$

27. $2x^2 - 18$

28. $16x^2 + 40x + 25$

29. $3x^2 + 10x - 8$

30. $m^4 + 2m^3 - 3m - 6$

31. $12t - 4t^2 - 48t^4$

32. $16y^4 - 81$

33. $6x^2 - 28x + 16$

34. $3 - 12x^6$

35. $4x^4 - 12x^2y + 9y^2$

Perform the indicated operations.

36. $(5xy^2 - 6x^2y^2 - 3xy^3) - (-4xy^3 + 7xy^2 - 2x^2y^2)$

37. $(3x^2 + 4y)(3x^2 - 4y)$

38. $(2a^2b - 5ab^2)^2$

Graph.

39. $y = \dfrac{1}{2}x$

40. $3x - 5y = 15$

41. $y = -x - 2$

42. $y = 1$

43. $x = -3$

Find an equation of variation where y varies directly as x, and the following are true.

44. $y = 8$ when $x = 12$

45. $y = 2.4$ when $x = 12$

Find the slope, if it exists, of the line containing each pair of points.

46. $(-2, 6)$ and $(-2, -1)$

47. $(-4, 1)$ and $(3, -2)$

48. $(2, 3)$ and $(-1, 3)$

49. Find the slope and the y-intercept: $4x - 3y = 6$.

Find an equation for the line containing the given point and having the given slope.

50. $(2, -3)$, $m = -4$

51. $(0, -3)$, $m = 6$

Find an equation of the line that contains the given pair of points.

52. $(-1, -3)$ and $(5, -2)$

53. $(-5, 6)$ and $(2, -4)$

Solve.

54. The product of a number and one more than the number is 20. Find the number.

55. If the sides of a square are increased by 2 ft, the sum of the areas of the two squares is 452 ft². Find the length of a side of the original square.

56. A person's salary varies directly as the number of hours worked. For working 9 hr the salary is $117. Find the salary for working 6 hr.

⭐ **EXTENSION**

57. Compute: $(x + 7)(x - 4) - (x + 8)(x - 5)$.

58. Multiply: $[4y^3 - (y^2 - 3)][4y^3 + (y^2 - 3)]$.

59. Factor: $2a^{32} - 13,122b^{40}$.

60. Solve: $(x - 4)(x + 7)(x - 12) = 0$.

61. Find an equation of the line that contains the point $(-3, -2)$ and has the same slope as the line $2x - 3y = -12$.

AN APPLICATION
There were 411 people at a play. Admission was $1.00 for adults and $0.75 for children. The receipts were $395.75. How many adults and how many children attended?

THE MATHEMATICS
We let x = the number of adults and y = the number of children. The problem translates to:

$$\left.\begin{array}{l} x + y = 411 \\ 100x + 75y = 39{,}575 \end{array}\right\}$$ ◄—— This is a system of equations.

A solution of a system of equations is a point at which two lines intersect. Methods of solving systems of equations are presented in this chapter. We will see that a problem that might otherwise be considered difficult is easy when translated to a system of equations.

The review sections to be tested in addition to the material in this chapter are 3.7, 4.7, 5.7, and 6.3.

Systems of Equations

7.1

Translate to a system of equations. Do not attempt to solve. Save for later use.

1. The sum of two numbers is 115. The difference is 21. Find the numbers.

Translating Problems to Equations

| • | The first and often hardest part of solving a problem is translating it to mathematical language. Translating becomes easier in many cases if we translate to more than one equation. When we do this, we use more than one variable.

Example 1 The sum of two numbers is 15. One number is four times the other. Find the numbers.

There are two statements in this problem. We translate the first one.

$$\underbrace{\text{The sum of two numbers}}_{x + y} \; \underbrace{\text{is}}_{} \; \underbrace{15.}_{= 15}$$

We have used x and y for the numbers. Now we translate the second statement, remembering to use x and y.

$$\underbrace{\text{One number}}_{y} \; \underbrace{\text{is}}_{=} \; \underbrace{\text{four}}_{4} \; \underbrace{\text{times}}_{\cdot} \; \underbrace{\text{the other.}}_{x}$$

For the second statement we could have also translated to $x = 4y$. This would also have been correct. The problem has been translated to a pair or *system of equations*. We list what the variables represent and then list the equations.

Let x represent one number and y represent the other number.

$$x + y = 15$$
$$y = 4x$$

DO EXERCISE 1.

Example 2 Badger Rent-a-Car rents compact cars at a daily rate of $43.95 plus 40¢ per mile. Thirsty Rent-a-Car rents compact cars at a daily rate of $42.95 plus 42¢ per mile. For what mileage is the cost the same?

We translate the first statement, using $0.40 for 40¢.

$$\underbrace{43.95}_{43.95} \; \underbrace{\text{plus}}_{+} \; \underbrace{40¢}_{0.40} \; \underbrace{\text{times}}_{\cdot} \; \underbrace{\text{the number of miles driven}}_{m} \; \underbrace{\text{is}}_{=} \; \underbrace{\text{cost.}}_{c}$$

We have let m represent the mileage and c the cost. We translate the second statement, but again it helps to reword it first.

$$\underbrace{42.95}_{42.95} \; \underbrace{\text{plus}}_{+} \; \underbrace{42¢}_{0.42} \; \underbrace{\text{times}}_{\cdot} \; \underbrace{\text{the number of miles driven}}_{m} \; \underbrace{\text{is}}_{=} \; \underbrace{\text{cost.}}_{c}$$

We have now translated to a system of equations. We then let m represent the mileage and c represent the cost.

$$43.95 + 0.40m = c$$
$$42.95 + 0.42m = c$$

DO EXERCISE 2.

> The problem in Example 2 can also be translated to one equation with one variable, as follows:
>
> $$43.95 + 0.40m = 42.95 + 0.42m.$$
>
> When solving applied problems, use one variable if it is reasonable. Use two variables when it makes the translation easier.

Making a drawing is often helpful. Whenever it makes sense to do so you should make a drawing before trying to translate.

Example 3 The perimeter of a rectangle is 90 cm. The length is 20 cm greater than the width. Find the length and the width.

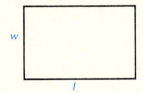

From the drawing we see that the perimeter (the distance around) of the rectangle is $l + l + w + w$, or $2l + 2w$. We translate the first statement.

The perimeter is 90 cm.

$$2l + 2w \quad = \quad 90$$

We translate the second statement.

The length is 20 cm greater than the width.

$$l \quad = \quad 20 + w$$

We have translated to a system of equations. We have let w represent the width and l represent the length.

$$2w + 2l = 90$$
$$l = 20 + w$$

DO EXERCISE 3.

Translate to a system of equations. Do not attempt to solve. Save for later use.

2. Acme Rent-a-Car rents a car at a daily rate of $31.95 plus 33¢ per mile. Speedo Rentzit rents a car for $34.95 plus 29¢ per mile. For what mileage is the cost the same?

Translate to a system of equations. Do not attempt to solve. Draw a picture if helpful. Save for later use.

3. The perimeter of a rectangle is 76 cm. The length is 17 cm more than the width. Find the length and the width.

ANSWERS ON PAGE A-16

CALCULATOR CORNER

Find the shortest path from *A* to *G*.

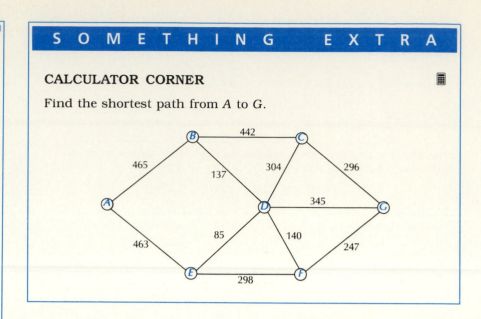

• Translate to a system of equations. Do not attempt to solve. Save for later use.

1. The sum of two numbers is 58. The difference is 16. Find the numbers.

2. The sum of two numbers is 26.4. One number is five times the other. Find the numbers.

3. The perimeter of a rectangle is 400 m. The width is 40 m less than the length. Find the length and the width.

4. The perimeter of a rectangle is 76 cm. The width is 17 cm less than the length. Find the length and the width.

5. Badger Rent-a-Car rents a compact car at a daily rate of $53.95 plus 30¢ per mile. Hartz Rent-a-Car rents a compact car at a daily rate of $54.95 plus 20¢ per mile. For what mileage is the cost the same?

6. Badger rents a basic car at a daily rate of $45.95 plus 40¢ per mile. Hartz rents a basic car at a daily rate of $46.95 plus 20¢ per mile. For what mileage is the cost the same?

7. The difference between two numbers is 16. Three times the larger number is seven times the smaller. What are the numbers?

8. The difference between two numbers is 18. Twice the smaller number plus three times the larger is 74. What are the numbers?

9. Two angles are supplementary. One is 8° more than three times the other. Find the angles. (Supplementary angles are angles whose sum is 180°.)

10. Two angles are supplementary. One is 30° more than two times the other. Find the angles.

ANSWERS
1.
2.
3.
4.
5.
6.
7.
8.
9.
10.

11. Two angles are complementary. Their difference is 34°. Find the angles. (Complementary angles are angles whose sum is 90°.)

12. Two angles are complementary. One angle is 42° more than $\frac{1}{2}$ the other. Find the angles.

13. In a vineyard a vintner uses 820 hectares to plant Chardonnay and Riesling grapes. The vintner knows that profits will be greatest by planting 140 hectares more of Chardonnay than of Riesling. How many hectares of each grape should be planted? (1 hectare is about 2.47 acres.)

14. The Hayburner Horse Farm allots 650 hectares to plant hay and oats. The owners know that their needs are best met if they plant 180 hectares more of hay than oats. How many hectares of each should they plant?

 SKILL MAINTENANCE

15. Evaluate $3x^2 - x + 8$ for $x = 4$.

16. Add: $(5x^3 - 4x^2 + 12x + 45) + (-8x^3 + 28x^2 - 49x - 63)$.

17. Multiply: $(9x^{-5})(12x^{-8})$.

18. Divide: $\dfrac{9x^{-5}}{3x^{-8}}$.

⭐ **EXTENSION**

Translate to a system of equations. Do not attempt to solve.

19. Patrick's age is 20% of his father's age. Twenty years from now, Patrick's age will be 52% of his father's age. How old are Patrick and his father?

20. If 5 is added to a mother's age and the total is then divided by 5, the result will be her daughter's age. Five years ago the mother's age was 8 times the daughter's age. Find their present ages.

21. When the base of a triangle is increased by 2 ft and the height decreased by 1 ft, the height becomes $\frac{1}{3}$ of the base and the area becomes 24 ft². Find the original dimensions of the triangle.

Systems of Equations

⬤ IDENTIFYING SOLUTIONS

After finishing Section 7.2, you
should be able to:

⬤ Determine whether an ordered
pair is a solution of a system of
equations.

⬤⬤ Solve systems of equations by
graphing.

A set of equations such as

$$x + y = 8$$
$$2x - y = 1$$

is called a *system of equations*. A *solution* of a system of two equations
is an ordered pair that makes both equations true. Consider the system
listed above. Look at the graphs. Recall that a graph of an equation is a
picture of its solution set. Each point on a graph corresponds to a
solution. Which points (ordered pairs) are solutions of *both* equations?
The graph shows that there is only one. It is the point P where the
graphs cross. This point looks as if its coordinates are $(3, 5)$. We check:

$$\begin{array}{c|c} x + y = 8 \\ \hline 3 + 5 & 8 \\ 8 & \end{array}$$

$$\begin{array}{c|c} 2x - y = 1 \\ \hline 2 \cdot 3 - 5 & 1 \\ 6 - 5 & \\ 1 & \end{array}$$

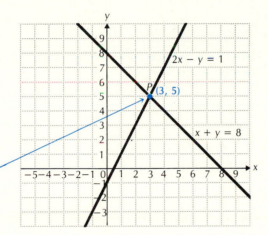

The point P is a
picture of the set
of common
solutions.

There is just one solution of the system of equations. It is $(3, 5)$. In other
words, $x = 3$ and $y = 5$.

Example 1 Determine whether $(1, 2)$ is a solution of the system

$$y = x + 1$$
$$2x + y = 4.$$

We check:
$$\begin{array}{c|c} y = x + 1 \\ \hline 2 & 1 + 1 \\ 2 & 2 \end{array}$$
$$\begin{array}{c|c} 2x + y = 4 \\ \hline 2 \cdot 1 + 2 & 4 \\ 2 + 2 & \\ 4 & \end{array}$$

$(1, 2)$ is a solution of the system.

Example 2 Determine whether $(-3, 2)$ is a solution of the system

$$a + b = -1$$
$$b + 3a = 4.$$

Determine whether the given ordered pair is a solution of the system of equations.

1. $(2, -3)$; $x = 2y + 8$
   $2x + y = 1$

2. $(20, 40)$; $a = \frac{1}{2}b$
   $b - a = 60$

Solve by graphing.

3. $2x + y = 1$
   $x = 2y + 8$

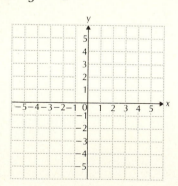

4. $y + 4 = x$
   $x - y = -2$

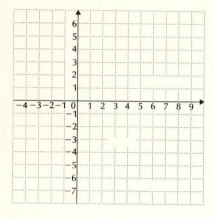

ANSWERS ON PAGE A-16

We check:

$$\begin{array}{c|c} a + b = -1 \\ \hline -3 + 2 & -1 \\ -1 & \end{array} \qquad \begin{array}{c|c} b + 3a = 4 \\ \hline 2 + 3(-3) & 4 \\ 2 - 9 & \\ -7 & \end{array}$$

The point $(-3, 2)$ is not a solution of $b + 3a = 4$. Thus it is not a solution of the system.

DO EXERCISES 1 AND 2.

•• SOLVING SYSTEMS BY GRAPHING

To solve a system of equations by graphing, we graph both equations and find coordinates of the point(s) of intersection. Then we check. If the lines are parallel, there is no solution.

Example 3 Solve by graphing: $x + 2y = 7$
   $x = y + 4$.

We graph the equations. Point P looks as if it has coordinates $(5, 1)$.

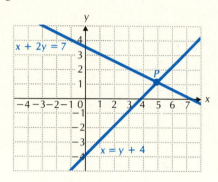

We check:

$$\begin{array}{c|c} x + 2y = 7 \\ \hline 5 + 2 \cdot 1 & 7 \\ 5 + 2 & \\ 7 & \end{array} \qquad \begin{array}{c|c} x = y + 4 \\ \hline 5 & 1 + 4 \\ 5 & 5 \end{array}$$

The solution is $(5, 1)$.

Example 4 Solve by graphing: $y = 3x + 4$
   $y = 3x - 3$.

The graphs are parallel. There is no point at which they cross, so the system has no solution.

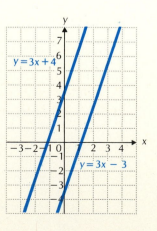

DO EXERCISES 3 AND 4.

· Determine whether the given ordered pair is a solution of the given system of equations. Remember to use alphabetical order of variables.

1. $(3, 2)$; $2x + 3y = 12$
             $x - 4y = -5$

2. $(1, 5)$; $5x - 2y - -5$
             $3x - 7y = -32$

3. $(-3, 4)$; $2x = -y - 2$
              $y + 7x = 9$

4. $(3, -2)$; $3t - 2s = 0$
              $t + 2s = 15$

5. $(2, -2)$; $b + 2a = 2$
             $b - a = -4$

6. $(-1, -3)$; $3r + s = -6$
              $2r = 1 + s$

· · Use graph paper. Solve by graphing. Check on this paper.

7. $x + 2y = 10$
   $3x + 4y = 8$
          Check: $x + 2y = 10$ $3x + 4y = 8$

8. $u = v$
   $4u = 2v - 6$
          Check: $u = v$ $4u = 2v - 6$

9. $8x - y = 29$
   $2x + y = 11$
          Check: $8x - y = 29$ $2x + y = 11$

ANSWERS
1. _____
2. _____
3. _____
4. _____
5. _____
6. _____
7. _____
8. _____
9. _____

10.

11.

12.

13.

14.

15.

16.

17.

18.

19.

20.

21.

22.

10. $4x - y = 10$ Check: $\underline{4x - y = 10}$ $\underline{3x + 5y = 19}$
    $3x + 5y = 19$

11. $a - b = 6$ Check: $\underline{a - b = 6}$ $\underline{a - 7 = b}$
    $a - 7 = b$

12. $x - 2y = 6$ Check: $\underline{x - 2y = 6}$ $\underline{2x - 3y = 5}$
    $2x - 3y = 5$

13. $y = 3$ **14.** $x + y = 4$
    $x = 5$ $x + y = -4$

SKILL MAINTENANCE

15. Simplify: $(5x^{-2}y^5)^{-4}$.

16. Divide: $\dfrac{-124}{-62}$.

17. Multiply: $(5x - 6)^2$.

18. Solve: $x^2 - 4x - 21 = 0$.

⭐ **EXTENSION**

19. The solution of the following system is $(2, -3)$. Find A and B.

$$Ax - 3y = 13$$
$$x - By = 8$$

20. Determine whether the ordered pair $(-1, -5)$ is a solution of the following system of three equations.

$$4a - b = 1$$
$$-a + b = -4$$
$$2a + 3b = -17$$

21. Describe in words the graph of the system in Exercise 20. Describe the solution.

22. Solve this system by graphing. What happens when you check your possible solution?

$$3x + 7y = 5$$
$$6x - 7y = 1$$

Solving by Substitution

• THE SUBSTITUTION METHOD

After finishing Section 7.3, you should be able to:

• Solve a system of two equations by the substitution method when one of them has a variable alone on one side.

•• Solve a system of two equations by the substitution method when neither equation has a variable alone on one side.

Graphing helps picture the solution of a system of equations, but solving by graphing is not fast or accurate. Let's learn better ways.

Example 1 Solve the system: $x + y = 6$
$$x = y + 2.$$

The second equation says that x and $y + 2$ name the same number. Thus in the first equation, we can substitute $y + 2$ for x:

$$x + y = 6$$
$$(y + 2) + y = 6. \qquad \text{Substituting } y + 2 \text{ for } x$$

This last equation has only one variable. We solve it:

$$(y + 2) + y = 6$$
$$2y + 2 = 6 \qquad \text{Collecting like terms}$$
$$2y = 4$$
$$y = 2.$$

We return to the original pair of equations. We substitute 2 for y in either of them. We use the second equation since x is alone on one side:

$$x = y + 2$$
$$x = 2 + 2 \qquad \text{Substituting 2 for } y$$
$$x = 4.$$

The ordered pair (4, 2) may be a solution. We check.

Check:

$$\begin{array}{c|c} x + y = 6 \\ \hline 4 + 2 & 6 \\ 6 & \end{array} \qquad \begin{array}{c|c} x = y + 2 \\ \hline 4 & 2 + 2 \\ & 4 \end{array}$$

Since (4, 2) checks, we have the solution. We could also express the answer as $x = 4$, $y = 2$.

DO EXERCISE 1.

Example 2 Solve: $s = 13 - 3t$
$$s + t = 5.$$

We substitute $13 - 3t$ for s in the second equation:

$$s + t = 5$$
$$(13 - 3t) + t = 5. \qquad \text{Substituting } 13 - 3t \text{ for } s$$

Now we solve for t:

$$13 - 2t = 5 \qquad \text{Collecting like terms}$$
$$-2t = -8 \qquad \text{Adding } -13$$
$$t = \frac{-8}{-2}, \text{ or } 4 \qquad \text{Multiplying by } \frac{1}{-2}$$

Solve by the substitution method. Do not graph.

1. $x + y = 5$
$x = y + 1$

ANSWER ON PAGE A-16

Next we substitute 4 for t in the first equation of the original system:

$$s = 13 - 3t$$
$$s = 13 - 3(4) \qquad \text{Substituting 4 for } t$$
$$s = 1.$$

Check:

$s = 13 - 3t$		$s + t = 5$	
1	$13 - 3 \cdot 4$	$1 + 4$	5
	$13 - 12$		5
	1		

Since (1, 4) checks, it is the solution.

DO EXERCISE 2.

•• SOLVING FOR THE VARIABLE FIRST

Sometimes neither equation of a pair has a variable alone on one side. Then we solve one equation for one of the variables and proceed as before.

Example 3 Solve: $x - 2y = 6$
              $3x + 2y = 4.$

We solve one equation for one variable. Since the coefficient of x is 1 in the first equation, it is easier to solve it for x:

$$x - 2y = 6$$
$$x = 6 + 2y.$$

We substitute $6 + 2y$ for x in the second equation of the original pair and solve:

$$3x + 2y = 4$$
$$3(6 + 2y) + 2y = 4 \qquad \text{Substituting } 6 + 2y \text{ for } x$$
$$18 + 6y + 2y = 4$$
$$18 + 8y = 4$$
$$8y = -14$$
$$y = \frac{-14}{8}, \text{ or } -\frac{7}{4}.$$

We go back to either of the original equations and substitute $-\frac{7}{4}$ for y. It will be easier to solve for x in the first equation:

$$x - 2y = 6$$
$$x - 2(-\tfrac{7}{4}) = 6$$
$$x + \tfrac{7}{2} = 6$$
$$x = 6 - \tfrac{7}{2}$$
$$x = \tfrac{5}{2}$$

Check:

$x - 2y = 6$		$3x + 2y = 4$	
$\frac{5}{2} - 2(-\frac{7}{4})$	6	$3 \cdot \frac{5}{2} + 2(-\frac{7}{4})$	4
$\frac{5}{2} + \frac{7}{2}$		$\frac{15}{2} - \frac{7}{2}$	
$\frac{12}{2}$		$\frac{8}{2}$	
6		4	

Since $(\frac{5}{2}, -\frac{7}{4})$ checks, it is the solution.

DO EXERCISE 3.

Chapter 7   Systems of Equations

• Solve by the substitution method.

1. $x + y = 4$
$y = 2x + 1$

2. $x + y = 10$
$y = x + 8$

3. $y = x + 1$
$2x + y - 4$

4. $y = x - 6$
$x + y = -2$

5. $y = 2x - 5$
$3y - x = 5$

6. $y = 2x + 1$
$x + y = -2$

7. $x = -2y$
$x + 4y = 2$

8. $r = -3s$
$r + 4s = 10$

•• Solve by the substitution method. Get one variable alone first.

9. $s + t = -4$
$s - t = 2$

10. $x - y = 6$
$x + y = -2$

11. $y - 2x = -6$
$2y - x = 5$

12. $x - y = 5$
$x + 2y = 7$

13. $2x + 3y = -2$
$2x - y = 9$

14. $x + 2y = 10$
$3x + 4y = 8$

ANSWERS
1.
2.
3.
4.
5.
6.
7.
8.
9.
10.
11.
12.
13.
14.

15. $x - y = -3$
    $2x + 3y = -6$

16. $3b + 2a = 2$
    $-2b + a = 8$

17. $r - 2s = 0$
    $4r - 3s = 15$

18. $y - 2x = 0$
    $3x + 7y = 17$

19. $x - 3y = 7$
    $-4x + 12y = 28$

20. $8x + 2y = 6$
    $4x = 3 - y$

✓ SKILL MAINTENANCE

Multiply.

21. $(7x^2 - 4)(7x^2 + 4)$

22. $(5t - 3m)(5t + 3m)$

Solve.

23. $3x + 5x - 4 = 2(x - 7)$

24. $25t^2 - 49 = 0$

⭐ EXTENSION

Solve by the substitution method.

25. ▦ $y - 2.35x = -5.97$
    $2.14y - x = 4.88$

26. $\frac{1}{4}(a - b) = 2$
    $\frac{1}{6}(a + b) = 1$

27. $\dfrac{x}{2} + \dfrac{3y}{2} = 2$

    $\dfrac{x}{5} - \dfrac{y}{2} = 3$

28. $0.4x + 0.7y = 0.1$
    $0.5x - 0.1y = 1.1$

The Addition Method

• SOLVING BY THE ADDITION METHOD

Another method of solving systems of equations is called the *addition method.*

Example 1 Solve: $x + y = 5$
$\qquad\qquad x - y = 1.$

We will use the addition principle for equations. According to the second equation, $x - y$ and 1 are the same number. Thus we can add $x - y$ to the left side of the first equation and 1 to the right side:

$$\begin{array}{r} x + y = 5 \\ \underline{x - y = 1} \\ 2x + 0y = 6. \end{array}$$

We have made one variable "disappear." We now have an equation with just one variable:

$$2x = 6.$$

We solve for x: $x = 3$. Next we substitute 3 for x in either of the original equations:

$$\begin{array}{ll} x + y = 5 & \\ 3 + y = 5 & \text{Substituting 3 for } x \text{ in the first equation} \\ \qquad y = 2. & \text{Solving for } y \end{array}$$

Check:

$$\begin{array}{c|c} x + y = 5 & x - y = 1 \\ \hline 3 + 2 \;\big|\; 5 & 3 - 2 \;\big|\; 1 \\ 5 \;\big| & 1 \;\big| \end{array}$$

Since (3, 2) checks, it is the solution.

DO EXERCISES 1 AND 2.

• • USING THE MULTIPLICATION PRINCIPLE FIRST

The addition method allows us to eliminate a variable. We may need to multiply by -1 to make this happen.

Example 2 Solve: $2x + 3y = 8$
$\qquad\qquad x + 3y = 7.$

If we add, we do not eliminate a variable. However, if the $3y$ were $-3y$ in one equation we could. We multiply on both sides of the second equation by -1 and then add:

$$\begin{array}{ll} 2x + 3y = 8 & \\ \underline{-x - 3y = -7} & \text{Multiplying by } -1 \\ x = 1. & \text{Adding} \end{array}$$

• Solve a system of two equations using the addition method when no multiplication is necessary.

• • Solve a system of two equations using the addition method when the multiplication principle must be used.

Solve using the addition method.

1. $x + y = 5$
$ 2x - y = 4$

2. $3x - 3y = 6$
$ 3x + 3y = 0$

ANSWERS ON PAGE A-16

3. $5x + 3y = 17$
$\quad 5x - 2y = -3$

Solve.

4. $4a + 7b = 11$
$\quad 2a + 3b = 5$

Now we substitute 1 for x in one of the original equations:

$$x + 3y = 7$$
$$1 + 3y = 7 \qquad \text{Substituting 1 for } x \text{ in the second equation}$$
$$3y = 6$$
$$y = 2. \qquad \text{Solving for } y$$

Check:

$2x + 3y = 8$	
$2 \cdot 1 + 3 \cdot 2$	8
$2 + 6$	
8	

$x + 3y = 7$	
$1 + 3 \cdot 2$	7
$1 + 6$	
7	

Since $(1, 2)$ checks, it is the solution.

DO EXERCISE 3.

In Example 2 we used the multiplication principle, multiplying by -1. We often need to multiply by something other than -1.

Example 3 Solve: $3x + 6y = -6$
$\qquad\qquad\qquad 5x - 2y = 14.$

This time we multiply by 3 on both sides of the second equation. Then we add:

$$3x + 6y = -6$$
$$\underline{15x - 6y = 42} \qquad \text{Multiplying by 3}$$
$$18x \quad\quad = 36 \qquad \text{Adding}$$
$$x \quad\quad = 2. \qquad \text{Solving for } x$$

We go back to the first equation and substitute 2 for x:

$$3 \cdot 2 + 6y = -6 \qquad \text{Substituting}$$
$$6 + 6y = -6$$
$$6y = -12$$
$$y = -2. \qquad \text{Solving for } y$$

Check:

$3x + 6y = -6$	
$3 \cdot 2 + 6 \cdot (-2)$	-6
$6 + (-12)$	
-6	

$5x - 2y = 14$	
$5 \cdot 2 - 2 \cdot (-2)$	14
$10 - (-4)$	
14	

Since $(2, -2)$ checks, it is the solution.

Caution! Solving a *system* of equations in two variables requires finding an ordered *pair* of numbers. Once you have solved for one variable, don't forget the other.

DO EXERCISE 4.

Chapter 7 Systems of Equations

Example 4 Solve: $3x + 5y = 6$
$5x + 3y = 4.$

We use the multiplication principle with both equations:

$3x + 5y = 6$
$5x + 3y = 4$

$15x + 25y = 30$ Multiplying on both sides of the first equation by 5

$-15x - 9y = -12$ Multiplying on both sides of the second equation by -3

$\overline{ }$

$16y = 18$ Adding

$y = \dfrac{18}{16}, \text{ or } \dfrac{9}{8}.$

We substitute $\dfrac{9}{8}$ for y in one of the original equations:

$3x + 5y = 6$

$3x + 5 \cdot \dfrac{9}{8} = 6$ Substituting $\dfrac{9}{8}$ for y in the first equation

$3x + \dfrac{45}{8} = 6$

$3x = 6 - \dfrac{45}{8}$

$3x = \dfrac{48}{8} - \dfrac{45}{8}$

$3x = \dfrac{3}{8}$

$x = \dfrac{3}{8} \cdot \dfrac{1}{3}, \text{ or } \dfrac{1}{8}.$ Solving for x

Check:

$$\begin{array}{c|c}
3x + 5y = 6 & \\
\hline
3 \cdot \dfrac{1}{8} + 5 \cdot \dfrac{9}{8} & 6 \\
\dfrac{3}{8} + \dfrac{45}{8} & \\
\dfrac{48}{8} & \\
6 &
\end{array}$$

$$\begin{array}{c|c}
5x + 3y = 4 & \\
\hline
5 \cdot \dfrac{1}{8} + 3 \cdot \dfrac{9}{8} & 4 \\
\dfrac{5}{8} + \dfrac{27}{8} & \\
\dfrac{32}{8} & \\
4 &
\end{array}$$

The solution is $\left(\dfrac{1}{8}, \dfrac{9}{8} \right).$

DO EXERCISES 5 AND 6.

Solve.

5. $5x + 3y = 2$
$3x + 5y = -2$

6. $2x + 3y = 1$
$3x - 2y = 5$

ANSWERS ON PAGE A-16

Solve.

7. $2x + y = 15$
$ 4x + 2y = 23$

Solve.

8. $\dfrac{1}{2}x + \dfrac{3}{10}y = \dfrac{1}{5}$
$ \dfrac{3}{10}x + \dfrac{1}{2}y = -\dfrac{1}{5}$

Some systems have no solution.

Example 5 Solve: $y - 3x = 2$
$ y - 3x = 1.$

We multiply by -1 on both sides of the second equation:

$$
\begin{aligned}
y - 3x &= 2 \\
-y + 3x &= -1 \qquad \text{Multiplying by } -1 \\
\hline
0 &= 1. \qquad \text{Adding}
\end{aligned}
$$

We obtain a false equation $0 = 1$, so there is no solution. The graphs of the equations are parallel lines. They do not intersect.

DO EXERCISE 7.

When decimals or fractions appear, multiply to clear them. Then proceed as before.

Example 6 Solve: $\dfrac{1}{4}x + \dfrac{5}{12}y = \dfrac{1}{2}$

$ \dfrac{5}{12}x + \dfrac{1}{4}y = \dfrac{1}{3}.$

The number 12 is a multiple of all the denominators. We multiply on both sides of each equation by 12:

$$12\left(\frac{1}{4}x + \frac{5}{12}y\right) = 12 \cdot \frac{1}{2} \qquad\qquad 12\left(\frac{5}{12}x + \frac{1}{4}y\right) = 12 \cdot \frac{1}{3}$$

$$12 \cdot \frac{1}{4}x + 12 \cdot \frac{5}{12}y = 6 \qquad\qquad 12 \cdot \frac{5}{12}x + 12 \cdot \frac{1}{4}y = 4$$

$$3x + 5y = 6; \qquad\qquad 5x + 3y = 4.$$

The resulting system is

$$3x + 5y = 6$$
$$5x + 3y = 4.$$

The solution of this system is given in Example 4.

WHICH METHOD IS BETTER?

Although there are exceptions, the substitution method is probably better when a variable has a coefficient of 1, as in the following examples:

$$8x + 4y = 180 \qquad\qquad x + y = 561$$
$$-x + y = 3; \qquad\qquad x = 2y.$$

Otherwise, the addition method is better. Both methods work. When in doubt, use the addition method.

DO EXERCISE 8.

Solve using the addition method.

1. $x + y = 10$
$x - y = 8$

2. $x - y = 7$
$x + y = 3$

3. $x + y = 8$
$2x - y = 7$

4. $x + y = 6$
$3x - y = -2$

5. $3a + 4b = 7$
$a - 4b = 5$

6. $7c + 5d = 18$
$c - 5d = -2$

7. $8x - 5y = -9$
$3x + 5y = -2$

8. $7a - 3b = -12$
$-4a + 3b = -3$

Solve using the addition method. Use the multiplication principle first.

9. $-x - y = 8$
$2x - y = -1$

10. $x + y = -7$
$3x + y = -9$

11. $x + 3y = 19$
$x + 3y = -1$

12. $4x - y = 1$
$4x - y = 7$

13. $3x - 2y = 10$
$5x + 3y = 4$

14. $2p + 5q = 9$
$3p - 2q = 4$

ANSWERS
1.
2.
3.
4.
5.
6.
7.
8.
9.
10.
11.
12.
13.
14.

15. $2a + 3b = -1$
$3a + 5b = -2$

16. $5x - 2y = 0$
$2x - 3y = -11$

17. $0.3x + 0.2y = 0$
$x + 0.5y = -0.5$

18. $0.4x + 0.1y = 0.1$
$0.6x + 0.2y = 0.3$

19. $\dfrac{3}{4}x + \dfrac{1}{3}y = 8$
$\dfrac{1}{2}x - \dfrac{5}{6}y = -1$

20. $\dfrac{2}{3}r - \dfrac{1}{5}s = 2$
$\dfrac{4}{3}r + 4s = -4$

21. $m - n = 32$
$3m - 8n - 6 = 0$

22. $x - \dfrac{3}{2}y = 13$
$\dfrac{3}{2}x - y = 17$

23. $0.06x + 0.05y = 0.07$
$0.04x - 0.03y = 0.11$

Copyright © 1987 Addison-Wesley Publishing Co., Inc.

✓ **SKILL MAINTENANCE**

24. Identify the degree of the polynomial: $5x^4 - 7x^3 + 3x^2 - 8x^5$.

25. Collect like terms: $5x^4 - 7x^3 + 9x^4 + 16x^3 - 19x^2$.

26. 35 is what percent of 105?

27. Find the LCM: 15, 45, 60.

⭐ **EXTENSION**

Solve.

28. $4.05x + 2.53y = 1.23$
$1.12x + 1.43y = 2.81$

29. $3(x - y) = 9$
$x + y = 7$

30. $2(x - y) = 3 + x$
$x = 3y + 4$

31. $\dfrac{x}{3} + \dfrac{y}{2} = 1\dfrac{1}{3}$
$x + 0.05y = 4$

Solving Problems

• When solving problems involving systems, any method, substitution or addition, can be used.

Example 1 Badger Rent-a-Car rents compact cars at a daily rate of $43.95 plus 40¢ per mile. Thirsty Rent-a-Car rents compact cars at a daily rate of $42.95 plus 42¢ per mile. For what mileage is the cost the same?

1. Acme rents a car at a daily rate of $31.95 plus 33¢ per mile. Speedo Rentzit rents a car for $34.95 plus 29¢ per mile. For what mileage is the cost the same?

(See Example 2 in Section 7.1.) If we let m represent the mileage and c represent the cost, we see that the cost for Badger will be

$$43.95 + 0.40m = c.$$

The cost for Thirsty will be

$$42.95 + 0.42m = c.$$

We solve the system

$$43.95 + 0.40m = c$$
$$42.95 + 0.42m = c.$$

We use substitution since there is a variable alone on one side of an equation—in fact, both:

$$43.95 + 0.40m = 42.95 + 0.42m$$
$$100(43.95 + 0.40m) = 100(42.95 + 0.42m) \quad \text{Multiplying by 100 to clear of decimals}$$
$$4395 + 40m = 4295 + 42m \quad \text{Simplifying}$$
$$100 + 40m = 42m \quad \text{Adding } -4295$$
$$100 = 2m \quad \text{Adding } -40m$$
$$\frac{100}{2} = m$$
$$50 = m.$$

Thus if the cars are driven 50 miles, the cost will be the same.

DO EXERCISE 1.

Example 2 Howie is 21 years older than Izzi. In six years Howie will be twice as old as Izzi. How old are they now?

Let us consider some conditions of the problem. We let x represent Howie's age now and y Izzi's age now. We know that now Howie is 21 years older than Izzi. We make a table to organize our information. How do the ages relate in six years? In six years Izzi will be $y + 6$ and Howie will be $x + 6$, or $2(y + 6)$.

	Age now	Age in 6 years
Howie	x, or $y + 21$	$x + 6$, or $2(y + 6)$
Izzi	y	$y + 6$

From the present ages we get the following rewording and translation.

Howie's age is 21 more than Izzi's age.   Rewording
$\downarrow$　$\downarrow\ \downarrow$　$\downarrow$　$\downarrow$
x　$= 21$　$+$　y Translating

2. Person A is 26 years older than Person B. In five years, A will be twice as old as B. How old are they now?

From their ages in six years we get the following rewording and translation.

$$\underbrace{\text{Howie's age in six years}}_{x + 6} \underbrace{\text{will be}}_{=} \underbrace{\text{twice}}_{2 \cdot} \underbrace{\text{Izzi's age in six years}}_{(y + 6)} \quad \text{Rewording / Translating}$$

We solve the system of equations

$$x = 21 + y$$
$$x + 6 = 2(y + 6).$$

We use the substitution method since there is a variable alone on one side of the equation:

$$(21 + y) + 6 = 2(y + 6) \quad \text{Substituting } 21 + y \text{ for } x \text{ in the second equation}$$

$$y + 27 = 2(y + 6) \quad \text{Collecting like terms}$$

$$y + 27 = 2y + 12 \quad \text{Removing parentheses}$$

$$15 = y. \quad \text{Adding } -12 \text{ and } -y$$

We find x by substituting 15 for y in $x = 21 + y$:

$$x = 21 + 15$$
$$x = 36.$$

We check in the original problem. Howie's age is 36, which is 21 more than 15, Izzi's age. In six years Howie will be 42 and Izzi 21. Howie's age will be twice Izzi's age. The answer is that Howie is 36 and Izzi is 15.

DO EXERCISE 2.

3. The perimeter of a rectangle is 76 cm. The length is 17 cm more than the width. Find the length and the width.

Example 3 The perimeter of a rectangle is 90 cm. The length is 20 cm greater than the width. Find the length and the width.

In Example 3 in Section 7.1, we translated this problem to the system

$$2l + 2w = 90$$
$$l = 20 + w,$$

where we have let l represent the length and w represent the width. We solve, first adding $-w$ on both sides of $l = 20 + w$:

$$2l + 2w = 90 \qquad 2l + 2w = 90$$
$$l - w = 20 \qquad \underline{2l - 2w = 40} \quad \text{Multiplying by 2}$$
$$4l \qquad = 130 \quad \text{Adding}$$
$$l = 32.5.$$

$$32.5 - w = 20 \quad \text{Substituting in the equation } l - w = 20$$
$$-w = -12.5$$
$$w = 12.5$$

We leave the check to the student. The length is 32.5 cm and the width is 12.5 cm.

DO EXERCISE 3.

ANSWERS ON PAGE A-16

Example 4 A student has some nickels and dimes. The value of the coins is $1.65. There are 12 more nickels than dimes. How many of each kind of coin are there?

We will use d to represent the number of dimes and n to represent the number of nickels. We have one equation at once:

$$d + 12 = n.$$

The value of the nickels, in cents, is $5n$, since each is worth 5¢. The value of the dimes, in cents, is $10d$, since each is worth 10¢. The total value is given as $1.65. Since we have the values of the nickels and dimes *in cents*, we must use cents for the total value. This is 165. Now we have another equation:

$$10d + 5n = 165.$$

Thus we have a system of equations to solve:

$$d + 12 = n$$
$$10d + 5n = 165.$$

Since we have n alone on one side of one equation, we use the substitution method. We substitute $d + 12$ for n in the second equation:

$$10d + 5n = 165$$
$$10d + 5(d + 12) = 165 \qquad \text{Substituting } d + 12 \text{ for } n$$
$$10d + 5d + 60 = 165 \qquad \text{Multiplying to remove parentheses}$$
$$15d + 60 = 165 \qquad \text{Collecting like terms}$$
$$15d = 105 \qquad \text{Adding } -60$$
$$d = \frac{105}{15}, \quad \text{or } 7. \qquad \text{Multiplying by } \frac{1}{15}$$

We substitute 7 for d in either of the original equations to find n. We use the first equation:

$$d + 12 = n$$
$$7 + 12 = n$$
$$19 = n.$$

The solution of this system is $d = 7$ and $n = 19$. This checks, so the student has 7 dimes and 19 nickels.

DO EXERCISE 4.

Example 5 There were 411 people at a play. Admission was $1.00 for adults and $0.75 for children. The receipts were $395.75. How many adults and how many children attended?

Let's organize the information in a table. This is often helpful in translating.

People	Paid	Number attending	Money taken in
Adults	$1.00	x	$1.00x$
Children	$0.75	y	$0.75y$
	Totals	411	$395.75

$$x + y = 411 \qquad 1.00x + 0.75y = 395.75$$

4. On a table are 20 coins, quarters and dimes. Their value is $3.05. How many of each kind of coin are there?

ANSWER ON PAGE A-16

5. There were 166 paid admissions to a game. The price was $2.10 for adults and $0.75 for children. The amount taken in was $293.25. How many adults and how many children attended?

We will use x for the number of adults and y for the number of children. The total number of people attending was 411, so

$$x + y = 411.$$

The amount taken in from the adults was $1.00x$, and the amount taken in from the children was $0.75y$. These amounts are in dollars. The total was $395.75, so we have

$$1.00x + 0.75y = 395.75.$$

We can multiply on both sides by 100 to clear of decimals. Thus we have the translation, a system of equations:

$$x + y = 411$$
$$100x + 75y = 39,575.$$

We solve the system of equations using the addition method. We multiply on both sides of the first equation by -100 and then add:

$$-100x - 100y = -41,100 \qquad \text{Multiplying by } -100$$
$$\underline{100x + \ 75y = \ \ 39,575}$$
$$-25y = -1525 \qquad \text{Adding}$$
$$y = \frac{-1525}{-25} \qquad \text{Dividing by } -25$$
$$y = 61.$$

We go back to the first equation and substitute 61 for y:

$$x + y = 411$$
$$x + 61 = 411$$
$$x = 350.$$

The solution of the system is $x = 350$ and $y = 61$. This checks, so we know that 350 adults and 61 children attended.

DO EXERCISE 5.

Example 6 A chemist has one solution that is 80% acid (and the rest water) and another solution that is 30% acid. What is needed is 200 L of a solution that is 62% acid. The chemist will prepare it by mixing the two solutions on hand. How much of each should be used?

We can draw a picture of the situation. The chemist uses x liters of the first solution and y liters of the second solution.

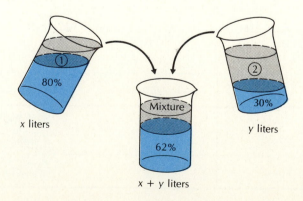

We can arrange the information in a table.

Type of solution	Amount of solution	Percent of acid	Amount of acid in solution
1 2	x y	80% 30%	80%x 30%y
Mixture	200 liters	62%	62% × 200, or 124 liters

$x + y = 200$ ←   80%x + 30%y = 124

The chemist uses x liters of the first solution and y liters of the second. Since the total is to be 200 liters, we have

$$x + y = 200.$$

The *amount* of acid in the new mixture is to be 62% of 200 liters, or 124 liters. The amounts of acid from the two solutions are 80%x and 30%y. Thus,

$$80\%x + 30\%y = 124, \quad \text{or} \quad 0.8x + 0.3y = 124.$$

We clear the decimals by multiplying on both sides by 10:

$$10(0.8x + 0.3y) = 10 \cdot 124$$
$$8x + 3y \quad = 1240.$$

We solve the system

$$x + \quad y = \quad 200$$
$$8x + 3y = 1240.$$

We use the addition method. We multiply on both sides of the first equation by -3 and then add:

$-3x - 3y = -600$	Multiplying by -3
$\underline{8x + 3y = 1240}$	
$5x \quad\quad = \quad 640$	Adding
$x = \dfrac{640}{5}$	Multiplying by $\dfrac{1}{5}$
$x = 128.$	Solving for x

We go back to the first equation and substitute 128 for x:

$$x + y = 200$$
$$128 + y = 200$$
$$y = 72.$$

The solution is $x = 128$ and $y = 72$.

Check: The sum of 128 and 72 is 200. Now 80% of 128 is 102.4 and 30% of 72 is 21.6. These add up to 124. Thus the chemist should use 128 liters of the stronger acid and 72 liters of the other.

DO EXERCISE 6.

6. One solution is 50% alcohol and a second is 70% alcohol. How much of each should be used to make 30 L of a solution that is 55% alcohol?

ANSWER ON PAGE A-16

7. Grass seed A is worth $1.00 per pound and seed B is worth $1.35 per pound. How much of each would you use to make 50 lb of a mixture worth $1.14 per pound?

Example 7 A grocer wishes to mix some candy worth 45¢ per pound and some worth 80¢ per pound to make 350 lb of a mixture worth 65¢ per pound. How much of each should be used?

Arranging information in a table will help.

Type of candy	Cost of candy	Amount (in lb)	Value (in ¢)
1	45¢	x	$45x$
2	80¢	y	$80y$
Mixture	65¢	350	65¢(350), or 22,750

$$x + y = 350 \qquad 45x + 80y = 22{,}750$$

We use x for the amount of 45¢ candy and y for the amount of 80¢ candy. Then

$$x + y = 350.$$

Our second equation will come from the values. The value of the first candy, in cents, is $45x$ (x pounds at 45¢ per pound). The value of the second is $80y$, and the value of the mixture is 350×65. Thus we have

$$45x + 80y = 350 \times 65, \quad \text{or} \quad 22{,}750.$$

We solve the system

$$x + y = 350$$
$$45x + 80y = 22750.$$

We use the addition method. We multiply on both sides of the first equation by -45 and then add:

$$
\begin{array}{rl}
-45x - 45y = -15750 & \textbf{Multiplying by } -45 \\
\underline{45x + 80y = 22750} & \\
35y = 7000 & \textbf{Adding} \\
y = \dfrac{7000}{35} & \\
y = 200. &
\end{array}
$$

We go back to the first equation and substitute 200 for y:

$$x + y = 350$$
$$x + \mathbf{200} = 350$$
$$x = 150.$$

We get $x = 150$ lb and $y = 200$ lb. This checks, so the grocer mixes 150 lb of the 45¢ candy with 200 lb of the 80¢ candy.

DO EXERCISE 7.

ANSWER ON PAGE A-16

Chapter 7   Systems of Equations

• Translate to systems of equations and solve. Many problems have already been translated in Exercise Set 7.1.

1. Badger rents an intermediate-size car at a daily rate of $53.95 plus 30¢ per mile. Another company rents an intermediate-size car for $54.95 plus 20¢ per mile. For what mileage is the cost the same?

2. Badger rents a basic car at a daily rate of $45.95 plus 40¢ per mile. Another company rents a basic car for $46.95 plus 20¢ per mile. For what mileage is the cost the same?

1. _____

2. _____

3. Sammy Tary is twice as old as his daughter. In four years Sammy's age will be three times what his daughter's age was six years ago. How old is each at present?

4. Ann Shent is eighteen years older than her son. One year ago she was three times as old as her son was. How old is each?

3. _____

4. _____

5. The difference between two numbers is 16. Three times the larger number is seven times the smaller. What are the numbers?

6. The difference between two numbers is 18. Twice the smaller number plus three times the larger is 74. What are the numbers?

5. _____

6. _____

7. Two angles are supplementary. One is 8° more than three times the other. Find the angles. (Supplementary angles are angles whose sum is 180°.)

8. Two angles are supplementary. One is 30° more than two times the other. Find the angles.

9. Two angles are complementary. Their difference is 34°. Find the angles. (Complementary angles are angles whose sum is 90°.)

10. Two angles are complementary. One angle is 42° more than one half the other. Find the angles.

11. In a vineyard a vintner uses 820 hectares to plant Chardonnay and Riesling grapes. The vintner knows that profits will be greatest by planting 140 hectares more of Chardonnay than Riesling. How many hectares of each grape should be planted? (A *hectare* is a metric unit of area, about 2.47 acres.)

12. The Hayburner Horse Farm allots 650 hectares to plant hay and oats. The owners know that their needs are best met if they plant 180 hectares more of hay than oats. How many hectares of each should they plant?

Chapter 7   Systems of Equations

13. The perimeter of a rectangle is 400 m. The width is 40 m less than the length. Find the length and the width.

14. The perimeter of a rectangle is 76 cm. The width is 17 cm less than the length. Find the length and the width.

15. A collection of dimes and quarters is worth $15.25. There are 103 coins in all. How many of each kind of coin are there?

16. A collection of quarters and nickels is worth $1.25. There are 13 coins in all. How many of each kind of coin are there?

17. A collection of nickels and dimes is worth $25. There are three times as many nickels as dimes. How many of each kind of coin are there?

18. A collection of nickels and dimes is worth $2.90. There are 19 more nickels than dimes. How many of each kind of coin are there?

13. _____

14. _____

15. _____

16. _____

17. _____

18. _____

19. There were 429 people at a play. Admission was $1 for adults and 75¢ for children. The receipts were $372.50. How many adults and how many children attended?

20. The attendance at a school concert was 578. Admission was $2 for adults and $1.50 for children. The receipts were $985. How many adults and how many children attended?

19. _____

20. _____

21. There were 200 tickets sold for a college basketball game. Tickets for students were $0.50 and for adults were $0.75. The total amount of money collected was $132.50. How many of each type of ticket were sold?

22. There were 203 tickets sold for a wrestling match. For activity card holders the price was $1.25 and for non-card holders the price was $2. The total amount of money collected was $310. How many of each type of ticket were sold?

21. _____

22. _____

23. Solution A is 50% acid and solution B is 80% acid. How much of each should be used to make 100 milliliters (mL) of a solution that is 68% acid? (*Hint:* 68% of what is acid?)

24. Solution A is 30% alcohol and solution B is 75% alcohol. How much of each should be used to make 100 L of a solution that is 50% alcohol?

23. _____

24. _____

Chapter 7   Systems of Equations

25. Farmer Jones has milk A, which is 1% butterfat, and milk B, which is 5% butterfat. How many gallons of each kind of milk are needed to make 400 gal of a mixture that is 2% butterfat?

26. Tank A contains a solution that is 40% acid. Tank B contains a solution that is 90% acid. How many gallons of each mixture are needed to make 1000 gal of a solution that is 70% acid?

27. A solution containing 30% insecticide is to be mixed with a solution containing 50% insecticide to make 200 L of a solution containing 42% insecticide. How much of each solution should be used?

28. A solution containing 28% fungicide is to be mixed with a solution containing 40% fungicide to make 300 L of a solution containing 36% fungicide. How much of each solution should be used?

29. The Nuthouse has 10 kg of mixed cashews and pecans worth $8.40 per kilogram. Cashews alone sell for $8.00 per kilogram and pecans sell for $9.00 per kilogram. How many kilograms of each are in the mixture?

30. A coffee shop mixed Brazilian coffee worth $5 per kilogram with Turkish coffee worth $8 per kilogram. The mixture is to sell for $7 per kilogram. How much of each type of coffee should be used to make a mixture of 300 kg?

Graph.

31. $y = -\frac{1}{4}x$

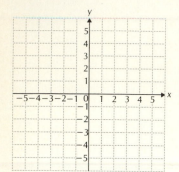

32. $y = 2x + 1$

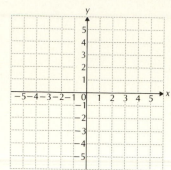

33. $2x - 3y = 6$

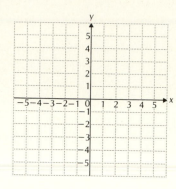

34. $2x - 5y = 10$

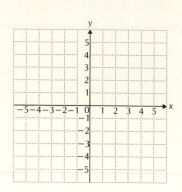

ANSWERS
31.
32.
33.
34.
35.
36.
37.
38.
39.
40.

⭐ **EXTENSION**

35. A wine maker ferments a batch of grapes and makes 1000 L of wine. A test reveals that the alcohol content of the wine is 5%. Consumer preference is for wine that is 12% alcohol. How many liters of a solution of alcohol and water that is 90% alcohol should be added to the wine in order to raise the alcohol level to the desired 12%?

36. A total of $27,000 is invested, part of it at 12% and part of it at 13%. The total yield after one year is $3385. How much was invested at each rate?

37. A two-digit number is six times the sum of its digits. The ten's digit is one more than the unit's digit. Find the number.

38. Farmer Jones has 100 L of milk that is 4.6% butterfat. How much skim milk (no butterfat) should be mixed with it in order to make milk that is 3.2% butterfat?

39. In Lewis Carroll's "Through the Looking Glass" Tweedledum says to Tweedledee, "The sum of your weight and twice mine is 361 pounds." Then Tweedledee says to Tweedledum, "Contrariwise, the sum of your weight and twice mine is 362 pounds." Find the weight of Tweedledum and Tweedledee.

40. Several ancient Chinese books included problems that can be solved by translating to systems of equations. *Arithmetical Rules in Nine Sections* is a book of 246 problems compiled by a Chinese mathematician, Chang Tsang, who died in 152 B.C. One of the problems is: Suppose there are a number of rabbits and pheasants confined in a cage. In all there are 35 heads and 94 feet. How many rabbits and how many pheasants are there? Solve the problem.

Problems Involving Motion

• Many problems deal with distance, time, and speed. A basic formula comes from the definition of speed.

$$\text{Speed} = \frac{\text{Distance}}{\text{Time}}, \quad r = \frac{d}{t}$$

From $r = \dfrac{d}{t}$, we can obtain two other formulas by solving. They are

$$d = rt \qquad \text{Multiplying by } t$$

and

$$t = \frac{d}{r}. \qquad \text{Multiplying by } t \text{ and dividing by } r$$

In most problems involving motion, you will use one of these formulas. It is probably easiest to remember the definition of speed, $r = d/t$. You can easily obtain either of the other formulas as you need them.

In Section 3.5 we considered some general steps for problem solving. The following are also helpful when solving motion problems.

1. Organize the information in a chart.
2. Look for as many things as you can that are the same, so you can write equations.

Example 1 A train leaves Podunk traveling east at 35 kilometers per hour (km/h). An hour later another train leaves Podunk and travels east on a parallel track at 40 km/h. How far from Podunk will the trains meet?

First make a drawing.

From the drawing we see that the distances are the same. Let's call the distance d. We don't know the times. Let t represent the time for the faster train. Then the time for the slower train will be $t + 1$. We can organize the information in a chart. In this case the distances are the same, so we will use the formula $d = rt$.

1. A car leaves Hereford traveling north at 56 km/h. Another car leaves Hereford one hour later traveling north at 84 km/h. How far from Hereford will the second car overtake the first? (*Hint:* The cars travel the same distance.)

	Distance	Speed	Time
Slow train	d	35	$t + 1$
Fast train	d	40	t

In these problems we look for things that are the same, so we can find equations. From each row of the chart we get an equation, $d = rt$. Thus we have two equations:

$$d = 35(t + 1)$$
$$d = 40t.$$

We solve the system:

$35(t + 1) = 40t$ Using the substitution method, substituting $35(t + 1)$ for d in the second equation

$35t + 35 = 40t$ Multiplying

$35 = 5t$ Adding $-35t$

$\dfrac{35}{5} = t$ Multiplying by $\dfrac{1}{5}$

$7 = t.$

The problem asks us to find how far from Podunk the trains meet. Thus we need to find d. We can do this by substituting 7 for t in the equation $d = 40t$:

$$d = 40(7) = 280.$$

Thus the trains meet 280 km from Podunk. Note also that $280 = 35(7 + 1)$, so 280 checks in the problem.

DO EXERCISE 1.

Example 2 A motorboat took 3 hr to make a downstream trip with a 6-km/h current. The return trip against the same current took 5 hr. Find the speed of the boat in still water.

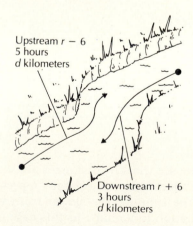

Upstream $r - 6$
5 hours
d kilometers

Downstream $r + 6$
3 hours
d kilometers

ANSWER ON PAGE A-17

First make a drawing. From the drawing we see that the distances are the same. Let's call the distance d. Let r represent the speed of the boat in still water. Then, when the boat is traveling downstream, its speed is $r + 6$ (the current helps the boat along). When it is traveling upstream, its speed is $r - 6$ (the current holds the boat back some). We can organize the information in a chart. In this case the distances are the same, so we will use the formula $d = rt$.

	Distance	Speed	Time
Downstream	d	$r + 6$	3
Upstream	d	$r - 6$	5

From each row of the chart, we get an equation, $d = rt$:

$$d = (r + 6)3,$$
$$d = (r - 6)5.$$

We solve the system using substitution:

$(r + 6)3 = (r - 6)5$ Using substitution, substituting $(r + 6)3$ for d in the second equation

$3r + 18 = 5r - 30$ Multiplying

$-2r + 18 = -30$ Adding $-5r$

$-2r = -48$ Adding -18

$r = \dfrac{-48}{-2}, \quad or \quad 24.$ Multiplying by $-\dfrac{1}{2}$

We check in the original problem. When $r = 24$, then $r + 6 = 30$, and $30 \cdot 3 = 90$, the distance. When $r = 24$, $r - 6 = 18$, and $18 \cdot 5 = 90$. In both cases we get the same distance. Thus the speed in still water is 24 km/h.

DO EXERCISE 2.

Example 3 Two cars leave town at the same time going in opposite directions. One of them travels 60 mph and the other 30 mph. In how many hours will they be 150 miles apart?

We first make a drawing.

From the wording of the problem and the drawing, we see that the distances may *not* be the same. But the times the cars travel are the same, so we can just use t for time. We can organize the information in a chart.

2. An airplane flew for 5 hr with a 25-km/h tail wind. The return flight against the same wind took 6 hr. Find the speed of the airplane in still air. (*Hint:* The distance is the same both ways. The speeds are $r + 25$ and $r - 25$, where r is the speed in still air.)

ANSWER ON PAGE A-17

3. Two cars leave town at the same time traveling in opposite directions. One travels 48 mph and the other 60 mph. How far apart will they be 3 hr later? (*Hint:* The times are the same. Be *sure* to make a drawing.)

	Distance	Speed	Time
Fast car	Distance of fast car	60	t
Slow car	Distance of slow car	30	t

From the drawing we see that

$$\text{(Distance of fast car)} + \text{(Distance of slow car)} = 150.$$

Then using $d = rt$ in each row of the table we get

$$60t + 30t = 150.$$

We solve the equation:

$90t = 150$ Collecting like terms

$t = \dfrac{150}{90}$, or $\dfrac{5}{3}$, or $1\dfrac{2}{3}$ hours.   Multiplying by $\dfrac{1}{90}$

This checks in the original problem, so in $1\frac{2}{3}$ hours the cars will be 150 miles apart.

DO EXERCISES 3 AND 4.

4. Two cars leave town at the same time in the same direction. One travels 35 mph and the other 40 mph. In how many hours will they be 15 mi apart? (*Hint:* The times are the same. Be *sure* to make a drawing.)

ANSWERS ON PAGE A-17

Chapter 7   Systems of Equations

• Solve.

1. Two cars leave town at the same time going in opposite directions. One travels 55 mph and the other travels 48 mph. In how many hours will they be 206 mi apart?

2. Two cars leave town at the same time going in opposite directions. One travels 44 mph and the other travels 55 mph. In how many hours will they be 297 mi apart?

3. Two cars leave town at the same time going in the same direction. One travels 30 mph and the other travels 46 mph. In how many hours will they be 72 mi apart?

4. Two cars leave town at the same time going in the same direction. One travels 32 mph and the other travels 47 mph. In how many hours will they be 69 mi apart?

ANSWERS

1. _____

2. _____

3. _____

4. _____

5. A train leaves a station and travels east at 72 km/h. Three hours later a second train leaves on a parallel track and travels east at 120 km/h. When will it overtake the first train?

6. A private airplane leaves an airport and flies due south at 192 km/h. Two hours later a jet leaves the same airport and flies due south at 960 km/h. When will the jet overtake the private plane?

5. _____

6. _____

7. A canoeist paddled for 4 hr with a 6-km/h current in order to reach a campsite. The return trip against the same current took 10 hr. Find the speed of the canoe in still water.

8. An airplane flew for 4 hr with a 20-km/h tail wind. The return flight against the same wind took 5 hr. Find the speed of the plane in still air.

7. _____

8. _____

Chapter 7   Systems of Equations

9. It takes a passenger train 2 hr less time than it takes a freight train to make the trip from Central City to Clear Creek. The passenger train averages 96 km/h while the freight train averages 64 km/h. How far is it from Central City to Clear Creek?

10. It takes a small jet plane 4 hr less time than it takes a propeller-driven plane to travel from Glen Rock to Oakville. The jet plane averages 637 km/h while the propeller plane averages 273 km/h. How far is it from Glen Rock to Oakville?

11. An airplane took 2 hr to fly 600 km against a head wind. The return trip with the wind took $1\frac{2}{3}$ hr. Find the speed of the plane in still air.

12. It took 3 hr to row a boat 18 km against the current. The return trip with the current took $1\frac{1}{2}$ hr. Find the speed of the rowboat in still water.

ANSWERS

9. _____

10. _____

11. _____

12. _____

13. A motorcycle breaks down and the rider has to walk the rest of the way to work. The motorcycle was being driven at 45 mph and the rider walks at a speed of 6 mph. The distance from home to work is 25 mi and the total time for the trip was 2 hr. How far did the motorcycle go before it broke down?

14. A student walks and jogs to college each day. The student averages 5 mph walking and 9 mph jogging. The distance from home to college is 8 mi and the student makes the trip in 1 hr. How far does the student jog?

13. _____

14. _____

15. _____

⭐ **EXTENSION**

16. _____

15. 🖩 An airplane flew for 4.23 hr with a 25.5-km/h tail wind. The return flight against the same wind took 4.97 hr. Find the speed of the plane in still air.

16. An airplane took $2\frac{1}{2}$ hr to fly 625 mi with the wind. It took 4 hr and 10 min to make the return trip against the same wind. Find the wind speed and the speed of the plane in still air.

17. To deliver a package, a messenger must travel at a speed of 60 mph on land and then use a motorboat whose speed is 20 mph in still water. While delivering the package, the messenger goes by land to a dock and then travels on a river against a current of 4 mph. The messenger reaches the destination in 4.5 hr and then returns to the starting point in 3.5 hr. How far did the messenger travel by land and how far by water?

18. Charles Lindbergh flew the Spirit of St. Louis in 1927 from New York to Paris at an average speed of 107.4 mph. Eleven years later, Howard Hughes flew the same route, averaged 217.1 mph, and took 16 hr, 57 min less time. Find the length of their route.

17. _____

18. _____

Summary and Review

The following contains a summary of what you should be able to do after completing this chapter. The review exercises are for practice. Answers are at the back of the book. If you miss an exercise, restudy the section and objective indicated alongside the answer.

The review sections to be tested in addition to the material in this chapter are 3.7, 4.7, 5.7, and 6.3.

You should be able to:

Determine whether an ordered pair is a solution of a system of equations.

Determine whether the given ordered pair is a solution of the system.

1. $(6, -1)$; $x - y = 3$
$2x + 5y = 6$

2. $(2, -3)$; $2x + y = 1$
$x - y = 5$

3. $(-2, 1)$; $x + 3y = 1$
$2x - y = -5$

4. $(-4, -1)$; $x - y = 3$
$x + y = -5$

Solve systems of equations using graphing and the substitution and addition methods.

Use graph paper. Solve each system graphically.

5. $x + y = 4$
$x - y = 8$

6. $a + 3b = 12$
$2a - 4b = 4$

7. $y = 5 - x$
$3x - 4y = -20$

8. $3x - 2y = -4$
$2y - 3x = -2$

Solve using the substitution method.

9. $y = 5 - x$
$3x - 4y = -20$

10. $x + 2y = 6$
$2x + 3y = 8$

11. $3x + y = 1$
$x - 2y = 5$

12. $x + y = 6$
$y = 3 - 2x$

13. $s + t = 5$
$s = 13 - 3t$

14. $x - y = 4$
$y = 2 - x$

Solve using the addition method.

15. $x + y = 4$
$2x - y = 5$

16. $x + 2y = 9$
$3x - 2y = -5$

17. $x - y = 8$
$2x + y = 7$

18. $\frac{2}{3}x + y = -\frac{5}{3}$
$x - \frac{1}{3}y = -\frac{13}{3}$

19. $2x + 3y = 8$
$5x + 2y = -2$

20. $5x - 2y = 2$
$3x - 7y = 36$

21. $-x - y = -5$
$2x - y = 4$

22. $6x + 2y = 4$
$10x + 7y = -8$

Solve problems using the substitution and addition methods for solving systems of equations.

Solve using the substitution and addition methods.

23. The sum of two numbers is 8. Their difference is 12. Find the numbers.

24. The sum of two numbers is 27. One half of the first number plus one third of the second number is 11. Find the numbers.

25. The perimeter of a rectangle is 98 cm. The length is 28 cm more than the width. Find the length and the width.

26. An airplane flew for 4 hr with a 15 km/h tail wind. The return flight against the same wind took 5 hr. Find the speed of the airplane in still air.

27. There were 508 people at an organ recital. Orchestra seats cost $5.00 per person with balcony seats costing $3.00. The total receipts were $2118. Find the number of orchestra and the number of balcony seats sold.

28. Solution A is 30% alcohol and solution B is 60% alcohol. How much of each is needed in order to make 80 L of a solution that is 45% alcohol?

29. Jeff is three times as old as his son. In nine years, Jeff will be twice as old as his son. How old is each now?

✔ **SKILL MAINTENANCE**

30. Multiply: $t^{-5} \cdot t^{13}$.

31. Divide: $\dfrac{t^{-5}}{t^{13}}$.

32. Simplify: $(x^4 y^{-5})^3$.

Multiply.

33. $(3t - 8)(3t + 8)$

34. $(5y^2 + 10)^2$

35. $(4y - 7)(5y + 8)$

Solve.

36. $20y^2 - 3y - 56 = 0$

37. $9x^2 - 64 = 0$

Graph.

38. $y = \frac{1}{2}x - 3$

39. $4y - x = 8$

⭐ **EXTENSION**

40. The solution of the following system is (6, 2). Find C and D.

$$2x - Dy = 6$$
$$Cx + 4y = 14$$

41. Solve.

$$3(x - y) = 4 + x$$
$$x = 5y + 2$$

42. For a two-digit number, the sum of the unit's digit and the ten's digit is 6. When the digits are reversed, the new number is 18 more than the original number. Find the original number.

43. A stable boy agreed to work for one year. At the end of that time he was to receive $240 and one horse. After 7 months the boy quit the job, but still received the horse and $100. What was the value of the horse?

Test: Chapter 7

1. Determine whether the given ordered pair is a solution of the system of equations:

$$(2, -1); \quad x = 4 + 2y$$
$$2y - 3x = 4$$

1. _____

Solve graphically.

2. $x - y = 3$

$\quad x - 2y = 4$

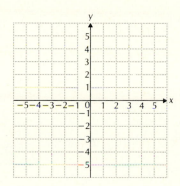

2. _____

3. _____

Solve using the substitution method.

3. $y = 6 - x$

$\quad 2x - 3y = 22$

4. $x + 2y = 5$

$\quad x + y = 2$

4. _____

5. _____

Solve using the addition method.

5. $x - y = 6$

$\quad 3x + y = -2$

6. $\frac{1}{2}x - \frac{1}{3}y = 8$

$\quad \frac{2}{3}x + \frac{1}{2}y = 5$

6. _____

7. $4x + 5y = 5$

$\quad 6x + 7y = 7$

8. $2x + 3y = 13$

$\quad 3x - 5y = 10$

7. _____

8. _____

Solve.

9. The sum of two numbers is 2. The difference is 8. Find the numbers.

10. A motorboat traveled for 2 hr with an 8 km/h current. The return trip against the same current took 3 hr. Find the speed of the motorboat in still water.

11. Solution A is 25% acid and solution B is 40% acid. How much of each is needed in order to make 60 L of a solution that is 30% acid?

 SKILL MAINTENANCE

12. Simplify: $(x^{-2}y^7)^{-3}$.

13. Multiply: $(4 - 5y)^2$.

14. Solve: $x^2 - 3x - 10 = 0$.

15. Graph: $3x + 4y = -12$.

⭐ EXTENSION

16. You are in line at a ticket window. There are two more people ahead of you in line than there are behind you. In the entire line there are three times as many people as there are behind you. How many are ahead of you in line?

17. The solution of the following system is $(-5, 1)$. Find A and B.

$$Ax - 2y = 7$$
$$3x + By = 8$$

AN APPLICATION
A student is taking an algebra course in which four tests are to be given. To average a B, a total of 320 points is needed. The student got scores of 81, 76, and 73 on the first three tests. Determine (in terms of an inequality) those scores that will allow the student to get a B.

THE MATHEMATICS
Let S = the score on the next test. Then a solution of the following will allow the student to get a B:

$$81 + 76 + 73 + S \geq 320$$

↑_____ This is an inequality.

The solution of inequalities is presented in this chapter. The principles used are very much like those used to solve equations, but the multiplication principle must be adapted for multiplication by a negative number.

The review sections to be tested in addition to the material in this chapter are 2.1, 3.6, 5.8, and 7.4.

Inequalities and Sets

Determine whether each number is a solution of the inequality.

1. $x < 3$

  a) 2   **b)** 0

  c) -5 **d)** 15

  e) 3

2. $x \geq 6$

  a) 6   **b)** 0

  c) -4 **d)** 25

  e) -6

Using the Addition Principle

In this chapter we learn principles for solving inequalities, which are similar to those we learned for equations.

• SOLUTIONS OF INEQUALITIES

In Section 2.1 we defined the symbols $>$ (greater than) and $<$ (less than). The symbol $\geq$ means *greater than or equal to*. The symbol $\leq$ means *less than or equal to*. For example, $-2 \leq 3$ and $3 \leq 3$ are both true, but $-3 \leq -5$ and $0 \geq 2$ are both false. An *inequality* is a number sentence with $>$, $<$, $\geq$, or $\leq$ for its verb. For example,

$$-4 < 5, \quad x > 3, \quad 2x + 5 \geq 0, \quad \text{and} \quad -3y + 7 \leq -8$$

are all inequalities. Some replacements for a variable in an inequality may make it true and some may make it false.

Example 1 Determine whether each number is a solution of $x < 3$.

a) -2;   Since $-2 < 3$ is true, -2 is a solution.

b) 5;   Since $5 < 3$ is false, 5 is not a solution.

c) $\frac{1}{4}$;   Since $\frac{1}{4} < 3$ is true, $\frac{1}{4}$ is a solution.

When we have found *all* the numbers that make an inequality true, we have *solved* the inequality.

> The replacements that make an inequality true are called its *solutions*. To *solve* an inequality means to find all its solutions.

DO EXERCISES 1 AND 2.

•• THE ADDITION PRINCIPLE

Consider the true inequality

$$-4 < 7.$$

Add 2 on both sides and we get another true inequality:

$$-4 + 2 < 7 + 2 \quad \text{or} \quad -2 < 9.$$

Similarly, if we add -3 on both sides we get another true inequality:

$$-4 + (-3) < 7 + (-3) \quad \text{or} \quad -7 < 4.$$

> **THE ADDITION PRINCIPLE FOR INEQUALITIES**
>
> If any number is added on both sides of a true inequality, we get another true inequality.

The addition principle holds whether the inequality contains $<$, $>$, $\leq$, or $\geq$.

Let's see how we use the addition principle to solve inequalities.

Example 2 Solve: $x + 2 > 8$.

We use the addition principle, adding -2:

$$x + 2 > 8$$
$$x + 2 + (-2) > 8 + (-2)$$
$$x > 6.$$

Using the addition principle, we get an inequality for which we can determine the solutions easily.

Any number greater than 6 makes the last sentence true, hence is a solution of that sentence. Any such number is also a solution of the original sentence. Thus we have solved the inequality.

We cannot check all the solutions of an inequality by substitution, as we can check solutions of equations. There are too many of them. A partial check can be done by substituting a number greater than 6, say 7, into the original inequality.

$$\frac{x + 2 > 8}{\begin{array}{c|c} 7 + 2 & 8 \\ 9 & \end{array}}$$

However, we really do not need to do such checks. Let us see why. Consider the first and last inequalities

$$x + 2 > 8 \quad \text{and} \quad x > 6.$$

Any number that makes the first one true must make the last one true. We know this by the addition principle. Now the question is, will any number that makes the last one true also be a solution of the first one? Let us use the addition principle again, adding 2:

$$x > 6$$
$$x + 2 > 6 + 2$$
$$x + 2 > 8.$$

Now we know that any number that makes $x > 6$ true also makes $x + 2 > 8$ true. Therefore, the sentences $x > 6$ and $x + 2 > 8$ have the same solutions. When two equations or inequalities have the same solutions, we say that they are *equivalent*. The addition principle tells us that $x > 6$ and $x + 2 > 8$ are equivalent. Whenever we use the addition principle with inequalities, the first and last sentences will be equivalent.

Solve. Write set notation for your answer.

3. $x + 3 > 5$

4. $x - 5 < 8$

5. $5x + 1 < 4x - 2$

Solve.

6. $x + \dfrac{2}{3} \leq \dfrac{4}{5}$

7. $5y + 2 \leq -1 + 4y$

ANSWERS ON PAGE A-17

Example 3 Solve: $3x + 1 < 2x - 3$.

$$3x + 1 < 2x - 3$$
$$3x + 1 - 1 < 2x - 3 - 1 \qquad \text{Adding } -1$$
$$3x < 2x - 4 \qquad \text{Simplifying}$$
$$3x - 2x < 2x - 4 - 2x \qquad \text{Adding } -2x$$
$$x < -4 \qquad \text{Simplifying}$$

Any number less than -4 is a solution. The following are some of the solutions:

$$-5, \quad -6, \quad -4.1, \quad -2045, \quad -18\pi, \quad -\sqrt{30}.$$

To describe all the solutions we use the set notation

$$\{x | x < -4\},$$

which is read:

The set of all x such that x is less than -4.

DO EXERCISES 3–5.

Example 4 Solve: $x + \dfrac{1}{3} \geq \dfrac{5}{4}$.

$$x + \frac{1}{3} \geq \frac{5}{4}$$
$$x + \frac{1}{3} - \frac{1}{3} \geq \frac{5}{4} - \frac{1}{3} \qquad \text{Adding } -\frac{1}{3}$$
$$x \geq \frac{5}{4} \cdot \frac{3}{3} - \frac{1}{3} \cdot \frac{4}{4} \qquad \text{Finding a common denominator}$$
$$x \geq \frac{15}{12} - \frac{4}{12}$$
$$x \geq \frac{11}{12}$$

Any number greater than or equal to $\frac{11}{12}$ is a solution. We say that the *solution set* is

$$\left\{ x | x \geq \frac{11}{12} \right\}.$$

Here is a handy way to help read and write set notation:

$$\left\{ x | x \geq \frac{11}{12} \right\}$$

The set of ←

all x ←

such that ←

x is greater than or equal to $\dfrac{11}{12}$ ←

DO EXERCISES 6 AND 7.

Chapter 8   Inequalities and Sets

• Determine whether each number is a solution of the inequality.

1. $x > -4$ **2.** $y < 5$ **3.** $x \geq 8$ **4.** $x \leq -10$ **5.** $t < -8$ **6.** $a \geq 0$

   a) 4   a) 0   a) -6 a) 4   a) 0   a) 2

   b) 0   b) 5   b) 0   b) -10 b) -8 b) -3

   c) -4 c) -1 c) 60   c) 0   c) -9 c) 0

   d) 6   d) -5 d) 8   d) -27 d) -7 d) 3

•• Solve using the addition principle. Write set notation for answers.

7. $x + 7 > 2$ **8.** $x + 6 > 3$ **9.** $y + 5 > 8$

10. $y + 7 > 9$ **11.** $x + 8 \leq -10$ **12.** $a + 12 < 6$

13. $x - 7 \leq 9$ **14.** $x - 6 > 2$ **15.** $y - 7 > -12$

16. $2x + 3 > x + 5$ **17.** $2x + 4 > x + 7$ **18.** $3x + 9 \leq 2x + 6$

19. $3x - 6 \geq 2x + 7$ **20.** $3x - 9 \geq 2x + 11$ **21.** $5x - 6 < 4x - 2$

22. $6x - 8 < 5x - 9$ **23.** $-7 + c > 7$

ANSWERS
1.
2.
3.
4.
5.
6.
7.
8.
9.
10.
11.
12.
13.
14.
15.
16.
17.
18.
19.
20.
21.
22.
23.

24. $-9 + b > 9$

25. $y + \dfrac{1}{4} \le \dfrac{1}{2}$

26. $y + \dfrac{1}{3} \le \dfrac{5}{6}$

27. $x - \dfrac{1}{3} > \dfrac{1}{4}$

28. $x - \dfrac{1}{8} > \dfrac{1}{2}$

29. $-14x + 21 > 21 - 15x$

30. $-10x + 15 > 18 - 11x$

31. $3(r + 2) < 2r + 4$

32. $4(r + 5) \ge 3r + 7$

33. $0.8x + 5 \ge 6 - 0.2x$

34. $0.7x + 6 \le 7 - 0.3x$

SKILL MAINTENANCE

35. Graph $\frac{5}{3}$ on a number line.

36. Find the absolute value: $|-8|$.

37. Solve: $-2y - 3 = 11$.

38. The length of a rectangle is 4 in. greater than the width. The area of the rectangle is 21 in². Find the perimeter of the rectangle.

EXTENSION

Solve.

39. $17x + 9{,}479{,}756 \le 16x - 8{,}579{,}243$

40. $9(m + 2) < 8(m + 5)$

41. Suppose $2x - 5 \ge 9$ is true for some number x. Determine whether $2x - 5 \ge 8$ is true for that same x.

42. A student is taking an introductory algebra course in which four tests are to be given. To average a B, a total of 320 points is needed. The student got scores of 81, 76, and 73 on the first three tests. Determine (in terms of an inequality) those scores that will allow the student to get a B.

Using the Multiplication Principle

◦ THE MULTIPLICATION PRINCIPLE

Consider the true inequality

$$3 < 7.$$

Multiply both numbers by 2 and we get another true inequality:

$$3 \cdot 2 < 7 \cdot 2$$

or

$$6 < 14.$$

Multiply both numbers by -3 and we get the false inequality

$$3 \cdot (-3) < 7 \cdot (-3)$$

or

$$-9 < -21. \quad \text{False}$$

However, if we reverse the inequality symbol we get a true inequality:

$$-9 > -21. \quad \text{True}$$

> **THE MULTIPLICATION PRINCIPLE FOR INEQUALITIES**
>
> If we multiply on both sides of a true inequality by a positive number, we get another true inequality. If we multiply by a negative number and the inequality symbol is reversed, we get another true inequality.

The multiplication principle holds whether the inequality contains $\geq$, $\leq$, $<$, or $>$.

Example 1 Solve: $4x < 28$.

$$\frac{1}{4} \cdot 4x < \frac{1}{4} \cdot 28 \quad \text{Multiplying by } \tfrac{1}{4}$$

> The symbol stays the same.

$$x < 7 \quad \text{Simplifying}$$

The solution set is $\{x|x < 7\}$.

DO EXERCISES 1 AND 2.

Example 2 Solve: $-2y < 18$.

$$-\frac{1}{2}(-2y) > -\frac{1}{2} \cdot 18 \quad \text{Multiplying by } -\tfrac{1}{2}$$

> The symbol must be reversed!

$$y > -9 \quad \text{Simplifying}$$

The solution set is $\{y|y > -9\}$.

DO EXERCISES 3 AND 4.

Solve.

1. $8x < 64$

2. $5y \geq 160$

Solve.

3. $-4x \leq 24$

4. $-5y > 13$

ANSWERS ON PAGE A-17

Solve.

5. $7 - 4x < 8$

Solve.

6. $13x + 5 \leq 12x + 4$

Solve.

7. $24 - 7y \geq 11y - 14$

● ● **USING THE PRINCIPLES TOGETHER**

We use the addition and multiplication principles together in solving inequalities in much the same way as in solving equations. We generally use the addition principle first.

Example 3 Solve: $6 - 5y > 7$.

$$-6 + 6 - 5y > -6 + 7 \qquad \text{Adding } -6$$
$$-5y > 1 \qquad \text{Simplifying}$$
$$-\frac{1}{5} \cdot (-5y) < -\frac{1}{5} \cdot 1 \qquad \text{Multiplying by } -\frac{1}{5}$$

The symbol must be reversed!

$$y < -\frac{1}{5}$$

The solution set is $\{y | y < -\frac{1}{5}\}$.

DO EXERCISE 5.

Example 4 Solve: $5x + 9 \leq 4x + 3$.

$$5x + 9 - 9 \leq 4x + 3 - 9 \qquad \text{Adding } -9$$
$$5x \leq 4x - 6 \qquad \text{Simplifying}$$
$$5x - 4x \leq 4x - 6 - 4x \qquad \text{Adding } -4x$$
$$x \leq -6 \qquad \text{Simplifying}$$

The solution set is $\{x | x \leq -6\}$.

DO EXERCISE 6.

Example 5 Solve: $8y - 5 > 17 - 5y$.

$$-17 + 8y - 5 > -17 + 17 - 5y \qquad \text{Adding } -17$$
$$8y - 22 > -5y \qquad \text{Simplifying}$$
$$-8y + 8y - 22 > -8y - 5y \qquad \text{Adding } -8y$$
$$-22 > -13y \qquad \text{Simplifying}$$
$$-\frac{1}{13} \cdot (-22) < -\frac{1}{13} \cdot (-13y) \qquad \text{Multiplying by } -\frac{1}{13}$$

The symbol must be reversed.

$$\frac{22}{13} < y$$

The solution set is $\{y | \frac{22}{13} < y\}$. Since $\frac{22}{13} < y$ has the same meaning as $y > \frac{22}{13}$, we can also describe the solution set as $\{y | y > \frac{22}{13}\}$.

DO EXERCISE 7.

• Solve using the multiplication principle.

1. $5x < 35$

2. $8x \geq 32$

3. $9y \leq 81$

4. $10x > 240$

5. $7x < 13$

6. $8y < 17$

7. $12x > -36$

8. $16x < -64$

9. $5y \geq -2$

10. $7x > -4$

11. $-2x \leq 12$

12. $-3y \leq 15$

13. $-4y \geq -16$

14. $-7x < -21$

15. $-3x < -17$

16. $-5y > -23$

17. $-2y > \frac{1}{7}$

18. $-4x \leq \frac{1}{0}$

19. $-\frac{6}{5} \leq -4x$

20. $-\frac{7}{8} > -56t$

•• Solve using the addition and multiplication principles.

21. $2x + 5 < 3$

22. $3x - 4 > 5$

ANSWERS
1.
2.
3.
4.
5.
6.
7.
8.
9.
10.
11.
12.
13.
14.
15.
16.
17.
18.
19.
20.
21.
22.

23.

24.

25.

26.

27.

28.

29.

30.

31.

32.

33.

34.

35.

36.

37.

38.

39.

40.

23. $-3x + 7 \geq -2$

24. $-4x - 3 \leq -5$

25. $6t - 8 \leq 4t + 1$

26. $5b + 7 \geq b - 1$

27. $3 - 6c > 15$

28. $5 - 3y < -4$

29. $15x - 21 \geq 8x + 7$

30. $21 - 15x < -8x - 7$

✔ SKILL MAINTENANCE

31. Find $-x$ when x is -17.

32. Solve $A = \pi r^2$ for r^2.

33. Solve: $x - y = 7$
$x + y = 9$.

34. Use the proper symbol $<$, $>$, or $=$.

$$\frac{4}{9} \qquad \frac{5}{11}$$

★ EXTENSION

Solve.

35. $5(12 - 3t) \geq 15(t + 4)$

36. $\dfrac{-x}{4} - \dfrac{3x}{8} + 2 > 3 - x$

37. $x^2 > 0$

38. Determine whether the following statement is true for all rational numbers a and b: If $a^2 < b^2$, then $a < b$.

39. Badger Rent-a-Car rents compact cars for \$53.95 plus 40¢ per mile. Beaver Rent-a-Car rents compact cars for \$52.95 plus 42¢ per mile. For what mileages is the cost of renting a car cheaper at Badger?

40. A salesperson can choose to be paid in one of two ways.

Plan A: A salary of \$600 per month, plus a commission of 4% of gross sales.

Plan B: A salary of \$800 per month, plus a commission of 6% of gross sales over \$100,000.

For what gross sales is plan A better than plan B, assuming that gross sales are always more than \$100,000?

Graphs of Inequalities

• INEQUALITIES IN ONE VARIABLE

We graph inequalities in one variable on a number line.

Example 1 Graph $x < 2$.

The solutions of $x < 2$ are those numbers less than 2. They are shown on the graph by shading all points to the left of 2.

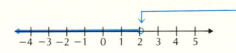

Since the inequality $x < 2$ involves $<$, we use an open circle to show that 2 is *not* part of the graph.

Example 2 Graph $y \geq -3$.

The solutions of $y \geq -3$ are shown by shading the point for -3 and all points to the right of -3.

Since the inequality $y \geq -3$ involves $\geq$, we use a solid circle to show that -3 *is* part of the graph.

DO EXERCISES 1 AND 2.

Example 3 Graph $3x + 2 < 5x - 1$.

First solve:

$$3x + 2 < 5x - 1$$
$$2 < 2x - 1 \qquad \text{Adding } -3x$$
$$3 < 2x \qquad \text{Adding 1}$$
$$\tfrac{3}{2} < x \qquad \text{Multiplying by } \tfrac{1}{2}$$

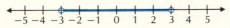

DO EXERCISES 3 AND 4.

•• GRAPHING INEQUALITIES WITH ABSOLUTE VALUE

Example 4 Graph $|x| < 3$.

The absolute value of a number is its distance from 0 on a number line. For the absolute value of a number to be less than 3 it must be between 3 and -3. Therefore, we shade the points between these two numbers. The open circles show that -3 and 3 are not part of the graph.

DO EXERCISES 5 AND 6.

After finishing Section 8.3, you should be able to:

• Graph inequalities in one variable on a number line.

•• Graph inequalities that contain absolute values on a number line.

••• Graph linear inequalities in two variables on a plane.

Graph on a number line.

1. $x < 4$

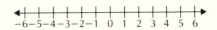

2. $y \geq -5$

Graph on a number line.

3. $x + 2 > 1$

4. $4x + 6 < 7x - 3$

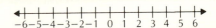

Graph on a number line.

5. $|x| < 5$

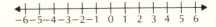

6. $|x| \leq 4$

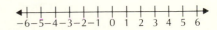

ANSWERS ON PAGE A-17

Graph on a number line.

7. $|x| \geq 3$

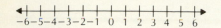

8. $|y| > 5$

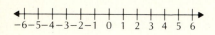

9. Determine whether $(4, 3)$ is a solution of $3x - 2y < 1$.

Example 5 Graph $|x| \geq 2$.

For the absolute value of a number to be greater than or equal to 2 its distance from 0 must be 2 or more. Thus the number must be 2 or greater, or it must be less than or equal to -2. Therefore, we shade the point for 2 and all points to its right. We also shade the point for -2 and all points to its left.

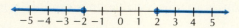

DO EXERCISES 7 AND 8.

••• INEQUALITIES IN TWO VARIABLES

The solutions of inequalities in two variables are ordered pairs.

Example 6 Determine whether $(-3, 2)$ is a solution of $5x - 4y < 13$.

We use alphabetical order of variables. We replace x by -3 and y by 2:

$$
\begin{array}{c|c}
5x - 4y & < 13 \\
\hline
5(-3) - 4 \cdot 2 & 13 \\
-15 - 8 & \\
-23 &
\end{array}
$$

Since $-23 < 13$ is true, $(-3, 2)$ is a solution.

DO EXERCISE 9.

Example 7 Graph $y > x$.

We first graph the line $y = x$ for comparison. Every solution of $y = x$ is an ordered pair such as $(3, 3)$. The first and second coordinates are the same. The graph of $y = x$ is shown to the left below.

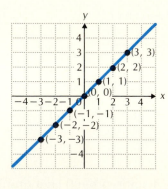 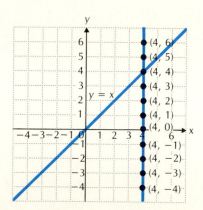

Now look at the graph to the right above. We consider a vertical line and ordered pairs on it. For all points above $y = x$, the second coordinate is greater than the first, $y > x$. For all points below the line, $y < x$. The same thing happens for any vertical line. Then for all points above $y = x$, the ordered pairs are solutions. We shade the half-plane above $y = x$.

This is the graph of $y > x$. Points on $y = x$ are not in the graph, so we draw it dashed.

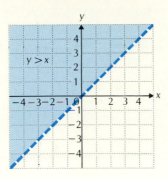

DO EXERCISE 10.

Example 8 Graph $y \leq x - 1$.

First sketch the line $y = x - 1$. Points on the line $y = x - 1$ are also in the graph of $y \leq x - 1$, so we draw the line solid. For points above the line, $y > x - 1$. These points are not in the graph. For points below the line, $y < x - 1$. These are in the graph, so we shade the lower half-plane.

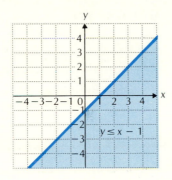

DO EXERCISE 11.

Example 9 Graph $6x - 2y < 10$.

We first solve for y, getting y on the left.

$$-2y < -6x + 10 \qquad \text{Adding } -6x$$

$$y > -\frac{1}{2}(-6x + 10) \qquad \text{Multiplying by } -\frac{1}{2}$$

Here the symbol must be reversed.

$$y > 3x - 5$$

We now graph the line $y = 3x - 5$. The intercepts are $(0, -5)$ and $(\frac{5}{3}, 0)$. The point $(3, 4)$ is also on the line. This line forms the boundary of the solutions of the inequality. In this case points on the line are not solutions of the inequality.

10. Graph $y < x$.

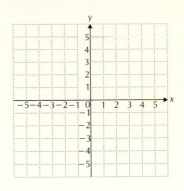

11. Graph $y \geq x + 2$.

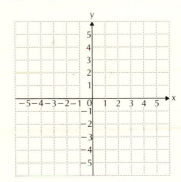

ANSWERS ON PAGE A-18

12. Graph $2x + 4y < 8$.

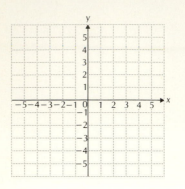

13. Graph $3x - 5y < 15$.

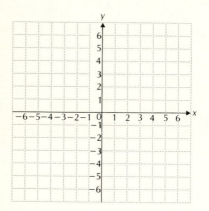

14. Graph $2x + 3y \geq 12$.

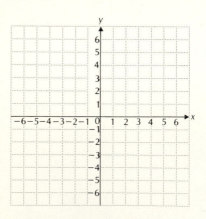

We shade the half-plane above the line.

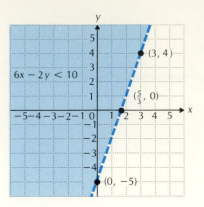

DO EXERCISE 12.

A *linear inequality* is one that we can get from a linear equation by changing the equals symbol to an inequality symbol. Every linear equation has a graph that is a straight line. The graph of a linear inequality is a half-plane, sometimes including the line along the edge. In the following example we give a different method of graphing. We graph the line using intercepts. Then we determine which side to shade by substituting a point from either half-plane.

Example 10 Graph $2x + 3y < 6$.

a) First graph the line $2x + 3y = 6$. The intercepts are $(3, 0)$ and $(0, 2)$. We use a dashed line for the graph since we have $<$.

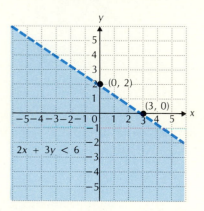

b) Pick a point that does not belong to the line. Substitute to determine whether this point is a solution. The origin $(0, 0)$ is usually an easy one to use: $2 \cdot 0 + 3 \cdot 0 < 6$ is true, so the origin is a solution. This means we shade the lower half-plane. Had the substitution given us a false inequality we would have shaded the other half-plane.

If the line goes through the origin, then we must test some other point not on the line. The point $(1, 1)$ is often a good one to try.

DO EXERCISES 13 AND 14.

Chapter 8   Inequalities and Sets

■ Graph on a number line.

1. $x < 5$

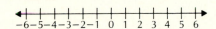

2. $x < 3$

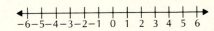

3. $y \geq -4$

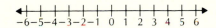

4. $x \geq -7$

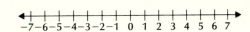

5. $t - 3 \leq -7$

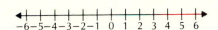

6. $x - 4 \leq -8$

7. $2x + 6 < 14$

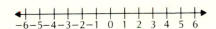

8. $4x - 8 \geq 12$

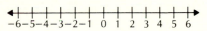

9. $4y + 9 > 11y - 12$

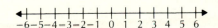

10. $6x + 11 \leq 14x + 7$

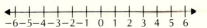

■■ Graph on a number line.

11. $|x| < 2$

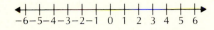

12. $|t| \leq 1$

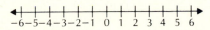

13. $|a| \geq 4$

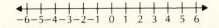

14. $|y| > 5$

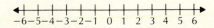

••• Graph on a plane.

15. $y > x - 2$

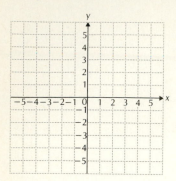

16. $y \leq x - 3$

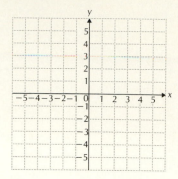

17. $6x - 2y \leq 12$

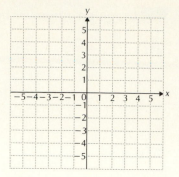

18. $2x + 3y < 12$

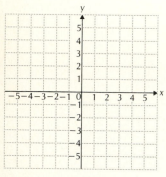

19. $3x - 5y \geq 15$

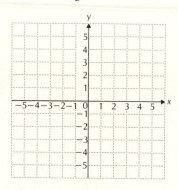

20. $5x + 2y > 10$

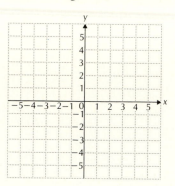

21. $y - 2x < 4$

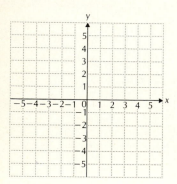

22. $2x - y \leq 4$

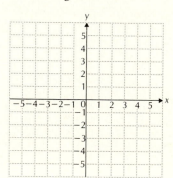

✓ **SKILL MAINTENANCE**

23. The base of a triangle is 5 m greater than the height. The area is 7 m². Find the base and the height.

☆ **EXTENSION**

Graph on a plane.

24. $y > 2$
   (*Hint:* Think of this as $0 \cdot x + y > 2$.)

25. $x \geq 3$

26. $x > 0$

27. $y \leq 0$

Sets

NAMING SETS

We have discussed notation like the following:

$$\{x|x < 6\},$$

where we were considering the rational numbers. If we were just considering whole numbers, the above set could also be named, or symbolized,

$$\{0, 1, 2, 3, 4, 5\}.$$

This way of naming a set is known as the *roster method*. In words, it is

The set consisting of 0, 1, 2, 3, 4, and 5.

DO EXERCISES 1 AND 2.

MEMBERSHIP

The symbol $\in$ means *is a member of* or *belongs to*. Thus

$$x \in A$$

means "x is a member of A" or "x belongs to A."

Examples Determine whether true or false.

1. $1 \in \{1, 2, 3\}$ True
2. $1 \in \{2, 3\}$ False
3. $4 \in \{x|x \text{ is an even whole number}\}$ True
4. $5 \in \{x|x \text{ is an even whole number}\}$ False

Set membership can be illustrated with a diagram such as the one at the right.

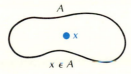

DO EXERCISES 3–6.

INTERSECTIONS

The *intersection* of two sets A and B is the set of members common to both sets and is indicated by the symbol

$$A \cap B.$$

Thus,

$$\{0, 1, 3, 5, 25\} \cap \{2, 3, 4, 5, 6, 7, 9\}$$

represents the set

$$\{3, 5\}.$$

Name each set using the roster method.

1. The set of whole numbers 2 through 10.

2. The set of even whole numbers between 30 and 40.

Determine whether true or false.

3. $4 \in \{a, b, 3, 4, 5\}$

4. $c \in \{a, b, 3, 4, 5\}$

5. $\dfrac{2}{3} \in \{x|x \text{ is a rational number}\}$

6. Heads $\in$ The set of outcomes of flipping a penny

ANSWERS ON PAGE A-18

Find.

7. $\{a, 1, 0, t, 6, 9, 14\} \cap \{2, 1, s, a, 9\}$

8. $\{1, 2\} \cap \{2, 3\}$

Find.

9. $\{1, 2\} \cap \{3, 4\}$

10. $A \cap B$, where

$A = \{x | x \text{ is an even integer}\}$
and
$B = \{x | x \text{ is an odd integer}\}$

Find.

11. $\{0, 1, 2, 3, 4\} \cup \{2, 3, 4, 5, 6, 7\}$

12. $A \cup B$, where

$A = \{x | x \text{ is an even integer}\}$
and
$B = \{x | x \text{ is an odd integer}\}$

ANSWERS ON PAGE A-18

Set intersection is illustrated by the diagram below.

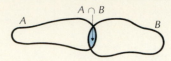

$A \cap B$

The solution of the system of equations

$$x + 2y = 7$$
$$x - y = 4$$

is the ordered pair $(5, 1)$. It is the intersection of the graphs of the two lines.

DO EXERCISES 7 AND 8.

The set without members is known as the *empty set*, and is often named 0. Each of the following is a description of the empty set:

The set of all six-eyed algebra teachers;

$\{2, 3\} \cap \{5, 6, 7\}$;

$\{x | x \text{ is an even whole number}\} \cap \{x | x \text{ is an odd whole number}\}$.

The system of equations

$$x + 2y = 7$$
$$x + 2y = 9$$

has no solution. The lines are parallel. Their intersection is empty.

DO EXERCISES 9 AND 10.

∷ UNIONS

Two sets A and B may be combined to form a new set that contains the members of A as well as those of B. The new set is called the *union* of A and B, and is represented by the symbol

$$A \cup B.$$

Thus,

$$\{0, 5, 7, 13, 27\} \cup \{0, 2, 3, 4, 5\}$$

represents the set

$$\{0, 2, 3, 4, 5, 7, 13, 27\}.$$

Set union is illustrated by the diagram below.

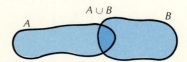

$A \cup B$

The solution set of the equation $(x - 3)(x + 2) = 0$ is $\{3, -2\}$. This set is the union of the solution sets of the equations $x - 3 = 0$ and $x + 2 = 0$, which are $\{3\}$ and $\{-2\}$.

DO EXERCISES 11 AND 12.

Chapter 8   Inequalities and Sets

● Name each set using the roster method.

1. The set of whole numbers 3 through 8

2. The set of whole numbers 101 through 107

3. The set of odd numbers between 40 and 50

4. The set of multiples of 5 between 10 and 40

5. $\{x|x$ times x is 9$\}$

6. $\{x|x$ is the cube of 0.2$\}$

●● Determine whether true or false.

7. $2 \in \{x|x$ is an odd number$\}$

8. $7 \in \{x|x$ is an odd number$\}$

9. Bruce Springsteen $\in$ The set of all rock stars

10. Apple $\in$ The set of all fruit

11. $-3 \in \{-4, -3, 0, 1\}$

12. $0 \in \{-4, -3, 0, 1\}$

ANSWERS
1.
2.
3.
4.
5.
6.
7.
8.
9.
10.
11.
12.

••• Find each intersection.

13. {a, b, c, d, e} ∩ {c, d, e, f, g}

14. {a, e, i, o, u} ∩ {q, u, i, c, k}

15. {1, 2, 5, 10} ∩ {0, 1, 7, 10}

16. {0, 1, 7, 10} ∩ {0, 1, 2, 5}

17. {1, 2, 5, 10} ∩ {3, 4, 7, 8}

18. {a, e, i, o, u} ∩ {m, n, f, g, h}

Find each union.

19. {a, e, i, o, u} ∪ {q, u, i, c, k}

20. {a, b, c, d, e} ∪ {c, d, e, f, g}

21. {0, 1, 7, 10} ∪ {0, 1, 2, 5}

22. {1, 2, 5, 10} ∪ {0, 1, 7, 10}

23. {a, e, i, o, u} ∪ {m, n, f, g, h}

24. {1, 2, 5, 10} ∪ {a, b}

 SKILL MAINTENANCE

25. Solve: $3x + 5y = 6$
$5x + 3y = 4$.

26. Solve $T = 3a + 4b$ for b.

27. Find $-(-x)$ when x is 9.

28. Find the absolute value: $|23|$.

29. When the sides of a square are lengthened by 0.2 km, the area becomes 0.64 km². Find the length of a side of the original square.

⭐ EXTENSION

30. For a set A, find the following.

a) $A \cup \emptyset$

b) $A \cup A$

c) $A \cap A$

d) $A \cap \emptyset$

31. Find the intersection of the set of even integers and the set of positive rational numbers.

32. A set is *closed* under an operation if, when the operation is performed on any of its members, the result is in the set. For example, the set of rational numbers is closed under the operation of addition since the sum of any two rational numbers is a rational number. Is the set of:

a) even whole numbers closed under addition?

b) odd whole numbers closed under addition?

c) {0, 1} closed under addition?

d) {0, 1} closed under multiplication?

e) integers closed under division?

Summary and Review

The following contains a summary of what you should be able to do after completing this chapter. The review exercises are for practice. Answers are at the back of the book. If you miss an exercise, restudy the section and objective indicated alongside the answer.

The review sections to be tested in addition to the material in this chapter are 2.1, 3.6, 5.8, and 7.4.

You should be able to:

Determine whether a number is a solution of an inequality.

Determine whether each number is a solution of the inequality $x \leq 4$.

1. -3 **2.** 7   **3.** 4

Solve inequalities using the addition principle, the multiplication principle, and both principles together.

Solve. Write set notation for the answers.

4. $y + \dfrac{2}{3} \geq \dfrac{1}{6}$ **5.** $9x \geq 63$

6. $2 + 6y > 14$ **7.** $7 - 3y \geq 27 + 2y$

8. $3x + 5 < 2x - 6$ **9.** $-4y < 28$

10. $3 - 4x < 27$ **11.** $4 - 8x < 13 + 3x$

12. $-3y \geq -21$ **13.** $-4x \leq \dfrac{1}{3}$

Graph inequalities on a number line and a plane.

Graph on a number line.

14. $y \leq 4$ **15.** $6x - 3 < x + 2$ **16.** $|x| \leq 2$

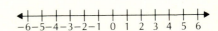

Graph on a plane.

17. $x < y$ **18.** $x + 2y \geq 4$

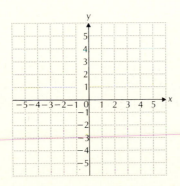 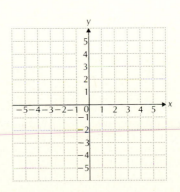

Name sets using the roster method and determine whether a given object is a member of a set.

19. Name the set of multiples of 4 between 30 and 60 using the roster method.

Determine whether true or false.

20. $91 \in \{x | x$ is a prime number$\}$

21. $-4 \in \{-6, -5, -3, -1, 0\}$

Find the intersection and union of sets.

Find each intersection.

22. $\{2, 4, 12, 14\} \cap \{-4, -2, 0, 2, 4\}$

23. $\{a, f, g\} \cap \{b, h, q, r, s\}$

Find each union.

24. $\{0, 1\} \cup \{1, 2, 3\}$

25. $\{A, W, R\} \cup \{E, F, A, B\}$

 SKILL MAINTENANCE

26. Find $-x$ when x is -10.

27. Find $-(-x)$ when x is -33.

Find the absolute value.

28. $|-56|$

29. $|0|$

30. $|76|$

31. Solve $d = rt$ for t.

32. Solve $Q = 5p - 7q$ for p.

Solve.

33. $3x + 5y = -2$
     $5x + 3y = 2$

34. $a - b = -12$
     $a + b = 16$

35. The length of a rectangle is 2 m greater than the width. The area is 35 m². Find the perimeter.

36. The product of two consecutive even integers is 360. Find the integers.

 EXTENSION

Solve.

37. $3[2 - 4(y - 3)] < 6(y - 1)$

38. $-\dfrac{x}{5} - \dfrac{4x}{15} + 6 \leq 10 - x$

39. Your quiz grades are 71, 75, 82, and 86. What is the lowest grade you can get on the last quiz to maintain an average of 80 or more?

40. The length of a rectangle is 43 cm. What widths will make the perimeter greater than 120 cm?

Graph on a plane.

41. $x < -3$

42. $y \geq -2$

43. Find the intersection of the set of odd integers greater than 6 and the set of odd integers less than 15.

Test: Chapter 8

Determine whether each number is a solution of the inequality $x > -2$.

1. -3 **2.** 0 **3.** 4

Solve. Write set notation for the answer.

4. $x + 6 \leq 2$ **5.** $14x + 9 > 13x - 4$

6. $12x \leq 60$ **7.** $-2y \geq 26$

8. $-4y \leq -32$ **9.** $-5x \geq \dfrac{1}{4}$

10. $4 - 6x > 40$ **11.** $5 - 9x \geq 19 + 5x$

Graph on a number line.

12. $4x - 6 < x + 3$ **13.** $|x| \geq 5$

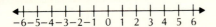   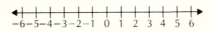

Graph on a plane.

14. $y > x - 1$ **15.** $2x - y \leq 6$

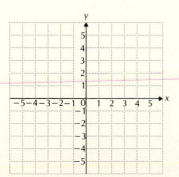   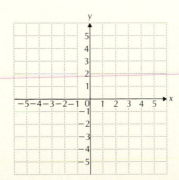

ANSWERS
1.
2.
3.
4.
5.
6.
7.
8.
9.
10.
11.
12.
13.
14.
15.

16. Name the set of prime numbers between 15 and 29 using the roster method.

16. _____

17. _____

18. _____

Find each intersection.

17. $\{0, 1\} \cap \{-1, 0\}$

18. $\{q, r, t, s, w\} \cap \{r, s, t, u, v\}$

19. _____

20. _____

21. _____

Find each union.

19. $\{-2, -1, 0, 1, 2\} \cup \{-4, -2, 0, 2, 4\}$

20. $\{a\} \cup \{b\}$

22. _____

23. _____

24. _____

 SKILL MAINTENANCE

21. Solve $A = \dfrac{c - d}{4}$ for c.

22. Solve: $5x - 2y = 14$
$3x + 6y = -6$.

25. _____

23. Find the absolute value: $|-37|$.

24. Find $-(-x)$ when x is -19.

25. The product of two consecutive integers is 306. Find the integers.

26. _____

 EXTENSION

26. Solve: $x^2 + 8 > 0$.

27. The length of a rectangle is 4 cm. Determine (in terms of an inequality) those widths for which the area will be less than 86 cm^2.

27. _____

Chapter 8   Inequalities and Sets

9

AN APPLICATION
One of these typists gets a certain typing job done in 4 hr, and the other gets the same job done in 6 hr. How much time would it take if they worked together?

THE MATHEMATICS
Let t = the time it takes them to get the job done if they work together. Then we can find t by solving the fractional equation

$$\frac{1}{4} + \frac{1}{6} = \frac{1}{t}.\left.\right\} \leftarrow \text{This is a fractional equation.}$$

In this chapter we learn to manipulate fractional expressions and solve fractional equations. The end result will be an ability to solve problems like the one above and others that we could not have solved before now.

The review sections to be tested in addition to the material in this chapter are 4.4, 5.6, 7.5, and 8.2.

Fractional Expressions and Equations

Multiply.

1. $\dfrac{x+3}{5} \cdot \dfrac{x+2}{x+4}$

2. $\dfrac{-3}{2x+1} \cdot \dfrac{4}{2x-1}$

Multiplying and Simplifying

These are *fractional expressions*:

$$\frac{3}{4}, \qquad \frac{5}{x+2}, \qquad \frac{x^2+3x-10}{7x^2-4}.$$

A fractional expression is a quotient of two polynomials and, therefore, indicates division. For example,

$$\frac{3}{4} \quad \text{means} \quad 3 \div 4$$

and

$$\frac{x^2+3x-10}{7x^2-4} \quad \text{means} \quad (x^2+3x-10) \div (7x^2-4).$$

We first study multiplying and simplifying fractional expressions. What we do is similar to multiplying and simplifying fractional expressions in arithmetic. But instead of simplifying an expression like

$$\frac{10}{25},$$

we may simplify an expression like

$$\frac{x^2-25}{x-5}.$$

Just as factoring is important in simplifying fractional notation in arithmetic, so too is it important in simplifying fractional expressions in algebra. The factoring we use the most is the factoring of polynomials, which we studied in Chapter 5.

• MULTIPLYING

For fractional expressions, multiplication is done as in arithmetic.

> To multiply two fractional expressions, multiply numerators and multiply denominators.

Examples Multiply.

1. $\dfrac{x-2}{3} \cdot \dfrac{x+2}{x+3} = \dfrac{(x-2)(x+2)}{3(x+3)}$ Multiplying numerators and multiplying denominators

   We could multiply out the numerator and the denominator to get $\dfrac{x^2-4}{3x+9}$, but it is best *not* to do it yet. We will see why later.

2. $\dfrac{-2}{2y+3} \cdot \dfrac{3}{y-5} = \dfrac{-2 \cdot 3}{(2y+3)(y-5)}$

DO EXERCISES 1 AND 2.

•• MULTIPLYING BY 1

Any fractional expression with the same numerator and denominator is a symbol for 1:

$$\frac{x + 2}{x + 2} = 1, \qquad \frac{3x^2 - 4}{3x^2 - 4} = 1, \qquad \frac{-1}{-1} = 1.$$

It should be noted that certain replacements are not sensible. For example, in

$$\frac{x + 2}{x + 2}$$

we should not substitute -2 for x. We would get 0 for the denominator.

Expressions that have the same value for all sensible replacements are called *equivalent expressions*. We can multiply by 1 to obtain an equivalent expression.

Examples Multiply.

3. $\dfrac{3x + 2}{x + 1} \cdot \dfrac{2x}{2x} = \dfrac{(3x + 2)2x}{(x + 1)2x}$

4. $\dfrac{x + 2}{x - 1} \cdot \dfrac{x + 1}{x + 1} = \dfrac{(x + 2)(x + 1)}{(x - 1)(x + 1)}$

5. $\dfrac{2 + x}{2 - x} \cdot \dfrac{-1}{-1} = \dfrac{(2 + x)(-1)}{(2 - x)(-1)}$

Note in Example 4 that the original expression

$$\frac{x + 2}{x - 1}$$

has 1 as a replacement, which is not sensible because it results in division by 0, which is not possible. The resulting expression

$$\frac{(x + 2)(x + 1)}{(x - 1)(x + 1)}$$

has 1 and -1 as replacements, which are not sensible. The two expressions are still equivalent. They have the same value for all of the *sensible* replacements.

DO EXERCISES 3–5.

Multiply.

3. $\dfrac{2x + 1}{3x - 2} \cdot \dfrac{x}{x}$

4. $\dfrac{x + 1}{x - 2} \cdot \dfrac{x + 2}{x + 2}$

5. $\dfrac{x - 8}{x - y} \cdot \dfrac{-1}{-1}$

ANSWERS ON PAGE A-19

Simplify by removing a factor of 1.

6. $\dfrac{5y}{y}$

7. $\dfrac{8x^2}{24x}$

••• SIMPLIFYING FRACTIONAL EXPRESSIONS

To simplify, we can do the reverse of multiplying. We factor numerator and denominator and "remove" a factor of 1.

Example 6 Simplify by removing a factor of 1: $\dfrac{3x}{x}$.

$$\dfrac{3x}{x} = \dfrac{3 \cdot x}{1 \cdot x} \qquad \text{Factoring numerator and denominator}$$

$$= \dfrac{3}{1} \cdot \dfrac{x}{x} \qquad \text{Factoring the fractional expression}$$

$$= \dfrac{3}{1} \cdot 1 \qquad \dfrac{x}{x} = 1$$

$$= 3 \qquad \text{We "removed" a factor of 1.}$$

In this example we supplied a 1 in the denominator. This can always be done whenever it is helpful.

Example 7 Simplify by removing a factor of 1: $\dfrac{7x^2}{14x}$.

$$\dfrac{7x^2}{14x} = \dfrac{7x \cdot x}{7x \cdot 2} \qquad \text{Factoring numerator and denominator}$$

$$= \dfrac{7x}{7x} \cdot \dfrac{x}{2} \qquad \text{Factoring the fractional expression}$$

$$= \dfrac{x}{2} \qquad \text{"Removing" a factor of 1}$$

DO EXERCISES 6 AND 7.

Examples Simplify by removing a factor of 1.

8. $\dfrac{6a + 12}{7(a + 2)} = \dfrac{6(a + 2)}{7(a + 2)}$

$$= \dfrac{6}{7} \cdot \dfrac{a + 2}{a + 2}$$

$$= \dfrac{6}{7} \qquad \text{"Removing" the factor } \dfrac{a + 2}{a + 2}$$

9. $\dfrac{6x^2 + 4x}{2x^2 + 2x} = \dfrac{2x(3x + 2)}{2x(x + 1)} \qquad \text{Factoring numerator and denominator}$

$$= \dfrac{2x}{2x} \cdot \dfrac{3x + 2}{x + 1} \qquad \text{Factoring the fractional expression}$$

$$= \dfrac{3x + 2}{x + 1} \qquad \text{"Removing" a factor of 1}$$

10. $\dfrac{x^2 + 3x + 2}{x^2 - 1} = \dfrac{(x + 2)(x + 1)}{(x + 1)(x - 1)}$

$\qquad = \dfrac{x + 1}{x + 1} \cdot \dfrac{x + 2}{x - 1}$

$\qquad = \dfrac{x + 2}{x - 1}$

11. $\dfrac{5a + 15}{10} = \dfrac{5(a + 3)}{5 \cdot 2}$

$\qquad = \dfrac{5}{5} \cdot \dfrac{a + 3}{2}$

$\qquad = \dfrac{a + 3}{2}$

DO EXERCISES 8–11.

⁚⁚ MULTIPLYING AND SIMPLIFYING

We try to simplify after we multiply. That is why we do not multiply out the numerator and the denominator too soon. We would need to factor them anyway in order to simplify.

Example 12 Multiply and simplify: $\dfrac{x^2 + 6x + 9}{x^2 - 4} \cdot \dfrac{x - 2}{x + 3}$.

$\qquad \dfrac{x^2 + 6x + 9}{x^2 - 4} \cdot \dfrac{x - 2}{x + 3}$

$\qquad = \dfrac{(x^2 + 6x + 9)(x - 2)}{(x^2 - 4)(x + 3)}$ Multiplying numerators and denominators

$\qquad = \dfrac{(x + 3)(x + 3)(x - 2)}{(x + 2)(x - 2)(x + 3)}$ Factoring numerator and denominator

> Note that if we had multiplied out the numerator and the denominator, we would have had to factor them again in order to simplify.

$\qquad = \dfrac{(x + 3)(x - 2)}{(x + 3)(x - 2)} \cdot \dfrac{x + 3}{x + 2}$ Factoring the fractional expression

$\qquad = \dfrac{x + 3}{x + 2}$ Simplifying by removing a factor of 1

Simplify by removing a factor of 1.

8. $\dfrac{2x^2 + x}{3x^2 + 2x}$

9. $\dfrac{x^2 - 1}{2x^2 - x - 1}$

10. $\dfrac{7x + 14}{7}$

11. $\dfrac{12y + 24}{48}$

ANSWERS ON PAGE A-19

Multiply and simplify.

12. $\dfrac{a^2 - 4a + 4}{a^2 - 9} \cdot \dfrac{a + 3}{a - 2}$

Example 13 Multiply and simplify: $\dfrac{x^2 + x - 2}{15} \cdot \dfrac{5}{2x^2 - 3x + 1}$.

$$\dfrac{x^2 + x - 2}{15} \cdot \dfrac{5}{2x^2 - 3x + 1}$$

$$= \dfrac{(x^2 + x - 2)5}{15(2x^2 - 3x + 1)} \qquad \text{Multiplying numerators and denominators}$$

$$= \dfrac{(x + 2)(x - 1)5}{(5)(3)(x - 1)(2x - 1)} \qquad \text{Factoring numerator and denominator}$$

$$= \dfrac{(x - 1)5}{(x - 1)5} \cdot \dfrac{x + 2}{3(2x - 1)} \qquad \text{Factoring the fractional expression}$$

$$= \dfrac{x + 2}{3(2x - 1)} \qquad \text{Simplifying by removing a factor of 1}$$

You need not carry out this multiplication.

DO EXERCISES 12 AND 13.

13. $\dfrac{x^2 - 25}{6} \cdot \dfrac{3}{x + 5}$

ANSWERS ON PAGE A-19

Chapter 9   Fractional Expressions and Equations

· Multiply. Do not carry out multiplications in the numerator and denominator.

1. $\dfrac{x-2}{x-5} \cdot \dfrac{x-2}{x+5}$

2. $\dfrac{x-1}{x+2} \cdot \dfrac{x+1}{x+2}$

3. $\dfrac{c-3d}{c+d} \cdot \dfrac{c+3d}{c-d}$

4. $\dfrac{a+2b}{a+b} \cdot \dfrac{a-2b}{a-b}$

· · Multiply.

5. $\dfrac{2a-1}{2a-1} \cdot \dfrac{3a-1}{3a+2}$

6. $\dfrac{3x-2}{x+7} \cdot \dfrac{2x+5}{2x+5}$

· · · Simplify by removing a factor of 1.

7. $\dfrac{x(3x+2)(x+1)}{x(x+1)(3x-2)}$

8. $\dfrac{x(3x+4)(5x+7)}{x(3x-4)(5x+7)}$

*9. $\dfrac{8a+8b}{8a-8b}$

*10. $\dfrac{6x-6y}{6x+6y}$

11. $\dfrac{t^2-25}{t^2+t-20}$

12. $\dfrac{a^2-9}{a^2+5a+6}$

13. $\dfrac{2x^2+6x+4}{4x^2-12x-16}$

14. $\dfrac{x^2-3x-4}{2x^2+10x+8}$

15. $\dfrac{a^2-10a+21}{a^2-11a+28}$

16. $\dfrac{x^2-3x-18}{x^2-2x-15}$

*Indicates exercises that can be omitted if Section 5.9 was not studied.

ANSWERS
1.
2.
3.
4.
5.
6.
7.
8.
9.
10.
11.
12.
13.
14.
15.
16.

17. $\dfrac{6x + 12}{x^2 - x - 6}$

18. $\dfrac{5y + 5}{y^2 + 7y + 6}$

19. $\dfrac{a^2 + 1}{a + 1}$

20. $\dfrac{t^2 - 1}{t + 1}$

Multiply and simplify.

21. $\dfrac{t^2}{t^2 - 4} \cdot \dfrac{t^2 - 5t + 6}{t^2 - 3t}$

(*Hint:* Factor t^2 in the numerator of $\dfrac{t^2}{t^2 - 4}$ as $t \cdot t$.)

22. $\dfrac{x^2 - 3x - 10}{(x - 2)^2} \cdot \dfrac{x - 2}{x - 5}$

[*Hint:* Factor $(x - 2)^2$ as $(x - 2)(x - 2)$.]

23. $\dfrac{24a^2}{3(a^2 - 4a + 4)} \cdot \dfrac{3a - 6}{2a}$

24. $\dfrac{5v + 5}{v - 2} \cdot \dfrac{v^2 - 4v + 4}{v^2 - 1}$

25. $\dfrac{ab - b^2}{2a} \cdot \dfrac{2a + 2b}{a^2b - b^3}$

26. $\dfrac{c^2 - 6c}{c - 6} \cdot \dfrac{c + 3}{c}$

✔ SKILL MAINTENANCE

27. Multiply and simplify: $\dfrac{1}{6} \cdot \dfrac{3}{8}$.

28. Divide and simplify: $\dfrac{2}{5} \div \dfrac{4}{15}$.

29. Add and simplify: $\dfrac{1}{2} + \dfrac{2}{5}$.

30. Subtract and simplify: $\dfrac{13}{15} - \dfrac{2}{45}$.

⭐ EXTENSION

Simplify.

31. $\dfrac{x^4 - 16y^4}{(x^2 + 4y^2)(x - 2y)}$

32. $\dfrac{(a - b)^2}{b^2 - a^2}$

33. $\dfrac{(t + 2)^3}{(t + 1)^3} \cdot \dfrac{t^2 + 2t + 1}{t^2 + 4t + 4} \cdot \dfrac{t + 1}{t + 2}$

Determine those replacements in the following that are not sensible.

34. $\dfrac{x + 1}{x^2 + 4x + 4}$

35. $\dfrac{x - 7}{x^3 - 9x^2 + 14x}$

36. $\dfrac{y^3 + 47}{y^2 - 49}$

Division and Reciprocals

There is a similarity throughout this chapter between what we do with fractional expressions and what we do with rational numbers. In fact, after replacements of variables by rational numbers, a fractional expression represents a rational number.

• FINDING RECIPROCALS

Two expressions are reciprocals of each other if their product is 1. The reciprocal of a fractional expression is found by interchanging numerator and denominator.

Examples

1. The reciprocal of $\frac{2}{5}$ is $\frac{5}{2}$. $\left(\text{This is because } \frac{2}{5} \cdot \frac{5}{2} = \frac{10}{10} = 1.\right)$

2. The reciprocal of $\frac{2x^2 - 3}{x + 4}$ is $\frac{x + 4}{2x^2 - 3}$.

3. The reciprocal of $x + 2$ is $\frac{1}{x + 2}$. $\left(\text{Think of } x + 2 \text{ as } \frac{x + 2}{1}.\right)$

DO EXERCISES 1–4.

• • DIVISION

> **To divide, we multiply by a reciprocal and simplify the result.**

(To review the reason for this, see Section 1.7.)

Examples Divide.

4. $\dfrac{3}{4} \div \dfrac{2}{5} = \dfrac{3}{4} \cdot \dfrac{5}{2}$ Multiplying by the reciprocal

$\phantom{\dfrac{3}{4} \div \dfrac{2}{5}} = \dfrac{3 \cdot 5}{4 \cdot 2}$

$\phantom{\dfrac{3}{4} \div \dfrac{2}{5}} = \dfrac{15}{8}$

5. $\dfrac{x + 1}{x + 2} \div \dfrac{x - 1}{x + 3} = \dfrac{x + 1}{x + 2} \cdot \dfrac{x + 3}{x - 1}$ Multiplying by the reciprocal

$\phantom{\dfrac{x + 1}{x + 2}} = \dfrac{(x + 1)(x + 3)}{(x + 2)(x - 1)}$ You need not carry out the multiplications in the numerator and denominator.

DO EXERCISES 5 AND 6.

Find the reciprocal.

1. $\dfrac{7}{2}$

2. $\dfrac{x^2 + 5}{2x^3 - 1}$

3. $x - 5$

4. $\dfrac{1}{x^2 - 3}$

Divide.

5. $\dfrac{3}{5} \div \dfrac{7}{2}$

6. $\dfrac{x - 3}{x + 5} \div \dfrac{x + 5}{x - 2}$

ANSWERS ON PAGE A-19

Divide and simplify.

7. $\dfrac{x-3}{x+5} \div \dfrac{x+2}{x+5}$

8. $\dfrac{x^2-5x+6}{x+5} \div \dfrac{x+2}{x+5}$

9. $\dfrac{y^2-1}{y+1} \div \dfrac{y^2-2y+1}{y+1}$

Example 6 Divide and simplify: $\dfrac{x+1}{x^2-1} \div \dfrac{x+1}{x^2-2x+1}$.

$$\dfrac{x+1}{x^2-1} \div \dfrac{x+1}{x^2-2x+1}$$

$= \dfrac{x+1}{x^2-1} \cdot \dfrac{x^2-2x+1}{x+1}$ Multiplying by the reciprocal

$= \dfrac{(x+1)(x^2-2x+1)}{(x^2-1)(x+1)}$

$= \dfrac{(x+1)(x-1)(x-1)}{(x-1)(x+1)(x+1)}$ Factoring numerator and denominator

$= \dfrac{(x+1)(x-1)}{(x+1)(x-1)} \cdot \dfrac{x-1}{x+1}$ Factoring the fractional expression

$= \dfrac{x-1}{x+1}$ Simplifying

Example 7 Divide and simplify: $\dfrac{x^2-2x-3}{x^2-4} \div \dfrac{x+1}{x+5}$.

$$\dfrac{x^2-2x-3}{x^2-4} \div \dfrac{x+1}{x+5}$$

$= \dfrac{x^2-2x-3}{x^2-4} \cdot \dfrac{x+5}{x+1}$ Multiplying by the reciprocal

$= \dfrac{(x^2-2x-3)(x+5)}{(x^2-4)(x+1)}$

$= \dfrac{(x-3)(x+1)(x+5)}{(x-2)(x+2)(x+1)}$ Factoring numerator and denominator

$= \dfrac{x+1}{x+1} \cdot \dfrac{(x-3)(x+5)}{(x-2)(x+2)}$ Factoring the fractional expression

$= \dfrac{(x-3)(x+5)}{(x-2)(x+2)}$ Simplifying

> You need not carry out the multiplications in the numerator and denominator.

DO EXERCISES 7–9.

• Find the reciprocal.

1. $\dfrac{4}{x}$

2. $\dfrac{a+3}{a-1}$

3. $x^2 - y^2$

4. $\dfrac{1}{a+b}$

5. $\dfrac{x^2 + 2x - 5}{x^2 - 4x + 7}$

6. $\dfrac{x^2 - 3xy + y^2}{x^2 + 7xy - y^2}$

•• Divide and simplify.

7. $\dfrac{2}{5} \div \dfrac{4}{3}$

8. $\dfrac{5}{6} \div \dfrac{2}{3}$

9. $\dfrac{2}{x} \div \dfrac{8}{x}$

10. $\dfrac{x}{2} \div \dfrac{3}{x}$

*11. $\dfrac{x^2}{y} \div \dfrac{x^3}{y^3}$

*12. $\dfrac{a}{b^2} \div \dfrac{a^2}{b^3}$

13. $\dfrac{a+2}{a-3} \div \dfrac{a-1}{a+3}$

14. $\dfrac{y+2}{4} \div \dfrac{y}{2}$

15. $\dfrac{x^2 - 1}{x} \div \dfrac{x+1}{x-1}$

16. $\dfrac{4y-8}{y+2} \div \dfrac{y-2}{y^2-4}$

17. $\dfrac{x+1}{6} \div \dfrac{x+1}{3}$

*18. $\dfrac{a}{a-b} \div \dfrac{b}{a-b}$

19. $\dfrac{x^2 - 9}{4x+12} \div \dfrac{x-3}{6}$

20. $\dfrac{c^2 + 3c}{c^2 + 2c - 3} \div \dfrac{c}{c+1}$

ANSWERS
1.
2.
3.
4.
5.
6.
7.
8.
9.
10.
11.
12.
13.
14.
15.
16.
17.
18.
19.
20.

*Indicates exercises that can be omitted if Section 5.9 was not studied.

*21. $\dfrac{x + y}{x - y} \div \dfrac{x^2 + y}{x^2 - y^2}$

*22. $\dfrac{x - b}{2x} \div \dfrac{x^2 - b^2}{5x^2}$

23. $\dfrac{x^2 - x - 20}{x^2 + 7x + 12} \div \dfrac{x^2 - 10x + 25}{x^2 + 6x + 9}$

24. $\dfrac{2y^2 - 7y + 3}{2y^2 + 3y - 2} \div \dfrac{6y^2 - 5y + 1}{3y^2 + 5y - 2}$

25. $\dfrac{c^2 + 10c + 21}{c^2 - 2c - 15} \div (c^2 + 2c - 35)$

26. $(1 - z) \div \dfrac{1 - z}{1 + 2z - z^2}$

27. $\dfrac{(t + 5)^3}{(t - 5)^3} \div \dfrac{(t + 5)^2}{(t - 5)^2}$

28. $\dfrac{(y - 3)^3}{(y + 3)^3} \div \dfrac{(y - 3)^2}{(y + 3)^2}$

 SKILL MAINTENANCE

29. Find the reciprocal of $\dfrac{2}{5}$.

30. Subtract:

$$(x^2 + 6x + 8) - (x^2 - 3x - 4).$$

Factor.

31. $x^2 + 3x + 2$

32. $25t^2 - 4$

⭐ **EXTENSION**

Simplify.

33. $\dfrac{x^2 - x + xy - y}{x^2 + 6x - 7} \div \dfrac{x^2 + 2xy + y^2}{4x + 4y}$

34. $\dfrac{3x^2 - 2xy - y^2}{x^2 - y^2} \div (3x^2 + 4xy + y^2)$

35. $\left[\dfrac{y^2 + 5y + 6}{y^2} \cdot \dfrac{3y^3 + 6y^2}{y^2 - y - 12}\right] \div \dfrac{y^2 - y}{y^2 - 2y - 8}$

36. $\dfrac{z^2 - 8z + 16}{z^2 + 8z + 16} \div \dfrac{(z - 4)^2}{(z + 4)^2}$

Addition and Subtraction

• ADDITION WHEN DENOMINATORS ARE THE SAME

After finishing Section 9.3, you should be able to:

•	Add fractional expressions having the same denominator.
• •	Add fractional expressions whose denominators are additive inverses of each other.
• • •	Subtract fractional expressions having the same denominator.
• • / • •	Subtract fractional expressions whose denominators are additive inverses of each other.

Addition is done as in arithmetic.

> When denominators are the same, we add the numerators and keep the denominator.

Examples Add.

1. $\dfrac{x}{x+1} + \dfrac{2}{x+1} = \dfrac{x+2}{x+1}$

2. $\dfrac{2x^2 + 3x - 7}{2x + 1} + \dfrac{x^2 + x - 8}{2x + 1} = \dfrac{(2x^2 + 3x - 7) + (x^2 + x - 8)}{2x + 1}$

$$= \dfrac{3x^2 + 4x - 15}{2x + 1}$$

DO EXERCISES 1–3.

• • ADDITION WHEN DENOMINATORS ARE ADDITIVE INVERSES

When one denominator is the additive inverse of the other, we first multiply one expression by $-1/-1$ to obtain equivalent expressions with the same denominator.

Examples Add.

3. $\dfrac{x}{2} + \dfrac{3}{-2} = \dfrac{x}{2} + \dfrac{-1}{-1} \cdot \dfrac{3}{-2}$ **Multiplying by** $\dfrac{-1}{-1}$

$$= \dfrac{x}{2} + \dfrac{(-1)3}{(-1)(-2)} = \dfrac{x}{2} + \dfrac{-3}{2}$$ **Denominators are now the same**

$$= \dfrac{x + (-3)}{2} = \dfrac{x - 3}{2}$$

4. $\dfrac{3x + 4}{x - 2} + \dfrac{x - 7}{2 - x} = \dfrac{3x + 4}{x - 2} + \dfrac{-1}{-1} \cdot \dfrac{x - 7}{2 - x}$

> We could have chosen to multiply this expression by $-1/-1$. This would also work.

$$= \dfrac{3x + 4}{x - 2} + \dfrac{-1(x - 7)}{-1(2 - x)}$$ *Note:* $-1(2 - x) = -2 + x$
$$= x - 2$$

$$= \dfrac{3x + 4}{x - 2} + \dfrac{7 - x}{x - 2}$$

$$= \dfrac{(3x + 4) + (7 - x)}{x - 2} = \dfrac{2x + 11}{x - 2}$$

DO EXERCISES 4 AND 5.

Add.

1. $\dfrac{5}{9} + \dfrac{2}{9}$

2. $\dfrac{3}{x - 2} + \dfrac{x}{x - 2}$

3. $\dfrac{4x + 5}{x - 1} + \dfrac{2x - 1}{x - 1}$

Add.

4. $\dfrac{x}{4} + \dfrac{5}{-4}$

5. $\dfrac{2x + 1}{x - 3} + \dfrac{x + 2}{3 - x}$

ANSWERS ON PAGE A-19

Subtract.

6. $\dfrac{7}{11} - \dfrac{3}{11}$

7. $\dfrac{2x^2 + 3x - 7}{2x + 1} - \dfrac{x^2 + x - 8}{2x + 1}$

Subtract.

8. $\dfrac{x}{3} - \dfrac{2x - 1}{-3}$

9. $\dfrac{3x}{x - 2} - \dfrac{x - 3}{2 - x}$

$\bullet\bullet\bullet$ **SUBTRACTION WHEN DENOMINATORS ARE THE SAME**

Subtraction is done as in arithmetic.

> When denominators are the same, we subtract the numerators and keep the denominator.

Example 5 Subtract: $\dfrac{3x}{x + 2} - \dfrac{x - 2}{x + 2}$.

$$\dfrac{3x}{x + 2} - \dfrac{x - 2}{x + 2} = \dfrac{3x - (x - 2)}{x + 2}$$

> The parentheses are important to make sure that you subtract the entire numerator.

$$= \dfrac{3x - x + 2}{x + 2}$$

$$= \dfrac{2x + 2}{x + 2}$$

DO EXERCISES 6 AND 7.

$\vcenter{\hbox{$\begin{smallmatrix}\bullet&\bullet\\\bullet&\bullet\end{smallmatrix}$}}$ **SUBTRACTION WHEN DENOMINATORS ARE ADDITIVE INVERSES**

When one denominator is the additive inverse of the other, we can first multiply one expression by $-1/-1$.

Example 6 Subtract: $\dfrac{x}{5} - \dfrac{3x - 4}{-5}$.

$$\dfrac{x}{5} - \dfrac{3x - 4}{-5} = \dfrac{x}{5} - \dfrac{-1}{-1} \cdot \dfrac{3x - 4}{-5}$$

$$= \dfrac{x}{5} - \dfrac{(-1)(3x - 4)}{(-1)(-5)} = \dfrac{x}{5} - \dfrac{-3x + 4}{5}$$

$$= \dfrac{x - (-3x + 4)}{5}$$

> Remember the parentheses!

$$= \dfrac{x + 3x - 4}{5} = \dfrac{4x - 4}{5}$$

Example 7 Subtract: $\dfrac{5y}{y - 5} - \dfrac{2y - 3}{5 - y}$.

$$\dfrac{5y}{y - 5} - \dfrac{2y - 3}{5 - y} = \dfrac{5y}{y - 5} - \dfrac{-1}{-1} \cdot \dfrac{2y - 3}{5 - y}$$

$$= \dfrac{5y}{y - 5} - \dfrac{(-1)(2y - 3)}{(-1)(5 - y)} = \dfrac{5y}{y - 5} - \dfrac{-2y + 3}{y - 5}$$

$$= \dfrac{5y - (-2y + 3)}{y - 5}$$

> Remember the parentheses!

$$= \dfrac{5y + 2y - 3}{y - 5} = \dfrac{7y - 3}{y - 5}$$

DO EXERCISES 8 AND 9.

• Add. Simplify if possible.

1. $\dfrac{5}{12} + \dfrac{7}{12}$

2. $\dfrac{3}{14} + \dfrac{5}{14}$

3. $\dfrac{1}{3+x} + \dfrac{5}{3+x}$

4. $\dfrac{4x+1}{6x+5} + \dfrac{3x-7}{5+6x}$

5. $\dfrac{x^2+7x}{x^2-5x} + \dfrac{x^2-4x}{x^2-5x}$

*6. $\dfrac{a}{x+y} + \dfrac{b}{y+x}$

•• Add. Simplify if possible.

7. $\dfrac{7}{8} + \dfrac{5}{-8}$

8. $\dfrac{11}{6} + \dfrac{5}{-6}$

9. $\dfrac{3}{t} + \dfrac{4}{-t}$

10. $\dfrac{5}{-a} + \dfrac{8}{a}$

11. $\dfrac{2x+7}{x-6} + \dfrac{3x}{6-x}$

12. $\dfrac{3x-2}{4x-3} + \dfrac{2x-5}{3-4x}$

13. $\dfrac{y^2}{y-3} + \dfrac{9}{3-y}$

14. $\dfrac{t^2}{t-2} + \dfrac{4}{2-t}$

15. $\dfrac{b-7}{b^2-16} + \dfrac{7-b}{16-b^2}$

16. $\dfrac{a-3}{a^2-25} + \dfrac{a-3}{25-a^2}$

*17. $\dfrac{z}{(y+z)(y-z)} + \dfrac{y}{(z+y)(z-y)}$

[Hint: Multiply by $-1/-1$. Note that $(z+y)(z-y)(-1) = (z+y)(y-z)$.]

ANSWERS
1.
2.
3.
4.
5.
6.
7.
8.
9.
10.
11.
12.
13.
14.
15.
16.
17.

*Indicates exercises that can be omitted if Section 5.9 was not studied.

*18. $\dfrac{a^2}{a-b} + \dfrac{b^2}{b-a}$

19. $\dfrac{x+3}{x-5} + \dfrac{2x-1}{5-x} + \dfrac{2(3x-1)}{x-5}$

18. _____

20. $\dfrac{3(x-2)}{2x-3} + \dfrac{5(2x+1)}{2x-3} + \dfrac{3(x-1)}{3-2x}$

21. $\dfrac{2(4x+1)}{5x-7} + \dfrac{3(x-2)}{7-5x} + \dfrac{-10x-1}{5x-7}$

19. _____

20. _____

22. $\dfrac{5(x-2)}{3x-4} + \dfrac{2(x-3)}{4-3x} + \dfrac{3(5x+1)}{4-3x}$

21. _____

23. $\dfrac{x+1}{(x+3)(x-3)} + \dfrac{4(x-3)}{(x-3)(x+3)} + \dfrac{(x-1)(x-3)}{(3-x)(x+3)}$

22. _____

23. _____

24. $\dfrac{2(x+5)}{(2x-3)(x-1)} + \dfrac{3x+4}{(2x-3)(1-x)} + \dfrac{x-5}{(3-2x)(x-1)}$

24. _____

●●● Subtract. Simplify if possible.

25. $\dfrac{7}{8} - \dfrac{3}{8}$

26. $\dfrac{5}{y} - \dfrac{7}{y}$

25. _____

26. _____

27. $\dfrac{x}{x-1} - \dfrac{1}{x-1}$

28. $\dfrac{x^2}{x+4} - \dfrac{16}{x+4}$

29. $\dfrac{x+1}{x^2-2x+1} - \dfrac{5-3x}{x^2-2x+1}$

30. $\dfrac{2x-3}{x^2+3x-4} - \dfrac{x-7}{x^2+3x-4}$

Subtract. Simplify if possible.

31. $\dfrac{11}{6} - \dfrac{5}{-6}$

32. $\dfrac{7}{8} - \dfrac{5}{-8}$

33. $\dfrac{5}{a} - \dfrac{8}{-a}$

34. $\dfrac{3}{t} - \dfrac{4}{-t}$

35. $\dfrac{x}{4} - \dfrac{3x-5}{-4}$

36. $\dfrac{2}{x-1} - \dfrac{2}{1-x}$

37. $\dfrac{3-x}{x-7} - \dfrac{2x-5}{7-x}$

38. $\dfrac{t^2}{t-2} - \dfrac{4}{2-t}$

39. $\dfrac{x-8}{x^2-16} - \dfrac{x-8}{16-x^2}$

40. $\dfrac{x-2}{x^2-25} - \dfrac{6-x}{25-x^2}$

41. $\dfrac{4-x}{x-9} - \dfrac{3x-8}{9-x}$

42. $\dfrac{3-x}{x-7} - \dfrac{2x-5}{7-x}$

ANSWERS

27.

28.

29.

30.

31.

32.

33.

34.

35.

36.

37.

38.

39.

40.

41.

42.

43. $\dfrac{2(x-1)}{2x-3} - \dfrac{3(x+2)}{2x-3} - \dfrac{x-1}{3-2x}$

44. $\dfrac{3(x-2)}{2x-3} - \dfrac{5(2x+1)}{2x-3} - \dfrac{3(x-1)}{3-2x}$

Perform the indicated operations and simplify.

45. $\dfrac{3(2x+5)}{x-1} - \dfrac{3(2x-3)}{1-x} + \dfrac{6x-1}{x-1}$

46. $\dfrac{2x-y}{x-y} + \dfrac{x-2y}{y-x} - \dfrac{3x-3y}{x-y}$

47. $\dfrac{x-y}{x^2-y^2} + \dfrac{x+y}{x^2-y^2} - \dfrac{2x}{x^2-y^2}$

48. $\dfrac{x+y}{2(x-y)} - \dfrac{2x-2y}{2(x-y)} + \dfrac{x-3y}{2(y-x)}$

49. $\dfrac{10}{2y-1} - \dfrac{6}{1-2y} + \dfrac{y}{2y-1} + \dfrac{y-4}{1-2y}$

50. $\dfrac{(x+3)(2x-1)}{(2x-3)(x-3)} - \dfrac{(x-3)(x+1)}{(3-x)(3-2x)} + \dfrac{(2x+1)(x+3)}{(3-2x)(x-3)}$

 SKILL MAINTENANCE

Factor.

51. $2x^2 - 3x + 1$

52. $x^5 + x^4 - 2x^3$

Find the LCM.

53. 12, 30

54. 16, 18

⭐ EXTENSION

Simplify.

55. $\dfrac{x^2}{3x^2-5x-2} - \dfrac{2x}{3x+1} \cdot \dfrac{1}{x-2}$

56. $\dfrac{x}{x-y} + \dfrac{y}{y-x} + \dfrac{x+y}{x-y} + \dfrac{x-y}{y-x}$

57. $\dfrac{x}{(x-y)(y-z)} - \dfrac{x}{(y-x)(z-y)}$

58. $\dfrac{3}{x+4} \cdot \dfrac{2x+11}{x-3} - \dfrac{-1}{4+x} \cdot \dfrac{6x+3}{3-x}$

Least Common Multiples

• LEAST COMMON MULTIPLES

To add when denominators are different, we first find a common denominator. For example, to add $\frac{5}{12}$ and $\frac{7}{30}$ we first look for a common multiple of both 12 and 30. Any multiple will do, but we prefer the smallest such number, the *Least Common Multiple* (LCM). To find the LCM, we factor.

$$12 = 2 \cdot 2 \cdot 3$$
$$30 = 2 \cdot 3 \cdot 5$$

The LCM is the number that has 2 as a factor twice, 3 as a factor once, and 5 as a factor once:

$$LCM = 2 \cdot 2 \cdot 3 \cdot 5, \quad \text{or} \quad 60.$$

> To find the LCM, we use each factor the greatest number of times that it appears in any one factorization.

Example 1 Find the LCM of 24 and 36.

$$\left.\begin{array}{l} 24 = 2 \cdot 2 \cdot 2 \cdot 3 \\ 36 = 2 \cdot 2 \cdot 3 \cdot 3 \end{array}\right\} \quad LCM = 2 \cdot 2 \cdot 2 \cdot 3 \cdot 3, \quad \text{or} \quad 72.$$

DO EXERCISES 1–4.

•• ADDING USING THE LCM

Let us finish adding $\frac{5}{12}$ and $\frac{7}{30}$.

$$\frac{5}{12} + \frac{7}{30} = \frac{5}{2 \cdot 2 \cdot 3} + \frac{7}{2 \cdot 3 \cdot 5}$$

The LCM is $2 \cdot 2 \cdot 3 \cdot 5$. To get the LCM in the first denominator we need a 5. To get the LCM in the second denominator we need another 2. We get these numbers by multiplying by 1.

$$\frac{5}{12} + \frac{7}{30} = \frac{5}{2 \cdot 2 \cdot 3} \cdot \frac{5}{5} + \frac{7}{2 \cdot 3 \cdot 5} \cdot \frac{2}{2} \quad \text{Multiplying by 1}$$

$$= \frac{25}{2 \cdot 2 \cdot 3 \cdot 5} + \frac{14}{2 \cdot 3 \cdot 5 \cdot 2} \quad \begin{array}{l}\text{Denominators are now} \\ \text{the LCM}\end{array}$$

$$= \frac{39}{2 \cdot 2 \cdot 3 \cdot 5} \quad \begin{array}{l}\text{Adding the numerators and keeping} \\ \text{the LCM}\end{array}$$

$$= \frac{13}{20} \quad \text{Simplifying}$$

Find the LCM by factoring.

1. 16, 18

2. 6, 12

3. 2, 5

4. 24, 30, 20

ANSWERS ON PAGE A-19

Add, first finding the LCM of the denominators. Simplify if possible.

5. $\dfrac{3}{16} + \dfrac{1}{18}$

6. $\dfrac{1}{6} + \dfrac{1}{12}$

7. $\dfrac{1}{2} + \dfrac{3}{5}$

8. $\dfrac{1}{24} + \dfrac{1}{30} + \dfrac{3}{20}$

Find the LCM.

9. $12xy^2,\ 15x^3y$

10. $y^2 + 5y + 4,\ y^2 + 2y + 1$

11. $t^2 + 16,\ t - 2,\ 7$

12. $x^2 + 2x + 1,\ 3x - 3x^2,\ x^2 - 1$

ANSWERS ON PAGE A-19

Example 2 Add: $\frac{5}{12} + \frac{11}{18}$.

$$\left.\begin{array}{l} 12 = 2 \cdot 2 \cdot 3 \\[2mm] 18 = 2 \cdot 3 \cdot 3 \end{array}\right\} \quad \text{LCM} = 2 \cdot 2 \cdot 3 \cdot 3, \quad \text{or} \quad 36.$$

$$\frac{5}{12} + \frac{11}{18} = \frac{5}{2 \cdot 2 \cdot 3} \cdot \frac{3}{3} + \frac{11}{2 \cdot 3 \cdot 3} \cdot \frac{2}{2} = \frac{37}{2 \cdot 2 \cdot 3 \cdot 3} = \frac{37}{36}$$

DO EXERCISES 5–8.

●●● LCM'S OF ALGEBRAIC EXPRESSIONS

To find the LCM of two or more algebraic expressions, we factor them. Then we use each factor the greatest number of times it occurs in any one expression.

Example 3 Find the LCM of $12x$, $16y$, and $8xyz$.

$$\left.\begin{array}{l} 12x = 2 \cdot 2 \cdot 3 \cdot x \\ 16y = 2 \cdot 2 \cdot 2 \cdot 2 \cdot y \\ 8xyz = 2 \cdot 2 \cdot 2 \cdot x \cdot y \cdot z \end{array}\right\} \quad \begin{array}{l} \text{LCM} = 2 \cdot 2 \cdot 2 \cdot 2 \cdot 3 \cdot x \cdot y \cdot z \\ = 48xyz \end{array}$$

Example 4 Find the LCM of $x^2 + 5x - 6$ and $x^2 - 1$.

$$\left.\begin{array}{l} x^2 + 5x - 6 = (x + 6)(x - 1) \\ x^2 - 1 = (x + 1)(x - 1) \end{array}\right\} \quad \text{LCM} = (x + 6)(x - 1)(x + 1)$$

Example 5 Find the LCM of $x^2 + 4$, $x + 1$, and 5.

These expressions are not factorable, so the LCM is their product: $5(x^2 + 4)(x + 1)$.

The additive inverse of an LCM is also an LCM. For example, if $(x + 2)(x - 5)$ is an LCM, then $-(x + 2)(x - 5)$ is an LCM. We can name the latter $(x + 2)(-1)(x - 5)$, or $(x + 2)(5 - x)$. When finding LCMs, if factors that are additive inverses occur, we do not use them both. For example, if $(x - 5)$ occurs in one factorization and $(5 - x)$ occurs in another, since these are additive inverses, we do not use them both.

Example 6 Find the LCM of $x^2 - 25$ and $10 - 2x$.

$$\left.\begin{array}{l} x^2 - 25 = (x + 5)(x - 5) \\ 10 - 2x = 2\,(5 - x) \end{array}\right\} \quad \begin{array}{l} \text{LCM} = 2(x + 5)(x - 5) \longleftarrow \\ \text{or } 2(x + 5)(5 - x) \longleftarrow \end{array}$$

We can use $x - 5$ or $5 - x$ but not both.

Example 7 Find the LCM of $x^2 - 4y^2$, $x^2 - 4xy + 4y^2$, and $x - 2y$.

$$\left.\begin{array}{r} x^2 - 4y^2 = (x - 2y)(x + 2y) \\ x^2 - 4xy + 4y^2 = (x - 2y)(x - 2y) \\ x - 2y = x - 2y \end{array}\right\} \quad \begin{array}{l} \text{LCM} = \\ (x + 2y)(x - 2y)(x - 2y) \\ = (x + 2y)(x - 2y)^2 \end{array}$$

DO EXERCISES 9–12.

Chapter 9   Fractional Expressions and Equations

⬤ Find the LCM.

1. 12, 27

2. 10, 15

3. 8, 9

4. 12, 15

5. 6, 9, 21

6. 8, 36, 40

7. 24, 36, 40

8. 3, 4, 5

9. 28, 42, 60

⬤⬤ Add, first finding the LCM of the denominators. Simplify if possible.

10. $\dfrac{7}{24} + \dfrac{11}{18}$

11. $\dfrac{7}{60} + \dfrac{6}{75}$

12. $\dfrac{1}{6} + \dfrac{3}{40} + \dfrac{2}{75}$

13. $\dfrac{5}{24} + \dfrac{3}{20} + \dfrac{7}{30}$

14. $\dfrac{2}{15} + \dfrac{5}{9} + \dfrac{3}{20}$

15. $\dfrac{1}{20} + \dfrac{1}{30} + \dfrac{2}{45}$

⬤⬤⬤ Find the LCM.

16. $6x^2,\ 12x^3$

*17. $2a^2b,\ 8ab^2$

*18. $2x^2,\ 6xy,\ 18y^2$

*19. $c^2d,\ cd^2,\ c^3d$

20. $2(y-3),\ 6(3-y)$

21. $4(x-1),\ 8(1-x)$

*Indicates exercises that can be omitted if Section 5.9 was not studied.

ANSWERS
1.
2.
3.
4.
5.
6.
7.
8.
9.
10.
11.
12.
13.
14.
15.
16.
17.
18.
19.
20.
21.

*22. $x^2 - y^2$, $2x + 2y$, $x^2 + 2xy + y^2$

23. $a + 1$, $(a - 1)^2$, $a^2 - 1$

24. $m^2 - 5m + 6$, $m^2 - 4m + 4$

25. $2 + 3k$, $9k^2 - 4$, $2 - 3k$

26. $10v^2 + 30v$, $-5v^2 - 35v - 60$

27. $9x^3 - 9x^2 - 18x$, $6x^5 - 24x^4 + 24x^3$

28. $x^5 + 4x^4 + 4x^3$, $3x^2 - 12$, $2x + 4$

29. $x^5 + 2x^4 + x^3$, $2x^3 - 2x$, $5x - 5$

 SKILL MAINTENANCE

Factor.

30. $x^2 - 6x + 9$

31. $6x^2 + 4x$

Solve.

32. $3t + 2t = 12$

33. $x^2 - 5x + 6 = 0$

EXTENSION

34. When is the LCM of two expressions the same as their product?

35. If the LCM of two expressions is the same as one of the expressions, what is their relationship?

Addition with Different Denominators

Now that we know how to find LCM's, we can add fractional expressions with different denominators. We first find the LCM of the denominators (the least common denominator) and then add.

Example 1 Add: $\dfrac{3}{x+1} + \dfrac{5}{x-1}$.

The denominators do not factor, so the LCM is their product. We multiply by 1 to get the LCM in each expression.

$$\frac{3}{x+1} \cdot \frac{x-1}{x-1} + \frac{5}{x-1} \cdot \frac{x+1}{x+1} = \frac{3(x-1) + 5(x+1)}{(x-1)(x+1)}$$

$$= \frac{3x - 3 + 5x + 5}{(x-1)(x+1)}$$

$$= \frac{8x + 2}{(x-1)(x+1)}$$

The numerator and denominator have no common factor (other than 1), so we cannot simplify.

DO EXERCISE 1.

Add.

1. $\dfrac{x}{x-2} + \dfrac{4}{x+2}$

Example 2 Add: $\dfrac{5}{x^2 + x} + \dfrac{4}{2x + 2}$.

First find the LCM of the denominators.

$$\left. \begin{array}{l} x^2 + x = x(x+1) \\ 2x + 2 = 2(x+1) \end{array} \right\} \quad \text{LCM} = 2x(x+1)$$

Multiply by 1 to get the LCM in each expression. Then add and simplify.

$$\frac{5}{x(x+1)} \cdot \frac{2}{2} + \frac{4}{2(x+1)} \cdot \frac{x}{x} = \frac{10}{2x(x+1)} + \frac{4x}{2x(x+1)} \quad \text{Multiplying by 1}$$

$$= \frac{10 + 4x}{2x(x+1)} \quad \text{Adding}$$

$$= \frac{2(5 + 2x)}{2x(x+1)} \quad \text{Factoring the numerator}$$

$$= \frac{2}{2} \cdot \frac{5 + 2x}{x(x+1)} \quad \text{Factoring the fractional expression}$$

$$= \frac{5 + 2x}{x(x+1)}$$

DO EXERCISES 2 AND 3.

Add.

2. $\dfrac{3}{x^3 - x} + \dfrac{4}{x^2 + 2x + 1}$

3. $\dfrac{5}{x^2 + 17x + 16} + \dfrac{3}{x^2 + 9x + 8}$

ANSWERS ON PAGE A-20

Add.

4. $\dfrac{a-1}{a+2} + \dfrac{a+3}{a-9}$

Example 3 Add: $\dfrac{x+4}{x-2} + \dfrac{x-7}{x+5}$.

First find the LCM of the denominators. It is just the product.

$$\text{LCM} = (x-2)(x+5)$$

Multiply by 1 to get the LCM in each expression. Then add and simplify.

$$\dfrac{x+4}{x-2} \cdot \dfrac{x+5}{x+5} + \dfrac{x-7}{x+5} \cdot \dfrac{x-2}{x-2} = \dfrac{(x+4)(x+5)}{(x-2)(x+5)} + \dfrac{(x-7)(x-2)}{(x-2)(x+5)}$$

$$= \dfrac{x^2+9x+20}{(x-2)(x+5)} + \dfrac{x^2-9x+14}{(x-2)(x+5)}$$

$$= \dfrac{x^2+9x+20+x^2-9x+14}{(x-2)(x+5)}$$

$$= \dfrac{2x^2+34}{(x-2)(x+5)}$$

DO EXERCISE 4.

Example 4 Add: $\dfrac{x}{x^2+11x+30} + \dfrac{-5}{x^2+9x+20}$.

Add.

5. $\dfrac{x}{x^2+5x+6} + \dfrac{-2}{x^2+3x+2}$

$$\dfrac{x}{x^2+11x+30} + \dfrac{-5}{x^2+9x+20}$$

$$= \dfrac{x}{(x+5)(x+6)} + \dfrac{-5}{(x+5)(x+4)} \qquad \text{Factoring denominators in order to find the LCM. The LCM is } (x+4)(x+5)(x+6).$$

$$= \dfrac{x}{(x+5)(x+6)} \cdot \dfrac{x+4}{x+4} + \dfrac{-5}{(x+5)(x+4)} \cdot \dfrac{x+6}{x+6} \qquad \text{Multiplying by 1}$$

$$= \dfrac{x(x+4)}{(x+5)(x+6)(x+4)} + \dfrac{-5(x+6)}{(x+5)(x+4)(x+6)}$$

$$= \dfrac{x(x+4)+(-5)(x+6)}{(x+4)(x+5)(x+6)} \qquad \text{Adding}$$

$$= \dfrac{x^2+4x-5x-30}{(x+4)(x+5)(x+6)}$$

$$= \dfrac{x^2-x-30}{(x+4)(x+5)(x+6)}$$

$$= \dfrac{(x-6)(x+5)}{(x+4)(x+5)(x+6)}$$

$$= \dfrac{x+5}{x+5} \cdot \dfrac{x-6}{(x+4)(x+6)} \qquad \longleftarrow \boxed{\text{Always simplify at the end if possible.}}$$

$$= \dfrac{x-6}{(x+4)(x+6)}$$

DO EXERCISE 5.

Chapter 9 Fractional Expressions and Equations

● Add, and simplify if possible.

1. $\dfrac{2}{x} + \dfrac{5}{x^2}$

2. $\dfrac{5}{6r} + \dfrac{7}{8r}$

*3. $\dfrac{x+y}{xy^2} + \dfrac{3x+y}{x^2y}$

*4. $\dfrac{2c-d}{c^2d} + \dfrac{c+d}{cd^2}$

5. $\dfrac{2}{x-1} + \dfrac{2}{x+1}$

6. $\dfrac{3}{x+1} + \dfrac{2}{3x}$

7. $\dfrac{2x}{x^2-16} + \dfrac{x}{x-4}$

8. $\dfrac{5}{z+4} + \dfrac{3}{3z+12}$

9. $\dfrac{3}{x-1} + \dfrac{2}{(x-1)^2}$

10. $\dfrac{4a}{5a-10} + \dfrac{3a}{10a-20}$

11. $\dfrac{x}{x^2+2x+1} + \dfrac{1}{x^2+5x+4}$

12. $\dfrac{7}{a^2+a-2} + \dfrac{5}{a^2-4a+3}$

ANSWERS
1.
2.
3.
4.
5.
6.
7.
8.
9.
10.
11.
12.

*Indicates exercises that can be omitted if Section 5.9 was not studied.

13. $\dfrac{x+3}{x-5} + \dfrac{x-5}{x+3}$

*14. $\dfrac{3x}{2y-3} + \dfrac{2x}{3y-2}$

15. $\dfrac{a}{a^2-1} + \dfrac{2a}{a^2-a}$

16. $\dfrac{3x+2}{3x+6} + \dfrac{x-2}{x^2-4}$

*17. $\dfrac{6}{x-y} + \dfrac{4x}{y^2-x^2}$

18. $\dfrac{a-2}{3-a} + \dfrac{4-a^2}{a^2-9}$

19. $\dfrac{10}{x^2+x-6} + \dfrac{3x}{x^2-4x+4}$

20. $\dfrac{2}{z^2-z-6} + \dfrac{3}{z^2-9}$

✓ SKILL MAINTENANCE

21. The perimeter of a rectangle is 642 ft. The length is 15 ft greater than the width. Find the area of the rectangle.

Solve.

22. $-3x + 14 \geq -10$

23. $5t - 45 - 8t < -4t + 67$

24. Subtract: $(4y^3 - 5y^2 + 7y - 24) - (-9y^3 + 9y^2 - 5y + 49)$.

⭐ EXTENSION

Find the perimeter and area of each figure.

25.

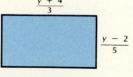

26.

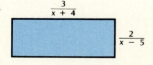

Add, and simplify if possible.

27. $\dfrac{5}{z+2} + \dfrac{4z}{z^2-4} + 2$

28. $\dfrac{-2}{y^2-9} + \dfrac{4y}{(y-3)^2} + \dfrac{6}{3-y}$

Chapter 9 Fractional Expressions and Equations

Subtraction with Different Denominators

⚫ SUBTRACTION

Subtraction is like addition, except that we subtract numerators.

Example 1 Subtract: $\dfrac{x+2}{x-4} - \dfrac{x+1}{x+4}$.

$LCM = (x-4)(x+4)$

$\dfrac{x+2}{x-4} \cdot \dfrac{x+4}{x+4} - \dfrac{x+1}{x+4} \cdot \dfrac{x-4}{x-4}$

$= \dfrac{(x+2)(x+4)}{(x-4)(x+4)} - \dfrac{(x+1)(x-4)}{(x-4)(x+4)}$

$= \dfrac{x^2+6x+8}{(x-4)(x+4)} - \dfrac{x^2-3x-4}{(x-4)(x+4)}$

$= \dfrac{x^2+6x+8-(x^2-3x-4)}{(x-4)(x+4)}$ Subtracting numerators

$= \dfrac{x^2+6x+8-x^2+3x+4}{(x-4)(x+4)}$ Don't forget parentheses.

$= \dfrac{9x+12}{(x-4)(x+4)}$

$= \dfrac{3(3x+4)}{(x-4)(x+4)}$

DO EXERCISE 1.

⚫⚫ SIMPLIFYING COMBINED ADDITIONS AND SUBTRACTIONS

Example 2 Simplify: $\dfrac{1}{x} - \dfrac{1}{x^2} + \dfrac{2}{x+1}$.

$LCM = x^2(x+1)$

$\dfrac{1}{x} \cdot \dfrac{x(x+1)}{x(x+1)} - \dfrac{1}{x^2} \cdot \dfrac{x+1}{x+1} + \dfrac{2}{x+1} \cdot \dfrac{x^2}{x^2}$

$= \dfrac{x(x+1)}{x^2(x+1)} - \dfrac{x+1}{x^2(x+1)} + \dfrac{2x^2}{x^2(x+1)}$

$= \dfrac{x(x+1) - (x+1) + 2x^2}{x^2(x+1)}$

$= \dfrac{x^2+x-x-1+2x^2}{x^2(x+1)}$

$= \dfrac{3x^2-1}{x^2(x+1)}$

DO EXERCISE 2.

Subtract.

1. $\dfrac{x-2}{3x} - \dfrac{2x-1}{5x}$

Simplify.

2. $\dfrac{1}{x} - \dfrac{5}{3x} + \dfrac{2x}{x+1}$

ANSWERS ON PAGE A-20

AN APPLICATION: HANDLING DIMENSION SYMBOLS (PART 2)

We can treat dimension symbols much like numerals and variables, since correct results can thus be obtained.

Example 1 Compare

$$4 \text{ m} \cdot 3 \text{ m} = 4 \cdot 3 \cdot \text{m} \cdot \text{m} = 12 \text{ m}^2 \text{ (square meters)}$$

with

$$4x \cdot 3x = 4 \cdot 3 \cdot x \cdot x = 12x^2.$$

Example 2 Compare

$$5 \text{ persons} \cdot 8 \text{ hr} = 5 \cdot 8 \text{ person-hr} = 40 \text{ person-hr}$$

with

$$5x \cdot 8y = 5 \cdot 8 \cdot x \cdot y = 40xy.$$

Example 3 Compare

$$480 \text{ cm} \cdot \frac{1 \text{ m}}{100 \text{ cm}} = \frac{480}{100} \text{ cm} \cdot \frac{\text{m}}{\text{cm}} = 4.8 \cdot \frac{\text{cm}}{\text{cm}} \cdot \text{m} = 4.8 \text{ m}$$

with

$$480x \cdot \frac{y}{100x} = \frac{480}{100}x \cdot \frac{y}{x} = 4.8 \cdot \frac{x}{x} \cdot y = 4.8y.$$

In each example, dimension symbols are treated as though they are variables or numerals, and as though a symbol like "3 ft" represents a product of "3" by "ft." A symbol like km/hr (standard notation is "km/h") is treated as if it represents a division of kilometers by hours.

Any two measures can be "multiplied" or "divided." For example,

$$6 \text{ ft} \cdot 4 \text{ lb} = 24 \text{ ft-lb}, \qquad 3 \text{ km} \cdot 4 \text{ sec} = 12 \text{ km-sec},$$

$$\frac{8 \text{ grams}}{4 \text{ min}} = 2\frac{\text{g}}{\text{min}}, \qquad \frac{3 \text{ in.} \cdot 8 \text{ days}}{6 \text{ lb}} = 4\frac{\text{in.-day}}{\text{lb}}.$$

EXERCISES

Perform these calculations and simplify if possible. *Do not* make unit changes.

1. $12 \text{ ft} \cdot \dfrac{1 \text{ yd}}{3 \text{ ft}}$

2. $6 \text{ lb} \cdot \dfrac{16 \text{ oz}}{1 \text{ lb}}$

3. $9\dfrac{\text{km}}{\text{hr}} \cdot 3 \text{ hr}$

4. $12\dfrac{\text{m}}{\text{sec}} \cdot 5 \text{ sec}$

5. $3 \text{ cm} \cdot \dfrac{2 \text{ g}}{2 \text{ cm}}$

6. $\dfrac{9 \text{ mi}}{3 \text{ days}} \cdot 6 \text{ days}$

7. $2347 \text{ m} \cdot \dfrac{1 \text{ km}}{1000 \text{ m}}$

8. $55 \text{ cm} \cdot \dfrac{10 \text{ mm}}{1 \text{ cm}}$

9. $\dfrac{3 \text{ kg}}{5 \text{ m}} \cdot \dfrac{7 \text{ kg}}{6 \text{ m}}$

10. $\dfrac{2000 \text{ lb} \cdot (6 \text{ mi/hr})^2}{100 \text{ ft}}$

11. $\dfrac{7 \text{ m} \cdot 8 \text{ kg/sec}}{4 \text{ sec}}$

• Subtract, and simplify if possible.

1. $\dfrac{x-2}{6} - \dfrac{x+1}{3}$

2. $\dfrac{a+2}{2} - \dfrac{a-4}{4}$

3. $\dfrac{4z-9}{3z} - \dfrac{3z-8}{4z}$

4. $\dfrac{x-1}{4x} - \dfrac{2x+3}{x}$

*5. $\dfrac{4x+2t}{3xt^2} - \dfrac{5x-3t}{x^2t}$

*6. $\dfrac{5x+3y}{2x^2y} - \dfrac{3x+4y}{xy^2}$

7. $\dfrac{5}{x+5} - \dfrac{3}{x-5}$

8. $\dfrac{2z}{z-1} - \dfrac{3z}{z+1}$

9. $\dfrac{3}{2t^2-2t} - \dfrac{5}{2t-2}$

10. $\dfrac{8}{x^2-4} - \dfrac{3}{x+2}$

*11. $\dfrac{2s}{t^2-s^2} - \dfrac{s}{t-s}$

12. $\dfrac{3}{12+x-x^2} - \dfrac{2}{x^2-9}$

ANSWERS
1.
2.
3.
4.
5.
6.
7.
8.
9.
10.
11.
12.

*Indicates exercises that can be omitted if Section 5.9 was not studied.

•• Simplify.

13. $\dfrac{4y}{y^2 - 1} - \dfrac{2}{y} - \dfrac{2}{y + 1}$

14. $\dfrac{x + 6}{4 - x^2} - \dfrac{x + 3}{x + 2} + \dfrac{x - 3}{2 - x}$

13. _____

14. _____

15. _____

16. _____

17. _____

18. _____

19. _____

20. _____

21. _____

22. _____

23. _____

24. _____

25. _____

26. _____

27. _____

28. _____

15. $\dfrac{2z}{1 - 2z} + \dfrac{3z}{2z + 1} - \dfrac{3}{4z^2 - 1}$

*16. $\dfrac{1}{x + y} + \dfrac{1}{x - y} - \dfrac{2x}{x^2 - y^2}$

17. $\dfrac{5}{3 - 2x} + \dfrac{3}{2x - 3} - \dfrac{x - 3}{2x^2 - x - 3}$

*18. $\dfrac{2r}{r^2 - s^2} + \dfrac{1}{r + s} - \dfrac{1}{r - s}$

19. $\dfrac{3}{2c - 1} - \dfrac{1}{c + 2} - \dfrac{5}{2c^2 + 3c - 2}$

20. $\dfrac{3y - 1}{2y^2 + y - 3} - \dfrac{2 - y}{y - 1}$

✓ SKILL MAINTENANCE

Simplify.

21. $x^{-7} \div x^{-8}$

22. $(a^2 b^{-5})^{-4}$

23. $3x^4 \cdot 10x^8$

24. A collection of dimes and nickels is worth $15.25. There are 157 coins in all. How many of each kind are there?

☆ EXTENSION

Simplify.

25. $\dfrac{1}{2xy - 6x + ay - 3a} - \dfrac{ay + xy}{(a^2 - 4x^2)(y^2 - 6y + 9)}$

26. $\dfrac{x}{x^4 - y^4} - \dfrac{1}{x^2 + 2xy + y^2}$

27. $\dfrac{3z^2}{z^4 - 4} - \dfrac{3 - 5z^2}{2z^4 + z^2 - 6}$

28. Find an expression equivalent to

$$\dfrac{5x^2 + 2xy}{x^2 + y^2}$$

that is a difference of two fractional expressions. Answers may vary.

9.7

Solving Fractional Equations

A *fractional equation* is an equation containing one or more fractional expressions. Here are some examples:

$$\frac{2}{3} + \frac{5}{6} = \frac{x}{9}, \qquad x + \frac{6}{x} = -5, \qquad \frac{x^2}{x-1} = \frac{1}{x-1}.$$

> To solve a fractional equation, multiply on both sides by the LCM of all the denominators. This is called *clearing of fractions*.

We have used clearing of fractions in Section 3.3 and Section 7.4.

Example 1 Solve: $\frac{2}{3} + \frac{5}{6} = \frac{x}{9}$.

The LCM of all denominators is 18, or $2 \cdot 3 \cdot 3$. We multiply on both sides by $2 \cdot 3 \cdot 3$.

$$2 \cdot 3 \cdot 3\left(\frac{2}{3} + \frac{5}{6}\right) = 2 \cdot 3 \cdot 3 \cdot \frac{x}{9} \qquad \text{Multiplying on both sides by the LCM}$$

$$2 \cdot 3 \cdot 3 \cdot \frac{2}{3} + 2 \cdot 3 \cdot 3 \cdot \frac{5}{6} = 2 \cdot 3 \cdot 3 \cdot \frac{x}{9} \qquad \text{Multiplying to remove parentheses}$$

$$2 \cdot 3 \cdot 2 + 3 \cdot 5 = 2 \cdot x \qquad \text{Simplifying}$$

$$12 + 15 = 2x$$

$$27 = 2x$$

$$\frac{27}{2} = x$$

> *Caution!* Note that we did not use the LCM to add or subtract fractional expressions. We used it in such a way that all the denominators disappeared and the resulting equation was easier to solve.

> When clearing an equation of fractions, be sure to multiply *all* terms in the equation by the LCM.

Example 2 Solve: $\frac{1}{x} = \frac{1}{4-x}$.

The LCM is $x(4-x)$. We multiply on both sides by $x(4-x)$.

$$x(4-x) \cdot \frac{1}{x} = x(4-x) \cdot \frac{1}{4-x} \qquad \text{Multiplying on both sides by the LCM}$$

$$4 - x = x \qquad \text{Simplifying}$$

$$4 = 2x$$

$$x = 2$$

This checks, so 2 is the solution.

DO EXERCISES 1 AND 2.

OBJECTIVES

After finishing Section 9.7, you should be able to:

• Solve fractional equations.

Solve.

1. $\frac{3}{4} + \frac{5}{8} = \frac{x}{12}$

2. $\frac{1}{x} = \frac{1}{6-x}$

ANSWERS ON PAGE A-20

Solve.

3. $\dfrac{x}{4} - \dfrac{x}{6} = \dfrac{1}{8}$

The following examples show how important it is to multiply *all* terms in an equation by the LCM.

Example 3 Solve: $\dfrac{x}{6} - \dfrac{x}{8} = \dfrac{1}{12}$.

The LCM is 24. We multiply on both sides by 24.

$$24\left(\dfrac{x}{6} - \dfrac{x}{8}\right) = 24 \cdot \dfrac{1}{12} \qquad \text{Multiplying on both sides by the LCM}$$

$$24 \cdot \dfrac{x}{6} - 24 \cdot \dfrac{x}{8} = 24 \cdot \dfrac{1}{12} \qquad \text{Multiplying to remove parentheses}$$

Be sure to multiply *every* term by the LCM.

$$4x - 3x = 2 \qquad \text{Simplifying}$$
$$x = 2$$

Check:

$$\begin{array}{c|c} \dfrac{x}{6} - \dfrac{x}{8} & \dfrac{1}{12} \\ \hline \dfrac{2}{6} - \dfrac{2}{8} & \dfrac{1}{12} \\ \dfrac{1}{3} - \dfrac{1}{4} & \\ \dfrac{4}{12} - \dfrac{3}{12} & \\ \dfrac{1}{12} & \end{array}$$

This checks, so the solution is 2.

DO EXERCISE 3.

Solve.

4. $\dfrac{1}{2x} + \dfrac{1}{x} = -12$

Example 4 Solve: $\dfrac{2}{3x} + \dfrac{1}{x} = 10$.

The LCM is $3x$. We multiply on both sides by $3x$.

$$3x\left(\dfrac{2}{3x} + \dfrac{1}{x}\right) = 3x \cdot 10 \qquad \text{Multiplying on both sides by the LCM}$$

$$3x \cdot \dfrac{2}{3x} + 3x \cdot \dfrac{1}{x} = 3x \cdot 10 \qquad \text{Multiplying to remove parentheses}$$

$$2 + 3 = 30x \qquad \text{Simplifying}$$
$$5 = 30x$$
$$\dfrac{5}{30} = x$$
$$\dfrac{1}{6} = x$$

This checks, so the solution is $\frac{1}{6}$.

DO EXERCISE 4.

Chapter 9   Fractional Expressions and Equations

Example 5 Solve: $x + \dfrac{6}{x} = -5$.

The LCM is x. We multiply by x.

$$x\left(x + \frac{6}{x}\right) = -5x \qquad \text{Multiplying on both sides by } x$$

$$x^2 + x \cdot \frac{6}{x} = -5x \qquad \begin{array}{l}\text{Note that } each\ term \text{ on the}\\ \text{left is now multiplied by } x.\end{array}$$

$$x^2 + 6 = -5x \qquad \text{Simplifying}$$

$$x^2 + 5x + 6 = 0 \qquad \text{Adding } 5x \text{ to get a 0 on one side}$$

$$(x + 3)(x + 2) = 0 \qquad \text{Factoring}$$

$$x + 3 = 0 \quad \text{or} \quad x + 2 = 0 \qquad \text{Principle of zero products}$$

$$x = -3 \quad \text{or} \qquad x = -2$$

Check:

$$\begin{array}{c|c} x + \dfrac{6}{x} = -5 & \\ \hline -3 + \dfrac{6}{-3} \ \bigg|\ -5 & \\ -5 \ \bigg| & \end{array} \qquad \begin{array}{c|c} x + \dfrac{6}{x} = -5 & \\ \hline -2 + \dfrac{6}{-2} \ \bigg|\ -5 & \\ -5 \ \bigg| & \end{array}$$

Both of these check, so the solutions are -3 and -2.

DO EXERCISE 5.

> It is important *always* to check when solving fractional equations.

Example 6 Solve: $\dfrac{x^2}{x - 1} = \dfrac{1}{x - 1}$.

The LCM is $x - 1$. We multiply by $x - 1$.

$$(x - 1) \cdot \frac{x^2}{x - 1} = (x - 1) \cdot \frac{1}{x - 1} \qquad \begin{array}{l}\text{Multiplying on both}\\ \text{sides by } x - 1\end{array}$$

$$x^2 = 1 \qquad \text{Simplifying}$$

$$x^2 - 1 = 0 \qquad \text{Adding } -1 \text{ to get a 0 on one side}$$

$$(x - 1)(x + 1) = 0 \qquad \text{Factoring}$$

$$x - 1 = 0 \quad \text{or} \quad x + 1 = 0 \qquad \text{Principle of zero products}$$

$$x = 1 \quad \text{or} \qquad x = -1$$

Possible solutions are 1 and -1.

Solve.

5. $x + \dfrac{1}{x} = 2$

ANSWER ON PAGE A-20

Solve.

6. $\dfrac{x^2}{x+2} = \dfrac{4}{x+2}$

Solve.

7. $\dfrac{4}{x-2} + \dfrac{1}{x+2} = \dfrac{26}{x^2-4}$

Check:

$\dfrac{x^2}{x-1} = \dfrac{1}{x-1}$		$\dfrac{x^2}{x-1} = \dfrac{1}{x-1}$	
$\dfrac{1^2}{1-1}$	$\dfrac{1}{1-1}$	$\dfrac{(-1)^2}{-1-1}$	$\dfrac{1}{-1-1}$
$\dfrac{1}{0}$	$\dfrac{1}{0}$	$-\dfrac{1}{2}$	$-\dfrac{1}{2}$

The number -1 is a solution, but 1 is not because it makes a denominator zero.

DO EXERCISE 6.

Example 7 Solve: $\dfrac{3}{x-5} + \dfrac{1}{x+5} = \dfrac{2}{x^2-25}$.

The LCM is $(x-5)(x+5)$. We multiply by $(x-5)(x+5)$.

$$(x-5)(x+5)\left[\dfrac{3}{x-5} + \dfrac{1}{x+5}\right]$$

$$= (x-5)(x+5)\cdot\left[\dfrac{2}{x^2-25}\right] \quad \text{Multiplying on both sides by the LCM}$$

$$(x-5)(x+5)\cdot\dfrac{3}{x-5} + (x-5)(x+5)\cdot\dfrac{1}{x+5}$$

$$= (x-5)(x+5)\cdot\dfrac{2}{x^2-25}$$

$$3(x+5) + (x-5) = 2 \quad \text{Simplifying}$$

$$3x + 15 + x - 5 = 2 \quad \text{Removing parentheses}$$

$$4x + 10 = 2$$

$$4x = -8$$

$$x = -2$$

Check:

$\dfrac{3}{x-5} + \dfrac{1}{x+5} = \dfrac{2}{x^2-25}$	
$\dfrac{3}{-2-5} + \dfrac{1}{-2+5}$	$\dfrac{2}{(-2)^2-25}$
$\dfrac{3}{-7} + \dfrac{1}{3}$	$\dfrac{2}{4-25}$
$\dfrac{9}{-21} + \dfrac{-7}{-21}$	$\dfrac{2}{-21}$
$\dfrac{2}{-21}$	

This checks, so the solution is -2.

DO EXERCISE 7.

• Solve.

1. $\dfrac{1}{4} + \dfrac{1}{6} = \dfrac{1}{t}$

2. $\dfrac{1}{3} - \dfrac{1}{4} = \dfrac{1}{x}$

3. $\dfrac{2}{3} - \dfrac{4}{5} = \dfrac{x}{15}$

4. $\dfrac{3}{4} + \dfrac{7}{8} = \dfrac{x}{2}$

5. $\dfrac{5}{x} = \dfrac{6}{x} - \dfrac{1}{3}$

6. $\dfrac{5}{3x} + \dfrac{3}{x} = 1$

7. $\dfrac{x-7}{x+2} = \dfrac{1}{4}$

8. $\dfrac{a-2}{a+3} = \dfrac{3}{8}$

9. $\dfrac{2}{x+1} = \dfrac{1}{x-2}$

10. $\dfrac{5}{x-1} = \dfrac{3}{x+2}$

11. $\dfrac{x}{8} - \dfrac{x}{12} = \dfrac{1}{8}$

12. $\dfrac{t}{3} + \dfrac{t}{9} = \dfrac{5}{6}$

11. _____

12. _____

13. $\dfrac{a-3}{3a+2} = \dfrac{1}{5}$

14. $\dfrac{2}{x+3} = \dfrac{5}{x}$

13. _____

14. _____

15. $\dfrac{x-1}{x-5} = \dfrac{4}{x-5}$

16. $\dfrac{x-7}{x-9} = \dfrac{2}{x-9}$

15. _____

16. _____

17. $x + \dfrac{4}{x} = -5$

18. $y + \dfrac{5}{y} = -6$

17. _____

18. _____

19. $\dfrac{x+1}{3} - \dfrac{x-1}{2} = 1$

20. $\dfrac{y-1}{4} - \dfrac{y+2}{5} = 3$

21. $\dfrac{1}{x} + \dfrac{2}{x} + \dfrac{3}{x} = 2$

22. $4 - \dfrac{1}{y} - \dfrac{2}{y} = \dfrac{6}{y}$

23. $\dfrac{y+3}{y} = \dfrac{5}{4}$

24. $\dfrac{t-7}{t} = \dfrac{3}{4}$

25. $\dfrac{x-2}{x-3} = \dfrac{x-1}{x+1}$

26. $\dfrac{2b-3}{3b+2} = \dfrac{2b+1}{3b-2}$

ANSWERS

27.

28.

29.

30.

31.

32.

33.

34.

35.

36.

37.

38.

39.

40.

27. $\dfrac{6x - 2}{2x - 1} = \dfrac{9x}{3x + 1}$

28. $\dfrac{2a}{a + 1} = 2 - \dfrac{5}{2a}$

29. $\dfrac{1}{x + 3} + \dfrac{1}{x - 3} = \dfrac{1}{x^2 - 9}$

30. $\dfrac{4}{x - 3} + \dfrac{2x}{x^2 - 9} = \dfrac{1}{x + 3}$

31. $\dfrac{x}{x + 4} - \dfrac{4}{x - 4} = \dfrac{x^2 + 16}{x^2 - 16}$

32. $\dfrac{5}{y - 3} - \dfrac{30}{y^2 - 9} = 1$

 SKILL MAINTENANCE

Graph on a plane.

33. $2x + y = 6$

34. $2x + y > 6$

35. Solve:

$$42 - 30x < -16x - 14.$$

36. Dried apricots are 5% protein, and dried prunes are 2% protein. How much of each type of fruit should be mixed in order to make a 100-oz mixture that is 3% protein?

⭐ **EXTENSION**

Solve.

37. $\dfrac{x}{x^2 + 3x - 4} + \dfrac{x + 1}{x^2 + 6x + 8} = \dfrac{2x}{x^2 + x - 2}$

38. $\dfrac{y}{y + 0.2} - 1.2 = \dfrac{y - 0.2}{y + 0.2}$

39. $\dfrac{3a - 5}{a^2 + 4a + 3} + \dfrac{2a + 2}{a + 3} = \dfrac{a - 3}{a + 1}$

40. $\dfrac{n}{n - \frac{4}{9}} - \dfrac{n}{n + \frac{4}{9}} = \dfrac{1}{n}$

Chapter 9   Fractional Expressions and Equations

Applied Problems and Proportions

APPLIED PROBLEMS

We now solve applied problems using fractional equations.

Example 1 The reciprocal of 2 less than a certain number is twice the reciprocal of the number itself. What is the number?

First translate to an equation. Let x = the number. Then 2 less than the number is $x - 2$, and the reciprocal of the number is $1/x$.

$$\left(\begin{array}{c}\text{Reciprocal of 2} \\ \text{less than number}\end{array}\right) \text{ is } \left(\begin{array}{c}\text{Twice the reciprocal} \\ \text{of the number}\end{array}\right)$$

$$\frac{1}{x-2} = 2 \cdot \frac{1}{x} \qquad \text{Translating}$$

Now we solve:

$$\frac{1}{x-2} = \frac{2}{x} \qquad \text{The LCM is } x(x-2).$$

$$\frac{x(x-2)}{x-2} = \frac{2\,x(x-2)}{x} \qquad \text{Multiplying by LCM}$$

$$x = 2(x-2) \qquad \text{Simplifying}$$
$$x = 2x - 4$$
$$x = 4.$$

Check: Go to the original problem. The number to be checked is 4. Two less than 4 is 2. The reciprocal of 2 is $\frac{1}{2}$. The reciprocal of the number itself is $\frac{1}{4}$. Now $\frac{1}{2}$ is twice $\frac{1}{4}$, so the conditions are satisfied. Thus the solution is 4.

DO EXERCISE 1.

Example 2 One car travels 20 km/h faster than another. While one of them goes 240 km, the other goes 160 km. Find their speeds.

First make a drawing. We really do not know the directions in which the cars are traveling, but it does not matter.

Slow car   160 km r km/h

Fast car   240 km $r + 20$ km/h

Let r represent the speed of the slow car. Then $r + 20$ is the speed of the fast car. The cars travel the same length of time, so we can just use t for time. We can organize the information in a chart, keeping in mind the formula $d = rt$.

After finishing Section 9.8, you should be able to:
- Solve applied problems using fractional equations.
- Find the ratio of one quantity to another.
- Solve proportion problems.

Solve.

1. The reciprocal of two more than a number is three times the reciprocal of the number. Find the number.

ANSWER ON PAGE A-20

2. One car goes 10 km/h faster than another. While one car goes 120 km, the other goes 150 km. How fast is each car?

	Distance	Speed	Time
Slow car	160	r	t
Fast car	240	$r + 20$	t

If we solve the formula $d = rt$ for t, we get $t = d/r$. Then from the rows of the table, we get

$$t = \frac{160}{r} \quad \text{and} \quad t = \frac{240}{r + 20}.$$

Since these times are the same, we get the following equation:

$$\frac{160}{r} = \frac{240}{r + 20}.$$

We solve:

$$\text{LCM} = r(r + 20)$$

$$\frac{160 \cdot r(r + 20)}{r} = \frac{240 \cdot r(r + 20)}{r + 20} \qquad \text{Multiplying on both sides by the LCM, } r(r + 20)$$

$$160(r + 20) = 240r \qquad \text{Simplifying}$$

$$160r + 3200 = 240r \qquad \text{Removing parentheses}$$

$$3200 = 80r \qquad \text{Adding } -160r$$

$$\frac{3200}{80} = r \qquad \text{Multiplying by } \frac{1}{80}$$

$$40 = r$$

$$60 = r + 20.$$

The speeds of 40 km/h for the slow car and 60 km/h for the fast car check in the problem.

DO EXERCISE 2.

Suppose it takes a person 4 hours to do a certain job. Then in 1 hour $\frac{1}{4}$ of the job is done.

> **If a job can be done in t hours (or days, or some other unit of time), then $1/t$ of it can be done in 1 hour (or day).**

Example 3 The head of a secretarial pool examines work records and finds that it takes Helen Huntinpeck 4 hr to type a certain report. It takes Willie Typitt 6 hr to type the same report. How long would it take them, working together, to type the same report?

We list the facts.

Helen: Can do the job in 4 hr.

Willie: Can do the job in 6 hr.

We want to know how long it will take them working together. Let's let t be that number.

ANSWER ON PAGE A-20

Chapter 9   Fractional Expressions and Equations

We might try a guess. Suppose we add the two times: 4 hr + 6 hr = 10 hr. Nice try, but we are way off. Either can do the job alone in less than 10 hr.

Let's try organizing the information in some kind of table. Helen should type $\frac{1}{4}$ of the report in 1 hr, since she can do it all in 4 hr.

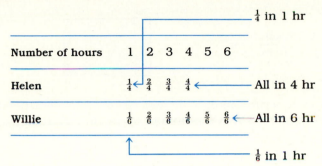

$\frac{1}{4}$ in 1 hr

Number of hours	1	2	3	4	5	6
Helen		$\frac{1}{4}$	$\frac{2}{4}$	$\frac{3}{4}$	$\frac{4}{4}$		— All in 4 hr
Willie		$\frac{1}{6}$	$\frac{2}{6}$	$\frac{3}{6}$	$\frac{4}{6}$	$\frac{5}{6}$	$\frac{6}{6}$ — All in 6 hr

$\frac{1}{6}$ in 1 hr

In 2 hr, Helen can do $\frac{2}{4}$ of the job. In 4 hr, she does it all, or $\frac{4}{4}$ of the job. Similarly, Willie can do $\frac{1}{6}$ of the job in 1 hr, $\frac{2}{6}$ in 2 hr, and so on. He can do $\frac{6}{6}$, or all of it in 6 hr.

Since Helen can do $\frac{1}{4}$ of the job in 1 hr and Willie can do $\frac{1}{6}$ of it in 1 hr, then working together they should do

$$\frac{1}{4} + \frac{1}{6} \quad \text{of the job in 1 hr.}$$

We are supposing that it takes them t hours working together, so they should do $1/t$ of the job in 1 hr. Thus,

$$\frac{1}{4} + \frac{1}{6} = \frac{1}{t}.$$

We now solve:

$$12t \cdot \left(\frac{1}{4} + \frac{1}{6}\right) = 12t \cdot \frac{1}{t} \qquad \text{The LCM is } 2 \cdot 2 \cdot 3t, \text{ or } 12t.$$

$$\frac{12t}{4} + \frac{12t}{6} = \frac{12t}{t}$$

$$3t + 2t = 12$$

$$5t = 12$$

$$t = \frac{12}{5}, \quad \text{or} \quad 2\frac{2}{5}.$$

Thus it takes $2\frac{2}{5}$ hr.

> Note that this answer, $2\frac{2}{5}$ hr, is less than the time it takes either person to do the job alone, 4 hr and 6 hr.

DO EXERCISE 3.

3. By checking work records, a contractor finds that it takes Red Bryck 6 hr to construct a wall of a certain size. It takes Lotta Mudd 8 hr to construct the same wall. How long would it take if they worked together?

ANSWER ON PAGE A-20

4. Find the ratio of 145 km to 2.5 liters (L).

5. Recently, a baseball player got 7 hits in 25 times at bat. What was the rate, or batting average, in hits per times at bat?

6. Impulses in nerve fibers travel 310 km in 2.5 hr. What is the rate, or speed, in kilometers per hour?

7. A lake of area 550 yd² contains 1320 fish. What is the population density of the lake in fish per square yard?

ANSWERS ON PAGE A-20

•• RATIO

The *ratio* of two quantities is their quotient. For example, in the rectangle, the ratio of width to length is

$$\frac{2 \text{ cm}}{3 \text{ cm}}, \quad \text{or} \quad \frac{2}{3}.$$

An older way to write this is 2:3, read "2 to 3." Percent notation is a ratio. For example, 37% is the ratio of 37 to 100, 37/100. The ratio of two different kinds of measure is called a *rate*.

Example 4 Betty Cuthbert of Australia set a world record in the 60-m dash of 7.2 sec. What was her rate, or *speed*, in meters per second?

$$\frac{60 \text{ m}}{7.2 \text{ sec}}, \quad \text{or} \quad 8.3 \frac{\text{m}}{\text{sec}} \qquad \text{(Rounded to the nearest tenth)}$$

DO EXERCISES 4–7.

••• PROPORTIONS

In applied problems a single ratio is often expressed in two ways. For example, it takes 9 gal of gas to drive 120 mi, and we wish to find how much will be required to go 550 mi. We can set up ratios:

$$\frac{9 \text{ gal}}{120 \text{ mi}} \qquad \frac{x \text{ gal}}{550 \text{ mi}}.$$

If we assume that the car uses gas at the same rate throughout the trip, the ratios are the same.

$$\text{Gas} \longrightarrow \frac{9}{120} = \frac{x}{550} \longleftarrow \text{Gas}$$
$$\text{Miles} \longrightarrow \qquad \qquad \longleftarrow \text{Miles}$$

To solve, we multiply by 550 to get *x* alone on one side:

$$550 \cdot \frac{9}{120} = 550 \cdot \frac{x}{550}$$

$$\frac{550 \cdot 9}{120} = x$$

$$41.25 = x.$$

Thus 41.25 gal will be required. (Note that we could have multiplied by the LCM of 120 and 550, which is 6600, but in this case, that would have been more complicated.)

An equality of ratios, $A/B = C/D$, is called a *proportion*. The numbers named in a true proportion are said to be *proportional*.

Example 5 A pitcher gave up 71 earned runs in 285 innings in a recent year. At this rate, how many runs did the pitcher give up every 9 innings? (There are 9 innings in a baseball game.)

Earned runs each 9 innings $\longrightarrow$ $\dfrac{A}{9} = \dfrac{71}{285}$ $\longleftarrow$ Earned runs
$\longleftarrow$ Innings pitched

Solve:

$$9 \cdot \dfrac{A}{9} = 9 \cdot \dfrac{71}{285}$$

$$A = \dfrac{9 \cdot 71}{285}$$

$$A = 2.24. \quad \text{(Rounded to the nearest hundredth)}$$

A stands for the *earned run average*. This means that on the average the pitcher gave up 2.24 earned runs for every 9 innings pitched.

DO EXERCISES 8 AND 9.

Example 6 (*Estimating wildlife populations*). To determine the number of fish in a lake, a conservationist catches 225 fish, tags them, and throws them back into the lake. Later, 108 fish are caught. Fifteen of them are found to be tagged. Estimate how many fish are in the lake.

Let F = the number of fish in the lake. Then translate to a proportion.

Fish tagged originally $\longrightarrow$ $\dfrac{225}{F} = \dfrac{15}{108}$ $\longleftarrow$ Tagged fish caught later
Fish in lake $\longrightarrow$ $\longleftarrow$ Fish caught later

This time we multiply by the LCM, which is $108F$:

$$108F \cdot \dfrac{225}{F} = 108F \cdot \dfrac{15}{108}$$

$$108 \cdot 225 = F \cdot 15$$

$$\dfrac{108 \cdot 225}{15} = F$$

$$1620 = F.$$

Thus we estimate that there are about 1620 fish in the lake.

DO EXERCISE 10.

8. A pitcher for the California Angels gave up 76 earned runs in 198 innings of pitching in a recent year. What was the earned run average?

9. A sample of 184 light bulbs contained 6 defective bulbs. How many would you expect in 1288 bulbs?

10. To determine the number of deer in a forest, a conservationist catches 612 deer, tags them, and lets them loose. Later 244 deer are caught. Seventy-two of them are tagged. Estimate how many deer are in the forest.

ANSWERS ON PAGE A-20

AN APPLICATION

The linear equation

$$P = 0.792t + 23$$

can be used to estimate the average percentage P of field goal completions by college basketball players t years from 1940. Since 1980 is 40 years from 1940, the average percentage of field goal completions in 1980 is given by

$$P = 0.792(40) + 23$$
$$P = 54.68\%.$$

EXERCISES

Find the average percentage of field goal completions in:

1. 1984.   **2.** 1990.   **3.** 1995.   **4.** 2000.

NAME: _____

CLASS/SECTION: _____ DATE: _____

Solve.

1. The reciprocal of 4 plus the reciprocal of 5 is the reciprocal of what number?

2. The reciprocal of 3 plus the reciprocal of 8 is the reciprocal of what number?

3. One number is 5 more than another. The quotient of the larger divided by the smaller is $\frac{4}{3}$. Find the numbers.

4. One number is 4 more than another. The quotient of the larger divided by the smaller is $\frac{5}{2}$. Find the numbers.

5. One car travels 40 km/h faster than another. While one travels 150 km, the other goes 350 km. Find their speeds.

6. One car travels 30 km/h faster than another. While one goes 250 km, the other goes 400 km. Find their speeds.

7. A person traveled 120 mi in one direction. The return trip was accomplished at double the speed, and took 3 hr less time. Find the speed going.

8. After making a trip of 126 mi, a person found that the trip would have taken 1 hr less time by increasing the speed by 8 mph. What was the actual speed?

ANSWERS
1. _____
2. _____
3. _____
4. _____
5. _____
6. _____
7. _____
8. _____

9. The speed of a freight train is 14 km/h slower than the speed of a passenger train. The freight train travels 330 km in the same time that it takes the passenger train to travel 400 km. Find the speed of each train.

10. The speed of a freight train is 15 km/h slower than the speed of a passenger train. The freight train travels 390 km in the same time that it takes the passenger train to travel 480 km. Find the speed of each train.

9. _____

10. _____

11. It takes painter A 4 hr to paint a certain area of a house. It takes painter B 5 hr to do the same job. How long would it take them, working together, to do the painting job?

12. By checking work records a carpenter finds that worker A can build a certain type of garage in 12 hr. Worker B can do the same job in 16 hr. How long would it take if they worked together?

11. _____

12. _____

13. By checking work records a plumber finds that worker A can do a certain job in 12 hr. Worker B can do the same job in 9 hr. How long would it take if they worked together?

14. A tank can be filled in 18 hr by pipe A alone and 24 hr by pipe B alone. How long would it take to fill the tank if both pipes were working?

13. _____

14. _____

Chapter 9   Fractional Expressions and Equations

Find the ratio of each of the following.

15. 54 days, 6 days

16. 800 mi, 50 gal

17. A black racer snake travels 4.6 km in 2 hr. What is the speed in kilometers per hour?

18. Light travels 558,000 mi in 3 sec. What is the speed in miles per second?

Solve.

19. The coffee beans from 14 trees are required to produce 7.7 kg of coffee (this is the average that each person in the United States drinks each year). How many trees are required to produce 320 kg of coffee?

20. Last season a minor league baseball player got 240 hits in 600 times at bat. This season his ratio of hits to number of times at bat is the same. He batted 500 times. How many hits has he had?

21. A student traveled 234 km in 14 days. At this same ratio, how far would the student travel in 42 days?

22. In a pancake recipe, the ratio of milk to flour is $\frac{4}{3}$. If 5 cups of milk are used, how many cups of flour are used?

23. 10 cm^3 of a normal specimen of human blood contains 1.2 g of hemoglobin. How many grams would 16 cm^3 of the same blood contain?

24. The winner of an election for class president won by a vote of 3 to 2, with 324 votes. How many votes did the loser get?

ANSWERS

15. _____

16. _____

17. _____

18. _____

19. _____

20. _____

21. _____

22. _____

23. _____

24. _____

ANSWERS

25. _____

26. _____

27. a) _____

b) _____

28. a) _____

b) _____

29. _____

30. _____

31. _____

32. _____

25. To determine the number of trout in a lake, a conservationist catches 112 trout, tags them, and throws them back into the lake. Later, 82 trout are caught; 32 of them are tagged. How many trout are in the lake?

26. To determine the number of deer in a game preserve, a conservationist catches 318 deer, tags them, and lets them loose. Later, 168 deer are caught; 56 of them are tagged. How many deer are in the preserve?

27. The ratio of the weight of an object on the moon to the weight of an object on earth is 0.16 to 1.

   a) How much would a 12-ton rocket weigh on the moon?

   b) How much would a 90-kg astronaut weigh on the moon?

28. The ratio of the weight of an object on Mars to the weight of an object on earth is 0.4 to 1.

   a) How much would a 12-ton rocket weigh on Mars?

   b) How much would a 90-kg astronaut weigh on Mars?

⭐ **EXTENSION**

29. The speed of a boat in still water is 10 mph. It travels 24 mi upstream and 24 mi downstream in a total time of 5 hr. What is the speed of the current?

30. Ann and Betty work together and complete a job in 4 hr. It would take Betty 6 hr longer, working alone, to do the job than it would Ann. How long would it take each of them to do the job working alone?

31. The denominator of a fraction is 1 more than the numerator. If 2 is subtracted from both the numerator and denominator, the resulting fraction is $\frac{1}{2}$. Find the original fraction.

32. In a proportion $A/B = C/D$ the numbers A and D are often called the *extremes* and the numbers B and C are called the *means*. Write four proportions. In each case, compare the product of the means with the product of the extremes.

Formulas

The use of formulas is important in many applications of mathematics. It is important to be able to solve a formula for a letter.

Example 1 (*Gravitational force*). The gravitational force f between planets of mass M and m, at a distance d from each other, is given by

$$f = \frac{kMm}{d^2},$$

where k is a constant. Solve for m.

$fd^2 = kMm$ Multiplying by the LCM, d^2

$\dfrac{fd^2}{kM} = m$ Multiplying by $\dfrac{1}{kM}$

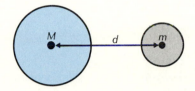

DO EXERCISE 1.

It is difficult, if not impossible, to look at a formula and tell which symbols are variables and which are constants. In fact, in a formula, a certain letter may be a variable at times and a constant at others, depending on how the formula is used.

Example 2 (*The area of a trapezoid*). The area A of a trapezoid is half the product of the height h and the sum of the lengths b_1 and b_2 of the parallel sides:

$$A = \frac{1}{2}(b_1 + b_2)h.$$

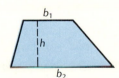

Solve for b_2.

We consider b_1 and b_2 to be different variables (or constants). The numbers 1 and 2 are called *subscripts*.

$2A = (b_1 + b_2)h$ Multiplying by 2 to clear of fractions

$2A = b_1h + b_2h$

$2A - b_1h = b_2h$ Adding $-b_1h$

$\dfrac{2A - b_1h}{h} = b_2$ Multiplying by $\dfrac{1}{h}$

Each of the following is also a correct answer:

$$\frac{2A}{h} - b_1 = b_2 \quad \text{and} \quad \frac{1}{h}(2A - b_1h) = b_2.$$

DO EXERCISE 2.

After finishing Section 9.9, you should be able to:

Solve a formula for a letter.

1. Solve $f = \dfrac{kMm}{d^2}$ for M.

2. Solve $V = \dfrac{1}{6}\pi h(h^2 + 3a^2)$ for a^2.

ANSWERS ON PAGE A-20

3. Solve for p: $\dfrac{n}{p} = 2 - m$.

In both of the examples, the letter for which we solved was on the right side of the equation. Ordinarily we put the letter for which we solve on the left. This is a matter of choice, since all equations are reversible.

Example 3 Solve for P: $A/P = 1 + r$. This is an interest formula.

The LCM is P. We multiply by this.

$$P \cdot \dfrac{A}{P} = P(1 + r) \qquad \text{Multiplying by } P$$

$$A = P(1 + r) \qquad \text{Simplifying the left side}$$

$$\dfrac{A}{1 + r} = \dfrac{P(1 + r)}{1 + r} \qquad \text{Multiplying by } \dfrac{1}{1 + r}$$

$$\dfrac{A}{1 + r} = P \qquad \text{Simplifying}$$

DO EXERCISE 3.

Example 4 If one person can do a job in a hr and another can do the same job in b hr, then working together they can do the same job in t hr, where a, b, and t are related by

$$\dfrac{1}{a} + \dfrac{1}{b} = \dfrac{1}{t}.$$

Solve for t.

4. Solve for f: $\dfrac{1}{p} + \dfrac{1}{q} = \dfrac{1}{f}$.

(This is an optics formula.)

The LCM is abt. We multiply by this.

$$abt \cdot \left(\dfrac{1}{a} + \dfrac{1}{b} \right) = abt \cdot \dfrac{1}{t} \qquad \text{Multiplying by } abt$$

$$abt \cdot \dfrac{1}{a} + abt \cdot \dfrac{1}{b} = \dfrac{abt}{t}$$

$$\dfrac{abt}{a} + \dfrac{abt}{b} = \dfrac{abt}{t}$$

$$bt + at = ab \qquad \text{Simplifying}$$

$$(b + a)t = ab \qquad \text{Factoring out } t$$

$$t = \dfrac{ab}{b + a} \qquad \text{Multiplying by } \dfrac{1}{b + a}$$

This answer can be used to find solutions to problems such as Example 3 in Section 9.8:

$$t = \dfrac{4 \cdot 6}{6 + 4} = \dfrac{24}{10} = 2\dfrac{2}{5}.$$

DO EXERCISE 4.

ANSWERS ON PAGE A-20

Chapter 9   Fractional Expressions and Equations

• Solve.

1. $S = 2\pi rh$ for r.

2. (*An interest formula*) $A = P(1 + rt)$ for t.

1. _____

3. (*The area of a triangle*) $A = \dfrac{1}{2}bh$ for b.

4. $s = \dfrac{1}{2}gt^2$ for g.

2. _____

3. _____

4. _____

5. $S = 180(n - 2)$ for n.

6. $S = \dfrac{n}{2}(a + l)$ for a.

5. _____

6. _____

7. $V = \dfrac{1}{3}k(B + b + 4M)$ for b.

8. $A = P + Prt$ for P.
(*Hint:* Factor the right-hand side.)

7. _____

8. _____

9. $S(r - 1) = rl - a$ for r.

10. $T = mg - mf$ for m.
(*Hint:* Factor the right-hand side.)

9. _____

10. _____

11. $A = \dfrac{1}{2}h(b_1 + b_2)$ for h.

12. (*The area of a right circular cylinder*)
$S = 2\pi r(r + h)$ for h.

11. _____

12. _____

13. $r = \dfrac{v^2 pL}{a}$ for a.

14. $L = \dfrac{Mt - g}{t}$ for M.

13. _____

14. _____

15. $\dfrac{1}{p} + \dfrac{1}{q} = \dfrac{1}{f}$ for p.

16. $\dfrac{1}{a} + \dfrac{1}{b} = \dfrac{1}{t}$ for b.

15. _____

16. _____

17. $\dfrac{A}{P} = 1 + r$ for A.

18. $\dfrac{2A}{h} = a + b$ for h.

19. (An electricity formula)

$\dfrac{1}{R} = \dfrac{1}{r_1} + \dfrac{1}{r_2}$ for R.

20. $\dfrac{1}{R} = \dfrac{1}{r_1} + \dfrac{1}{r_2}$ for r_1.

21. $\dfrac{A}{B} = \dfrac{C}{D}$ for D.

22. $\dfrac{A}{B} = \dfrac{C}{D}$ for C.

23. $h_1 = q\left(1 + \dfrac{h_2}{p}\right)$ for h_2.

24. $S = \dfrac{a - ar^n}{1 - r}$ for a.

25. $C = \dfrac{Ka - b}{a}$ for a.

26. $Q = \dfrac{Pt + h}{t}$ for t.

 SKILL MAINTENANCE

27. Subtract: $(5x^4 - 6x^3 + 23x^2 - 79x + 24) - (-18x^4 - 56x^3 + 84x - 17)$.

Multiply.

28. $(x - 2)(x + 2)$

29. $(6y^2 - 3)(5y^2 + 4)$

30. $(7m - 8n)^2$

⭐ **EXTENSION**

Solve.

31. $u = -F\left(E - \dfrac{P}{T}\right)$ for T

32. $1 = a + (n - 1)d$ for d.

33. The formula $C = \dfrac{5}{9}(F - 32)$ is used to convert Fahrenheit temperatures to Celsius temperatures. At what temperature are the Fahrenheit and Celsius readings the same?

34. In $N = \dfrac{a}{c}$, what is the effect on N when c increases? when c decreases? Assume that a, c, and N are positive.

Chapter 9   Fractional Expressions and Equations

Complex Fractional Expressions

·

A *complex fractional expression* is one that has a fractional expression in its numerator or its denominator or both.

Here are some examples of complex expressions.

$$\frac{1 + \dfrac{2}{x}}{3}, \qquad \frac{\dfrac{x+y}{2}}{\dfrac{2x}{x+1}}, \qquad \frac{\dfrac{1}{3} + \dfrac{1}{5}}{\dfrac{2}{x} - \dfrac{x}{y}}.$$

To simplify a complex fractional expression we first add or subtract, if necessary, to get a single fractional expression in both numerator and denominator. Then we divide by multiplying by the reciprocal of the denominator.

Example 1 Simplify.

$$\frac{\dfrac{1}{5} + \dfrac{2}{5}}{\dfrac{7}{3}} = \frac{\dfrac{3}{5}}{\dfrac{7}{3}} \qquad \text{Adding in the numerator}$$

$$= \frac{3}{5} \cdot \frac{3}{7} \qquad \text{Multiplying by the reciprocal of the denominator}$$

$$= \frac{9}{35}$$

Example 2 Simplify.

$$\frac{1 + \dfrac{2}{x}}{\dfrac{3}{4}} = \frac{1 \cdot \dfrac{x}{x} + \dfrac{2}{x}}{\dfrac{3}{4}} \qquad \text{Multiplying by } \dfrac{x}{x} \text{ to get a common denominator}$$

$$= \frac{\dfrac{x+2}{x}}{\dfrac{3}{4}} \qquad \text{Adding in the numerator}$$

$$= \frac{x+2}{x} \cdot \frac{4}{3} \qquad \text{Multiplying by the reciprocal of the denominator}$$

$$= \frac{4(x+2)}{3x}$$

DO EXERCISES 1 AND 2.

Simplify.

1. $\dfrac{\dfrac{2}{7} + \dfrac{3}{7}}{\dfrac{3}{4}}$

2. $\dfrac{3 + \dfrac{x}{2}}{\dfrac{5}{4}}$

ANSWERS ON PAGE A-21

Simplify.

3. $\dfrac{\dfrac{x}{2} + \dfrac{2x}{3}}{\dfrac{1}{x} - \dfrac{x}{2}}$

Example 3 Simplify.

$$\frac{\dfrac{3}{x} + \dfrac{1}{2x}}{\dfrac{1}{3x} - \dfrac{3}{4x}} = \frac{\dfrac{3}{x} \cdot \dfrac{2}{2} + \dfrac{1}{2x}}{\dfrac{1}{3x} \cdot \dfrac{4}{4} - \dfrac{3}{4x} \cdot \dfrac{3}{3}} \left.\begin{array}{l} \\ \\ \\ \\ \end{array}\right\} \leftarrow$$ Finding the LCM, $2x$, and multiplying by 1 in the numerator

Finding the LCM, $12x$, and multiplying by 1 in the denominator

$$= \frac{\dfrac{6}{2x} + \dfrac{1}{2x}}{\dfrac{4}{12x} - \dfrac{9}{12x}}$$

$$= \frac{\dfrac{7}{2x}}{\dfrac{-5}{12x}}$$ Adding in the numerator; subtracting in the denominator

$$= \frac{7}{2x} \cdot \frac{12x}{-5}$$ Multiplying by the reciprocal of the denominator

$$= \frac{2x}{2x} \cdot \frac{7 \cdot 6}{-5}$$ Factoring

$$= \frac{42}{-5} = -\frac{42}{5}$$ Simplifying

DO EXERCISE 3.

Example 4 Simplify.

Simplify.

4. $\dfrac{1 + \dfrac{1}{x}}{1 - \dfrac{1}{x^2}}$

$$\frac{1 - \dfrac{1}{x}}{1 - \dfrac{1}{x^2}} = \frac{\dfrac{x}{x} - \dfrac{1}{x}}{\dfrac{x^2}{x^2} - \dfrac{1}{x^2}} \left.\begin{array}{l} \\ \\ \\ \\ \end{array}\right\} \leftarrow$$ Finding the LCM, x, and multiplying by 1 in the numerator

Finding the LCM, x^2, and multiplying by 1 in the denominator

$$= \frac{\dfrac{x - 1}{x}}{\dfrac{x^2 - 1}{x^2}}$$ Subtracting in the numerator; subtracting in the denominator

$$= \frac{x - 1}{x} \cdot \frac{x^2}{x^2 - 1}$$ Multiplying by the reciprocal of the divisor

$$= \frac{(x - 1)x^2}{x(x^2 - 1)}$$

$$= \frac{x(x - 1)}{x(x - 1)} \cdot \frac{x}{x + 1}$$ Factoring

$$= \frac{x}{x + 1}.$$

DO EXERCISE 4.

ANSWERS ON PAGE A-21

• Simplify.

1. $\dfrac{1 + \dfrac{9}{16}}{1 - \dfrac{3}{4}}$

2. $\dfrac{9 - \dfrac{1}{4}}{3 + \dfrac{1}{2}}$

3. $\dfrac{1 - \dfrac{3}{5}}{1 + \dfrac{1}{5}}$

4. $\dfrac{\dfrac{5}{27} - 5}{\dfrac{1}{3} + 1}$

5. $\dfrac{\dfrac{1}{x} + 3}{\dfrac{1}{x} - 5}$

6. $\dfrac{\dfrac{3}{s} + s}{\dfrac{s}{3} + s}$

7. $\dfrac{\dfrac{1}{2} + \dfrac{3}{4}}{\dfrac{5}{8} - \dfrac{5}{6}}$

8. $\dfrac{\dfrac{2}{3} - \dfrac{5}{6}}{\dfrac{3}{4} + \dfrac{7}{8}}$

9. $\dfrac{\dfrac{2}{y} + \dfrac{1}{2y}}{y + \dfrac{y}{2}}$

ANSWERS
1. _____
2. _____
3. _____
4. _____
5. _____
6. _____
7. _____
8. _____
9. _____

10. $\dfrac{4 - \dfrac{1}{x^2}}{2 - \dfrac{1}{x}}$

11. $\dfrac{8 + \dfrac{8}{d}}{1 + \dfrac{1}{d}}$

12. $\dfrac{2 - \dfrac{3}{b}}{2 - \dfrac{b}{3}}$

13. $\dfrac{\dfrac{1}{5} - \dfrac{1}{a}}{\dfrac{5 - a}{5}}$

14. $\dfrac{2 - \dfrac{1}{x}}{\dfrac{2}{x}}$

*15. $\dfrac{\dfrac{x}{x - y}}{\dfrac{x^2}{x^2 - y^2}}$

*16. $\dfrac{\dfrac{x}{y} - \dfrac{y}{x}}{\dfrac{1}{y} + \dfrac{1}{x}}$

17. $\dfrac{x - 3 + \dfrac{2}{x}}{x - 4 + \dfrac{3}{x}}$

*18. $\dfrac{1 + \dfrac{a}{b - a}}{\dfrac{a}{a + b} - 1}$

 SKILL MAINTENANCE

19. Convert to decimal notation: 3.47×10^{-5}.

20. Convert to scientific notation: 0.0001209.

Factor.

21. $100m^2 - 81$

22. $64t^2 + 80t + 25$

⭐ EXTENSION

23. Find the reciprocal of $\dfrac{2}{x - 1} - \dfrac{1}{3x - 2}$.

Simplify.

24. $\left[\dfrac{\dfrac{x + 1}{x - 1} + 1}{\dfrac{x + 1}{x - 1} - 1} \right]^5$

25. $1 + \dfrac{1}{1 + \dfrac{1}{1 + \dfrac{1}{1 + \dfrac{1}{x}}}}$

26. $\dfrac{\dfrac{z}{1 - \dfrac{z}{2 + 2z}} - 2z}{\dfrac{2z}{5z - 2} - 3}$

*Indicates exercises that can be omitted if Section 5.9 was not studied.

Chapter 9   Fractional Expressions and Equations

Division of Polynomials

• DIVISOR A MONOMIAL

Division can be shown by a fractional expression.

Example 1 Divide $x^3 + 10x^2 + 8x$ by $2x$.

We write a fractional expression to show division:

$$\frac{x^3 + 10x^2 + 8x}{2x}.$$

This is equivalent to

$$\frac{x^3}{2x} + \frac{10x^2}{2x} + \frac{8x}{2x}. \quad \text{To see this, add and get the original expression.}$$

Next, we do the separate divisions:

$$\frac{x^3}{2x} + \frac{10x^2}{2x} + \frac{8x}{2x} = \frac{1}{2}x^2 + 5x + 4.$$

DO EXERCISES 1 AND 2.

Example 2 Divide and check: $(5y^2 - 2y + 3) \div 2$.

$$\frac{5y^2 - 2y + 3}{2} = \frac{5y^2}{2} - \frac{2y}{2} + \frac{3}{2} = \frac{5}{2}y^2 - y + \frac{3}{2}$$

Check:
$$\frac{5}{2}y^2 - y + \frac{3}{2}$$
$$\underline{ 2} \quad \text{We multiply.}$$
$$5y^2 - 2y + 3 \quad \text{The answer checks.}$$

Try to write only the answer.

> **To divide by a monomial, we can divide each term by the monomial.**

DO EXERCISES 3 AND 4.

•• DIVISOR NOT A MONOMIAL

When the divisor is not a monomial, we use long division very much as we do in arithmetic.

Example 3 Divide $x^2 + 5x + 6$ by $x + 2$.

$$
\begin{array}{r}
x \\
x + 2 \overline{)\, x^2 + 5x + 6} \\
\underline{x^2 + 2x} \\
3x
\end{array}
$$

Divide first term by first term, to get x. Ignore the term 2.
Multiply x by the divisor.
Subtract.

We now "bring down" the next term of the dividend, 6.

After finishing Section 9.11, you should be able to:

• Divide a polynomial by a monomial and check the result.

•• Divide a polynomial by a divisor that is not a monomial and, if there is a remainder, express the result in two ways.

Divide.

1. $\dfrac{2x^3 + 6x^2 + 4x}{2x}$

2. $(6x^2 + 3x - 2) \div 3$

Divide and check.

3. $(8x^2 - 3x + 1) \div 2$

4. $\dfrac{2x^4 - 3x^3 + 5x^2}{x^2}$

ANSWERS ON PAGE A-21

Divide and check.

5. $(x^2 + x - 6) \div (x + 3)$

6. $x - 2 \overline{)x^2 + 2x - 8}$

Divide and check. Express your answer in two ways.

7. $x + 3 \overline{)x^2 + 7x + 10}$

8. $(x^3 - 1) \div (x - 1)$

ANSWERS ON PAGE A-21

$$
\begin{array}{r}
x + 3 \\
x + 2 \overline{)x^2 + 5x + 6} \\
\underline{x^2 + 2x} \\
3x + 6 \\
\underline{3x + 6} \\
0
\end{array}
$$

— Divide first term by first term to get 3.

— Multiply 3 by the divisor.

Subtract.

The quotient is $x + 3$.

To check, multiply the quotient by the divisor and add the remainder, if any, to see if you get the dividend:

$$(x + 2)(x + 3) = x^2 + 5x + 6.$$ The division checks.

Example 4 Divide and check: $(x^2 + 2x - 12) \div (x - 3)$.

$$
\begin{array}{r}
x + 5 \\
x - 3 \overline{)x^2 + 2x - 12} \\
\underline{x^2 - 3x} \\
5x - 12 \\
\underline{5x - 15} \\
3
\end{array}
$$

Quotient

— Remainder

Check:
$(x - 3)(x + 5) + 3 = x^2 + 2x - 15 + 3$
$= x^2 + 2x - 12$

The quotient is $x + 5$ and the remainder is 3. We can write this as $x + 5$, R 3. The answer can also be given in this way: $x + 5 + \dfrac{3}{x - 3}$.

DO EXERCISES 5 AND 6.

Example 5 Divide: $(x^3 + 1) \div (x + 1)$.

$$
\begin{array}{r}
x^2 - x \quad\quad + 1 \\
x + 1 \overline{)x^3 \quad\quad\quad + 1} \\
\underline{x^3 + x^2} \\
- x^2 \\
\underline{- x^2 - x} \\
x + 1 \\
\underline{x + 1}
\end{array}
$$

— Leave space for missing terms.

The answer is $x^2 - x + 1$.
You need not write a 0 remainder.

Example 6 Divide: $(x^4 - 3x^2 + 1) \div (x - 4)$.

$$
\begin{array}{r}
x^3 + 4x^2 + 13x \ + 52 \\
x - 4 \overline{)x^4 \quad\quad - 3x^2 \quad\quad\quad + 1} \\
\underline{x^4 - 4x^3} \\
4x^3 - 3x^2 \\
\underline{4x^3 - 16x^2} \\
13x^2 \\
\underline{13x^2 - 52x} \\
52x + 1 \\
\underline{52x - 208} \\
209
\end{array}
$$

The answer can be expressed as $x^3 + 4x^2 + 13x + 52$, R 209, or

$$x^3 + 4x^2 + 13x + 52 + \frac{209}{x - 4}.$$

DO EXERCISES 7 AND 8.

Chapter 9   **Fractional Expressions and Equations**

• Divide and check.

1. $\dfrac{u - 2u^2 - u^5}{u}$

2. $\dfrac{50x^5 - 7x^4 + x^2}{x}$

3. $(15t^3 + 24t^2 - 6t) \div 3t$

4. $(25x^3 + 15x^2 - 30x) \div 5x$

5. $\dfrac{20x^6 - 20x^4 - 5x^2}{-5x^2}$

6. $\dfrac{24x^6 + 32x^5 - 8x^2}{-8x^2}$

7. $\dfrac{4x^4y - 8x^6y^2 + 12x^8y^6}{4x^4y}$

8. $\dfrac{9r^2s^2 + 3r^2s - 6rs^2}{-3rs}$

•• Divide and check.

9. $(x^2 - 10x - 25) \div (x - 5)$

10. $(x^2 + 8x - 16) \div (x + 4)$

11. $(x^2 + 4x + 4) \div (x + 2)$

12. $(x^2 - 6x + 9) \div (x - 3)$

ANSWERS

1. _____

2. _____

3. _____

4. _____

5. _____

6. _____

7. _____

8. _____

9. _____

10. _____

11. _____

12. _____

13. $(x^2 + 4x - 14) \div (x + 6)$

14. $(x^2 + 5x - 9) \div (x - 2)$

15. $(x^5 - 1) \div (x - 1)$

16. $(x^5 + 1) \div (x + 1)$

17. $(t^3 - t^2 + t - 1) \div (t - 1)$

18. $(t^3 - t^2 + t - 1) \div (t + 1)$

19. $(x^6 - 13x^3 + 42) \div (x^3 - 7)$

20. $(x^6 + 5x^3 - 24) \div (x^3 - 3)$

 SKILL MAINTENANCE

Solve.

21. $x = y + 2$
$\quad x + y = 6$

22. $5t^2 + 7t = 0$

23. $-4t + 9 + 8t > -23$

24. Divide: $\left(-\dfrac{5}{8}\right) \div \dfrac{3}{10}$.

⭐ **EXTENSION**

Divide.

25. $(5a^3 + 8a^2 - 23a - 1) \div (5a^2 - 7a - 2)$

26. $(5x^7 - 3x^4 + 2x^2 - 10x + 2) \div (x^2 - x + 1)$

27. $(y^4 + a^2) \div (y + a)$

28. $(a^6 - b^6) \div (a - b)$

If the remainder is 0 when one polynomial is divided by another, then the divisor is a factor of the dividend. Find the value(s) of c for which $x - 1$ is a factor of each polynomial.

29. $x^2 + 4x + c$

30. $c^2x^2 - 2cx + 1$

Chapter 9   Fractional Expressions and Equations

Inverse Variation

· EQUATIONS OF INVERSE VARIATION

A car is traveling a distance of 10 km. At a speed of 10 km/h it will take 1 hr. At 20 km/h, it will take $\frac{1}{2}$ hr. At 30 km/h it will take $\frac{1}{3}$ hr, and so on. This determines a set of numbers, all having the same product:

$$(10, 1), \quad (20, \tfrac{1}{2}), \quad (30, \tfrac{1}{3}), \quad (40, \tfrac{1}{4}), \quad \text{and so on.}$$

Note that as the first number gets larger, the second number gets smaller. Whenever a situation produces pairs of numbers whose product is constant, we say that there is *inverse variation*. Here the time *varies inversely* as the speed:

$$rt = 10 \text{ (a constant)}, \quad \text{or} \quad t = \frac{10}{r}.$$

> If a situation translates to an equation $y = k/x$, where k is a positive constant, $y = k/x$ is called an *equation of inverse variation*. We say that y *varies inversely* as x.

Although we will not study such graphs in this text, it is helpful to look at the graph of $y = k/x$, $k > 0$. The graph is shaped like the following for positive values of x. Note that as x increases, y decreases; and as x decreases, y increases.

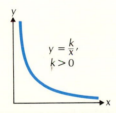

$y = \dfrac{k}{x}$, $k > 0$

Example 1 Find an equation of variation where y varies inversely as x and $y = 145$ when $x = 0.8$.

We substitute to find k:

$$y = \frac{k}{x}$$

$$145 = \frac{k}{0.8} \qquad \text{Substituting 145 for } y \text{ and 0.8 for } x$$

$$(0.8)145 = k \qquad \text{Multiplying on both sides by 0.8}$$

$$116 = k.$$

The equation of variation is $y = 116/x$.

DO EXERCISES 1 AND 2.

After finishing Section 9.12, you should be able to:

· Find an equation of inverse variation given a pair of values of the variables.

·· Solve problems involving inverse variation.

Find an equation of variation where y varies inversely as x.

1. $y = 105$ when $x = 0.6$

2. $y = 45$ when $x = 20$

ANSWERS ON PAGE A-21

3. In Example 2, how long would it take 10 people to do the job?

4. The time required to drive a fixed distance varies inversely as the speed r. It takes 5 hr at 60 km/h to drive a fixed distance. How long would it take at 40 km/h?

• • SOLVING PROBLEMS INVOLVING INVERSE VARIATION

Example 2 The time T to do a certain job varies inversely as the number N working (assuming all work at the same rate). It takes 4 hr for 20 people to wash and wax the floors in a building. How long would it take 25 people to do the job?

The problem states that we have inverse variation between the variables T and N. Thus the equation $T = k/N$, $k > 0$, applies. As the number of people increases, the time it takes to do the job decreases.

a) First find an equation of variation.

$$T = \frac{k}{N}$$

$$4 = \frac{k}{20} \qquad \text{Substituting 4 for } T \text{ and 20 for } N$$

$$20 \cdot 4 = k$$

$$80 = k$$

The equation of variation is $T = \dfrac{80}{N}$.

b) Use the equation to find the amount of time it takes 25 people to do the job.

$$T = \frac{80}{N}$$

$$T = \frac{80}{25}$$

$$= 3.2$$

It takes 3.2 hr for 25 people to do the job.

DO EXERCISES 3 AND 4.

• Find an equation of variation where y varies inversely as x and the following are true.

1. $y = 25$ when $x = 3$

2. $y = 45$ when $x = 2$

3. $y = 8$ when $x = 10$

4. $y = 7$ when $x = 10$

5. $y = 0.125$ when $x = 8$

6. $y = 6.25$ when $x = 0.16$

7. $y = 42$ when $x = 25$

8. $y = 42$ when $x = 50$

9. $y = 0.2$ when $x = 0.3$

10. $y = 0.4$ when $x = 0.6$

•• Solve.

11. It takes 16 hr for 2 people to resurface a gym floor. How long would it take 6 people to do the job?

12. It takes 4 hr for 9 cooks to prepare a school lunch. How long would it take 8 cooks to prepare the lunch?

A N S W E R S
1. _____
2. _____
3. _____
4. _____
5. _____
6. _____
7. _____
8. _____
9. _____
10. _____
11. _____
12. _____

13. The volume V of a gas varies inversely as the pressure P upon it. The volume of a gas is 200 cubic centimeters (cm^3) under a pressure of 32 kg/cm^3. What will be its volume under a pressure of 20 kg/cm^3?

14. The current I in an electrical conductor varies inversely as the resistance R of the conductor. The current is 2 amperes when the resistance is 960 ohms. What is the current when the resistance is 540 ohms?

15. The time t required to empty a tank varies inversely as the rate r of pumping. A pump can empty a tank in 90 min at the rate of 1200 gal/min. How long will it take the pump to empty the tank at 2000 gal/min?

16. The height H of a triangle of fixed area varies inversely as the base B. Suppose the height is 50 cm when the base is 40 cm. Find the height when the base is 8 cm. What is the fixed area?

17. The pitch P of a musical tone varies inversely as its wavelength W. One tone has a pitch of 660 vibrations per second and a wavelength of 1.6 ft. Find the wavelength of another tone that has a pitch of 440 vibrations per second.

18. The time t required to drive a fixed distance varies inversely as the speed r. It takes 6 hr at 55 km/h to drive a fixed distance. How long would it take at 40 km/h?

✔ **SKILL MAINTENANCE**

19. What is the degree of the polynomial

$$5x^3 - 7x^4 + 3x - x^2 - 1?$$

20. Solve: $-7y > 56$.

Factor.

21. $3x^3 + 21x^2 + 2x + 14$

22. $m^4 - n^4$

⭐ **EXTENSION**

Which of the following seem to vary inversely?

23. The cost of mailing a letter in the U.S. and the distance it travels

24. A runner's speed in a race and the time it takes to run the race

25. The weight of a turkey and the cooking time

Chapter 9   Fractional Expressions and Equations

The following contains a summary of what you should be able to do after completing this chapter. The review exercises are for practice. Answers are at the back of the book. If you miss an exercise, restudy the section and objective indicated alongside the answer.

The review sections to be tested in addition to the material in this chapter are 4.4, 5.6, 7.5, and 8.2.

You should be able to:

Simplify fractional expressions.

Simplify.

1. $\dfrac{4x^2 - 8x}{4x^2 + 4x}$

2. $\dfrac{14x^2 - x - 3}{2x^2 - 7x + 3}$

3. $\dfrac{(y - 5)^2}{y^2 - 25}$

Multiply and divide fractional expressions and simplify.

Multiply or divide and simplify.

4. $\dfrac{a^2 - 36}{10a} \cdot \dfrac{2a}{a + 6}$

5. $\dfrac{6t - 6}{2t^2 + t - 1} \cdot \dfrac{t^2 - 1}{t^2 - 2t + 1}$

6. $\dfrac{10 - 5t}{3} \div \dfrac{t - 2}{12t}$

7. $\dfrac{4x^4}{x^2 - 1} \div \dfrac{2x^3}{x^2 - 2x + 1}$

Find the least common multiple (LCM) of algebraic expressions.

Find the LCM.

8. $3x^2,\ 10xy,\ 15y^2$

9. $a - 2,\ 4(2 - a)$

10. $y^2 - y - 2,\ y^2 - 4$

Add and subtract fractional expressions and simplify.

Perform the indicated operations and simplify.

11. $\dfrac{x + 8}{x + 7} + \dfrac{10 - 4x}{x + 7}$

12. $\dfrac{3}{3x - 9} + \dfrac{x - 2}{3 - x}$

13. $\dfrac{6x - 3}{x^2 - x - 12} - \dfrac{2x - 15}{x^2 - x - 12}$

14. $\dfrac{3x - 1}{2x} - \dfrac{x - 3}{x}$

15. $\dfrac{x + 3}{x - 2} - \dfrac{x}{2 - x}$

16. $\dfrac{2a}{a + 1} + \dfrac{4a}{a^2 - 1}$

17. $\dfrac{d^2}{d - c} + \dfrac{c^2}{c - d}$

18. $\dfrac{1}{x^2 - 25} - \dfrac{x - 5}{x^2 - 4x - 5}$

19. $\dfrac{3x}{x + 2} - \dfrac{x}{x - 2} + \dfrac{8}{x^2 - 4}$

Solve fractional equations.

Solve.

20. $\dfrac{3}{y} - \dfrac{1}{4} = \dfrac{1}{y}$

21. $\dfrac{15}{x} - \dfrac{15}{x + 2} = 2$

Solve problems involving fractional equations, and solve proportion problems.

Solve.

22. In checking records a contractor finds that crew A can pave a certain length of highway in 9 hr. Crew B can do the same job in 12 hr. How long would it take if they worked together?

23. A lab is testing two high-speed trains. One train travels 40 km/h faster than the other. While one train travels 70 km, the other travels 60 km. Find their speeds.

24. The reciprocal of one more than a number is twice the reciprocal of the number itself. What is the number?

25. A sample of 250 batteries contained 8 defective batteries. How many defective batteries would you expect in a batch of 5000 batteries?

Solve a formula for a given letter.

Solve.

26. $\dfrac{1}{r} + \dfrac{1}{s} = \dfrac{1}{t}$ for s

27. $F = \dfrac{9C + 160}{5}$ for C

Simplify complex fractional expressions.

Simplify.

28. $\dfrac{\dfrac{1}{z} + 1}{\dfrac{1}{z^2} - 1}$

29. $\dfrac{\dfrac{c}{d} - \dfrac{d}{c}}{\dfrac{1}{c} + \dfrac{1}{d}}$

Divide polynomials by monomials and by divisors that are not monomials and check the result.

Divide.

30. $(10x^3 - x^2 + 6x) \div 2x$

31. $(6x^3 - 5x^2 - 13x + 13) \div (2x + 3)$

Find equations of variation where y varies inversely as x, and solve problems involving inverse variation.

Find an equation of variation where y varies inversely as x and the following are true.

32. $y = 5$ when $x = 6$

33. $y = 1.3$ when $x = 0.5$

34. It takes 5 hr for 2 washing machines to wash a fixed amount of clothing. How long would it take 10 washing machines? (The number of hours T varies inversely as the number of washing machines W.)

 **SKILL MAINTENANCE**

35. Factor: $5x^3 + 20x^2 - 3x - 12$.

36. Solve: $5x - 16 - 9x \geq -8$.

37. Subtract: $(5x^3 - 4x^2 + 3x - 4) - (7x^3 - 7x^2 - 9x + 14)$.

38. There were 12,000 people at a rock concert. Admission was $7.00 at the door and $6.50 if bought in advance. Total receipts were $81,165. How many people bought their tickets in advance?

 EXTENSION

39. Determine those replacements in the following expression that are not sensible:

$$\frac{x - 5}{x^3 - 8x^2 + 15x}.$$

1. Simplify:

$$\frac{6x^2 + 17x + 7}{2x^2 + 7x + 3}.$$

2. Multiply and simplify:

$$\frac{a^2 - 25}{6a} \cdot \frac{3a}{a - 5}.$$

3. Divide and simplify:

$$\frac{25x^2 - 1}{9x^2 - 6x} \div \frac{5x^2 + 9x - 2}{3x^2 + x - 2}.$$

4. Find the LCM:

$$y^2 - 9, \ y^2 + 10y + 21, \ y^2 + 4y - 21.$$

Perform the indicated operations and simplify.

5. $\dfrac{16 + x}{x^3} + \dfrac{7 - 4x}{x^3}$

6. $\dfrac{5 - t}{t^2 + 1} - \dfrac{t - 3}{t^2 + 1}$

7. $\dfrac{x - 4}{x - 3} + \dfrac{x - 1}{3 - x}$

8. $\dfrac{x - 4}{x - 3} - \dfrac{x - 1}{3 - x}$

9. $\dfrac{5}{t - 1} + \dfrac{3}{t}$

10. $\dfrac{1}{x^2 - 16} - \dfrac{x + 4}{x^2 - 3x - 4}$

11. $\dfrac{1}{x - 1} + \dfrac{4}{x^2 - 1} - \dfrac{2}{x^2 - 2x + 1}$

Solve.

12. $\dfrac{7}{y} - \dfrac{1}{3} = \dfrac{1}{4}$

13. $\dfrac{15}{x} - \dfrac{15}{x - 2} = -2$

14. The reciprocal of three less than a number is four times the reciprocal of the number itself. Find the number.

15. A sample of 125 spark plugs contained 4 defective. How many defective spark plugs would you expect in a batch of 500?

ANSWERS
1.
2.
3.
4.
5.
6.
7.
8.
9.
10.
11.
12.
13.
14.
15.

16. One car travels 20 km/h faster than another. While one goes 225 km, the other goes 325 km. Find their speeds.

17. Solve for t: $L = \dfrac{Mt - g}{t}$.

Divide.

18. $(12x^4 + 9x^3 - 15x^2) \div 3x^2$

19. $(6x^3 - 8x^2 - 14x + 13) \div (3x + 2)$

20. Simplify: $\dfrac{9 - \dfrac{1}{y^2}}{3 - \dfrac{1}{y}}$.

21. Find an equation of variation where y varies inversely as x, and $y = 6$ when $x = 2$.

22. It takes 3 hr for 2 cement mixers to mix a certain amount. The number T of hours varies inversely as the number N of cement mixers. How long would it take 5 cement mixers to do the job?

✓ SKILL MAINTENANCE

23. Factor: $16a^2 - 49$.

24. Solve: $-12 + 3x > -5x + 24$.

25. Subtract: $(5x^2 - 19x + 34) - (-8x^2 + 10x - 42)$.

26. The perimeter of a rectangle is 118 yd. The width is 18 yd less than the length. Find the area of the rectangle.

⭐ EXTENSION

27. Team A and team B work together and complete a job in $2\frac{6}{7}$ hr. It would take team B 6 hr longer, working alone, to do the job than it would team A. How long would it take each of them to do the job working alone?

28. Simplify: $1 + \dfrac{1}{1 + \dfrac{1}{1 + \dfrac{1}{a}}}$.

ANSWERS column: 16. 17. 18. 19. 20. 21. 22. 23. 24. 25. 26. 27. 28.

Cumulative Review

Solve.

1. $y = 2x - 9$
   $2x + 3y = -3$

2. $6x + 3y = -6$
   $-2x + 5y = 14$

3. $x = y + 3$
   $3y - 4x = -13$

4. $2x - 3y = 30$
   $5y - 2x = -46$

5. $8x - 2 \geq 7x + 5$

6. $-4x \geq 24$

7. $-3x < 30 + 2x$

8. $5x + 3 \geq 6(x - 4) + 7$

9. $\dfrac{5x - 2}{4} - \dfrac{4x - 5}{3} = 1$

10. $\dfrac{2x}{x - 3} - \dfrac{6}{x} = \dfrac{18}{x^2 - 3x}$

Solve.

11. $T = Rn + \dfrac{mn}{p}$ for p

12. $\dfrac{E}{r} = \dfrac{R + r}{R}$ for R

Graph on a number line.

13. $5x - 1 < 24$

14. $|x| \geq 3$

Graph on a plane.

15. $y \leq 5x$

16. $2y - 3x > -6$

Find each of the following.

17. $\{1, 4, 7, 11, 15, 19, 23, 25, 27, 31\} \cup \{3, 6, 9, 12, 15, 18, 21, 24, 27, 30\}$

18. $\{0, 1, 3\} \cap \{-1, 0, 1\}$

Find an equation of variation where y varies inversely as x and the following are true.

19. $y = 1$ when $x = 15$

20. $y = \dfrac{1}{700}$ when $x = 35$

Solve.

21. The second angle of a triangle is twice as large as the first. The third angle is 48° less than the sum of the other two angles. Find the measures of the angles.

22. The cost of 6 hamburgers and 4 milkshakes is $11.40. Three hamburgers and 1 milkshake cost $4.80. Find the cost of a hamburger and the cost of a milkshake.

23. The perimeter of a rectangle is 220 ft. One-fourth the length is 10 ft less than the width. Find the length and the width.

24. A tank contains 200 L of a 30% salt solution. How much pure water should be added to make a solution that is 12% salt?

25. The height h of a parallelogram of fixed area varies inversely as the base b. Suppose the height is 24 ft when the base is 15 ft. Find the height when the base is 5 ft. What is the variation constant?

26. One number is 7 more than another number. The quotient of the larger divided by the smaller is $\frac{5}{4}$. Find the numbers.

27. One car travels 105 mi in the same time that a car traveling 10 mph slower travels 75 mi. Find the speed of each.

Simplify.

28. $\dfrac{x^2 - 9}{2x^2 - 7x + 3}$

29. $\dfrac{t^2 - 16}{(t + 4)^2}$

Compute and simplify.

30. $\dfrac{2y + 4}{21} \cdot \dfrac{7}{y^2 + 4y + 4}$

31. $\dfrac{x^2 - 9}{x^2 + 8x + 15} \div \dfrac{x - 3}{2x + 10}$

32. $\dfrac{x^2}{x - 4} + \dfrac{16}{4 - x}$

33. $\dfrac{7}{5x - 25} + \dfrac{x + 7}{5 - x}$

34. $\dfrac{2x - 1}{x - 2} - \dfrac{2x}{2 - x}$

35. $\dfrac{5x}{x^2 - 4} - \dfrac{-3}{2 - x}$

36. $\dfrac{\dfrac{4}{x} - \dfrac{6}{x^2}}{\dfrac{5}{x} + \dfrac{7}{2x}}$

37. Divide: $(3x^3 - 2x^2 + x - 5) \div (x - 2)$.

⭐ **EXTENSION**

38. Divide:

$$(a^3 - b^3) \div (a - b).$$

39. Suppose that

$$x = \frac{ab}{a + b} \text{ and } y = \frac{ab}{a - b}.$$

Show that

$$\frac{y^2 - x^2}{y^2 + x^2} = \frac{2ab}{a^2 + b^2}.$$

40. Solve: $x^2 + 2 < 0$.

41. Graph on a plane: $x < 3$.

42. The solution of the following system is $(-5, 2)$. Find A and B.

$$3x - Ay = -7$$
$$Bx + 4y = 15$$

43. Solve:

$$2(5a - 5b) = 10$$
$$-5(6a + 2b) = 10.$$

10

AN APPLICATION
The *period T* of a pendulum is the time it takes to make a move from one side to the other and back. Find the period of the pendulum of the armswing of a bowler that is 3 ft long.

THE MATHEMATICS
Suppose *L* is the length of the pendulum. Then the period is given by

$$T = 2\pi\sqrt{\frac{L}{32}}.$$

⎵——— This is a radical expression.

We substitute 3 for *L* and compute *T*.

A square root of a number c is a number which when multiplied by itself is c. This chapter deals with expressions involving square roots. These are called radical expressions. We learn to solve equations involving radical expressions and to use these equations to solve problems.

The review sections to be tested in addition to the material in this chapter are 6.4, 7.6, 9.6, and 9.7.

Radical Expressions and Equations

Square Roots and Real Numbers

• SQUARE ROOTS

When we raise a number to the second power we have squared the number. Sometimes we may need to find the number that was squared. We call this process *finding the square root* of a number.

> The number c is a *square root* of a if $c^2 = a$.

Every positive number has two square roots. For example, the square roots of 25 are 5 and −5 because $5^2 = 25$ and $(-5)^2 = 25$.

Example 1 Find the square roots of 81.

The square roots are 9 and −9, because $9^2 = 81$ and $(-9)^2 = 81$.

DO EXERCISES 1–4.

The nonnegative square root is called the *principal square root*. The symbol $\sqrt{}$ is called a *radical*. The radical symbol refers only to the principal square root. Thus $\sqrt{25} = 5$. To name the negative square root of a number, we use $-\sqrt{}$. The number 0 has only one square root, 0.

Example 2 Find $\sqrt{225}$.

$$\sqrt{225} = 15 \qquad \text{Remember, } \sqrt{} \text{ means to take the principal, or}$$
nonnegative, square root.

Example 3 Find $\sqrt{0}$, $-\sqrt{25}$, and $-\sqrt{64}$.

$$\sqrt{0} = 0, \qquad -\sqrt{25} = -5, \qquad -\sqrt{64} = -8$$

> It would be most helpful for our work in this chapter to make a list of squares of numbers from 0 to 25 and to memorize the list.

DO EXERCISES 5–10.

• • IRRATIONAL NUMBERS

Recall that all rational numbers can be named by fractional notation a/b, where a and b are integers and $b \neq 0$. Rational numbers can be named in other ways, such as with decimal notation, but they can all be named with fractional notation. Suppose we try to find a rational number a/b for which $(a/b)^2 = 2$.

Find the square roots of each number.

1. 36

2. 64

3. 225

4. 100

Find the following.

5. $\sqrt{16}$

6. $\sqrt{49}$

7. $\sqrt{100}$

8. $\sqrt{441}$

9. $-\sqrt{49}$

10. $-\sqrt{169}$

ANSWERS ON PAGE A-22

We can find rational numbers whose square roots are quite close to 2. For example,

$$\left(\frac{14}{10}\right)^2 = (1.4)^2 = 1.96;$$

$$\left(\frac{141}{100}\right)^2 = (1.41)^2 = 1.9881;$$

$$\left(\frac{1414}{1000}\right)^2 = (1.414)^2 = 1.999396;$$

$$\left(\frac{14142}{10000}\right)^2 = (1.4142)^2 = 1.99996164.$$

Actually, we can never find a rational number whose square is 2. That can be proved but we will not do so here. Since $\sqrt{2}$ is not a rational number, we call it an *irrational number*.

> An *irrational number* is a number that cannot be named by fractional notation a/b, where a and b are integers and $b \neq 0$.

The square roots of most whole numbers are irrational. Only the perfect squares 0, 1, 4, 9, 16, 25, 36, 49, 64, 81, 100, and so on, have rational square roots.

Examples Identify each number as rational or irrational.

4. $\sqrt{3}$ $\sqrt{3}$ is irrational, since 3 is not a perfect square.

5. $\sqrt{25}$ $\sqrt{25}$ is rational, since 25 is a perfect square.

6. $-\sqrt{35}$ $-\sqrt{35}$ is irrational, since 35 is not a perfect square.

DO EXERCISES 11–14.

••• REAL NUMBERS

The rational numbers are very close together on the number line. Yet no matter how close together two rational numbers are, we can find many rational numbers between them. By averaging, we find the number halfway between.

Example 7 Find the number halfway between $\frac{1}{32}$ and $\frac{2}{32}$.

We average:

$$\frac{\frac{1}{32} + \frac{2}{32}}{2} = \frac{\frac{3}{32}}{2} = \frac{3}{64}.$$

Example 8 Find the number halfway between -0.52 and -0.523.

$$\frac{(-0.52) + (-0.523)}{2} = \frac{-1.043}{2} = -0.5215$$

DO EXERCISES 15–17.

Identify each number as rational or irrational.

11. $\sqrt{5}$

12. $-\sqrt{36}$

13. $-\sqrt{32}$

14. $\sqrt{101}$

Find the number halfway between each pair of numbers.

15. $\frac{3}{64}$ and $\frac{4}{64}$

16. -0.678 and -0.6782

17. 5.698 and 5.6999

ANSWERS ON PAGE A-22

This process can be repeated indefinitely. For example, $\frac{5}{128}$ is halfway between $\frac{1}{32}$ and $\frac{3}{64}$, and -0.52075 is halfway between -0.52 and -0.5215.

It looks like the rational numbers fill up the number line, but they do not. There are many points on the number line for which there are no rational numbers. These points correspond to irrational numbers.

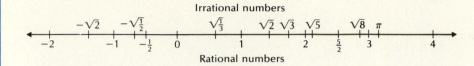

The *real numbers* consist of the rational numbers and the irrational numbers. For each point on a number line, there is a real number, and for each real number, there is a point on a number line.

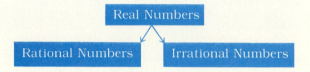

Decimal notation for a rational number either ends or repeats a group of digits. For example,

$\frac{1}{4} = 0.25$ The decimal ends.

$\frac{1}{3} = 0.3333\ldots$ The 3 repeats.

$\frac{5}{11} = 0.45\overline{45}$ The bar indicates that "45" repeats.

Decimal notation for an irrational number never ends and does not repeat. The number π is an example:

$$\pi = 3.1415926535.\ldots$$

Decimal notation for π never ends and never repeats. The numbers 3.14, 3.1416, and $\frac{22}{7}$ are only rational-number approximations for π. Decimal notation for $\frac{22}{7}$ is

$$3.\,142857\ 142857.\ldots$$

It repeats.

Here are some other examples of irrational numbers.

2.818118111811118111118 . . .   No group of digits repeats.

$-$ 0.040504550455504555550 . . .   No group of digits repeats.

Decimal notation for an irrational number is nonrepeating and nonending.

Examples Identify each number as rational or irrational.

9. $\frac{18}{37}$ Rational, since it is the ratio of two integers.

10. 2.565656 . . .

    (numeral repeats)   Rational, since the digits "56" repeat.

Chapter 10   Radical Expressions and Equations

11. 4.020020002 . . .
  (numeral does not repeat)   Irrational, since the decimal notation neither ends nor repeats.

DO EXERCISES 18–21.

⠂⠂ APPROXIMATING SQUARE ROOTS

We often need to use rational numbers to approximate square roots that are irrational. Such approximations can be found using a table such as Table 2 at the back of the book. They can also be found on a calculator with a square root key [$\sqrt{}$].

Example 12 Use your calculator or Table 2 to approximate $\sqrt{14}$. Round to three decimal places.

  You will need to consult the instruction manual for your calculator to find square roots. Calculators vary in their methods of operation.

  $\sqrt{14} \approx 3.741657387$ Using a calculator with a 10-digit readout

Different calculators give different numbers of digits in their readouts. This may cause some variance in answers. We round to three decimal places. Then

  $\sqrt{14} \approx 3.742.$ This can also be found in Table 2.

The symbol $\approx$ means "is approximately equal to."

DO EXERCISES 22 AND 23.

⠒⠒ SOLVING PROBLEMS INVOLVING SQUARE ROOTS

Example 13 (*An application: Parking lot arrival spaces*). When a parking lot has attendants to park cars, it uses spaces for drivers to pull in and leave cars before they are taken to permanent parking stalls. The number N of such spaces needed is approximated by the formula

$$N = 2.5\sqrt{A},$$

where A is the average number of arrivals in peak hours. Find the number of spaces needed when the average number of arrivals in peak hours is 49 and 77.

  We substitute 49 into the formula:

$$N = 2.5\sqrt{49} = 2.5(7) = 17.5 \approx 18.$$

Note that we round up to 18 spaces because 17.5 spaces would give us a half space, which we could not use. To ensure that we have enough spaces we need 18.

  We substitute 77 into the formula. We use a calculator or Table 2 to find an approximation:

$$N = 2.5\sqrt{77} \approx 2.5(8.775) \approx 21.938 \approx 22.$$

When the average number of arrivals is 49, about 18 spaces are needed. When the average number of arrivals is 77, about 22 spaces are needed.

DO EXERCISE 24.

Identify whether the number is rational or irrational.

18. $-\dfrac{95}{37}$ **19.** 433

20. 0.535353 . . .
  (numeral repeats)

21. 2.01001000100001000001 . . .
  (numeral does not repeat)

Use your calculator or Table 2 to approximate these square roots. Round to three decimal places.

22. $\sqrt{8}$

23. $\sqrt{62}$

24. Referring to Example 13, find the number of spaces needed when the average number of arrivals in peak hours is 64; when it is 39.

ANSWERS ON PAGE A-22

EXERCISES

Approximate these square roots to three decimal places using a calculator.

1. $\sqrt{17}$

2. $\sqrt{80}$

3. $\sqrt{110}$

4. $\sqrt{69}$

5. $\sqrt{200}$

6. $\sqrt{10.5}$

7. $\sqrt{890}$

8. $\sqrt{265.78}$

9. $\sqrt{2}$

10. $\sqrt{0.2344}$

11. $\sqrt{\pi}$

12. $\sqrt{2 + \sqrt{3}}$

CALCULATOR CORNER: FINDING SQUARE ROOTS ON A CALCULATOR THAT DOES NOT HAVE A SQUARE ROOT KEY

Some calculators have a square root key. If yours does not, you can use Table 2 if the number occurs there, or you can use the following method.

1. **Make a guess.**
2. **Square the guess on a calculator.**
3. **Adjust the guess.**

Example Approximate $\sqrt{27}$ to three decimal places using a calculator.

First we make a guess. Let's guess 4. Using a calculator we find 4^2, which is 16. Since $16 < 27$, we know that the square root is at *least* 4.

Next we try 5. We find 5^2, which is 25. This is also less than 27, so the square root is at least 5.

Next we try 6: $6^2 = 36$, which is greater than 27, so we know that $\sqrt{27}$ is between 5 and 6. Thus $\sqrt{27}$ is 5 plus some fraction: $\sqrt{27} = 5. \ldots$

Now we find the tenths digit by trying 5.1, 5.2, 5.3, and so on.

$$(5.1)^2 = 26.01 \qquad \text{This is less than 27.}$$
$$(5.2)^2 = 27.04 \qquad \text{This is greater than 27.}$$

The tenths digit is 1, so $\sqrt{27} = 5.1. \ldots$

Now we find the hundredths digit by trying 5.11, 5.12, 5.13, and so on.

$$(5.11)^2 = 26.1121$$
$$(5.12)^2 = 26.2144$$
$$\vdots \qquad\qquad \vdots$$
$$(5.19)^2 = 26.9361 \qquad \text{All of these are less than 27, so the hundredths digit is 9.}$$

The hundredths digit is 9, so $\sqrt{27} = 5.19. \ldots$

Now we find the thousandths digit by trying 5.191, 5.192, and so on.

$$(5.196)^2 = 26.998416 \qquad \text{This is less than 27.}$$
$$(5.197)^2 = 27.008809 \qquad \text{This is greater than 27.}$$

The thousandths digit is 6, so $\sqrt{27} \approx 5.196. \ldots$

Now we find the ten-thousandths digit by trying 5.1961, 5.1962, and so on.

$$(5.1961)^2 = 26.99945521 \qquad \text{This is less than 27.}$$
$$(5.1962)^2 = 27.00049444 \qquad \text{This is greater than 27.}$$

$\sqrt{27} \approx 5.1961 \ldots$

How far do we continue? It depends on the desired accuracy. We decide to stop here and round back to three decimal places. Thus we get

$$\sqrt{27} \approx 5.196$$

as a final answer. Check this in Table 2.

Tables can be made this way, but there are other methods.

• Find the square roots of each number.

1. 1   2. 4   3. 16   4. 9

5. 100   6. 121   7. 169   8. 144

Find each of the following.

9. $\sqrt{4}$ 10. $\sqrt{1}$ 11. $-\sqrt{9}$ 12. $-\sqrt{25}$

13. $-\sqrt{64}$ 14. $-\sqrt{81}$ 15. $\sqrt{225}$ 16. $\sqrt{400}$

17. $\sqrt{361}$ 18. $\sqrt{625}$ 19. $\sqrt{324}$ 20. $\sqrt{196}$

•• Identify each square root as rational or irrational.

21. $\sqrt{2}$ 22. $\sqrt{6}$ 23. $-\sqrt{8}$ 24. $-\sqrt{10}$

25. $\sqrt{49}$ 26. $\sqrt{100}$ 27. $\sqrt{98}$ 28. $-\sqrt{12}$

••• Identify each number as rational or irrational.

29. $-\dfrac{2}{3}$ 30. $\dfrac{136}{51}$ 31. 23   32. 4.23

33. 0.424242 . . .   34. 0.1565656 . . .
    (numeral repeats)   (numeral repeats)

35. 4.28228222822228 . . .   36. 7.76776777677776 . . .
    (numeral does not repeat)   (numeral does not repeat)

37. −1   38. 0

ANSWERS
1.
2.
3.
4.
5.
6.
7.
8.
9.
10.
11.
12.
13.
14.
15.
16.
17.
18.
19.
20.
21.
22.
23.
24.
25.
26.
27.
28.
29.
30.
31.
32.
33.
34.
35.
36.
37.
38.

39. −45.6919119111911119 . . .
(numeral does not repeat)

40. −63.03003000300003 . . .
(numeral does not repeat)

41. 14.678989‾

42. −8.4253253‾

 Use your calculator or Table 2 to approximate these square roots. Round to three decimal places.

43. $\sqrt{5}$

44. $\sqrt{6}$

45. $\sqrt{17}$

46. $\sqrt{19}$

47. $\sqrt{43}$

48. $\sqrt{99}$

Use the formula $N = 2.5\sqrt{A}$ of Example 13.

49. Find the number of spaces needed when the average number of arrivals is 25 and 89.

50. Find the number of spaces needed when the average number of arrivals is 62 and 100.

SKILL MAINTENANCE

51. What is the meaning of 5^2?

52. Multiply: $(-5)(-5)$.

53. Find the absolute value: $|-8|$.

54. Multiply and simplify: $x^3 \cdot x$.

EXTENSION

Simplify.

55. $\sqrt{\sqrt{16}}$

56. $\sqrt{3^2 + 4^2}$

57. Between what two consecutive integers is $-\sqrt{33}$?

Use a calculator to approximate these square roots. Round to three decimal places.

58. $\sqrt{12.8}$

59. $\sqrt{4230}$

60. $\sqrt{1057.61}$

Radical Expressions

• RADICANDS

After finishing Section 10.2, you should be able to:

• Identify radicands.

•• Identify which are meaningless as real numbers, determine whether a given number is a sensible replacement in a radical expression, and determine the sensible replacements in a radical expression.

••• Simplify radical expressions with a perfect-square radicand.

When an expression is written under a radical, we have a *radical expression*. These are radical expressions:

$$\sqrt{14}, \qquad \sqrt{x}, \qquad \sqrt{x^2 + 4}, \qquad \sqrt{\frac{x^2 - 5}{2}}.$$

The expression written under the radical is called the *radicand*.

Examples Identify the radicand in each expression.

1. $\sqrt{x}$ The radicand is x. **2.** $\sqrt{y^2 - 5}$ The radicand is $y^2 - 5$.

DO EXERCISES 1 AND 2.

•• MEANINGLESS EXPRESSIONS

The square of any number is always positive. For example, $8^2 = 64$ and $(-11)^2 = 121$. There are no real numbers that can be squared to get negative numbers.

Radical expressions with negative radicands have *no* meaning in the real-number system.

Thus the following expressions do not represent real numbers.

$$\sqrt{-100}, \qquad \sqrt{-49}, \qquad -\sqrt{-3}.$$

Later in your mathematical studies you may encounter a number system in which negative numbers do have square roots.

Example 3 Determine whether 6 is a sensible replacement in the expression $\sqrt{1 - y}$.

If we replace y by 6, we get $\sqrt{1 - 6} = \sqrt{-5}$, which is meaningless because the radicand is negative. Thus 6 is not a sensible replacement.

Examples Determine the sensible replacements in each expression.

4. $\sqrt{x}$ Any number x for which $x \geq 0$ is sensible.

5. $\sqrt{x + 2}$ Any number x for which $x + 2 \geq 0$ is sensible. We solve this inequality and get $x \geq -2$. Thus any number x for which $x \geq -2$ is sensible.

6. $\sqrt{x^2}$ Squares of numbers are never negative. All real-number replacements are sensible.

7. $\sqrt{x^2 + 1}$ Since x^2 is never negative, $x^2 + 1$ is never negative. All real-number replacements are sensible.

DO EXERCISES 3–12.

Identify the radicand in each expression.

1. $\sqrt{45 + x}$

2. $\sqrt{\dfrac{x}{x + 2}}$

Which of these expressions are meaningless as real numbers? Write "yes" or "no."

3. $-\sqrt{25}$ **4.** $\sqrt{-25}$

5. $-\sqrt{-36}$ **6.** $-\sqrt{36}$

7. Determine whether 8 is a sensible replacement in $\sqrt{x}$.

8. Determine whether 10 is a sensible replacement in $\sqrt{4 - x}$.

Determine the sensible replacements in each expression.

9. $\sqrt{a}$ **10.** $\sqrt{x - 3}$

11. $\sqrt{2x - 5}$ **12.** $\sqrt{x^2 + 3}$

ANSWERS ON PAGE A-22

Simplify. Assume that expressions represent any real number.

13. $\sqrt{(xy)^2}$ **14.** $\sqrt{x^2y^2}$

15. $\sqrt{(x-1)^2}$

16. $\sqrt{x^2 + 8x + 16}$

Simplify. Assume that expressions represent nonnegative real numbers.

17. $\sqrt{(xy)^2}$

18. $\sqrt{x^2y^2}$

19. $\sqrt{(x-1)^2}$

20. $\sqrt{x^2 + 8x + 16}$

21. $\sqrt{25y^2}$ **22.** $\sqrt{\frac{1}{4}t^2}$

ANSWERS ON PAGE A-22

The expression $\sqrt{x^2}$ is somewhat troublesome. Note that since squares are never negative, all replacements are sensible.

Suppose that $x = 3$.   Then we have $\sqrt{3^2}$, which is $\sqrt{9}$, or 3.

Suppose that $x = -3$.   Then we have $\sqrt{(-3)^2}$, which is $\sqrt{9}$, or 3.

In either case, when replacements for x are considered to be any real number, it follows that $\sqrt{x^2} = |x|$.

For any real-number radicand A,

$$\sqrt{A^2} = |A|.$$

(For any real-number radicand A, the principal square root of A squared is the absolute value of A.)

Examples Simplify. Assume that expressions represent any numbers.

8. $\sqrt{(3x)^2} = |3x|$

9. $\sqrt{a^2b^2} = \sqrt{(ab)^2} = |ab|$

10. $\sqrt{x^2 + 2x + 1} = \sqrt{(x+1)^2} = |x+1|$

DO EXERCISES 13–16.

Fortunately, in most uses of radicals it can be assumed that radicands are nonnegative, or positive. Indeed, many computers are programmed to consider only nonnegative radicands. Suppose that $x \geq 0$. Then

$$\sqrt{x^2} = |x| = x, \quad \text{since } x \text{ is nonnegative.}$$

For any nonnegative real-number radicand A,

$$\sqrt{A^2} = A.$$

(For any nonnegative real-number radicand A, the principal square root of A squared is A.)

Examples Simplify. Assume that expressions represent nonnegative real numbers.

11. $\sqrt{(3x)^2} = 3x$ Since $3x$ is assumed to be nonnegative

12. $\sqrt{a^2b^2} = \sqrt{(ab)^2} = ab$ Since ab is assumed to be nonnegative

13. $\sqrt{x^2 + 2x + 1} = \sqrt{(x+1)^2} = x + 1$ Since $x + 1$ is nonnegative

DO EXERCISES 17–22.

Henceforth in this text we will assume that *radicands represent nonnegative real numbers*. We make this assumption in order to eliminate some confusion and because it is valid in many applications. As you study further in mathematics, however, you will often have to make a determination about radicands being nonnegative, or positive. This will often be necessary in calculus.

• Identify the radicand in each expression.

1. $\sqrt{a-4}$

2. $\sqrt{t+3}$

3. $5\sqrt{t^2+1}$

4. $8\sqrt{x^2+5}$

5. $x^2 y \sqrt{\dfrac{3}{x+2}}$

6. $ab^2 \sqrt{\dfrac{a}{a-b}}$

•• Which of these expressions are meaningless? Write "yes" or "no."

7. $\sqrt{-16}$

8. $\sqrt{-81}$

9. $-\sqrt{81}$

10. $-\sqrt{64}$

Determine whether the given number is a sensible replacement in the given radical expression.

11. 4; $\sqrt{y}$

12. -8; $\sqrt{m}$

13. -11; $\sqrt{1+x}$

14. -11; $\sqrt{2-x}$

Determine the sensible replacements in each expression.

15. $\sqrt{5x}$

16. $\sqrt{3y}$

17. $\sqrt{t-5}$

18. $\sqrt{y-8}$

19. $\sqrt{y+8}$

20. $\sqrt{x+6}$

21. $\sqrt{x+20}$

22. $\sqrt{m-18}$

23. $\sqrt{2y-7}$

24. $\sqrt{3x+8}$

25. $\sqrt{t^2+5}$

26. $\sqrt{y^2+1}$

••• Simplify.

27. $\sqrt{t^2}$

28. $\sqrt{x^2}$

29. $\sqrt{9x^2}$

30. $\sqrt{4a^2}$

ANSWERS
1.
2.
3.
4.
5.
6.
7.
8.
9.
10.
11.
12.
13.
14.
15.
16.
17.
18.
19.
20.
21.
22.
23.
24.
25.
26.
27.
28.
29.
30.

31. $\sqrt{(ab)^2}$

32. $\sqrt{(6y)^2}$

33. $\sqrt{(34d)^2}$

34. $\sqrt{(53b)^2}$

35. $\sqrt{(x + 3)^2}$

36. $\sqrt{(x - 7)^2}$

37. $\sqrt{a^2 - 10a + 25}$

38. $\sqrt{x^2 + 2x + 1}$

39. $\sqrt{4x^2 - 20x + 25}$

40. $\sqrt{9x^2 + 12x + 4}$

 SKILL MAINTENANCE

41. Simplify: $\dfrac{81}{27}$.

42. Solve: $x + 1 = 2x - 5$.

43. Simplify: $\dfrac{1}{x} - \dfrac{1}{x^2} + \dfrac{2}{x + 1}$.

44. The amount F that a family spends on food varies directly as its income I. A family making $19,600 a year will spend $5096 on food. At this rate, how much would a family making $20,500 spend on food?

⭐ **EXTENSION**

Solve.

45. $\sqrt{x^2} = 6$

46. $\sqrt{y^2} = -7$

47. $m^2 = 49$

Determine the sensible replacements in each expression.

48. $\sqrt{x^2(x - 3)}$

49. $\sqrt{t^2 - 4}$

Multiplication and Factoring

After finishing Section 10.3, you should be able to:

• Multiply with radical notation.

•• Factor radical expressions and where possible simplify.

••• Approximate square roots not in Table 2 using factoring and the square root table.

• MULTIPLICATION

To see how we can multiply with radical notation, look at the following examples.

Example 1 Simplify.

a) $\sqrt{9} \cdot \sqrt{4} = 3 \cdot 2 = 6$ This is a product of square roots.

b) $\sqrt{9 \cdot 4} = \sqrt{36} = 6$ This is the square root of a product.

Example 2 Simplify.

a) $\sqrt{4} \cdot \sqrt{25} = 2 \cdot 5 = 10$

b) $\sqrt{4 \cdot 25} = \sqrt{100} = 10$

Example 3

$$\sqrt{-9}\sqrt{-4} \quad \text{Meaningless (negative radicands)}$$

DO EXERCISE 1.

We can multiply radical expressions by multiplying the radicands* provided the radicands are not negative.

> For any nonnegative radicands A and B, $\sqrt{A} \cdot \sqrt{B} = \sqrt{A \cdot B}$. (The product of square roots, provided they exist, is the square root of the product of the radicands.)

Examples Multiply. Assume that all radicands are nonnegative.

4. $\sqrt{5}\sqrt{7} = \sqrt{5 \cdot 7} = \sqrt{35}$

5. $\sqrt{8}\sqrt{8} = \sqrt{8 \cdot 8} = \sqrt{64} = 8$

6. $\sqrt{\dfrac{2}{3}}\sqrt{\dfrac{4}{5}} = \sqrt{\dfrac{2}{3} \cdot \dfrac{4}{5}} = \sqrt{\dfrac{8}{15}}$

7. $\sqrt{2x}\sqrt{3x-1} = \sqrt{2x(3x-1)} = \sqrt{6x^2 - 2x}$

DO EXERCISES 2–5.

1. Simplify.

   a) $\sqrt{4} \cdot \sqrt{16}$

   b) $\sqrt{4 \cdot 16}$

Multiply. Assume that all radicands are nonnegative.

2. $\sqrt{3}\sqrt{7}$

3. $\sqrt{5}\sqrt{5}$

4. $\sqrt{x}\sqrt{x+1}$

5. $\sqrt{x+1}\sqrt{x-1}$

---

*A proof (optional). We consider a product $\sqrt{A}\sqrt{B}$, where A and B are not negative. We square this product to show that we get AB; thus the product is the square root of AB (or $\sqrt{AB}$).

$$(\sqrt{A}\sqrt{B})^2 = (\sqrt{A}\sqrt{B})(\sqrt{A}\sqrt{B})$$
$$= (\sqrt{A}\sqrt{A})(\sqrt{B}\sqrt{B})$$
$$= AB$$

ANSWERS ON PAGE A-22

6. $\sqrt{32}$

7. $\sqrt{x^2 + 14x + 49}$

8. $\sqrt{25x^2}$

9. $\sqrt{36m^2}$

10. $\sqrt{76}$

11. $\sqrt{x^2 - 8x + 16}$

12. $\sqrt{64t^2}$

13. $\sqrt{100a^2}$

Approximate these square roots. Round to three decimal places.

14. $\sqrt{275}$

15. $\sqrt{102}$

ANSWERS ON PAGE A-22

◦ ◦ FACTORING AND SIMPLIFYING

We know that for nonnegative radicands,

$$\sqrt{A}\sqrt{B} = \sqrt{AB}.$$

To factor radical expressions we can think of this equation in reverse:

$$\sqrt{AB} = \sqrt{A}\sqrt{B}.$$

In some cases we can simplify after factoring. A radical expression is simplified when its radicand has no factors that are perfect squares.

Examples Simplify by factoring.

8. $\sqrt{18} = \sqrt{9 \cdot 2}$ Factoring the radicand: 9 is a perfect square.

     $= \sqrt{9} \cdot \sqrt{2}$ Factoring the radical expression

     $= 3\sqrt{2}$ $3\sqrt{2}$ means $3 \cdot \sqrt{2}$ The radicand has no factors that are perfect squares.

9. $\sqrt{25x} = \sqrt{25} \cdot \sqrt{x} = 5\sqrt{x}$

10. $\sqrt{x^2 - 6x + 9} = x - 3$ 11. $\sqrt{36x^2} = \sqrt{36}\sqrt{x^2} = 6x$

DO EXERCISES 6–13.

◦ ◦ ◦ APPROXIMATING SQUARE ROOTS

If we are using a table to approximate square roots, there can be numbers too large to find in the table. For example, Table 2 goes only to 100. We may still be able to find approximate square roots for other numbers. We do this by first factoring out the largest perfect square, if there is one. If there is none, we use any factorization we can find that will give smaller factors.

Examples Approximate these square roots.

12. $\sqrt{160} = \sqrt{16 \cdot 10}$ Factoring the radicand (make one factor a perfect square, if you can)

     $= \sqrt{16}\sqrt{10}$ Factoring the radical expression

     $= 4\sqrt{10}$

     $\approx 4(3.162)$ From Table 2, $\sqrt{10} \approx 3.162$

     ≈ 12.648

Note that we can also approximate $\sqrt{160}$ directly using a calculator. Doing so may give us a different approximation. Using a calculator with a 10-digit readout and rounding to three decimal places, we get $\sqrt{160} \approx 12.649$.

13. $\sqrt{341} = \sqrt{11 \cdot 31}$ Factoring (there is no perfect-square factor.)

     $= \sqrt{11}\sqrt{31}$

     $\approx 3.317 \times 5.568$ Table 2

     ≈ 18.469 Rounded to 3 decimal places

DO EXERCISES 14 AND 15.

● Multiply. Assume that all radicands are nonnegative.

1. $\sqrt{2}\sqrt{3}$

2. $\sqrt{3}\sqrt{5}$

3. $\sqrt{3}\sqrt{3}$

4. $\sqrt{6}\sqrt{6}$

5. $\sqrt{7}\sqrt{2}$

6. $\sqrt{14}\sqrt{3}$

7. $\sqrt{\dfrac{2}{5}}\sqrt{\dfrac{3}{4}}$

8. $\sqrt{\dfrac{4}{5}}\sqrt{\dfrac{15}{8}}$

9. $\sqrt{2}\sqrt{x}$

10. $\sqrt{3}\sqrt{t}$

11. $\sqrt{x}\sqrt{x-3}$

12. $\sqrt{5}\sqrt{2x-1}$

13. $\sqrt{x+2}\sqrt{x+1}$

14. $\sqrt{x+4}\sqrt{x-4}$

15. $\sqrt{x+y}\sqrt{x-y}$

16. $\sqrt{a-b}\sqrt{a-b}$

17. $\sqrt{43}\sqrt{2x}$

18. $\sqrt{35}\sqrt{4x}$

●● Simplify by factoring.

19. $\sqrt{12}$

20. $\sqrt{8}$

21. $\sqrt{75}$

22. $\sqrt{50}$

23. $\sqrt{200x}$

24. $\sqrt{300x}$

25. $\sqrt{16a^2}$

26. $\sqrt{64y^2}$

ANSWERS
1.
2.
3.
4.
5.
6.
7.
8.
9.
10.
11.
12.
13.
14.
15.
16.
17.
18.
19.
20.
21.
22.
23.
24.
25.
26.

27. $\sqrt{49t^2}$ **28.** $\sqrt{81p^2}$ **29.** $\sqrt{x^3 - 2x^2}$ **30.** $\sqrt{t^3 + t^2}$

31. $\sqrt{4x^2 - 4x + 1}$ **32.** $\sqrt{x^2 - 12x + 36}$

33. $\sqrt{9a^2 - 18ab + 9b^2}$ **34.** $\sqrt{100t^2 + 40t + 4}$

••• Approximate these square roots using Table 2. Round to three decimal places.

35. $\sqrt{125}$ **36.** $\sqrt{180}$ **37.** $\sqrt{360}$ **38.** $\sqrt{105}$

39. $\sqrt{300}$ **40.** $\sqrt{143}$ **41.** $\sqrt{122}$ **42.** $\sqrt{2000}$

✓ SKILL MAINTENANCE

43. Solve: $\dfrac{1}{5} + \dfrac{1}{7} = \dfrac{1}{t}$.   **44.** Factor: $x^3 - x^2$.

45. Find the prime factorization of 50.

46. Two cars leave town at the same time going in opposite directions. One travels at 43 mph and the other travels at 54 mph. In how many hours will they be 291 miles apart?

☆ EXTENSION

Simplify.

47. $\sqrt{0.01}$ **48.** $\sqrt{0.25}$ **49.** $\sqrt{x^4}$ **50.** $\sqrt{x^{-2}}$

51. Find $\sqrt{49}$, $\sqrt{490}$, $\sqrt{4900}$, $\sqrt{49,000}$, and $\sqrt{490,000}$. What pattern do you see?

52. *Speed of a skidding car.* How do police determine the speed of a car after an accident? The formula

$$r = 2\sqrt{5L}$$

can be used to approximate the speed r, in mph, of a car that has left a skid mark of length L, in feet. What was the speed of a car that left skid marks of lengths 20 ft? 70 ft? 90 ft?

Simplifying Radical Expressions

● SIMPLIFYING

To simplify radical expressions we usually try to factor out as many perfect squares as possible. Compare the following.

$$\sqrt{50} = \sqrt{10 \cdot 5} = \sqrt{10}\sqrt{5}, \qquad \sqrt{50} = \sqrt{25 \cdot 2} = \sqrt{25} \cdot \sqrt{2} = 5\sqrt{2}$$

In the case on the right, we factored out the perfect square 25. If you do not recognize perfect squares, try factoring the radicand into its prime factors. For example,

$$\sqrt{50} = \sqrt{2 \cdot \underline{5 \cdot 5}} = 5\sqrt{2} \qquad \text{Perfect square}$$

Radical expressions, such as $5\sqrt{2}$, where the radicand has no perfect-square factors, are considered to be in simplest form.

Examples Simplify by factoring. Assume that expressions under radicals represent nonnegative numbers.

1. $\sqrt{48t} = \sqrt{16 \cdot 3t}$ Identifying perfect-square factors and factoring the radicand

    $= \sqrt{16}\sqrt{3t}$ Factoring into a product of radicals

    $= 4\sqrt{3t}$ Taking the square root

2. $\sqrt{20t^2} = \sqrt{4 \cdot t^2 \cdot 5}$ Identifying perfect-square factors and factoring the radicand

    $= \sqrt{4} \cdot \sqrt{t^2} \cdot \sqrt{5}$ Factoring into a product of radicals

    $= 2t\sqrt{5}$ Taking the square roots. Absolute-value notation is not necessary because expressions are not negative.

3. $\sqrt{3x^2 + 6x + 3} = \sqrt{3(x^2 + 2x + 1)}$ Factoring the radicand

    $= \sqrt{3}\sqrt{x^2 + 2x + 1}$ Factoring into a product of radicals

    $= \sqrt{3}\sqrt{(x + 1)^2}$

    $= \sqrt{3}(x + 1)$ Taking the square root. Absolute-value notation is not necessary because expressions are not negative.

DO EXERCISES 1–3.

●● SIMPLIFYING SQUARE ROOTS OF POWERS

To take the square root of an even power such as x^8, we note that $x^8 = (x^4)^2 = x^4 \cdot x^4$. Then

$$\sqrt{x^8} = \sqrt{(x^4)^2} = \sqrt{x^4 \cdot x^4} = x^4.$$

We can find the answer by taking half the exponent. That is,

$$\sqrt{x^8} = x^4. \longleftarrow \qquad \tfrac{1}{2}(8) = 4$$

Simplify by factoring. Assume that all expressions under radicals represent nonnegative numbers.

1. $\sqrt{32}$

2. $\sqrt{50h^2}$

3. $\sqrt{3x^2 - 6x + 3}$

ANSWERS ON PAGE A-23

Simplify by factoring.

4. $\sqrt{x^{10}}$

5. $\sqrt{(x+2)^{14}}$

6. $\sqrt{x^{15}}$

Multiply and simplify.

7. $\sqrt{3}\sqrt{6}$

8. $\sqrt{2}\sqrt{50}$

Multiply and simplify.

9. $\sqrt{2x^3}\sqrt{8x^3y^4}$

10. $\sqrt{10xy^2}\sqrt{5x^2y^3}$

ANSWERS ON PAGE A-23

Absolute-value notation is again not necessary because expressions are not negative.

Examples Simplify by factoring.

4. $\sqrt{x^6} = x^3$ ⟵————— $\frac{1}{2}(6) = 3$ **Absolute-value notation is not necessary because expressions are not negative.**

5. $\sqrt{x^{10}} = x^5$
6. $\sqrt{t^{22}} = t^{11}$

When odd powers occur, express the power in terms of the next lower even power. Then simplify the even power.

Example 7 Simplify by factoring: $\sqrt{x^9}$.

$$\sqrt{x^9} = \sqrt{x^8 x}$$
$$= \sqrt{x^8}\sqrt{x}$$
$$= x^4\sqrt{x}$$

DO EXERCISES 4–6.

■■■ **MULTIPLYING AND SIMPLIFYING**

Sometimes we can simplify after multiplying. We factor the new radicand and look for perfect-square factors.

Example 8 Multiply and then simplify by factoring: $\sqrt{2}\sqrt{14}$.

$$\sqrt{2}\sqrt{14} = \sqrt{2 \cdot 14}$$ **Multiplying**
$$= \sqrt{2 \cdot 2 \cdot 7}$$ **Looking for perfect-square factors and factoring**

Perfect square

$$= \sqrt{2 \cdot 2}\,\sqrt{7}$$
$$= 2\sqrt{7}$$

DO EXERCISES 7 AND 8.

Example 9 Multiply and simplify by factoring: $\sqrt{3x^2}\sqrt{9x^3}$.

$$\sqrt{3x^2}\sqrt{9x^3} = \sqrt{3 \cdot 9x^5}$$ **Multiplying**
$$= \sqrt{3 \cdot 9 \cdot x^4 \cdot x}$$ **Looking for perfect-square factors and factoring**
$$= \sqrt{9 \cdot x^4 \cdot 3 \cdot x}$$ **Perfect squares**
$$= \sqrt{9}\sqrt{x^4}\sqrt{3x}$$
$$= 3x^2\sqrt{3x}$$

DO EXERCISES 9 AND 10.

Chapter 10   Radical Expressions and Equations

∙ Simplify by factoring.

1. $\sqrt{24}$

2. $\sqrt{12}$

3. $\sqrt{40}$

4. $\sqrt{200}$

5. $\sqrt{175}$

6. $\sqrt{243}$

7. $\sqrt{48x}$

8. $\sqrt{40m}$

9. $\sqrt{28x^2}$

10. $\sqrt{20x^2}$

11. $\sqrt{8x^2 + 8x + 2}$

12. $\sqrt{36y + 12y^2 + y^3}$

∙∙ Simplify by factoring.

13. $\sqrt{t^6}$

14. $\sqrt{x^4}$

15. $\sqrt{x^5}$

16. $\sqrt{t^{19}}$

17. $\sqrt{(y - 2)^8}$

18. $\sqrt{4(x + 5)^{10}}$

19. $\sqrt{36m^3}$

20. $\sqrt{8a^5}$

21. $\sqrt{448x^6y^3}$

22. $\sqrt{243x^5y^4}$

ANSWERS
1.
2.
3.
4.
5.
6.
7.
8.
9.
10.
11.
12.
13.
14.
15.
16.
17.
18.
19.
20.
21.
22.

Multiply and then simplify by factoring.

23. $\sqrt{3}\sqrt{18}$

24. $\sqrt{5}\sqrt{10}$

25. $\sqrt{18}\sqrt{14}$

26. $\sqrt{12}\sqrt{18}$

27. $\sqrt{10}\sqrt{10}$

28. $\sqrt{11}\sqrt{11}$

29. $\sqrt{5b}\sqrt{15b}$

30. $\sqrt{6a}\sqrt{18a}$

31. $\sqrt{ab}\sqrt{ac}$

32. $\sqrt{xy}\sqrt{xz}$

33. $\sqrt{18x^2y^3}\sqrt{6xy^4}$

34. $\sqrt{12x^3y^2}\sqrt{8xy}$

35. $\sqrt{50ab}\sqrt{10a^2b^4}$

36. $\sqrt{18xy}\sqrt{14x^3y^2}$

✓ **SKILL MAINTENANCE**

37. Solve: $\dfrac{3}{x-5} + \dfrac{1}{x+5} = \dfrac{2}{x^2-25}$.

38. Subtract: $\dfrac{x+2}{x-4} - \dfrac{x+1}{x+4}$.

39. Find an equation of variation where y varies directly as x and $y=7$ when $x=32$.

40. An airplane flew for 7 hr with a 5 km/h tailwind. The return flight against the wind took 8 hr. Find the speed of the plane in still air.

⭐ **EXTENSION**

Simplify.

41. $\sqrt{x}\sqrt{2x}\sqrt{10x^5}$

42. $\sqrt{x^{8n}}$

43. $\sqrt{0.04x^{4n}}$

44. $\sqrt{2^{109}}\sqrt{x^{306}}\sqrt{x^{11}}$

Simplifying Square Roots of Quotients

● SQUARE ROOTS OF QUOTIENTS

We now look at division instead of multiplication. Consider the expressions

$$\sqrt{\frac{25}{16}} \quad \text{and} \quad \frac{\sqrt{25}}{\sqrt{16}}.$$

Let us evaluate them separately.

a) $\sqrt{\frac{25}{16}} = \frac{5}{4}$ because $\frac{5}{4} \cdot \frac{5}{4} = \frac{25}{16}$.

b) $\frac{\sqrt{25}}{\sqrt{16}} = \frac{5}{4}$ since $\sqrt{25} = 5$ and $\sqrt{16} = 4$.

We see that both expressions represent the same number. This suggests that we can take the square root of a quotient by taking the square root of the numerator and denominator separately. That is true.

> **For any nonnegative number A and any positive number B,**
> $$\sqrt{\frac{A}{B}} = \frac{\sqrt{A}}{\sqrt{B}}.$$
>
> **(We can take the square root of the numerator and denominator separately.)**

For simplicity we usually consider both numerator and denominator to be positive.

Examples Simplify by taking square roots of the numerator and the denominator. Assume that all expressions under radicals represent positive numbers.

1. $\sqrt{\frac{25}{9}} = \frac{\sqrt{25}}{\sqrt{9}} = \frac{5}{3}$ Taking the square root of the numerator and the denominator

2. $\sqrt{\frac{49}{t^2}} = \frac{\sqrt{49}}{\sqrt{t^2}} = \frac{7}{t}$ Taking the square root of the numerator and the denominator

We are assuming that no expression represents 0 or a negative number. Thus we need not be concerned with zero denominators or absolute-value signs.

Sometimes a fractional expression can be simplified to one that has a perfect-square numerator and denominator.

Example 3 Simplify: $\sqrt{\frac{18}{50}}$.

$$\sqrt{\frac{18}{50}} = \sqrt{\frac{9 \cdot 2}{25 \cdot 2}} = \sqrt{\frac{9}{25} \cdot \frac{2}{2}} = \sqrt{\frac{9}{25} \cdot 1} = \sqrt{\frac{9}{25}} = \frac{3}{5}$$

Simplify.

1. $\sqrt{\frac{16}{9}}$

2. $\sqrt{\frac{1}{25}}$

3. $\sqrt{\frac{1}{9}}$

4. $\sqrt{\frac{18}{32}}$

5. $\sqrt{\frac{2250}{2560}}$

ANSWERS ON PAGE A-23

Rationalize the denominator.

6. $\sqrt{\dfrac{3}{5}}$

7. $\sqrt{\dfrac{5}{8}}$

$\left(Hint: \text{Multiply the radicand by } \dfrac{2}{2}.\right)$

Approximate to three decimal places.

8. $\sqrt{\dfrac{2}{7}}$

9. $\sqrt{\dfrac{5}{8}}$

ANSWERS ON PAGE A-23

Example 4 Simplify: $\sqrt{\dfrac{2560}{2890}}$.

$$\sqrt{\dfrac{2560}{2890}} = \sqrt{\dfrac{256 \cdot 10}{289 \cdot 10}} = \sqrt{\dfrac{256}{289} \cdot \dfrac{10}{10}} = \sqrt{\dfrac{256}{289} \cdot 1} = \sqrt{\dfrac{256}{289}} = \dfrac{16}{17}$$

DO EXERCISES 1–5 (ON THE PRECEDING PAGE).

•• RATIONALIZING DENOMINATORS

When neither the numerator nor the denominator is a perfect square, we can multiply by 1 to make the denominator a perfect square. Then we can take the square root of the denominator.

Example 5 Simplify: $\sqrt{\dfrac{2}{3}}$.

We multiply by 1, choosing 3/3 for 1. This makes the denominator a perfect square.

$$\sqrt{\dfrac{2}{3}} = \sqrt{\dfrac{2}{3} \cdot \dfrac{3}{3}} = \sqrt{\dfrac{6}{9}} = \dfrac{\sqrt{6}}{\sqrt{9}} = \dfrac{\sqrt{6}}{3}$$

We can always multiply by 1 to make the denominator a perfect square. Then we can take the square root of the denominator. This procedure is called *rationalizing the denominator*.

Example 6 Rationalize the denominator: $\sqrt{\dfrac{5}{12}}$.

$$\sqrt{\dfrac{5}{12}} = \sqrt{\dfrac{5}{12} \cdot \dfrac{3}{3}} = \sqrt{\dfrac{15}{36}} = \dfrac{\sqrt{15}}{\sqrt{36}} = \dfrac{\sqrt{15}}{6}$$

DO EXERCISES 6 AND 7.

••• APPROXIMATING SQUARE ROOTS OF FRACTIONS

We can use a calculator or Table 2 to approximate square roots of fractions. There are various ways to do it.

Example 7 Approximate $\sqrt{3/5}$ to three decimal places.

Method 1. Suppose we are using a calculator. Then we divide and approximate the square root.

$$\sqrt{\dfrac{3}{5}} = \sqrt{0.6} \approx 0.774596669 \approx 0.775$$

Method 2. Suppose we are using a table such as Table 2. We rationalize the denominator. Then we use a calculator or Table 2 to approximate the square root in the numerator. Then we divide and round.

$$\sqrt{\dfrac{3}{5}} = \sqrt{\dfrac{3}{5} \cdot \dfrac{5}{5}} = \sqrt{\dfrac{15}{25}} = \dfrac{\sqrt{15}}{\sqrt{25}} = \dfrac{\sqrt{15}}{5} \approx \dfrac{3.873}{5} \approx 0.775$$

DO EXERCISES 8 AND 9.

Chapter 10   Radical Expressions and Equations

Simplify. Assume that all expressions under radicals represent positive numbers.

1. $\sqrt{\dfrac{9}{49}}$

2. $\sqrt{\dfrac{16}{25}}$

3. $\sqrt{\dfrac{1}{36}}$

4. $\sqrt{\dfrac{1}{4}}$

5. $-\sqrt{\dfrac{16}{81}}$

6. $-\sqrt{\dfrac{25}{49}}$

7. $\sqrt{\dfrac{64}{289}}$

8. $\sqrt{\dfrac{81}{361}}$

9. $\sqrt{\dfrac{1690}{1960}}$

10. $\sqrt{\dfrac{1440}{6250}}$

11. $\sqrt{\dfrac{36}{a^2}}$

12. $\sqrt{\dfrac{25}{x^2}}$

13. $\sqrt{\dfrac{9a^2}{625}}$

14. $\sqrt{\dfrac{x^2y^2}{256}}$

Rationalize the denominator. Assume that all expressions under radicals represent positive numbers.

15. $\sqrt{\dfrac{2}{5}}$

16. $\sqrt{\dfrac{2}{7}}$

17. $\sqrt{\dfrac{3}{8}}$

18. $\sqrt{\dfrac{7}{8}}$

19. $\sqrt{\dfrac{1}{2}}$

20. $\sqrt{\dfrac{1}{3}}$

21. $\sqrt{\dfrac{3}{x}}$

22. $\sqrt{\dfrac{a}{b}}$

ANSWERS
1.
2.
3.
4.
5.
6.
7.
8.
9.
10.
11.
12.
13.
14.
15.
16.
17.
18.
19.
20.
21.
22.

Approximate to three decimal places.

23. $\sqrt{\dfrac{3}{7}}$ 24. $\sqrt{\dfrac{3}{2}}$ 25. $\sqrt{\dfrac{1}{3}}$ 26. $\sqrt{\dfrac{1}{5}}$

27. $\sqrt{\dfrac{7}{20}}$ 28. $\sqrt{\dfrac{3}{20}}$ 29. $\sqrt{\dfrac{12}{5}}$ 30. $\sqrt{\dfrac{8}{3}}$

✓ SKILL MAINTENANCE

31. Collect like terms: $2x^3 - 8x^2 - 10x^3 + 5x^2 + 7$.

32. Simplify: $\dfrac{2x^2 - x - 15}{x^2 - 9}$.   33. Solve: $x + 5 = -\dfrac{6}{x}$.

34. The weight J of an object on Jupiter varies directly as its weight E on earth. An object that weighs 225 kg on earth has a weight of 594 kg on Jupiter. What is the weight on Jupiter of an object that has a weight of 115 kg on earth?

★ EXTENSION

Rationalize the denominator.

35. $\sqrt{\dfrac{5}{1600}}$ 36. $\sqrt{\dfrac{3}{1000}}$ 37. $\sqrt{\dfrac{1}{5x^3}}$ 38. $\sqrt{\dfrac{3x^2y}{a^2x^5}}$

36

39. *Pendulums.* The *period* T of a pendulum is the time it takes to make a move from one side to the other and back. A formula for the period is

$$T = 2\pi\sqrt{\dfrac{L}{32}},$$

where T is in seconds and L is in feet. Find the periods of pendulums of lengths 2 ft, 8 ft, 64 ft, and 100 ft. (Use 3.14 for π.)

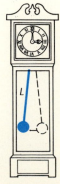

Division

• QUOTIENTS OF SQUARE ROOTS

We now find quotients of square roots, and we continue to study rationalizing denominators. We know that for positive radicands,

$$\sqrt{\frac{A}{B}} = \frac{\sqrt{A}}{\sqrt{B}}.$$

To find quotients we can use this equation. It may help to think of it in reverse.

> For any positive radicands A and B,
>
> $$\frac{\sqrt{A}}{\sqrt{B}} = \sqrt{\frac{A}{B}}.$$
>
> (The quotient of square roots, provided they exist, is the square root of the quotient of the radicands.)

Examples Divide and simplify. Assume that all expressions under radicals represent positive numbers.

1. $\dfrac{\sqrt{27}}{\sqrt{3}} = \sqrt{\dfrac{27}{3}} = \sqrt{9} = 3$

2. $\dfrac{\sqrt{7}}{\sqrt{14}} = \sqrt{\dfrac{7}{14}} = \sqrt{\dfrac{1}{2}} = \sqrt{\dfrac{1}{2} \cdot \dfrac{2}{2}} = \sqrt{\dfrac{2}{4}} = \dfrac{\sqrt{2}}{\sqrt{4}} = \dfrac{\sqrt{2}}{2}$

3. $\dfrac{\sqrt{30a^3}}{\sqrt{6a^2}} = \sqrt{\dfrac{30a^3}{6a^2}} = \sqrt{5a}$

DO EXERCISES 1–3.

•• RATIONALIZING DENOMINATORS

Expressions with radicals are considered simpler if there are no radicals in denominators. We can simplify by multiplying by 1, but this time we do it a bit differently. We use notation like $\sqrt{a}/\sqrt{a}$.

Example 4 Rationalize the denominator.

$$\frac{\sqrt{2}}{\sqrt{3}} = \frac{\sqrt{2}}{\sqrt{3}} \cdot \frac{\sqrt{3}}{\sqrt{3}} = \frac{\sqrt{2} \cdot \sqrt{3}}{\sqrt{3} \cdot \sqrt{3}} = \frac{\sqrt{6}}{3}, \quad \text{or} \quad \frac{1}{3}\sqrt{6}$$

After finishing Section 10.6, you should be able to:

- **•** Divide with radical notation.
- **••** Rationalize denominators.

Divide and simplify. Assume that all expressions under radicals represent positive numbers.

1. $\dfrac{\sqrt{5}}{\sqrt{45}}$

2. $\dfrac{\sqrt{2}}{\sqrt{6}}$

3. $\dfrac{\sqrt{42x^5}}{\sqrt{7x^2}}$

ANSWERS ON PAGE A-23

Rationalize the denominator. Assume that all expressions under radicals represent positive numbers.

4. $\dfrac{\sqrt{5}}{\sqrt{7}}$

5. $\dfrac{\sqrt{64y^3}}{\sqrt{7}}$

Rationalize the denominator. Assume that all expressions under radicals represent positive numbers.

6. $\dfrac{\sqrt{x}}{\sqrt{y}}$

ANSWERS ON PAGE A-23

Example 5 Rationalize the denominator. Assume that all expressions under radicals represent positive numbers.

$$\frac{\sqrt{49a^5}}{\sqrt{12}} = \frac{\sqrt{49a^5}}{\sqrt{12}} \cdot \frac{\sqrt{3}}{\sqrt{3}} = \frac{\sqrt{49a^5}\sqrt{3}}{\sqrt{12}\sqrt{3}} = \frac{\sqrt{49a^5}\sqrt{3}}{\sqrt{36}}$$

$$= \frac{\sqrt{49a^4 \cdot 3a}}{\sqrt{36}} = \frac{7a^2\sqrt{3a}}{6}$$

DO EXERCISES 4 AND 5.

Example 6 Rationalize the denominator. Assume that all expressions under radicands represent positive numbers.

$$\frac{\sqrt{5}}{\sqrt{x}} = \frac{\sqrt{5}}{\sqrt{x}} \cdot \frac{\sqrt{x}}{\sqrt{x}} \qquad \text{Multiplying by 1}$$

$$= \frac{\sqrt{5}\sqrt{x}}{\sqrt{x}\sqrt{x}}$$

$$= \frac{\sqrt{5x}}{x}$$

DO EXERCISE 6.

Chapter 10 Radical Expressions and Equations

• Divide and simplify. Assume that all expressions under radicals represent positive numbers.

1. $\dfrac{\sqrt{18}}{\sqrt{2}}$

2. $\dfrac{\sqrt{20}}{\sqrt{5}}$

3. $\dfrac{\sqrt{60}}{\sqrt{15}}$

4. $\dfrac{\sqrt{108}}{\sqrt{3}}$

5. $\dfrac{\sqrt{75}}{\sqrt{15}}$

6. $\dfrac{\sqrt{18}}{\sqrt{3}}$

7. $\dfrac{\sqrt{12}}{\sqrt{75}}$

8. $\dfrac{\sqrt{18}}{\sqrt{32}}$

9. $\dfrac{\sqrt{8x}}{\sqrt{2x}}$

10. $\dfrac{\sqrt{18b}}{\sqrt{2b}}$

11. $\dfrac{\sqrt{63y^3}}{\sqrt{7y}}$

12. $\dfrac{\sqrt{48x^3}}{\sqrt{3x}}$

13. $\dfrac{\sqrt{15x^5}}{\sqrt{3x}}$

14. $\dfrac{\sqrt{30a^5}}{\sqrt{5a}}$

15. $\dfrac{\sqrt{3x}}{\sqrt{\dfrac{3x}{4}}}$

16. $\dfrac{\sqrt{5x}}{\sqrt{\dfrac{5x}{9}}}$

•• Rationalize the denominator. Assume that all variables under radicals represent positive numbers.

17. $\dfrac{\sqrt{2}}{\sqrt{5}}$

18. $\dfrac{\sqrt{3}}{\sqrt{2}}$

19. $\dfrac{2}{\sqrt{2}}$

20. $\dfrac{3}{\sqrt{3}}$

ANSWERS
1.
2.
3.
4.
5.
6.
7.
8.
9.
10.
11.
12.
13.
14.
15.
16.
17.
18.
19.
20.

21. $\dfrac{\sqrt{48}}{\sqrt{32}}$

22. $\dfrac{\sqrt{56}}{\sqrt{40}}$

23. $\dfrac{\sqrt{450}}{\sqrt{18}}$

24. $\dfrac{\sqrt{224}}{\sqrt{14}}$

25. $\dfrac{\sqrt{3}}{\sqrt{x}}$

26. $\dfrac{\sqrt{2}}{\sqrt{y}}$

27. $\dfrac{4y}{\sqrt{3}}$

28. $\dfrac{8x}{\sqrt{5}}$

29. $\dfrac{\sqrt{a^3}}{\sqrt{8}}$

30. $\dfrac{\sqrt{x^3}}{\sqrt{27}}$

31. $\dfrac{\sqrt{16a^4b^6}}{\sqrt{128a^6b^6}}$

32. $\dfrac{\sqrt{45mn^2}}{\sqrt{32m}}$

 SKILL MAINTENANCE

33. Divide and simplify: $\dfrac{x^2 + 2x}{x^2 + 5x + 6} \div \dfrac{x}{3x + 9}$.

34. Subtract: $\dfrac{y}{y^2 - 9} - \dfrac{2}{y + 3}$.

35. Solve: $\dfrac{x^2}{x + 4} = \dfrac{16}{x + 4}$.

36. A train leaves a station and travels west at 70 km/h. Two hours later a second train leaves on a parallel track and travels west at 90 km/h. When will it overtake the first train?

⭐ EXTENSION

Simplify. Assume that all expressions under radicals represent positive numbers. Rationalize denominators where appropriate.

37. $\dfrac{3\sqrt{15}}{5\sqrt{32}}$

38. $\dfrac{4\sqrt{\dfrac{6}{7}}}{\sqrt{\dfrac{12}{63}}}$

39. $\dfrac{\sqrt{\dfrac{2}{3}}}{\sqrt{\dfrac{3}{2}}}$

40. $\dfrac{\sqrt{\dfrac{3x}{2}}}{\sqrt{\dfrac{x^3}{5}}}$

10.7

Addition and Subtraction

We can add any two real numbers. The sum of 5 and $\sqrt{2}$ can be expressed as

$$5 + \sqrt{2}.$$

We cannot simplify this unless we use rational approximations. When we have *like radicals*, however, a sum can be simplified using the distributive laws and collecting like terms. *Like radicals* have the same radicands.

Example 1 Add $3\sqrt{5}$ and $4\sqrt{5}$ and simplify by collecting like radical terms if possible.

$$3\sqrt{5} + 4\sqrt{5} = (3 + 4)\sqrt{5} \qquad \text{Using the distributive law to factor out } \sqrt{5}$$

$$= 7\sqrt{5}$$

To simplify like this, the radical expressions must be the same. Sometimes we can make them the same.

Examples Add or subtract. Simplify, if possible, by collecting like radical terms.

2. $\sqrt{2} - \sqrt{8} = \sqrt{2} - \sqrt{4 \cdot 2}$ \qquad Factoring 8

$\phantom{2.\ \sqrt{2} - \sqrt{8}} = \sqrt{2} - \sqrt{4}\sqrt{2}$

$\phantom{2.\ \sqrt{2} - \sqrt{8}} = \sqrt{2} - 2\sqrt{2}$

$\phantom{2.\ \sqrt{2} - \sqrt{8}} = 1\sqrt{2} - 2\sqrt{2}$

$\phantom{2.\ \sqrt{2} - \sqrt{8}} = (1 - 2)\sqrt{2}$ \qquad Using the distributive law to factor out the common factor $\sqrt{2}$

$\phantom{2.\ \sqrt{2} - \sqrt{8}} = -1 \cdot \sqrt{2}$

$\phantom{2.\ \sqrt{2} - \sqrt{8}} = -\sqrt{2}$

3. $\sqrt{x^3 - x^2} + \sqrt{4x - 4}$

$ = \sqrt{x^2(x - 1)} + \sqrt{4(x - 1)}$ \qquad Factoring radicands

$ = \sqrt{x^2}\sqrt{x - 1} + \sqrt{4}\sqrt{x - 1}$

$ = x\sqrt{x - 1} + 2\sqrt{x - 1}$ \qquad Assume that all expressions under radicals are nonnegative.

$ = (x + 2)\sqrt{x - 1}$ \qquad Using the distributive law to factor out the common factor $\sqrt{x - 1}$

> *Warning!* Do not make the mistake of thinking that the sum of square roots is the square root of a sum. For example,
>
> $$\sqrt{9} + \sqrt{16} = 3 + 4 = 7$$
>
> but
>
> $$\sqrt{9 + 16} = \sqrt{25} = 5.$$
>
> In general,
>
> $$\sqrt{a} + \sqrt{b} \neq \sqrt{a + b}.$$

DO EXERCISES 1–5.

OBJECTIVES

After finishing Section 10.7, you should be able to:

- Add or subtract with radical notation, using the distributive law to simplify.

Add or subtract and simplify by collecting like radical terms if possible.

1. $3\sqrt{2} + 9\sqrt{2}$

2. $8\sqrt{5} - 3\sqrt{5}$

3. $2\sqrt{10} - 7\sqrt{40}$

4. $\sqrt{24} + \sqrt{54}$

5. $\sqrt{9x + 9} - \sqrt{4x + 4}$

ANSWERS ON PAGE A-23

Add or subtract.

6. $\sqrt{2} + \sqrt{\dfrac{1}{2}}$

7. $\sqrt{\dfrac{5}{3}} - \sqrt{\dfrac{3}{5}}$

Sometimes rationalizing denominators will enable us to factor and then combine expressions.

Example 4 Add: $\sqrt{3} + \sqrt{\dfrac{1}{3}}$.

$$\sqrt{3} + \sqrt{\dfrac{1}{3}} = \sqrt{3} + \sqrt{\dfrac{1}{3} \cdot \dfrac{3}{3}} \qquad \text{Multiplying by 1}$$

$$= \sqrt{3} + \sqrt{\dfrac{3}{9}}$$

$$= \sqrt{3} + \dfrac{\sqrt{3}}{\sqrt{9}}$$

$$= \sqrt{3} + \dfrac{\sqrt{3}}{3}$$

$$= \left(1 + \dfrac{1}{3}\right)\sqrt{3} \qquad \text{Factoring and simplifying}$$

$$= \dfrac{4}{3}\sqrt{3}$$

DO EXERCISES 6 AND 7.

S O M E T H I N G   E X T R A

AN APPLICATION: WIND CHILL TEMPERATURE

In cold weather one feels colder when there is wind than when there is not. The *wind chill temperature* is what the temperature would have to be with no wind to give the same chilling effect as when there is wind. A formula for finding the wind chill temperature, T_W, is

$$T_W = 91.4 - \dfrac{(10.45 + 6.68\sqrt{v} - 0.447v)(457 - 5T)}{110},$$

where T is the actual temperature given by a thermometer, in degrees Fahrenheit, and v is the wind speed in mph.

EXERCISES

You will need a calculator. You can get the square roots from Table 2. Find the wind chill temperature in each case. Round to the nearest one degree.

1. $T = 30°F$, $v = 25$ mph
2. $T = 20°F$, $v = 20$ mph
3. $T = 20°F$, $v = 40$ mph
4. $T = -10°F$, $v = 30$ mph

Chapter 10   Radical Expressions and Equations

Add or subtract. Simplify by collecting like radical terms if possible.

1. $3\sqrt{2} + 4\sqrt{2}$

2. $7\sqrt{5} + 3\sqrt{5}$

3. $6\sqrt{a} - 14\sqrt{a}$

4. $10\sqrt{x} - 13\sqrt{x}$

5. $3\sqrt{12} + 2\sqrt{3}$

6. $5\sqrt{8} + 15\sqrt{2}$

7. $\sqrt{27} - 2\sqrt{3}$

8. $\sqrt{45} - \sqrt{20}$

9. $\sqrt{72} + \sqrt{98}$

10. $\sqrt{45} + \sqrt{80}$

11. $3\sqrt{18} - 2\sqrt{32} - 5\sqrt{50}$

12. $2\sqrt{12} + 4\sqrt{27} - 5\sqrt{48}$

13. $\sqrt{4x} + \sqrt{81x^3}$

14. $\sqrt{27} - \sqrt{12x^2}$

ANSWERS
1.
2.
3.
4.
5.
6.
7.
8.
9.
10.
11.
12.
13.
14.

15. $\sqrt{8x + 8} + \sqrt{2x + 2}$

16. $\sqrt{x^5 - x^2} + \sqrt{9x^3 - 9}$

17. $3x\sqrt{y^3x} - x\sqrt{yx^3} + y\sqrt{y^3x}$

18. $4a\sqrt{a^2b} + a\sqrt{a^2b^3} - 5\sqrt{b^3}$

19. $\sqrt{3} - \sqrt{\dfrac{1}{3}}$

20. $\sqrt{2} - \sqrt{\dfrac{1}{2}}$

21. $5\sqrt{2} + 3\sqrt{\dfrac{1}{2}}$

22. $\sqrt{\dfrac{2}{3}} - \sqrt{\dfrac{1}{6}}$

23. $\sqrt{\dfrac{1}{12}} - \sqrt{\dfrac{1}{27}}$

24. $\sqrt{\dfrac{2}{3}} + \sqrt{\dfrac{3}{2}}$

✓ SKILL MAINTENANCE

25. Multiply and simplify: $\dfrac{x^2 - 25}{3x} \cdot \dfrac{9x}{x + 5}$.

26. Add: $\dfrac{3}{x - 3} + \dfrac{2}{x + 3}$.

27. Solve: $-3x + 6y = -6$
$-2x + 7y = 2.$

28. Two cars leave town at the same time going in the same direction. One travels at 42 mph and the other travels at 55 mph. In how many hours will they be 52 mi apart?

★ EXTENSION

Perform the indicated operations and simplify.

29. $\sqrt{1 + x^2} + \dfrac{1}{\sqrt{1 + x^2}}$

30. $(\sqrt{a} - \sqrt{b})(\sqrt{a} + \sqrt{b})$

Rationalize the denominator. Use the result of Exercise 30.

31. $\dfrac{2}{\sqrt{3} - \sqrt{5}}$

32. $\dfrac{1 - \sqrt{7}}{3 + \sqrt{7}}$

33. Can you find any pairs of numbers a and b, for which $\sqrt{a} + \sqrt{b} = \sqrt{a + b}$? If so, name them.

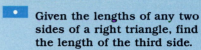
Right Triangles and Solving Problems

■ RIGHT TRIANGLES

In a right triangle, the longest side is called the *hypotenuse*. It is also called the side opposite the right angle. The other two sides are called the *legs*. We generally use the letters a and b for the lengths of the legs and c for the length of the hypotenuse. They are related as follows.

THE PYTHAGOREAN PROPERTY OF RIGHT TRIANGLES

In any right triangle, if a and b are the lengths of the legs and c is the length of the hypotenuse, then

$$a^2 + b^2 = c^2.$$

The equation $a^2 + b^2 = c^2$ is called *the Pythagorean equation*.

After finishing Section 10.8, you should be able to:

■ Given the lengths of any two sides of a right triangle, find the length of the third side.

■ ■ Solve applied problems involving right triangles.

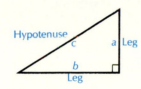

If we know the lengths of any two sides, we can find the length of the third side.

Example 1 Find the length of the hypotenuse of this right triangle. Give an exact answer and an approximation to three decimal places.

$4^2 + 5^2 = c^2$ Substituting in the Pythagorean equation

$16 + 25 = c^2$

$41 = c^2$

$c = \sqrt{41}$

$c \approx 6.403$ Using your calculator or Table 2

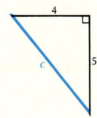

Example 2 Find the length of the leg of this right triangle. Give an exact answer and an approximation to three decimal places.

$10^2 + b^2 = 12^2$ Substituting in the Pythagorean equation

$100 + b^2 = 144$

$b^2 = 144 - 100$

$b^2 = 44$

$b = \sqrt{44}$

$b \approx 6.633$ Using your calculator or Table 2

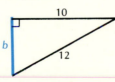

DO EXERCISES 1 AND 2.

1. Find the length of the hypotenuse in this right triangle. Give an exact answer and an approximation to three decimal places.

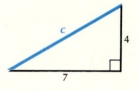

2. Find the length of the leg of this right triangle. Give an exact answer and an approximation to three decimal places.

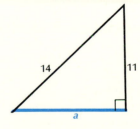

ANSWERS ON PAGE A-23

Find the length of the leg of each right triangle. Give an exact answer and an approximation to three decimal places.

3.

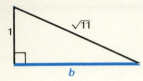

4.

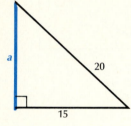

5. How long is a guy wire reaching from the top of a 15-ft pole to a point on the ground 10 ft from the pole? Give an exact answer and an approximation to three decimal places.

ANSWERS ON PAGE A-23

Example 3 Find the length of the leg of this right triangle. Give an exact answer and an approximation to three decimal places.

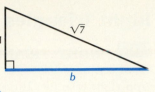

$1^2 + b^2 = (\sqrt{7})^2$ Substituting in the Pythagorean equation

$1 + b^2 = 7$

$b^2 = 7 - 1 = 6$

$b = \sqrt{6} \approx 2.449$ Using your calculator or Table 2

Example 4 Find the length of the leg of this right triangle. Give an exact answer and an approximation to three decimal places.

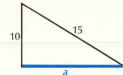

$a^2 + 10^2 = 15^2$

$a^2 + 100 = 225$

$a^2 = 225 - 100$

$a^2 = 125$

$a = \sqrt{125} \approx 11.180$

Use your calculator. If you use Table 2 to find an approximation, you will need to simplify before finding an approximation: $\sqrt{125} = \sqrt{25 \cdot 5} = 5\sqrt{5} \approx 5(2.236) = 11.180$. This can result in a different answer.

DO EXERCISES 3 AND 4.

●● APPLICATIONS

Example 5 A slow-pitch softball diamond is actually a square 65 ft on a side. How far is it from home to second base? (This can be helpful information when lining up the bases.) Give an exact answer and an approximation to three decimal places.

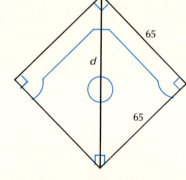

a) We first make a drawing. We note that the first and second base lines, together with a line from home to second, form a right triangle. We label the unknown distance d.

b) Now $65^2 + 65^2 = d^2$. We solve this equation:

$$4225 + 4225 = d^2$$
$$8450 = d^2.$$

Exact answer: $\sqrt{8450} = d$

Approximation: $91.924 \approx d$

If you use Table 2 to find an approximation, you will need to simplify before finding an approximation in the table:

$$d = \sqrt{8450} = \sqrt{25 \cdot 169 \cdot 2} = \sqrt{25}\sqrt{169}\sqrt{2}$$
$$\approx 5(13)(1.414) = 91.910.$$

Note that we get a variance in the last two decimal places.

DO EXERCISE 5.

   Chapter 10   Radical Expressions and Equations

● Find the length of the third side of each right triangle. Give an exact answer and an approximation to three decimal places.

1.

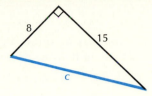

2.

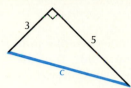

3.

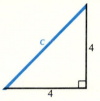

4.

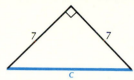

5.

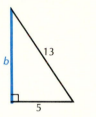

6.

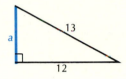

7.

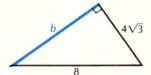

8.

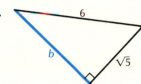

In a right triangle, find the length of the side not given. Give an exact answer and an approximation to three decimal places.

9. $a = 10$, $b = 24$

10. $a = 5$, $b = 12$

11. $a = 9$, $c = 15$

12. $a = 18$, $c = 30$

13. $b = 1$, $c = \sqrt{5}$

14. $b = 1$, $c = \sqrt{2}$

15. $a = 1$, $c = \sqrt{3}$

16. $a = \sqrt{3}$, $b = \sqrt{5}$

17. $c = 10$, $b = 5\sqrt{3}$

18. $a = 5$, $b = 5$

●● Solve. Don't forget to make drawings. Give an exact answer and an approximation to three decimal places.

19. A 10-m ladder is leaning against a building. The bottom of the ladder is 5 m from the building. How high is the top of the ladder?

20. Find the length of a diagonal of a square whose sides are 3 cm long.

ANSWERS
1.
2.
3.
4.
5.
6.
7.
8.
9.
10.
11.
12.
13.
14.
15.
16.
17.
18.
19.
20.

21. How long is a guy wire reaching from the top of a 12-ft pole to a point 8 ft from the pole?

22. How long must a wire be to reach from the top of a 13-m telephone pole to a point on the ground 9 m from the foot of the pole?

23. A little league baseball diamond is a square 60 ft on a side. How far is it from home to second base?

24. A baseball diamond is a square 90 ft on a side. How far is it from first to third base?

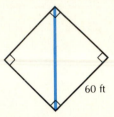

60 ft

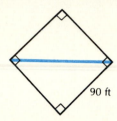

90 ft

✔ SKILL MAINTENANCE

Solve.

25. $\dfrac{12}{x} = \dfrac{48}{x + 9}$

26. $2x^2 - x - 15 = 0$

⭐ EXTENSION

27. The length and the width of a rectangle are given by consecutive integers. The area of the rectangle is 90 cm². Find the length of the diagonal of the rectangle.

28. Two cars leave a service station at the same time. One car travels east at a speed of 50 mph, and the other travels south at a speed of 60 mph. After one-half hour, how far apart are they?

Find x.

29.

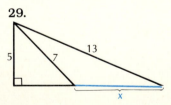

30.

31.

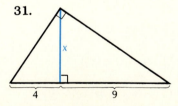

An *equilateral* triangle is shown at the right.

32. Find an expression for its height h in terms of a.

33. Find an expression for its area A in terms of a.

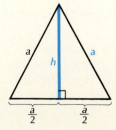

Copyright © 1987 Addison-Wesley Publishing Co., Inc.

Equations with Radicals

• SOLVING EQUATIONS WITH RADICALS

To solve equations with radicals, we first convert them to equations without radicals. We do this by squaring both sides of the equation. The following new principle is used.

THE PRINCIPLE OF SQUARING

If an equation $a = b$ is true, then the equation $a^2 = b^2$ is true.

Example 1 Solve: $\sqrt{2x} - 4 = 7$.

$$\sqrt{2x} - 4 = 7$$
$$\sqrt{2x} = 11 \qquad \text{Adding 4 to get the radical alone on one side}$$
$$(\sqrt{2x})^2 = 11^2 \qquad \text{Squaring both sides}$$
$$2x = 121$$
$$x = \frac{121}{2}$$

Check:

$$\begin{array}{c|c} \sqrt{2x} - 4 = 7 \\ \hline \sqrt{2 \cdot \dfrac{121}{2}} - 4 & 7 \\ \sqrt{121} - 4 \\ 11 - 4 \\ 7 \end{array}$$

It is important to check! When we square both sides of an equation, the new equation may have solutions that the first one does not. For example, the equation

$$x = 1$$

has just one solution, the number 1. When we square both sides we get

$$x^2 = 1,$$

which has two solutions, 1 and -1. The equations $x = 1$ and $x^2 = 1$ are not equivalent.

DO EXERCISE 1.

Example 2 Solve: $\sqrt{x + 1} = \sqrt{2x - 5}$.

$$\sqrt{x + 1} = \sqrt{2x - 5}$$
$$(\sqrt{x + 1})^2 = (\sqrt{2x - 5})^2 \qquad \text{Squaring both sides}$$
$$x + 1 = 2x - 5$$
$$x = 6$$

Since 6 checks, it is the solution.

DO EXERCISES 2 AND 3.

Solve.

1. $\sqrt{3x} - 5 = 3$

Solve.

2. $\sqrt{3x + 1} = \sqrt{2x + 3}$

3. $\sqrt{x - 2} - 5 = 3$
   (*Hint:* Get $\sqrt{x - 2}$ alone on one side.)

ANSWERS ON PAGE A-23

4. How far can you see to the horizon through an airplane window at a height of 8000 m?

APPLICATIONS

How far can you see from a given height? There is a formula for this. At a height of h meters you can see V kilometers to the horizon. These numbers are related as follows:

$$V = 3.5\sqrt{h} \qquad (1)$$

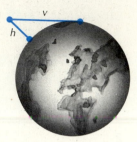

Earth

Example 3 How far to the horizon can you see through an airplane window at a height, or altitude, of 9000 m?

We substitute 9000 for h in equation (1) and find an approximation.

Method 1. We use a calculator and approximate $\sqrt{9000}$ directly:

$$V = 3.5\sqrt{9000} \approx 3.5(94.868) = 332.038.$$

Method 2. We simplify and then approximate:

$$V = 3.5\sqrt{9000} = 3.5\sqrt{900 \cdot 10} = 3.5 \times 30 \times \sqrt{10}$$
$$V \approx 3.5 \times 30 \times 3.162 \approx 332.010 \text{ km}.$$

You can see about 332 km at a height of 9000 m.

DO EXERCISES 4 AND 5.

5. How far can a sailor see to the horizon from the top of a 20-m mast?

Example 4 A person can see 50.4 km to the horizon from the top of a cliff. How high is the cliff?

We substitute 50.4 for V in equation (1) and solve:

$$50.4 = 3.5\sqrt{h}$$
$$\frac{50.4}{3.5} = \sqrt{h}$$
$$14.4 = \sqrt{h}$$
$$(14.4)^2 = (\sqrt{h})^2$$
$$(14.4)^2 = h$$
$$207.36 \text{ m} = h.$$

6. A sailor can see 91 km to the horizon from the top of a mast. How high is the mast?

50.4 km

The cliff is 207.36 m high.

ANSWERS ON PAGE A-23

DO EXERCISE 6.

Chapter 10   Radical Expressions and Equations

• Solve.

1. $\sqrt{x} = 5$

2. $\sqrt{x} = 7$

3. $\sqrt{x} = 6.2$

4. $\sqrt{x} = 4.3$

5. $\sqrt{x + 3} = 20$

6. $\sqrt{x + 4} = 11$

7. $\sqrt{2x + 4} = 25$

8. $\sqrt{2x + 1} = 13$

9. $3 + \sqrt{x - 1} = 5$

10. $4 + \sqrt{y - 3} = 11$

11. $6 - 2\sqrt{3n} = 0$

12. $8 - 4\sqrt{5n} = 0$

13. $\sqrt{5x - 7} = \sqrt{x + 10}$

14. $\sqrt{4x - 5} = \sqrt{x + 9}$

15. $\sqrt{x} = -7$

16. $\sqrt{x} = -5$

17. $\sqrt{2y + 6} = \sqrt{2y - 5}$

18. $2\sqrt{3x - 2} = \sqrt{2x - 3}$

ANSWERS
1. _____
2. _____
3. _____
4. _____
5. _____
6. _____
7. _____
8. _____
9. _____
10. _____
11. _____
12. _____
13. _____
14. _____
15. _____
16. _____
17. _____
18. _____

•• Solve.

Use $V = 3.5\sqrt{h}$ for Exercises 19–22.

19. How far can you see to the horizon through an airplane window at a height of 9800 m?

20. How far can a sailor see to the horizon from the top of a 24-m mast?

21. A person can see 371 km to the horizon from an airplane window. How high is the airplane?

22. A sailor can see 99.4 km to the horizon from the top of a mast. How high is the mast?

The formula $r = 2\sqrt{5L}$ can be used to approximate the speed r, in mph, of a car that has left a skid mark of length L, in feet.

23. How far will a car skid at 50 mph? at 70 mph?

24. How far will a car skid at 60 mph? at 100 mph?

✓ **SKILL MAINTENANCE**

25. Solve for E: $A = \dfrac{\pi r^2 E}{180}$.

26. Graph: $y = -\dfrac{1}{4}x - 3$.

27. The amount that a family spends on entertainment varies directly as its income. A family making $25,000 a year will spend $6000 for entertainment. How much will a family making $30,000 a year spend for entertainment?

⭐ **EXTENSION**

Solve.

28. $\sqrt{5x^2 + 5} = 5$

29. $\sqrt{x} = -x$

30. $x - 1 = \sqrt{x + 5}$

31. $\sqrt{x - 5} + \sqrt{x} = 5$

(*Hint:* Use the principle of squaring twice.)

Summary and Review

The following contains a summary of what you should be able to do after completing this chapter. The review exercises are for practice. Answers are at the back of the book. If you miss an exercise, restudy the section and objective indicated alongside the answer.

The review sections to be tested in addition to the material in this chapter are 6.4, 7.6, 9.6, and 9.7.

You should be able to:

Find the square roots of a number and simplify radical expressions with a perfect-square radicand.

Find the square roots of each number.

1. 64   **2.** 25   **3.** 196   **4.** 400

Simplify.

5. $\sqrt{36}$ **6.** $-\sqrt{81}$ **7.** $\sqrt{49}$ **8.** $-\sqrt{169}$

Identify a given square root or other real number as rational or irrational.

Identify each number as rational or irrational.

9. $\sqrt{3}$ **10.** $\sqrt{36}$ **11.** $-\sqrt{12}$ **12.** $-\sqrt{4}$ **13.** $-\dfrac{4}{5}$ **14.** 5.56

15. 0.272727 . . .   (numeral repeats)   **16.** 0.313313331 . . .   (numeral does not repeat)

Use a calculator or Table 2 to approximate expressions involving square roots to three decimal places.

Approximate these square roots to three decimal places.

17. $\sqrt{3}$ **18.** $\sqrt{108}$ **19.** $\sqrt{\dfrac{1}{8}}$ **20.** $\sqrt{\dfrac{11}{20}}$

Identify radicands of radical expressions, identify meaningless radical expressions, and determine the sensible replacements in a radical expression.

Identify the radicand in each expression.

21. $\sqrt{x^2 + 4}$ **22.** $\sqrt{5ab^3}$

Which of these expressions are meaningless? Write "yes" or "no."

23. $\sqrt{-22}$ **24.** $-\sqrt{49}$ **25.** $\sqrt{-36}$ **26.** $\sqrt{-100}$

Determine the sensible replacements in each expression.

27. $\sqrt{x + 7}$ **28.** $\sqrt{2y - 20}$

Simplify radical expressions with a perfect-square radicand.

Simplify.

29. $\sqrt{m^2}$ **30.** $\sqrt{49t^2}$ **31.** $\sqrt{p^2}$ **32.** $\sqrt{(x - 4)^2}$

Multiply with radical notation and, where possible, simplify the result.

Multiply.

33. $\sqrt{3}\sqrt{7}$ **34.** $\sqrt{a}\sqrt{t}$ **35.** $\sqrt{x - 3}\sqrt{x + 3}$ **36.** $\sqrt{2x}\sqrt{3y}$

Simplify radical expressions by factoring.

Simplify by factoring.

37. $-\sqrt{48}$ **38.** $\sqrt{32t^2}$ **39.** $\sqrt{x^2 + 16x + 64}$ **40.** $\sqrt{t^2 - 49}$

Simplify radical expressions where radicands are powers, and multiply and simplify radical expressions.

Simplify.

41. $\sqrt{x^8}$ **42.** $\sqrt{m^{15}}$

Multiply and simplify.

43. $\sqrt{6}\sqrt{10}$ 44. $\sqrt{5x}\sqrt{8x}$ 45. $\sqrt{5x}\sqrt{10xy^2}$ 46. $\sqrt{20a^3b}\sqrt{5a^2b^2}$

Simplify radical expressions with fractional radicands.

Simplify.

47. $\sqrt{\dfrac{25}{64}}$ 48. $\sqrt{\dfrac{20}{45}}$ 49. $\sqrt{\dfrac{49}{t^2}}$

Rationalize denominators of radical expressions, and approximate square roots of fractions.

Rationalize the denominator.

50. $\sqrt{\dfrac{1}{2}}$ 51. $\sqrt{\dfrac{1}{8}}$ 52. $\sqrt{\dfrac{5}{y}}$ 53. $\dfrac{2}{\sqrt{3}}$

Divide with radical notation and, where possible, simplify the result.

Divide and simplify.

54. $\dfrac{\sqrt{27}}{\sqrt{45}}$ 55. $\dfrac{\sqrt{45x^2y}}{\sqrt{54y}}$

Add or subtract with radical notation, using the distributive law to simplify.

Add or subtract.

56. $10\sqrt{5} + 3\sqrt{5}$ 57. $\sqrt{80} - \sqrt{45}$ 58. $3\sqrt{2} - 5\sqrt{\tfrac{1}{2}}$

Given the lengths of any two sides of a right triangle, find the length of the third side, and solve problems involving right triangles.

In a right triangle, find the length of the side not given. Find an exact answer and an approximation to three decimal places.

59. $a = 15$, $c = 25$ 60. $a = 1$, $b = \sqrt{2}$

61. Find the length of the diagonal of a square whose sides are 7 cm.

Solve equations with radicals.

62. $\sqrt{x - 3} = 7$ 63. $\sqrt{5x + 3} = \sqrt{2x - 1}$

Solve problems involving the solution of radical equations.

64. The formula $r = 2\sqrt{5L}$ can be used to approximate the speed r, in mph, of a car that has left a skid mark of length L, in ft. How far will a car skid at a speed of 90 mph?

 SKILL MAINTENANCE

65. Solve: $\dfrac{x - 4}{2x + 3} = \dfrac{1}{3}$.

66. Subtract: $\dfrac{x}{x^2 - 9} - \dfrac{x - 1}{x^2 - 5x + 6}$.

67. A student's test score varies directly as the time studied. A score of 90 is made after studying 18 hr. What score can be made after studying 20 hr?

68. A car and a truck leave town at the same time going in opposite directions. The car travels at 90 km/h and the truck travels at 60 km/h. In how many hours will they be 200 km apart?

 EXTENSION

69. The area of square $PQRS$ is 100 ft^2, and A, B, C, and D are midpoints of the sides on which they lie. Find the area of square $ABCD$.

70. Simplify: $\sqrt{\sqrt{\sqrt{625}}}$.

71. Solve: $\sqrt{x^2} = -10$.

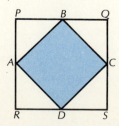

1. Find the square roots of 81.

Simplify.

2. $\sqrt{64}$ **3.** $-\sqrt{25}$

Identify each number as rational or irrational.

4. $-\sqrt{10}$ **5.** $\sqrt{16}$ **6.** $\dfrac{1}{4}$

7. 0.4324432444 . . .   **8.** 0.136136136 . . .
  (numeral does not repeat)   (numeral repeats)

Approximate these expressions involving square roots to three decimal places.

9. $\sqrt{116}$ **10.** $\sqrt{87}$ **11.** $\dfrac{3}{\sqrt{3}}$

12. Identify the radicand: $\sqrt{4 - y^3}$.

Which of these expressions are meaningless? Write "yes" or "no."

13. $\sqrt{24}$ **14.** $\sqrt{-23}$

15. Determine the sensible replacements in $\sqrt{8 - x}$.

Simplify.

16. $\sqrt{a^2}$ **17.** $\sqrt{36y^2}$

Multiply. Remember, radicands are nonnegative.

18. $\sqrt{5}\sqrt{6}$ **19.** $\sqrt{x-3}\sqrt{x+3}$

Simplify by factoring.

20. $\sqrt{27}$ **21.** $\sqrt{25x - 25}$ **22.** $\sqrt{t^5}$

Multiply and simplify.

23. $\sqrt{5}\sqrt{10}$ **24.** $\sqrt{3ab}\sqrt{6ab^3}$

ANSWERS
1. _____
2. _____
3. _____
4. _____
5. _____
6. _____
7. _____
8. _____
9. _____
10. _____
11. _____
12. _____
13. _____
14. _____
15. _____
16. _____
17. _____
18. _____
19. _____
20. _____
21. _____
22. _____
23. _____
24. _____

Simplify.

25. $\sqrt{\dfrac{27}{12}}$

26. $\sqrt{\dfrac{144}{a^2}}$

Rationalize the denominator.

27. $\sqrt{\dfrac{2}{5}}$

28. $\sqrt{\dfrac{2x}{y}}$

Divide and simplify.

29. $\dfrac{\sqrt{27}}{\sqrt{32}}$

30. $\dfrac{\sqrt{35x}}{\sqrt{80xy^2}}$

Add or subtract.

31. $3\sqrt{18} - 5\sqrt{18}$

32. $\sqrt{5} - \sqrt{\dfrac{1}{5}}$

33. In a right triangle, $a = 8$ and $b = 4$. Find c. Find an exact answer and an approximation to three decimal places.

34. Solve: $\sqrt{3x} + 2 = 14$.

35. A person can see 247.49 km to the horizon from an airplane window. How high is the airplane? Use the formula $V = 3.5\sqrt{h}$.

✔ SKILL MAINTENANCE

36. The weight of an object on planet X varies directly as its weight on earth. A person weighing 33 kg on planet X weighs 63 kg on earth. How much would an 84-kg person weigh on planet X?

37. Solve: $\dfrac{x}{x + 3} = \dfrac{5}{2}$.

38. Subtract: $\dfrac{1}{x^2 - 16} - \dfrac{x - 4}{x^2 - 3x - 4}$.

39. An airplane flew for 8 hr with a 29-mph tailwind. The return flight against the same wind took 10 hr. Find the speed of the airplane in still air.

★ EXTENSION

40. Simplify: $\sqrt{y^{16n}}$.

41. The diagonal of a square has length $8\sqrt{2}$ ft. Find the length of a side of the square.

AN APPLICATION
The Sears Tower in Chicago is 1451 ft tall. How long would it take an object to fall from the top?

THE MATHEMATICS
Let t = the time of the fall. To find the answer to the problem we solve the equation

$1451 = 16t^2$. ← This is a *quadratic equation*.

In this chapter equations containing polynomials of second degree are introduced. We began our study of these equations in Chapter 5 on factoring. Certain equations could not be factored easily. The quadratic formula, which can be used to solve any quadratic equation, is presented. Finally, we solve problems using quadratic equations.

The review sections to be tested in addition to the material in this chapter are 7.4, 9.8, 10.4, and 10.7.

Quadratic Equations

Introduction to Quadratic Equations

■ STANDARD FORM

The following are *quadratic equations*. They contain polynomials of second degree.

$$4x^2 + 7x - 5 = 0, \qquad 3t^2 - \tfrac{1}{4}t = 9, \qquad 5y^2 = -6y, \qquad 3m^2 = 0$$

The quadratic equation

$$4x^2 + 7x - 5 = 0$$

is said to be in *standard form*. The quadratic equation

$$4x^2 = 5 - 7x$$

is equivalent to the preceding, but it is not in standard form.

> An equation of the type $ax^2 + bx + c = 0$, where a, b, and c are real-number constants and $a > 0$, is called the *standard form of a quadratic equation*.

Suppose we are considering an equation like $-3x^2 + 8x - 2 = 0$. It is not in standard form. We can find an equivalent equation that *is* in standard form by multiplying on both sides by -1:

$$-1(-3x^2 + 8x - 2) = -1(0)$$
$$3x^2 - 8x + 2 = 0.$$

We require $a > 0$ in the standard form to make for a smoother proof of the quadratic formula, which we consider later.

To write standard form for a quadratic equation, we find an equivalent equation that *is* in standard form.

Examples Write standard form and determine a, b, and c.

1. $4x^2 + 7x - 5 = 0$ The equation is already in standard form.
     $a = 4$; $b = 7$; $c = -5$

2. $3x^2 - 0.5x = 9$
       $3x^2 - 0.5x - 9 = 0$ Adding -9. This is standard form.
     $a = 3$; $b = -0.5$; $c = -9$

3. $-4y^2 = 5y$
       $-4y^2 - 5y = 0$ Adding $-5y$
                 Not positive!

     $4y^2 + 5y = 0$ Multiplying by -1. This is standard form.
     $a = 4$; $b = 5$; $c = 0$

Write standard form and determine a, b, and c.

1. $x^2 = 7x$

2. $3 - x^2 = 9x$

3. $3x + 5x^2 = x^2 - 4 + x$

ANSWERS ON PAGE A-24

4. $4x^2 = 20$
         $4x^2 - 20 = 0$ Adding -20
    $a = 4; \; b = 0; \; c = -20$

DO EXERCISES 1–3 ON THE PRECEDING PAGE.

•• SOLVING EQUATIONS OF THE TYPE $ax^2 + c = 0$

When b is 0, we solve for x^2 and take the square roots.

Example 5 Solve: $5x^2 = 15$.

$$5x^2 = 15$$
$$x^2 = 3 \quad \text{Solving for } x^2\text{; multiplying by } \tfrac{1}{5}$$

The square roots of 3 are $\sqrt{3}$ and $-\sqrt{3}$. Thus,

$$x = \sqrt{3} \quad \text{or} \quad x = -\sqrt{3}.$$

Check: For $\sqrt{3}$:   For $-\sqrt{3}$:

$$\frac{5x^2 \;=\; 15}{\begin{array}{c|c} 5(\sqrt{3})^2 & 15 \\ 5 \cdot 3 & \\ 15 & \end{array}} \qquad \frac{5x^2 \;=\; 15}{\begin{array}{c|c} 5(-\sqrt{3})^2 & 15 \\ 5 \cdot 3 & \\ 15 & \end{array}}$$

The solutions are $\sqrt{3}$ and $-\sqrt{3}$.

DO EXERCISE 4.

Example 6 Solve: $\tfrac{1}{3}x^2 = 0$.

$$\frac{1}{3}x^2 = 0$$
$$x^2 = 0 \quad \text{Multiplying by 3}$$
$$x = 0$$

The only number whose square is 0 is 0. It checks, so it is the solution.

DO EXERCISE 5.

Example 7 Solve: $-3x^2 + 7 = 0$.

$$-3x^2 + 7 = 0$$
$$-3x^2 = -7 \quad \text{Adding } -7$$
$$x^2 = \frac{-7}{-3} \quad \text{Multiplying by } -\frac{1}{3}$$
$$x^2 = \frac{7}{3}$$
$$x = \sqrt{\frac{7}{3}} \quad \text{or} \quad x = -\sqrt{\frac{7}{3}}$$
$$x = \sqrt{\frac{7}{3} \cdot \frac{3}{3}} \quad \text{or} \quad x = -\sqrt{\frac{7}{3} \cdot \frac{3}{3}} \quad \begin{array}{l}\text{Rationalizing the}\\\text{denominators}\end{array}$$
$$x = \frac{\sqrt{21}}{3} \quad \text{or} \quad x = -\frac{\sqrt{21}}{3}$$

Solve.

4. $4x^2 = 20$

Solve.

5. $2x^2 = 0$

ANSWERS ON PAGE A-24

Solve.

6. $2x^2 - 3 = 0$

7. $4x^2 - 9 = 0$

8. The height of the World Trade Center in New York is 1377 ft (excluding TV towers and antennas). How long would it take an object to fall from the top?

Check: For $\dfrac{\sqrt{21}}{3}$:

$$\begin{array}{c|c} -3x^2 + 7 = 0 & \\ \hline -3\left(\dfrac{\sqrt{21}}{3}\right)^2 + 7 & 0 \\ -3 \cdot \dfrac{21}{9} + 7 & \\ -7 + 7 & \\ 0 & \end{array}$$

For $-\dfrac{\sqrt{21}}{3}$:

$$\begin{array}{c|c} -3x^2 + 7 = 0 & \\ \hline -3\left(-\dfrac{\sqrt{21}}{3}\right)^2 + 7 & 0 \\ -3 \cdot \dfrac{21}{9} + 7 & \\ -7 + 7 & \\ 0 & \end{array}$$

The solutions are $\dfrac{\sqrt{21}}{3}$ and $-\dfrac{\sqrt{21}}{3}$.

DO EXERCISES 6 AND 7.

••• APPLICATIONS

Example 8 The Sears Tower in Chicago is 1451 ft tall. How long would it take an object to fall from the top?

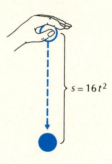

$s = 16t^2$

A formula that fits this situation is

$$s = 16t^2.$$

According to this formula, the distance s in feet that an object falls freely from rest in t seconds is approximated by $16t^2$. This formula is actually an approximation in that it does not account for air resistance. In this problem we know the distance s to be 1451 ft. We are looking for the time t. We substitute 1451 for s and solve for t:

$$1451 = 16t^2$$

$$\dfrac{1451}{16} = t^2 \qquad \text{Solving for } t^2$$

$$90.6875 = t^2 \qquad \text{Dividing on both sides by 16}$$

$$\sqrt{90.6875} = t \qquad \text{Taking the principal square root}$$

$$9.5 \approx t \qquad \text{Using a calculator to find the square root and rounding to the nearest tenth}$$

DO EXERCISE 8.

• Write standard form and determine a, b, and c.

1. $x^2 = 3x + 2$

2. $2x^2 = 3$

3. $7x^2 = 4x - 3$

4. $5 = -2x^2 + 3x$

5. $2x - 1 = 3x^2 + 7$

6. $4x^2 - 3x + 2 = 3x^2 + 7x - 5$

•• Solve.

7. $x^2 = 4$

8. $x^2 = 1$

9. $x^2 = 49$

10. $x^2 = 16$

11. $x^2 = 7$

12. $x^2 = 11$

13. $3x^2 = 30$

14. $5x^2 = 35$

ANSWERS
1. _____
2. _____
3. _____
4. _____
5. _____
6. _____
7. _____
8. _____
9. _____
10. _____
11. _____
12. _____
13. _____
14. _____

15. $3x^2 = 24$ **16.** $5x^2 = 60$ **17.** $4x^2 - 25 = 0$ **18.** $9x^2 - 4 = 0$

19. $3x^2 - 49 = 0$ **20.** $5y^2 - 16 = 0$ **21.** $4y^2 - 3 = 9$ **22.** $5x^2 - 100 = 0$

●●● Solve. Use $s = 16t^2$.

23. A body falls 1000 ft. How many seconds does this take?

24. A body falls 2496 ft. How many seconds does this take?

25. The world record for free-fall by a woman is 175 ft and is held by Kitty O'Neill. This kind of fall is from a certain height, say the top of a tower, and does not involve a parachute for any part of the fall. Approximately how long did the fall take?

26. The world record for free-fall by a man is 311 ft and is held by Dar Robinson. Approximately how long did the fall take?

✔ **SKILL MAINTENANCE**

Solve.

27. $x^2 - 6x = 0$

28. $x^2 - 5x + 6 = 0$

Simplify.

29. $\sqrt{20}$

30. $\sqrt{\dfrac{2890}{2560}}$

☆ **EXTENSION**

Solve.

31. $4.82x^2 = 12{,}000$ **32.** $\dfrac{x}{4} = \dfrac{9}{x}$ **33.** $\dfrac{4}{x^2 - 7} = \dfrac{6}{x^2}$

34. Solve for x: $3ax^2 - 9b = 3b^2$.

Solving by Factoring

◆ EQUATIONS OF THE TYPE $ax^2 + bx = 0$

When c is 0 (and $b \neq 0$), we can factor and use the principle of zero products.

Example 1 Solve: $7x^2 + 2x = 0$

$$7x^2 + 2x = 0$$
$$x(7x + 2) = 0 \qquad \text{Factoring}$$
$$x = 0 \quad \text{or} \quad 7x + 2 = 0 \qquad \text{Principle of zero products}$$
$$x = 0 \quad \text{or} \qquad 7x = -2$$
$$x = 0 \quad \text{or} \qquad x = -\tfrac{2}{7}$$

Check: For 0:   For $-\tfrac{2}{7}$:

$$
\begin{array}{c|c}
7x^2 + 2x = 0 & \\ \hline
7 \cdot 0^2 + 2 \cdot 0 & 0 \\
0 &
\end{array}
\qquad
\begin{array}{c|c}
7x^2 + 2x = 0 & \\ \hline
7(-\tfrac{2}{7})^2 + 2(-\tfrac{2}{7}) & 0 \\
7(\tfrac{4}{49}) - \tfrac{4}{7} & \\
\tfrac{4}{7} - \tfrac{4}{7} & \\
0 &
\end{array}
$$

The solutions are 0 and $-\tfrac{2}{7}$.

When we use the principle of zero products, we need not check except to detect errors in solving.

Example 2 Solve: $20x^2 - 15x = 0$.

$$20x^2 - 15x = 0$$
$$5x(4x - 3) = 0 \qquad \text{Factoring}$$
$$5x = 0 \quad \text{or} \quad 4x - 3 = 0 \qquad \text{Principle of zero products}$$
$$x = 0 \quad \text{or} \qquad 4x = 3$$
$$x = 0 \quad \text{or} \qquad x = \tfrac{3}{4}$$

The solutions are 0 and $\tfrac{3}{4}$.

A quadratic equation of this type will always have 0 as one solution and a nonzero number as the other solution.

DO EXERCISES 1 AND 2.

◆◆ EQUATIONS OF THE TYPE $ax^2 + bx + c = 0$

When neither b nor c is 0, we can sometimes solve by factoring.

Solve.

1. $3x^2 + 5x = 0$

2. $10x^2 - 6x = 0$

ANSWERS ON PAGE A-24

Solve.

3. $3x^2 + x - 2 = 0$

4. $(x - 1)(x + 1) = 5(x - 1)$

5. Use $d = \dfrac{n^2 - 3n}{2}$.

 a) A heptagon has 7 sides. How many diagonals does it have?

 b) A polygon has 44 diagonals. How many sides does it have?

ANSWERS ON PAGE A-24

Example 3 Solve: $5x^2 - 8x + 3 = 0$.

$$5x^2 - 8x + 3 = 0$$
$$(5x - 3)(x - 1) = 0 \qquad \text{Factoring}$$
$$5x - 3 = 0 \quad \text{or} \quad x - 1 = 0$$
$$5x = 3 \quad \text{or} \qquad x = 1$$
$$x = \tfrac{3}{5} \quad \text{or} \qquad x = 1$$

The solutions are $\tfrac{3}{5}$ and 1.

Example 4 Solve: $(y - 3)(y - 2) = 6(y - 3)$.

We write standard form and then try to factor.

$$y^2 - 5y + 6 = 6y - 18$$
$$y^2 - 11y + 24 = 0 \qquad \text{Standard form}$$
$$(y - 8)(y - 3) = 0$$
$$y - 8 = 0 \quad \text{or} \quad y - 3 = 0$$
$$y = 8 \quad \text{or} \qquad y = 3$$

The solutions are 8 and 3.

DO EXERCISES 3 AND 4.

••• SOLVING PROBLEMS

Example 5 The number of diagonals, d, of a polygon of n sides is given by

$$d = \frac{n^2 - 3n}{2}.$$

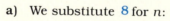

(a) An octagon has 8 sides. How many diagonals does it have? (b) A polygon has 27 diagonals. How many sides does it have?

a) We substitute **8** for n:

$$d = \frac{n^2 - 3n}{2} = \frac{8^2 - 3 \cdot 8}{2} = \frac{64 - 24}{2} = 20.$$

An octagon has 20 diagonals.

b) We substitute **27** for d and solve for n:

$$\frac{n^2 - 3n}{2} = d$$
$$\frac{n^2 - 3n}{2} = 27 \qquad \text{Substituting 27 for } d$$
$$n^2 - 3n = 54 \qquad \text{Multiplying by 2 to clear of fractions}$$
$$n^2 - 3n - 54 = 0$$
$$(n - 9)(n + 6) = 0$$
$$n - 9 = 0 \quad \text{or} \quad n + 6 = 0$$
$$n = 9 \quad \text{or} \qquad n = -6.$$

Since the number of sides cannot be negative, -6 cannot be a solution. But 9 checks, so the polygon has 9 sides (it is a nonagon).

DO EXERCISE 5.

Chapter 11   Quadratic Equations

⬤ Solve.

1. $x^2 + 7x = 0$ **2.** $x^2 - 5x = 0$ **3.** $3x^2 + 2x = 0$ **4.** $5x^2 + 2x = 0$

5. $4x^2 + 4x = 0$ **6.** $10x^2 - 30x = 0$ **7.** $55x^2 - 11x = 0$ **8.** $14x^2 - 3x = 0$

⬤⬤ Solve.

9. $x^2 - 16x + 48 = 0$ **10.** $x^2 + 7x + 6 = 0$

11. $x^2 + 4x - 21 = 0$ **12.** $x^2 - 9x + 14 = 0$

13. $x^2 + 10x + 25 = 0$ **14.** $x^2 - 2x + 1 = 0$

15. $2x^2 - 13x + 15 = 0$ **16.** $3a^2 - 10a - 8 = 0$

17. $3x^2 - 7x = 20$ **18.** $2x^2 + 12x = -10$

ANSWERS
1.
2.
3.
4.
5.
6.
7.
8.
9.
10.
11.
12.
13.
14.
15.
16.
17.
18.

19. _____

20. _____

21. _____

22. _____

23. _____

24. _____

25. _____

26. _____

27. _____

28. _____

29. _____

30. _____

31. _____

32. _____

33. _____

34. _____

35. _____

36. _____

19. $t(t - 5) = 14$

20. $3y^2 + 8y = 12y + 15$

21. $t(9 + t) = 4(2t + 5)$

22. $(2x - 3)(x + 1) = 4(2x - 3)$

 Solve. Use $d = \dfrac{n^2 - 3n}{2}$.

23. A hexagon has 6 sides. How many diagonals does it have?

24. A decagon has 10 sides. How many diagonals does it have?

25. A polygon has 14 diagonals. How many sides does it have?

26. A polygon has 9 diagonals. How many sides does it have?

✔ SKILL MAINTENANCE

27. Multiply: $(3x + 1)^2$.

28. Graph: $5x + 3y = 15$.

29. Approximate $\sqrt{17}$ to three decimal places using a calculator or Table 2.

30. Add: $\sqrt{2} + \sqrt{8}$.

☆ EXTENSION

Solve.

31. $4m^2 - (m + 1)^2 = 0$

32. $x^2 + \sqrt{3}x = 0$

33. 🖩 $0.0025x^2 + 70{,}400x = 0$

34. Solve for x: $ax^2 + bx = 0$.

35. Solve: $y^4 - 4y^2 + 4 = 0$. (*Hint*: Let $x^2 = y^4$. Write a quadratic equation in x and solve. Remember to solve for y after finding x.)

36. Solve: $z - 10\sqrt{z} + 9 = 0$. (*Hint*: Let $x = \sqrt{z}$.)

Completing the Square

• SOLVING EQUATIONS OF THE TYPE $(x + k)^2 = d$

In equations of the type $(x + k)^2 = d$, we have the square of a binomial equal to a constant. We can solve such an equation by taking the square roots.

Example 1 Solve: $(x - 5)^2 = 9$.

$$(x - 5)^2 = 9$$
$$x - 5 = 3 \quad \text{or} \quad x - 5 = -3$$
$$x = 8 \quad \text{or} \quad x = 2$$

The solutions are 8 and 2.

Example 2 Solve: $(x + 2)^2 = 7$.

$$(x + 2)^2 = 7$$
$$x + 2 = \sqrt{7} \quad \text{or} \quad x + 2 = -\sqrt{7}$$
$$x = -2 + \sqrt{7} \quad \text{or} \quad x = -2 - \sqrt{7}$$

The solutions are $-2 + \sqrt{7}$ and $-2 - \sqrt{7}$, or simply $-2 \pm \sqrt{7}$ (read "-2 plus or minus $\sqrt{7}$").

DO EXERCISES 1–3.

•• COMPLETING THE SQUARE

The following is the square of a binomial:

$$x^2 + 10x + 25.$$

An equivalent expression is $(x + 5)^2$. We could find the 25 from $10x$ by taking half the coefficient of x and squaring it.

From the preceding we can see how to make $x^2 + 10x$ the square of a binomial. We add the proper number to it. In this case that number is 25. This is called *completing the square*.

$$x^2 + 10x$$
$$\downarrow$$
$$\frac{10}{2} = 5 \qquad \text{Taking half the } x\text{-coefficient}$$
$$\downarrow$$
$$5^2 = 25 \qquad \text{Squaring}$$
$$\downarrow$$
$$x^2 + 10x + 25 \qquad \text{Adding}$$

The trinomial $x^2 + 10x + 25$ is the square of $x + 5$.

Solve.

1. $(x - 3)^2 = 16$

2. $(x + 3)^2 = 10$

3. $(x - 1)^2 = 5$

ANSWERS ON PAGE A-24

Complete the square.

4. $x^2 - 8x$

5. $x^2 - 10x$

6. $x^2 + 7x$

7. $x^2 - 3x$

Solve.

8. $1000 invested at 14% compounded annually for 2 years will grow to what amount?

Examples Complete the square.

3. $x^2 \underbrace{- 12x}$

$$\left(\frac{-12}{2}\right)^2 = (-6)^2 = 36 \qquad \text{Taking half the } x\text{-coefficient and squaring}$$

$$x^2 - 12x + 36$$

The trinomial $x^2 - 12x + 36$ is the square of $x - 6$.

4. $x^2 \underbrace{- 5x}$

$$\left(\frac{-5}{2}\right)^2 = \frac{25}{4}$$

$$x^2 - 5x + \frac{25}{4}$$

The trinomial $x^2 - 5x + \frac{25}{4}$ is the square of $x - \frac{5}{2}$.

DO EXERCISES 4–7.

● ● ● **APPLICATIONS: INTEREST PROBLEMS**

If you put money in a savings account, the bank will pay you interest. At the end of a year, the bank will start paying you interest on both the original amount and the interest. This is called *compounding interest annually.*

An amount of money P is invested at interest rate r, compounded annually. In t years it will grow to the amount A given by

$$A = P(1 + r)^t.$$

Example 5 $1000 invested at 16% for 2 years compounded annually will grow to what amount?

$$A = P(1 + r)^t$$
$$A = 1000(1 + 0.16)^2 \qquad \text{Substituting into the formula}$$
$$A = 1000(1.16)^2$$
$$A = 1000(1.3456)$$
$$A = 1345.60 \qquad \text{Computing}$$

The amount is $1345.60.

DO EXERCISE 8.

Example 6 $2560 is invested at interest rate r compounded annually. In 2 years it grows to $2890. What is the interest rate?

We substitute 2560 for P, 2890 for A, and 2 for t in the formula, and solve for r.

$$A = P(1 + r)^t$$
$$2890 = 2560(1 + r)^2$$
$$\frac{2890}{2560} = (1 + r)^2$$
$$\frac{289}{256} = (1 + r)^2$$
$$\sqrt{\frac{289}{256}} = 1 + r \qquad \text{Taking the principal square root. Since } r \text{ must be positive, } 1 + r \text{ must also be positive.}$$
$$\frac{17}{16} = 1 + r$$
$$-\frac{16}{16} + \frac{17}{16} = r$$
$$\frac{1}{16} = r$$
$$0.0625 = r \qquad \tfrac{1}{16} = 0.0625$$

or

$$6.25\% = r.$$

The interest rate must be 6.25% in order for $2560 to grow to $2890 in 2 years.

DO EXERCISE 9.

Example 7 For $2000 to double itself in 2 years, what would the interest rate have to be?

We substitute 2000 for P, 4000 for A, and 2 for t in the formula.

$$A = P(1 + r)^t$$
$$4000 = 2000(1 + r)^2$$
$$\frac{4000}{2000} = (1 + r)^2$$
$$2 = (1 + r)^2$$
$$\sqrt{2} = 1 + r \qquad \text{Taking the principal square root}$$
$$-1 + \sqrt{2} = r$$
$$-1 + (1.414) \approx r$$
$$0.414 \approx r \qquad \text{Using Table 2}$$
$$41.4\% \approx r$$

The interest rate would have to be 41.4% in order for the $2000 to double itself. Such a rate would be hard to get!

DO EXERCISE 10.

9. Suppose $2560 is invested at interest rate r compounded annually, and grows to $3240 in 2 years. What is the interest rate?

10. Suppose $1000 is invested at interest rate r compounded annually, and grows to $3000 (it triples) in 2 years. What is the interest rate?

ANSWERS ON PAGE A-24

CALCULATOR CORNER: COMPOUND INTEREST

We have considered the formula

$$A = P(1 + r)^t$$

for interest compounded annually. If interest is compounded more than once a year, say quarterly, we can find a formula like the one above as follows:

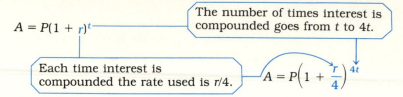

$$A = P(1 + r)^t$$

The number of times interest is compounded goes from t to $4t$.

Each time interest is compounded the rate used is $r/4$.

$$A = P\left(1 + \frac{r}{4}\right)^{4t}$$

In general:

> If principal P is invested at interest rate r, compounded n times a year, in t years it will grow to the amount A given by
>
> $$A = P\left(1 + \frac{r}{n}\right)^{nt}.$$

Example Suppose $1000 is invested at 16%, compounded quarterly. How much is in the account at the end of 2 years?

Substituting 1000 for P, 0.16 for r, 4 for n, and 2 for t, we get

$$A = 1000\left(1 + \frac{0.16}{4}\right)^{4 \cdot 2} = 1000(1.04)^8$$

$$\approx 1000(1.368569) \approx \$1368.57.$$

EXERCISES

Use your calculator. A power key $\boxed{a^b}$ will be needed for many of the exercises. Use 365 days for one year.

1. Suppose $1000 is invested at 16%, compounded semiannually ($n = 2$). How much is in the account at the end of 2 years?

2. Suppose $1000 is invested at 12%, compounded quarterly. How much is in the account at the end of 3 years?

3. Suppose $1000 is invested at 16%. How much is in the account at the end of 1 year, if interest is compounded (a) annually? (b) semiannually? (c) quarterly? (d) daily? (e) hourly?

4. Suppose $1 is invested at the interest rate of 100%, even though it would be hard to get such a rate. How much is in the account at the end of 1 year, if interest is compounded (a) annually? (b) semiannually? (c) quarterly? (d) daily? (e) hourly?

Solve.

1. $(x + 2)^2 = 25$ **2.** $(x - 2)^2 = 49$ **3.** $(x + 1)^2 = 6$

4. $(x + 3)^2 = 21$ **5.** $(x - 3)^2 = 6$ **6.** $(x + 13)^2 = 8$

Complete the square.

7. $x^2 - 2x$ **8.** $x^2 - 4x$ **9.** $x^2 + 18x$ **10.** $x^2 + 22x$

11. $x^2 - x$ **12.** $x^2 + x$ **13.** $x^2 + 5x$ **14.** $x^2 - 9x$

Solve. Use $A = P(1 + r)^t$. What is the interest rate?

15. \$1000 grows to \$1210 in 2 years.   **16.** \$1000 grows to \$1440 in 2 years.

ANSWERS
1. _____
2. _____
3. _____
4. _____
5. _____
6. _____
7. _____
8. _____
9. _____
10. _____
11. _____
12. _____
13. _____
14. _____
15. _____
16. _____

17. $2560 grows to $3610 in 2 years.

18. $4000 grows to $4410 in 2 years.

19. $6250 grows to $7290 in 2 years.

20. $6250 grows to $6760 in 2 years.

21. $2500 grows to $3600 in 2 years.

22. $1600 grows to $2500 in 2 years.

✓ SKILL MAINTENANCE

23. Solve: $2x + 5y = 3$
$3x + 2y = 10$

24. Multiply and simplify: $\sqrt{2}\,\sqrt{14}$.

25. Jack can paint a shed alone in 5 hr. Jill can paint the same shed alone in 10 hr. How long would it take both of them working together to paint the shed?

26. Rationalize the denominator: $\sqrt{\dfrac{7}{3}}$.

★ EXTENSION

Factor the left side. Then solve.

27. $x^2 + 2x + 1 = 81$

28. $y^2 - 16y + 64 = 15$

29. 🖩 $1000 is invested at interest rate r, compounded annually. In 2 years it grows to $1267.88. What is the interest rate?

30. 🖩 In two years you want $3000. How much do you need to invest now if you can get an interest rate of 15.75%, compounded annually?

Solving by Completing the Square

• We have seen that a quadratic equation $(x + k)^2 = d$ can be solved by taking the square roots. An equation such as $x^2 + 6x + 8 = 0$ can be put in this form by completing the square. Then we can solve as before.

Example 1 Solve: $x^2 + 6x + 8 = 0$.

$$x^2 + 6x + 8 = 0$$
$$x^2 + 6x \quad = -8 \qquad \text{Adding } -8$$

We take half of 6 and square it to get 9. Then we add 9 on *both* sides of the equation. This makes the left side the square of a binomial. We have *completed the square*.

$$x^2 + 6x + 9 = -8 + 9$$
$$(x + 3)^2 = 1$$
$$x + 3 = 1 \quad \text{or} \quad x + 3 = -1$$
$$x = -2 \quad \text{or} \quad x = -4$$

The solutions are -2 and -4.

This method of solving is called *completing the square*.

Example 2 Solve $x^2 - 4x - 7 = 0$ by completing the square.

$$x^2 - 4x - 7 = 0$$
$$x^2 - 4x \quad = 7 \qquad \text{Adding } 7$$
$$x^2 - 4x + 4 = 7 + 4 \qquad \text{Adding 4: } \left(\frac{-4}{2}\right)^2 = (-2)^2 = 4$$
$$(x - 2)^2 = 11$$
$$x - 2 = \sqrt{11} \quad \text{or} \quad x - 2 = -\sqrt{11}$$
$$x = 2 + \sqrt{11} \quad \text{or} \quad x = 2 - \sqrt{11}$$

The solutions are $2 \pm \sqrt{11}$.

DO EXERCISES 1–3.

Example 3 Solve $x^2 + 3x - 10 = 0$ by completing the square.

$$x^2 + 3x - 10 = 0$$
$$x^2 + 3x \quad = 10$$
$$x^2 + 3x + \frac{9}{4} = 10 + \frac{9}{4} \qquad \text{Adding } \frac{9}{4} : \left(\frac{3}{2}\right)^2 = \frac{9}{4}$$
$$\left(x + \frac{3}{2}\right)^2 = \frac{40}{4} + \frac{9}{4}$$
$$\left(x + \frac{3}{2}\right)^2 = \frac{49}{4}$$
$$x + \frac{3}{2} = \sqrt{\frac{49}{4}} \quad \text{or} \quad x + \frac{3}{2} = -\sqrt{\frac{49}{4}}$$

Solve by completing the square.

1. $x^2 + 8x + 12 = 0$

2. $x^2 - 10x + 22 = 0$

3. $x^2 + 6x - 1 = 0$

ANSWERS ON PAGE A-24

Solve by completing the square.

4. $x^2 - 3x - 10 = 0$

5. $x^2 + 5x - 14 = 0$

Solve by completing the square.

6. $2x^2 + 3x - 3 = 0$

7. $3x^2 - 2x - 3 = 0$

We then have

$$x + \frac{3}{2} = \frac{7}{2} \quad \text{or} \quad x + \frac{3}{2} = -\frac{7}{2}$$

$$x = \frac{4}{2} \quad \text{or} \quad x = -\frac{10}{2}$$

$$x = 2 \quad \text{or} \quad x = -5.$$

The solutions are 2 and -5.

DO EXERCISES 4 AND 5.

When the coefficient of x^2 is not 1, we can make it 1.

Example 4 Solve $2x^2 - 3x - 1 = 0$ by completing the square.

$$2x^2 - 3x - 1 = 0$$

$$\frac{1}{2}(2x^2 - 3x - 1) = \frac{1}{2} \cdot 0 \qquad \text{Multiplying on both sides by } \frac{1}{2}$$
$$\text{to make the } x^2\text{-coefficient 1}$$

$$x^2 - \frac{3}{2}x - \frac{1}{2} = 0$$

$$x^2 - \frac{3}{2}x = \frac{1}{2}$$

$$x^2 - \frac{3}{2}x + \frac{9}{16} = \frac{1}{2} + \frac{9}{16} \qquad \text{Adding } \frac{9}{16}\colon \left[\frac{1}{2}\left(-\frac{3}{2}\right)\right]^2 = \left[-\frac{3}{4}\right]^2 = \frac{9}{16}$$

$$\left(x - \frac{3}{4}\right)^2 = \frac{8}{16} + \frac{9}{16}$$

$$\left(x - \frac{3}{4}\right)^2 = \frac{17}{16}$$

$$x - \frac{3}{4} = \sqrt{\frac{17}{16}} \quad \text{or} \quad x - \frac{3}{4} = -\sqrt{\frac{17}{16}}$$

$$x - \frac{3}{4} = \frac{\sqrt{17}}{4} \quad \text{or} \quad x - \frac{3}{4} = -\frac{\sqrt{17}}{4}$$

$$x = \frac{3}{4} + \frac{\sqrt{17}}{4} \quad \text{or} \quad x = \frac{3}{4} - \frac{\sqrt{17}}{4}$$

The solutions are $\dfrac{3 \pm \sqrt{17}}{4}$.

DO EXERCISES 6 AND 7.

There are at least two reasons for learning to complete the square. One is to enhance your ability to graph certain second-degree equations, which you will encounter later in mathematics. The other is to prove a general formula that can be used to solve quadratic equations. We will prove this formula in the next section.

Solve by completing the square. Show your work.

1. $x^2 - 6x - 16 = 0$ **2.** $x^2 + 8x + 15 = 0$ **3.** $x^2 + 22x + 21 = 0$

4. $x^2 + 14x - 15 = 0$ **5.** $x^2 - 2x - 5 = 0$ **6.** $x^2 - 4x - 11 = 0$

7. $x^2 - 18x + 74 = 0$ **8.** $x^2 - 22x + 102 = 0$ **9.** $x^2 + 7x - 18 = 0$

10. $x^2 + 5x - 6 = 0$ **11.** $x^2 + x - 6 = 0$ **12.** $x^2 + 10x - 4 = 0$

ANSWERS
1.
2.
3.
4.
5.
6.
7.
8.
9.
10.
11.
12.

A N S W E R S
13.
14.
15.
16.
17.
18.
19.
20.
21.
22.
23.
24.
25.
26.
27.
28.
29.

13. $x^2 - 7x - 2 = 0$

14. $x^2 + 3x - 28 = 0$

15. $x^2 + \dfrac{3}{2}x - \dfrac{1}{2} = 0$

16. $2x^2 + 3x - 17 = 0$

17. $3x^2 + 4x - 1 = 0$

18. $2x^2 - 9x - 5 = 0$

19. $4x^2 + 12x - 7 = 0$

20. $9x^2 - 6x - 9 = 0$

21. $6x^2 + 11x - 10 = 0$

 SKILL MAINTENANCE

22. A tank can be filled in 6 hr by pipe A alone and in 9 hr by pipe B alone. How long would it take to fill an empty tank if both pipes were working?

23. Subtract: $\sqrt{54} - \sqrt{24}$.

24. Simplify: $\sqrt{80}$.

25. Solve: $7x - 2y = -31$
$4x - 3y = -27$.

⭐ **EXTENSION**

Find q such that each trinomial is a square.

26. $x^2 + qx + 36$

27. $x^2 + qx + 55$

28. $qx^2 + 16x + 16$

29. Solve for x by completing the square: $4x^2 + 4x + c = 0$.

The Quadratic Formula

● SOLVING USING THE QUADRATIC FORMULA

Each time you solve by completing the square, you do about the same thing. In situations like this in mathematics, when we do about the same kind of computation many times, we look for a formula so we can speed up our work. Consider any quadratic equation in standard form:

$$ax^2 + bx + c = 0, \qquad a > 0.$$

Let's solve by completing the square.

$$x^2 + \frac{b}{a}x + \frac{c}{a} = 0 \qquad \text{Multiplying by } \frac{1}{a}$$

$$x^2 + \frac{b}{a}x \quad = -\frac{c}{a} \qquad \text{Adding } -\frac{c}{a}$$

Half of $\dfrac{b}{a}$ is $\dfrac{b}{2a}$. The square is $\dfrac{b^2}{4a^2}$. We add $\dfrac{b^2}{4a^2}$ on both sides.

$$x^2 + \frac{b}{a}x + \frac{b^2}{4a^2} = -\frac{c}{a} + \frac{b^2}{4a^2}$$

$$\left(x + \frac{b}{2a}\right)^2 = -\frac{4ac}{4a^2} + \frac{b^2}{4a^2}$$

$$\left(x + \frac{b}{2a}\right)^2 = \frac{b^2 - 4ac}{4a^2}$$

$$x + \frac{b}{2a} = \sqrt{\frac{b^2-4ac}{4a^2}} \quad \text{or} \quad x + \frac{b}{2a} = -\sqrt{\frac{b^2-4ac}{4a^2}}$$

Since $a > 0$, $\sqrt{4a^2} = 2a$, so

$$x + \frac{b}{2a} = \frac{\sqrt{b^2 - 4ac}}{2a} \quad \text{or} \quad x + \frac{b}{2a} = -\frac{\sqrt{b^2 - 4ac}}{2a}.$$

Thus,

$$x + \frac{b}{2a} = \pm\frac{\sqrt{b^2 - 4ac}}{2a},$$

so

$$x = -\frac{b}{2a} + \frac{\sqrt{b^2 - 4ac}}{2a} \quad \text{or} \quad x = -\frac{b}{2a} - \frac{\sqrt{b^2 - 4ac}}{2a}.$$

The solutions are given by the following.

THE QUADRATIC FORMULA

$$x = \frac{-b \pm \sqrt{b^2 - 4ac}}{2a}$$

Example 1 Solve $5x^2 - 8x = -3$ using the quadratic formula.

First find standard form and determine a, b, and c.

$$5x^2 - 8x + 3 = 0$$
$$a = 5,\ b = -8,\ c = 3$$

Then use the quadratic formula:

$$x = \frac{-b \pm \sqrt{b^2 - 4ac}}{2a}$$

$$x = \frac{-(-8) \pm \sqrt{(-8)^2 - 4 \cdot 5 \cdot 3}}{2 \cdot 5}$$

$$x = \frac{8 \pm \sqrt{64 - 60}}{10}$$

$$x = \frac{8 \pm \sqrt{4}}{10}$$

$$x = \frac{8 \pm 2}{10}$$

$$x = \frac{8 + 2}{10} \quad \text{or} \quad x = \frac{8 - 2}{10}$$

$$x = \frac{10}{10} \quad \text{or} \quad x = \frac{6}{10}$$

$$x = 1 \quad \text{or} \quad x = \frac{3}{5}.$$

The solutions are 1 and $\frac{3}{5}$.

It turns out that the equation in Example 1 could have been solved by factoring. Actually, factoring would have been easier.

To solve a quadratic equation:

1. **Try factoring.**
2. **If it is not possible to factor or if factoring seems difficult, use the quadratic formula.**

The solutions of a quadratic equation can always be found using the quadratic formula. They cannot always be found by factoring.

DO EXERCISE 1.

When $b^2 - 4ac \geq 0$, the equation has real-number solutions. When $b^2 - 4ac < 0$, the equation has no real-number solutions. The expression $b^2 - 4ac$ is called the *discriminant*.

When using the quadratic formula, it is wise to compute the discriminant first. If it is negative, there are no real-number solutions and no further computation is necessary.

ANSWER ON PAGE A-24

Example 2 Solve $3x^2 = 7 - 2x$ using the quadratic formula.

Find standard form and determine a, b, and c.

$$3x^2 + 2x - 7 = 0$$
$$a = 3, \, b = 2, \, c = -7$$

We compute the discriminant:

$$b^2 - 4ac = 2^2 - 4 \cdot 3 \cdot (-7)$$
$$= 4 + 84$$
$$= 88.$$

This is positive, so there are real-number solutions. They are given by

$$x = \frac{-2 \pm \sqrt{88}}{6} \quad \text{Substituting into the quadratic formula}$$

$$x = \frac{-2 \pm \sqrt{4 \cdot 22}}{6}$$

$$x = \frac{-2 \pm 2\sqrt{22}}{6} \quad \text{Simplifying the radical}$$

$$x = \frac{2(-1 \pm \sqrt{22})}{2 \cdot 3} \quad \text{Factoring out 2 in the numerator and denominator}$$

$$x = \frac{-1 \pm \sqrt{22}}{3}.$$

The solutions are $\dfrac{-1 + \sqrt{22}}{3}$ and $\dfrac{-1 - \sqrt{22}}{3}$.

DO EXERCISE 2.

• • APPROXIMATING SOLUTIONS

A calculator or Table 2 can be used to approximate solutions.

Example 3 Use a calculator or Table 2 to approximate the solutions to the equation in Example 2 to the nearest tenth.

Using a calculator or Table 2, we see that $\sqrt{22} \approx 4.690$:

$$\frac{-1 + \sqrt{22}}{3} \approx \frac{-1 + 4.690}{3} \qquad \frac{-1 - \sqrt{22}}{3} \approx \frac{-1 - 4.690}{3}$$

$$\approx \frac{3.69}{3} \qquad\qquad\qquad \approx \frac{-5.69}{3}$$

$$\approx 1.2 \text{ to the} \qquad\qquad \approx -1.9 \text{ to the}$$
$$\text{nearest tenth;} \qquad\qquad \text{nearest tenth.}$$

DO EXERCISE 3.

Solve using the quadratic formula.

2. $5x^2 - 8x = 3$

3. Approximate the solutions to the equation in Exercise 2 above. Round to the nearest tenth.

ANSWERS ON PAGE A-24

Perform the following changes of unit using substitution or multiplying by 1.

1. 72 in., change to ft

2. 17 hr, change to min

3. 2 days, change to sec

4. 360 sec, change to hr

5. $60 \dfrac{kg}{m}$, change to $\dfrac{g}{cm}$

6. $44 \dfrac{ft}{sec}$, change to $\dfrac{mi}{hr}$

7. $216 \ m^2$, change to cm^2

8. $60 \dfrac{lb}{ft^3}$, change to $\dfrac{ton}{yd^3}$

9. $\dfrac{\$36}{day}$, change to $\dfrac{\cent}{hr}$

10. 1440 person-hr, change to person-days

11. $186{,}000 \dfrac{mi}{sec}$ (speed of light), change to $\dfrac{mi}{yr}$. Let 365 days = 1 yr.

12. $1100 \dfrac{ft}{sec}$ (speed of sound), change to $\dfrac{mi}{yr}$. Let 365 days = 1 yr.

S O M E T H I N G   E X T R A

**AN APPLICATION:
HANDLING DIMENSION SYMBOLS (PART 3)**

Changes of unit can be achieved by substitutions.

Example 1 Change to inches: 25 yd.

$$25 \text{ yd} = 25 \cdot 1 \text{ yd}$$
$$= 25 \cdot 3 \text{ ft} \qquad \text{Substituting 3 ft for 1 yd}$$
$$= 25 \cdot 3 \cdot 1 \text{ ft}$$
$$= 25 \cdot 3 \cdot 12 \text{ in.} \qquad \text{Substituting 12 in. for 1 ft}$$
$$= 900 \text{ in.}$$

Example 2 Change to meters: 4 km.

$$4 \text{ km} = 4 \cdot 1 \text{ km}$$
$$= 4 \cdot 1000 \text{ m} \qquad \text{Substituting 1000 m for 1 km}$$
$$= 4000 \text{ m}$$

The notion of "multiplying by 1" can also be used to change units.

Example 3 Change to yd: 7.2 in.

$$7.2 \text{ in.} = 7.2 \text{ in.} \cdot \frac{1 \text{ ft}}{12 \text{ in}} \cdot \frac{1 \text{ yd}}{3 \text{ ft}} \qquad \text{Both of these are equal to 1.}$$
$$= \frac{7.2}{12 \cdot 3} \cdot \frac{\text{in.}}{\text{in.}} \cdot \frac{\text{ft}}{\text{ft}} \cdot \text{yd} = 0.2 \text{ yd}$$

Example 4 Change to mm: 55 cm.

$$55 \text{ cm} = 55 \text{ cm} \cdot \frac{1 \text{ m}}{100 \text{ cm}} \cdot \frac{1000 \text{ mm}}{1 \text{ m}}$$
$$= \frac{55 \cdot 1000}{100} \cdot \frac{\text{cm}}{\text{cm}} \cdot \frac{\text{m}}{\text{m}} \cdot \text{mm}$$
$$= 550 \text{ mm}$$

Example 5 Change to $\dfrac{m}{sec}$: $95 \dfrac{km}{hr}$.

$$95 \frac{km}{hr} = 95 \frac{km}{hr} \cdot \frac{1000 \text{ m}}{1 \text{ km}} \cdot \frac{1 \text{ hr}}{60 \text{ min}} \cdot \frac{1 \text{ min}}{60 \text{ sec}}$$
$$= \frac{95 \cdot 1000}{60 \cdot 60} \cdot \frac{km}{km} \cdot \frac{hr}{hr} \cdot \frac{min}{min} \cdot \frac{m}{sec} = 26.4 \frac{m}{sec}$$

Below is a shortcut for the procedure in Example 5. It can also be used in Examples 3 and 4.

$$95 \frac{km}{hr} = 95 \frac{\overset{19}{\cancel{km}}}{\cancel{hr}} \cdot \frac{1000 \text{ m}}{1 \cancel{km}} \cdot \frac{1 \cancel{hr}}{\underset{12}{\cancel{60 \text{ min}}}} \cdot \frac{1 \cancel{min}}{60 \text{ sec}} = 26.4 \frac{m}{sec}$$

• Solve. Try factoring first. If factoring is not possible or is difficult, use the quadratic formula.

1. $x^2 - 4x - 21$

2. $x^2 + 7x = 18$

3. $x^2 = 6x - 9$

4. $x^2 = 8x - 16$

5. $3y^2 - 2y - 8 = 0$

6. $4y^2 + 12y = 7$

7. $x^2 - 9 = 0$

8. $x^2 - 4 = 0$

9. $y^2 - 10y + 26 = 4$

10. $x^2 + 4x + 4 = 7$

11. $x^2 - 2x = 2$

12. $x^2 + 6x = 1$

13. $4y^2 + 3y + 2 = 0$

14. $2t^2 + 6t + 5 = 0$

15. $3p^2 + 2p = 3$

ANSWERS
1.
2.
3.
4.
5.
6.
7.
8.
9.
10.
11.
12.
13.
14.
15.

16. $3z^2 - 2z = 2$ **17.** $(y + 4)(y + 3) = 15$ **18.** $x^2 + (x + 2)^2 = 7$

16. _____

17. _____

18. _____

19. $5x + x(x - 7) = 0$ **20.** $(x + 3)^2 + (x + 1)^2 = 0$

19. _____

20. _____

21. _____

 Solve using the quadratic formula. Use a calculator or Table 2 to approximate the solutions to the nearest tenth.

21. $x^2 - 4x - 7 = 0$ **22.** $x^2 = 5$ **23.** $y^2 - 6y - 1 = 0$

22. _____

23. _____

24. _____

24. $4x^2 + 4x = 1$ **25.** $3x^2 + 4x - 2 = 0$ **26.** $2y^2 + 2y - 3 = 0$

25. _____

26. _____

27. _____

✓ SKILL MAINTENANCE

28. _____

27. Multiply and simplify: $\sqrt{3x^2}\ \sqrt{9x^3}$.   **28.** Solve: $2x + 5y = -4$
$3x - 5y = 19.$

29. _____

29. Add: $\sqrt{3} + \sqrt{\dfrac{1}{3}}$.

30. _____

30. Crew A can paint a house in 10 hr. Crew B can paint the same house in 12 hr. How many hours would it take to paint the house if both crews worked together?

31. _____

☆ EXTENSION

32. _____

Solve for x.

31. $0.8x^2 + 0.16x - 0.09 = 0$ **32.** $\frac{1}{2}x^2 + bx + (b - \frac{1}{2}) = 0$

33. _____

33. Determine as an inequality the possible values of a for which $ax^2 + 2x = 3$ will have real-number solutions.   **34.** Determine as an inequality the possible values of c for which $-5x^2 + 4x + c = 0$ will have real-number solutions.

34. _____

Fractional and Radical Equations

Suppose we are solving fractional or radical equations. After a few steps we may obtain a quadratic equation. When that happens we can finish using the methods we have learned in this chapter.

● Solve certain fractional equations by first deriving a quadratic equation.

●● Solve certain radical equations by first using the principle of squaring to derive a quadratic equation.

● FRACTIONAL EQUATIONS

Recall that to solve a fractional equation we multiply on both sides by the LCM of all the denominators.

Example 1 Solve: $\dfrac{3}{x-1} + \dfrac{5}{x+1} = 2$.

The LCM is $(x-1)(x+1)$. We multiply by this.

$$(x-1)(x+1) \cdot \left(\frac{3}{x-1} + \frac{5}{x+1}\right) = 2 \cdot (x-1)(x+1)$$

We use the distributive law on the left.

$$(x-1)(x+1) \cdot \frac{3}{x-1} + (x-1)(x+1) \cdot \frac{5}{x+1} = 2(x-1)(x+1)$$

$$3(x+1) + 5(x-1) = 2(x-1)(x+1)$$

$$3x + 3 + 5x - 5 = 2(x^2 - 1)$$

$$8x - 2 = 2x^2 - 2$$

$$-2x^2 + 8x = 0$$

$$2x^2 - 8x = 0 \quad \text{Multiplying by } -1$$

$$2x(x-4) = 0 \quad \text{Factoring}$$

$$2x = 0 \quad \text{or} \quad x - 4 = 0$$

$$x = 0 \quad \text{or} \quad x = 4$$

Check:　For 0:

$$\frac{3}{x-1} + \frac{5}{x+1} = 2$$

$$\frac{3}{0-1} + \frac{5}{0+1} \;\Big|\; 2$$

$$\frac{3}{-1} + \frac{5}{1}$$

$$-3 + 5$$

$$2$$

For 4:

$$\frac{3}{x-1} + \frac{5}{x+1} = 2$$

$$\frac{3}{4-1} + \frac{5}{4+1} \;\Big|\; 2$$

$$\frac{3}{3} + \frac{5}{5}$$

$$1 + 1$$

$$2$$

Both numbers check. The solutions are 0 and 4.

DO EXERCISE 1.

Solve.

1. $\dfrac{20}{x+5} - \dfrac{1}{x-4} = 1$

ANSWER ON PAGE A-25

Solve.

2. $\sqrt{x + 2} = 4 - x$

● ● RADICAL EQUATIONS

We can solve some radical equations by first using the principle of squaring to find a quadratic equation. When we do this we must be sure to check.

Example 2 Solve: $x - 5 = \sqrt{x + 7}$.

$$x - 5 = \sqrt{x - 7}$$
$$(x - 5)^2 = (\sqrt{x + 7})^2 \qquad \text{Principle of squaring}$$
$$x^2 - 10x + 25 = x + 7$$
$$x^2 - 11x + 18 = 0$$
$$(x - 9)(x - 2) = 0$$
$$x - 9 = 0 \quad \text{or} \quad x - 2 = 0$$
$$x = 9 \quad \text{or} \quad x = 2$$

Check: For 9:

$$\begin{array}{c|c} x - 5 = \sqrt{x + 7} \\ \hline 9 - 5 & \sqrt{9 + 7} \\ 4 & 4 \end{array}$$

For 2:

$$\begin{array}{c|c} x - 5 = \sqrt{x + 7} \\ \hline 2 - 5 & \sqrt{2 + 7} \\ -3 & 3 \end{array}$$

The number 9 checks, but 2 does not. Thus the solution is 9.

DO EXERCISE 2.

Solve.

3. $\sqrt{30 - 3x} + 4 = x$

Example 3 Solve: $\sqrt{27 - 3x} + 3 = x$.

$$\sqrt{27 - 3x} + 3 = x$$
$$\sqrt{27 - 3x} = x - 3 \qquad \text{Adding } -3 \text{ to get the radical alone on one side}$$
$$(\sqrt{27 - 3x})^2 = (x - 3)^2 \qquad \text{Principle of squaring}$$
$$27 - 3x = x^2 - 6x + 9$$
$$0 = x^2 - 3x - 18 \qquad \text{We can have 0 on the left.}$$
$$0 = (x - 6)(x + 3) \qquad \text{Factoring}$$
$$x - 6 = 0 \quad \text{or} \quad x + 3 = 0$$
$$x = 6 \quad \text{or} \quad x = -3$$

Check: For 6:

$$\begin{array}{c|c} \sqrt{27 - 3x} + 3 = x \\ \hline \sqrt{27 - 3 \cdot 6} + 3 & 6 \\ \sqrt{9} + 3 & \\ 3 + 3 & \\ 6 & \end{array}$$

For −3:

$$\begin{array}{c|c} \sqrt{27 - 3x} + 3 = x \\ \hline \sqrt{27 - 3 \cdot (-3)} + 3 & -3 \\ \sqrt{27 + 9} + 3 & \\ \sqrt{36} + 3 & \\ 6 + 3 & \\ 9 & \end{array}$$

There is only one solution, 6.

DO EXERCISE 3.

ANSWERS ON PAGE A-25

Chapter 11   Quadratic Equations

• Solve.

1. $\dfrac{8}{x+2} + \dfrac{8}{x-2} = 3$

2. $\dfrac{24}{x-2} + \dfrac{24}{x+2} = 5$

3. $\dfrac{1}{x} + \dfrac{1}{x+6} = \dfrac{1}{4}$

4. $\dfrac{1}{x} + \dfrac{1}{x+9} = \dfrac{1}{20}$

5. $1 + \dfrac{12}{x^2-4} = \dfrac{3}{x-2}$

6. $\dfrac{5}{t-3} - \dfrac{30}{t^2-9} = 1$

7. $\dfrac{r}{r-1} + \dfrac{2}{r^2-1} = \dfrac{8}{r+1}$

8. $\dfrac{x+2}{x^2-2} = \dfrac{2}{3-x}$

9. $\dfrac{4-x}{x-4} + \dfrac{x+3}{x-3} = 0$

10. $\dfrac{y+2}{y} = \dfrac{1}{y+2}$

11. $\dfrac{x^2}{x-4} - \dfrac{7}{x-4} = 0$

12. $\dfrac{x^2}{x+3} - \dfrac{5}{x+3} = 0$

13. $x + 2 = \dfrac{3}{x+2}$

14. $x - 3 = \dfrac{5}{x-3}$

15. $\dfrac{1}{x} + \dfrac{1}{x+6} = \dfrac{1}{5}$

16. $\dfrac{1}{x} + \dfrac{1}{x+1} = \dfrac{1}{3}$

1. _____

2. _____

3. _____

4. _____

5. _____

6. _____

7. _____

8. _____

9. _____

10. _____

11. _____

12. _____

13. _____

14. _____

15. _____

16. _____

●● Solve.

17. $x - 7 = \sqrt{x - 5}$

18. $\sqrt{x + 7} = x - 5$

19. $\sqrt{x + 18} = x - 2$

20. $x - 9 = \sqrt{x - 3}$

21. $2\sqrt{x - 1} = x - 1$

22. $x + 4 = 4\sqrt{x + 1}$

23. $\sqrt{5x + 21} = x + 3$

24. $\sqrt{27 - 3x} = x - 3$

25. $x = 1 + 6\sqrt{x - 9}$

26. $\sqrt{2x - 1} + 2 = x$

27. $\sqrt{x^2 + 6} - x + 3 = 0$

28. $\sqrt{x^2 + 5} - x + 2 = 0$

29. $\sqrt{(p + 6)(p + 1)} - 2 = p + 1$

30. $\sqrt{(4x + 5)(x + 4)} = 2x + 5$

★ **EXTENSION**

Solve.

31. $\dfrac{7}{1 + x} - 1 = \dfrac{5x}{x^2 + 3x + 2}$

32. $\dfrac{x}{x + 1} = 4 + \dfrac{1}{3x^2 - 3}$

33. $x + 1 + 3\sqrt{x + 1} = 4$

34. $\dfrac{12}{\sqrt{5x + 6}} = \sqrt{2x + 5}$

Copyright © 1987 Addison-Wesley Publishing Co., Inc.

Chapter 11   Quadratic Equations

Formulas

• To solve a formula for a given letter, we try to get the letter alone on one side.

Example 1 Solve for h: $V = 3.5 \sqrt{h}$ (the distance to the horizon).

$$V^2 = (3.5 \sqrt{h})^2$$
$$V^2 = (3.5)^2 (\sqrt{h})^2 \qquad \text{Squaring both sides}$$
$$V^2 = 12.25h$$
$$\frac{V^2}{12.25} = h \qquad \text{Multiplying by } \frac{1}{12.25} \text{ to get } h \text{ alone}$$

DO EXERCISE 1.

Example 2 Solve for g: $T = 2\pi \sqrt{\dfrac{L}{g}}$ (the period of a pendulum).

$$T^2 = (2\pi)^2 \left(\sqrt{\frac{L}{g}} \right)^2 \qquad \text{Squaring both sides}$$

$$T^2 = 4\pi^2 \frac{L}{g}$$

$$T^2 = \frac{4\pi^2 L}{g}$$

$$gT^2 = 4\pi^2 L \qquad \text{Multiplying by } g \text{ to clear of fractions}$$

$$g = \frac{4\pi^2 L}{T^2} \qquad \text{Multiplying by } \frac{1}{T^2} \text{ to get } g \text{ alone}$$

DO EXERCISES 2 AND 3.

In most formulas the letters represent nonnegative numbers, so you don't need to use absolute values when taking square roots.

Example 3 (*Torricelli's theorem*). In hydrodynamics the speed v of a liquid leaving a tank from an orifice is related to the height h of the top of the water above the orifice by the formula

$$h = \frac{v^2}{2g}.$$

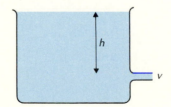

Solve for v:

$$2gh = v^2 \qquad \text{Multiplying by } 2g \text{ to clear of fractions}$$
$$\sqrt{2gh} = v. \qquad \text{Taking the square root. We assume that } v \text{ is nonnegative.}$$

DO EXERCISE 4.

1. Solve for L: $r = 2\sqrt{5L}$.
   (A formula for the speed of a skidding car)

2. Solve for L: $T = 2\pi \sqrt{\dfrac{L}{g}}$.

3. Solve for m: $c = \sqrt{\dfrac{E}{m}}$.

4. Solve for r: $A = \pi r^2$.
   (The area of a circle)

ANSWERS ON PAGE A-25

5. Solve for d: $C = P(d-1)^2$.

Example 4 Solve for r: $A = P(1 + r^2)$ (a compound interest formula).

$$A = P(1 + r)^2$$

$$\frac{A}{P} = (1 + r)^2 \qquad \text{Multiplying by } \frac{1}{P}$$

$$\sqrt{\frac{A}{P}} = 1 + r \qquad \begin{array}{l}\text{Taking the square root. We assume}\\ 1 + r \text{ is positive.}\end{array}$$

$$-1 + \sqrt{\frac{A}{P}} = r \qquad \text{Adding } -1 \text{ to get } r \text{ alone}$$

DO EXERCISE 5.

Sometimes you need to use the quadratic formula to solve a formula for a certain letter.

6. Solve for n: $N = n^2 - n$.

Example 5 Solve for n: $d = \dfrac{n^2 - 3n}{2}$ (the number of diagonals of a polygon).

$$d = \frac{n^2 - 3n}{2}$$

$$n^2 - 3n = 2d \qquad \text{Multiplying by 2 to clear of fractions}$$

$$n^2 - 3n - 2d = 0 \qquad \text{Finding standard form}$$

$$a = 1, \, b = -3, \, c = -2d \qquad \begin{array}{l}\text{All letters are considered}\\ \text{constants except } n.\end{array}$$

$$n = \frac{-b \pm \sqrt{b^2 - 4ac}}{2a}$$

$$n = \frac{-(-3) \pm \sqrt{(-3)^2 - 4 \cdot 1 \cdot (-2d)}}{2 \cdot 1} \qquad \begin{array}{l}\text{Substituting into the}\\ \text{quadratic formula}\end{array}$$

$$n = \frac{3 \pm \sqrt{9 + 8d}}{2}$$

DO EXERCISE 6.

7. Solve for t: $h = vt + 8t^2$.

Example 6 Solve for t: $S = gt + 16t^2$.

$$16t^2 + gt - S = 0 \qquad \text{Finding standard form}$$

$$a = 16, \, b = g, \, c = -S$$

$$t = \frac{-b \pm \sqrt{b^2 - 4ac}}{2a}$$

$$t = \frac{-g \pm \sqrt{g^2 - 4 \cdot 16 \cdot (-S)}}{2 \cdot 16} \qquad \begin{array}{l}\text{Substituting into the}\\ \text{quadratic formula}\end{array}$$

$$t = \frac{-g \pm \sqrt{g^2 + 64S}}{32}$$

DO EXERCISE 7.

ANSWERS ON PAGE A-25

Chapter 11   Quadratic Equations

Solve for the indicated letter.

1. $N = 2.5 \sqrt{A}$, for A.

2. $T - 2\pi \sqrt{\dfrac{L}{32}}$, for L.

3. $Q = \sqrt{\dfrac{aT}{c}}$, for T.

4. $v = \sqrt{\dfrac{2gE}{m}}$, for E.

5. $E = mc^2$, for c.

6. $S = 4\pi r^2$, for r.

7. $Q = ad^2 - cd$, for d.

8. $P = kA^2 + mA$, for A.

9. $c^2 = a^2 + b^2$, for a.

10. $c = \sqrt{a^2 + b^2}$, for b.

11. $S = \dfrac{1}{2}gt^2$, for t.

12. $V = \pi r^2 h$, for r.

ANSWERS
1. _____
2. _____
3. _____
4. _____
5. _____
6. _____
7. _____
8. _____
9. _____
10. _____
11. _____
12. _____

13. $A = \pi r^2 + 2\pi rh$, for r.

14. $A = 2\pi r^2 + 2\pi rh$, for r.

15. $A = \dfrac{\pi r^2 S}{360}$, for r.

16. $H = \dfrac{D^2 N}{2.5}$, for D.

17. $c = \sqrt{a^2 + b^2}$, for a.

18. $c^2 = a^2 + b^2$, for b.

19. $h = \dfrac{a}{2}\sqrt{3}$, for a.

(The height of an equilateral triangle with sides of length a)

20. $d = s\sqrt{2}$, for s.

(The hypotenuse of an isosceles right triangle for which s is the length of the two sides that have the same length)

✓ **SKILL MAINTENANCE**

21. Subtract: $\dfrac{x-7}{x^2-9} - \dfrac{x-7}{9-x^2}$.

22. Subtract: $\sqrt{5} - \sqrt{\dfrac{1}{5}}$.

23. Multiply and simplify: $\sqrt{8x^3}\,\sqrt{2x^3 y^4}$.

24. Simplify: $\sqrt{40t^2}$.

⭐ **EXTENSION**

25. Solve $n = aT^2 - 4T + m$ for T.

26. Solve $3ax^2 - x - 3ax + 1 = 0$ for x.

27. Solve $y = ax^2 + bx + c$ for x.

28. The circumference of a circle is given by $C = 2\pi r$ and the area is given by $A = \pi r^2$, where in both cases r is the radius. Express the area A in terms of the circumference without r in the formula.

Applied Problems

- We now use quadratic equations to solve more applied problems.

Example 1 A picture frame measures 20 cm by 14 cm. 160 square centimeters of picture shows. Find the width of the frame.

We first make a drawing. Let x = the width of the frame. To translate, we recall that area is length × width. Thus,

$$A = lw = 160$$
$$(20 - 2x)(14 - 2x) = 160.$$

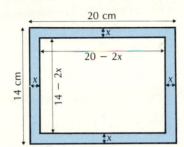

We solve.

$$280 - 68x + 4x^2 = 160$$
$$4x^2 - 68x + 120 = 0$$
$$x^2 - 17x + 30 = 0 \qquad \text{Multiplying by } \tfrac{1}{4}$$
$$(x - 15)(x - 2) = 0 \qquad \text{Factoring}$$
$$x - 15 = 0 \quad \text{or} \quad x - 2 = 0 \qquad \text{Principle of zero products}$$
$$x = 15 \quad \text{or} \quad x = 2$$

We check in the original problem. 15 is not a solution because when $x = 15$, $20 - 2x = -10$, and the length of the picture cannot be negative. When $x = 2$, $20 - 2x = 16$. This is the length. When $x = 2$, $14 - 2x = 10$. This is the width. The area is 16×10, or 160. This checks, so the width of the frame is 2 cm.

DO EXERCISE 1.

1. A rectangular garden is 80 m by 60 m. Part of the garden is torn up to install a strip of lawn around the garden. The new area of the garden is 800 m². How wide is the strip of lawn?

Example 2 The hypotenuse of a right triangle is 6 m long. One leg is 1 m longer than the other. Find the lengths of the legs. Round to the nearest tenth.

We first make a drawing. Let x = the length of one leg. Then $x + 1$ is the length of the other leg. To translate we use the Pythagorean equation.

$$x^2 + (x + 1)^2 = 6^2$$
$$x^2 + x^2 + 2x + 1 = 36$$
$$2x^2 + 2x - 35 = 0$$

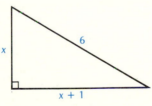

Since we cannot factor, we use the quadratic formula.

$$a = 2, \ b = 2, \ c = -35$$
$$x = \frac{-b \pm \sqrt{b^2 - 4ac}}{2a} = \frac{-2 \pm \sqrt{2^2 - 4 \cdot 2(-35)}}{2 \cdot 2}$$
$$= \frac{-2 \pm \sqrt{4 + 280}}{4} = \frac{-2 \pm \sqrt{284}}{4}$$
$$= \frac{-2 \pm \sqrt{4 \cdot 71}}{4} = \frac{-2 \pm 2 \cdot \sqrt{71}}{2 \cdot 2} = \frac{-1 \pm \sqrt{71}}{2}$$

ANSWER ON PAGE A-25

2. The hypotenuse of a right triangle is 4 cm long. One leg is
1 cm longer than the other.
Find the lengths of the legs.
Round to the nearest tenth.

Using a calculator or Table 2 gives an approximation: $\sqrt{71} \approx 8.426$.

$$\frac{-1 + \sqrt{71}}{2} \approx 3.7, \qquad \frac{-1 - \sqrt{71}}{2} \approx -4.7$$

Since the length of a leg cannot be negative, -4.7 does not check; 3.7
does: $3.7^2 + 4.7^2 = 13.69 + 22.09 = 35.78$ and $\sqrt{35.78} \approx 5.98 \approx 6$.
Thus one leg is about 3.7 m long and the other is about 4.7 m long.

DO EXERCISE 2.

Example 3 The current in a stream moves at a speed of 2 km/h. A boat
travels 24 km upstream and 24 km downstream in a total time of 5 hr.
What is the speed of the boat in still water?

First make a drawing. The distances are the same. Let r represent
the speed of the boat in still water. Then, when traveling upstream the
speed of the boat is $r - 2$. When traveling downstream, the speed of the
boat is $r + 2$. We let t_1 represent the time it takes the boat to go upstream, and t_2 the time it takes to go downstream. We summarize in a
table.

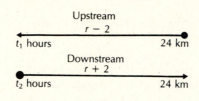

	d	r	t
Upstream	24	$r - 2$	t_1
Downstream	24	$r + 2$	t_2

3. The speed of a boat in still
water is 12 km/h. The boat travels 45 km upstream and 45 km
downstream in a total time of
8 hr. What is the speed of the
stream? (*Hint:* Let $s =$ the speed
of the stream. Then $12 - s$ is
the speed upstream and $12 + s$
is the speed downstream.)

Recall the basic formula for motion: $r = d/t$. From it we can obtain $t = d/r$. Then using the rows of the table, we have $t_1 = 24/(r - 2)$ and $t_2 = 24/(r + 2)$. Since the total time is 5 hours, $t_1 + t_2 = 5$, and we have

$$\frac{24}{r - 2} + \frac{24}{r + 2} = 5.$$

The translation is complete. Now we solve. The LCM $= (r - 2)(r + 2)$.

$$(r - 2)(r + 2) \cdot \left[\frac{24}{r - 2} + \frac{24}{r + 2} \right] = (r - 2)(r + 2) \cdot 5 \qquad \text{Multiplying by the LCM}$$

$$(r - 2)(r + 2) \cdot \frac{24}{r - 2} + (r - 2)(r + 2) \cdot \frac{24}{r + 2} = (r^2 - 4)5$$

$$24(r + 2) + 24(r - 2) = 5r^2 - 20$$

$$24r + 48 + 24r - 48 = 5r^2 - 20$$

$$-5r^2 + 48r + 20 = 0$$

$$5r^2 - 48r - 20 = 0 \qquad \text{Multiplying by } -1$$

$$(5r + 2)(r - 10) = 0 \qquad \text{Factoring}$$

$5r + 2 = 0 \quad$ or $\quad r - 10 = 0 \qquad$ Principle of zero products

$\qquad 5r = -2 \quad$ or $\qquad\quad r = 10$

$\qquad r = -\frac{2}{5} \quad$ or $\qquad\quad r = 10$

Since speed cannot be negative, $-\frac{2}{5}$ cannot be a solution. But 10 checks,
so the speed of the boat in still water is 10 km/h.

ANSWERS ON PAGE A-25

DO EXERCISE 3.

Chapter 11   Quadratic Equations

• Solve.

1. A picture frame measures 20 cm by 12 cm. There are 84 cm^2 of picture showing. Find the width of the frame.

2. A picture frame measures 18 cm by 14 cm. There are 192 cm^2 of picture showing. Find the width of the frame.

3. The hypotenuse of a right triangle is 25 ft long. One leg is 17 ft longer than the other. Find the lengths of the legs.

4. The hypotenuse of a right triangle is 26 yd long. One leg is 14 yd longer than the other. Find the lengths of the legs.

5. The length of a rectangle is 2 cm greater than the width. The area is 80 cm^2. Find the length and width.

6. The length of a rectangle is 3 m greater than the width. The area is 70 m². Find the length and the width.

7. The width of a rectangle is 4 cm less than the length. The area is 320 cm². Find the length and the width.

8. The width of a rectangle is 3 cm less than the length. The area is 340 cm². Find the length and the width.

9. The length of a rectangle is twice the width. The area is 50 m². Find the length and the width.

10. The length of a rectangle is twice the width. The area is 32 cm². Find the length and the width.

6. _____

7. _____

8. _____

9. _____

10. _____

Give approximate answers for Exercises 11–16. Round to the nearest tenth.

11. The hypotenuse of a right triangle is 8 m long. One leg is 2 m longer than the other. Find the lengths of the legs.

12. The hypotenuse of a right triangle is 5 cm long. One leg is 2 cm longer than the other. Find the lengths of the legs.

13. The length of a rectangle is 2 in. greater than the width. The area is 20 in^2. Find the length and the width.

14. The length of a rectangle is 3 ft greater than the width. The area is 15 ft^2. Find the length and the width.

15. The length of a rectangle is twice the width. The area is 10 m^2. Find the length and the width.

ANSWERS

11. _____

12. _____

13. _____

14. _____

15. _____

16. The length of a rectangle is twice the width. The area is 20 cm². Find the length and the width.

17. The current in a stream moves at a speed of 3 km/h. A boat travels 40 km upstream and 40 km downstream in a total time of 14 hr. What is the speed of the boat in still water?

16. _____

17. _____

18. The current in a stream moves at a speed of 3 km/h. A boat travels 45 km upstream and 45 km downstream in a total time of 8 hr. What is the speed of the boat in still water?

19. The current in a stream moves at a speed of 4 mph. A boat travels 4 mi upstream and 12 mi downstream in a total time of 2 hr. What is the speed of the boat in still water?

18. _____

20. The current in a stream moves at a speed of 4 mph. A boat travels 5 mi upstream and 13 mi downstream in a total time of 2 hr. What is the speed of the boat in still water?

19. _____

20. _____

Chapter 11   Quadratic Equations

21. The speed of a boat in still water is 10 km/h. The boat travels 12 km upstream and 28 km downstream in a total time of 4 hr. What is the speed of the stream?

22. The speed of a boat in still water is 8 km/h. The boat travels 60 km upstream and 60 km downstream in a total time of 16 hr. What is the speed of the stream?

23. An airplane flies 738 mi against the wind and 1062 mi with the wind in a total time of 9 hr. The speed of the airplane in still air is 200 mph. What is the speed of the wind?

24. An airplane flies 520 km against the wind and 680 km with the wind in a total time of 4 hr. The speed of the airplane in still air is 300 km/h. What is the speed of the wind?

21. _____

22. _____

23. _____

24. _____

25. Find r in this figure. Round to the nearest hundredth.

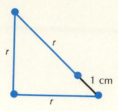

25. _____

26. A 20-ft pole is struck by lightning, and, while not completely broken, falls over and touches the ground 10 ft from the bottom of the pole. How high up the pole did the lightning strike?

26. _____

27. What should the diameter d of a pizza be so that it has the same area as two 10-in. pizzas? Do you get more to eat with a 13-in. pizza or two 10-in. pizzas?

27. _____

28. 🖩 In this figure, the area of the shaded region is 24 cm². Find r if $R = 6$ cm. Round to the nearest hundredth.

28. _____

Graphs of Quadratic Equations

• **GRAPHING QUADRATIC EQUATIONS,**
$y = ax^2 + bx + c$

We now learn to graph quadratic equations

$$y = ax^2 + bx + c, \qquad a \neq 0.$$

The polynomial above is of second degree, or *quadratic*. Examples of the types of equations we are going to graph are

$$y = x^2, \qquad y = x^2 + 2x - 3, \quad \text{and} \quad y = -2x^2 + 3.$$

Graphs of these equations are always cup-shaped. They all have a *line of symmetry* like the dashed line shown in these figures. If you fold on this line, the two halves will match exactly. The curve goes on forever.

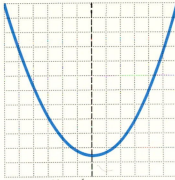

Line of symmetry

These curves are called *parabolas.* Some parabolas are thin and others are flat, but they all have the same general shape.

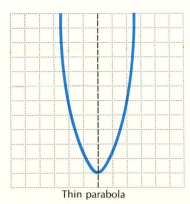

Thin parabola

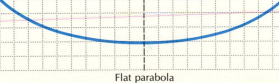

Flat parabola

After finishing Section 11.9, you should be able to:

• Without graphing, tell whether the graph of an equation of the type $y = ax^2 + bx + c$ opens upward or downward. Then graph the equation.

•• Approximate the solutions of $0 = ax^2 + bx + c$ by graphing.

••• Graph quadratic equations using intercepts.

1. a) Without graphing tell whether the graph of

$$y = x^2 - 3$$

opens upward or downward.

To graph a quadratic equation we begin by choosing some numbers for x and computing the corresponding values of y.

Example 1 Graph $y = x^2$.

We plot the ordered pairs resulting from the computations and connect them with a smooth curve.

x	y
-2	4
-1	1
0	0
1	1
2	4

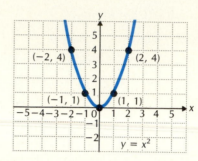

For $x = -2$,
$y = (-2)^2 = 4.$

Example 2 Graph $y = x^2 + 2x - 3$.

x	y
1	0
0	-3
-1	-4
-2	-3
-3	0
-4	5
2	5

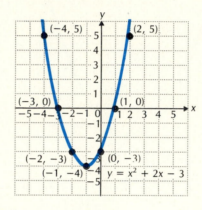

For $x = -1$,
$y = (-1)^2 + 2(-1) - 3$
$= 1 - 2 - 3$
$= -4$

b) Graph the equation.

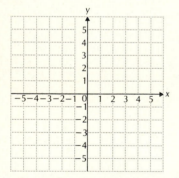

Example 3 Graph $y = -2x^2 + 3$.

x	y
0	3
1	1
-1	1
2	-5
-2	-5

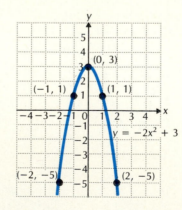

For $x = 2$,
$y = -2(2)^2 + 3$
$= -2 \cdot 4 + 3$
$= -8 + 3$
$= -5$

The graphs in Examples 1 and 2 open upward and the coefficients of x^2 are both 1, which is positive. The graph in Example 3 opens downward and the coefficient of x^2 is -2, which is negative.

ANSWERS ON PAGE A-25

Chapter 11   Quadratic Equations

Graphs of quadratic equations $y = ax^2 + bx + c$ are all parabolas. They are *smooth* cup-shaped symmetric curves, with no sharp points or kinks in them.

The graph of $y = ax^2 + bx + c$ opens upward if $a > 0$. It opens downward if $a < 0$.

In drawing parabolas, be sure to plot enough points to see the general shape of each graph.

If your graphs look like any of the following, they are incorrect.

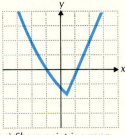

a) Sharp point is wrong.

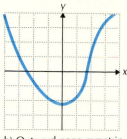

b) Outward nonsymmetric curve is wrong.

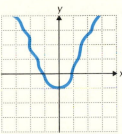

c) Kinks are wrong.

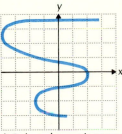

d) S-shaped curve is wrong.

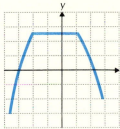

e) Flat nose is wrong.

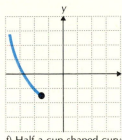

f) Half a cup-shaped curve is wrong.

DO EXERCISES 1–3. (EXERCISE 1 IS ON THE PRECEDING PAGE.)

•• APPROXIMATING SOLUTIONS OF $ax^2 + bx + c = 0$

We can use graphing to approximate the solutions of quadratic equations, $ax^2 + bx + c = 0$. We graph the equation $y = ax^2 + bx + c$. If the graph crosses the x-axis, the points of crossing are the x-intercepts. They will give us solutions. If the graph does *not* cross the x-axis, then there is no real-number solution.

Example 4 Approximate the solutions of

$$-2x^2 + 3 = 0$$

by graphing.

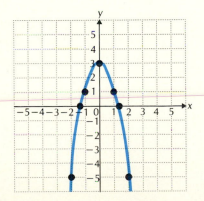

2. a) Without graphing tell whether the graph of

$$y = x^2 + 6x + 9$$

opens upward or downward.

b) Graph the equation.

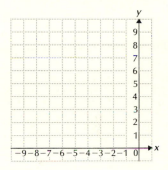

3. a) Without graphing tell whether the graph of

$$y = -3x^2 + 6x$$

opens upward or downward.

b) Graph the equation.

ANSWERS ON PAGE A-25

Approximate the solutions by graphing.

4. $x^2 - 4x + 4 = 0$

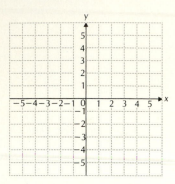

5. $-2x^2 - 4x + 1 = 0$

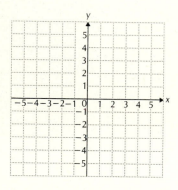

The graph was found in Example 3. It crosses the x-axis at about $(-1.2, 0)$ and $(1.2, 0)$. These are the x-intercepts. So the solutions are about -1.2 and 1.2.

DO EXERCISES 4 AND 5.

••• GRAPHING QUADRATIC EQUATIONS USING INTERCEPTS

The graph of a quadratic equation is symmetric about a line through its *vertex*, or turning point. This means that if we were to fold the graph on this line, the two parts of the graph would coincide.

Note also the location of the x-intercepts. They can be used to find the vertex.

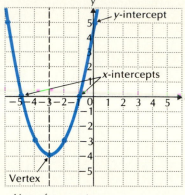

Line of symmetry

We are going to consider an alternative or more refined method for graphing quadratic equations. It works well when the x-intercepts can be found easily by factoring.

Example 5 Graph: $y = x^2 - 4x - 5$.

We start by finding the x- and y-intercepts. The y-intercept occurs when $x = 0$, so we substitute 0 for x and compute y:

$$y = x^2 - 4x - 5 = 0^2 - 4(0) - 5 = -5.$$

The x-intercepts occur when $y = 0$. To find them we substitute 0 for y and solve for x:

$$x^2 - 4x - 5 = 0$$
$$(x - 5)(x + 1) = 0$$
$$x - 5 = 0 \quad \text{or} \quad x + 1 = 0$$
$$x = 5 \quad \text{or} \quad x = -1.$$

We can now begin to make a table of solutions.

x	y	
0	−5	← y-intercept
−1	0	← x-intercepts
5	0	
		← Vertex?

How can we find the vertex? One way is to use the first coordinates of the x-intercepts. We average them to find the first coordinate of the vertex. Then we substitute to find the second coordinate:

$$\frac{-1 + 5}{2} = \frac{4}{2} = 2, \qquad y = x^2 - 4x - 5 = 2^2 - 4(2) - 5 = -9.$$

Chapter 11   Quadratic Equations

The coordinates of the vertex are $(2, -9)$.

We can come very close to completing the graph using the points we have found. To be more certain we can substitute to find one or two more points. Then we draw the graph.

x	y
0	-5
-1	0
5	0
2	-9
4	-5
-2	7

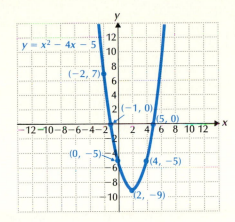

Below is the method we are using for graphing quadratic equations when it is easy to find the x-intercepts by factoring, that is, when it is not necessary to use the quadratic formula.

To graph a quadratic equation when the quadratic formula is not necessary to find the x-intercepts:

1. **Find the y-intercept by substituting 0 for x and computing y.**
2. **Find the x-intercepts by substituting 0 for y and solving for x.**
3. **Find the vertex by averaging the first coordinates of the x-intercepts and substituting to find the second coordinate.**
4. **Find one or two more solutions.**
5. **Plot the points and draw the graph.**

DO EXERCISE 6.

Example 6 Graph: $y = -x^2 - x + 6$.

This graph opens downward, since $-x^2 = -1 \cdot x^2$ and -1 is negative.

1. Find the y-intercept by substituting 0 for x and computing y:

$$y = -x^2 - x + 6 = -(0)^2 - 0 + 6 = 6.$$

This gives us $(0, 6)$ as a point on the graph.

2. Find the x-intercepts by substituting 0 for y and solving for x. After we have made the substitution, it eases factoring if we multiply on both sides by -1 to get rid of the minus sign for the x^2-term:

$$0 = -x^2 - x + 6 \qquad \text{Substituting 0 for } y$$
$$0 = x^2 + x - 6 \qquad \text{Multiplying on both sides by } -1$$
$$0 = (x + 3)(x - 2) \qquad \text{Factoring}$$
$$0 = x + 3 \quad \text{or} \quad 0 = x - 2$$
$$-3 = x \qquad \text{or} \quad 2 = x.$$

This gives us $(-3, 0)$ and $(2, 0)$ as points on the graph.

6. Consider $y = x^2 + 6x + 8$.

   a) Complete the following table.

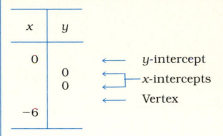

x	y	
0		← y-intercept
	0	← x-intercepts
	0	←
-6		← Vertex

   b) Graph the equation.

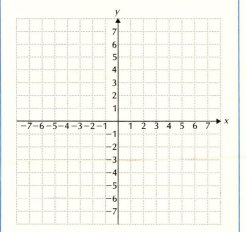

ANSWERS ON PAGE A-25

7. Graph $y = -x^2 + 3x + 4$.

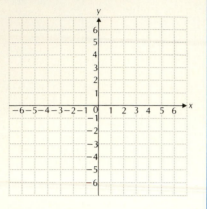

3. Find the vertex. We find the average of -3 and 2:

$$\frac{-3 + 2}{2} = \frac{-1}{2} = -\frac{1}{2}.$$

We substitute $-\frac{1}{2}$ for x and compute y:

$$y = -x^2 - x + 6 = -1(-\tfrac{1}{2})^2 - (-\tfrac{1}{2}) + 6 = -1(\tfrac{1}{4}) + \tfrac{1}{2} + 6 = \tfrac{25}{4}.$$

This gives us $(-\frac{1}{2}, \frac{25}{4})$ as the vertex, which is another point on the graph.

4. Find one or two more solutions.

5. Plot the points and draw the graph.

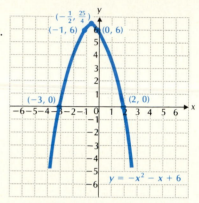

x	y	
0	6	← y-intercept
-3	0	←
2	0	← ┐ x-intercepts
$-\frac{1}{2}$	$\frac{25}{4}$	← Vertex
-1	6	

DO EXERCISE 7.

When the trinomial is a square, there is only one x-intercept and that intercept is also the vertex. In such a case, we can find other solutions, perhaps two, on each side of the vertex.

Example 7 Graph: $y = x^2 - 4x + 4$.

1. To find the y-intercept by substituting 0 for x and computing y, we have

$$y = 0^2 - 4(0) + 4 = 4.$$

2. Find the x-intercepts by substituting 0 for y and solving for x:

$$0 = x^2 - 4x + 4$$
$$0 = (x - 2)(x - 2)$$
$$0 = x - 2 \quad \text{or} \quad 0 = x - 2$$
$$2 = x \qquad \text{or} \quad 2 = x.$$

The solution is 2. There is only one x-intercept and it is also the vertex. We compute four other solutions and draw the graph.

8. Graph $y = x^2 + 4x + 4$.

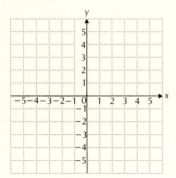

x	y	
0	4	← y-intercept
2	0	← x-intercept, vertex
3	1	
4	4	
1	1	
-1	9	

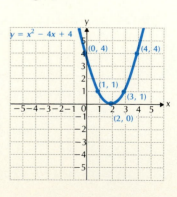

ANSWERS ON PAGE A-25

DO EXERCISE 8.

Chapter 11   Quadratic Equations

. , Without graphing, tell whether the graph of the equation opens upward or downward. Then graph the equation.

1. $y = x^2$

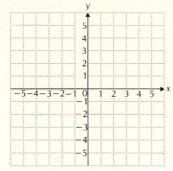

2. $y = 2x^2$

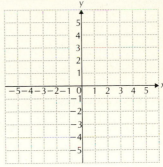

3. $y = -1 \cdot x^2$

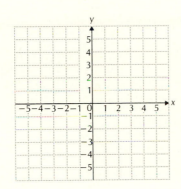

4. $y = x^2 - 1$

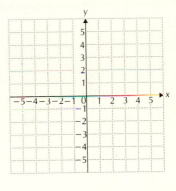

5. $y = -x^2 + 2x$

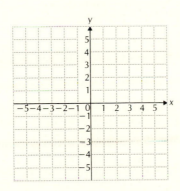

6. $y = x^2 + x - 6$

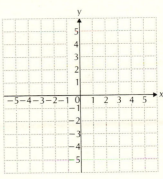

7. $y = 8 - x - x^2$

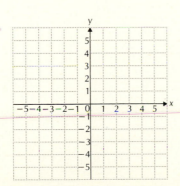

8. $y = x^2 + 2x + 1$

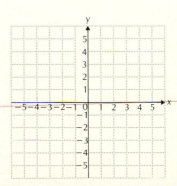

9. $y = x^2 - 2x + 1$

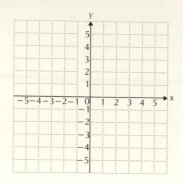

10. $y = -\frac{1}{2}x^2$

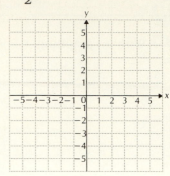

15. _____

16. _____

17. _____

11. $y = x^2 + 2x - 3$

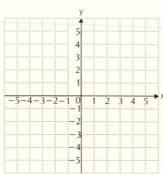

12. $y = -x^2 - 2x + 3$

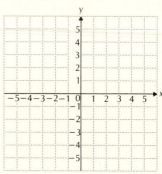

18. _____

19. _____

20. _____

13. $y = -2x^2 - 4x + 1$

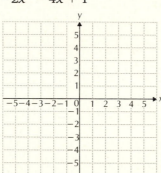

14. $y = 2x^2 + 4x - 1$

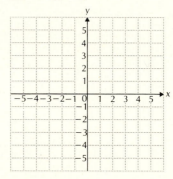

21. _____

22. _____

23. _____

 Approximate the solutions by graphing. Use graph paper.

15. $x^2 - 5 = 0$ 16. $x^2 + 2x = 0$ 17. $8 - x - x^2 = 0$

18. $x^2 + 10x + 25 = 0$ 19. $x^2 - 8x + 16 = 0$ 20. $x^2 + 3 = 0$

21. $x^2 + 8 = 0$ 22. $2x^2 + 4x - 1 = 0$ 23. $x^2 + 2x + 3 = 0$

24. _____

✔ SKILL MAINTENANCE

24. Add: $\sqrt{x^3 - x^2} + \sqrt{4x - 4}$.   25. Simplify: $\sqrt{20t^2}$.

25. _____

26. _____

☆ EXTENSION

26. Graph the equations $y = x^2 - x - 6$ and $y = 2$ using the same set of axes. Use your graphs to approximate the solutions of $x^2 - x - 6 = 2$.

27. What is the y-intercept of $y = ax^2 + bx + c$?

27. _____

The following contains a summary of what you should be able to do after completing this chapter. The review exercises are for practice. Answers are at the back of the book. If you miss an exercise, restudy the section and objective indicated alongside the answer.

The review sections to be tested in addition to the material in this chapter are 7.4, 9.8, 10.4, and 10.7.

You should be able to:

Solve quadratic equations by factoring or using the quadratic formula.

Solve.

1. $8x^2 = 24$

2. $5x^2 - 8x + 3 = 0$

3. $x^2 - 2x - 10 = 0$

4. $3y^2 + 5y = 2$

5. $(x + 8)^2 = 13$

6. $9x^2 = 0$

7. $5t^2 - 7t = 0$

8. $9x^2 - 6x - 9 = 0$

9. $x^2 + 6x = 9$

10. $1 + 4x^2 = 8x$

11. $6 + 3y = y^2$

12. $3m = 4 + 5m^2$

13. $3x^2 = 4x$

14. $40 = 5y^2$

Solve quadratic equations by completing the square.

Solve by completing the square. Show your work.

15. $3x^2 - 2x - 5 = 0$

16. $x^2 - 5x + 2 = 0$

Use a calculator or Table 2 to approximate the solutions of a quadratic equation.

Approximate the solutions to the nearest tenth.

17. $x^2 - 5x + 2 = 0$

18. $4y^2 + 8y + 1 = 0$

Solve certain fractional equations by first deriving a quadratic equation; solve certain radical equations by first using the principle of squaring to derive a quadratic equation; and solve a formula for a letter.

Solve.

19. $\dfrac{15}{x} - \dfrac{15}{x + 2} = 2$

20. $x + \dfrac{1}{x} = 2$

21. $\sqrt{x + 5} = x - 1$

22. $1 + 2x = \sqrt{1 + 5x}$

23. Solve for T: $V = \dfrac{1}{2} \sqrt{1 + \dfrac{L}{T}}$.

Without graphing, tell whether the graph of an equation of the type $y = ax^2 + bx + c$ opens upward or downward. Then graph the equation. Approximate the solutions of $0 = ax^2 + bx + c$ by graphing. Graph quadratic equations using intercepts.

Without graphing, tell whether the graph of the equation opens upward or downward. Then graph the equation.

24. $y = 2 - x^2$

25. $y = x^2 - 4x - 2$

26. Approximate the solutions of $x^2 - 4x - 2 = 0$ by graphing.

Solve problems involving the solution of quadratic equations.

Solve.

27. The hypotenuse of a right triangle is 5 m long. One leg is 3 m longer than the other. Find the lengths of the legs. Round to the nearest tenth.

28. $1000 is invested at interest rate r, compounded annually. In 2 years it grows to $1690. What is the interest rate?

29. The length of a rectangle is 3 m greater than the width. The area is 70 m². Find the length and the width.

30. The current in a stream moves at a speed of 2 km/h. A boat travels 56 km upstream and 64 km downstream in a total time of 4 hr. What is the speed of the boat in still water?

Solve.

31. $x - y = -2$
    $7x - 2y = 11$

32. $2x - 5y = -43$
    $3x + 5y = 23$

Multiply and simplify.

33. $\sqrt{18a} \ \sqrt{2}$

34. $\sqrt{12xy^2} \ \sqrt{5xy}$

Add or subtract.

35. $5\sqrt{11} + 7\sqrt{11}$

36. $2\sqrt{90} - \sqrt{40}$

37. Crew A can construct a specified length of road in 28 days. Crew B can construct the same length of road in 35 days. How many days would it take both crews working together to construct the road?

38. Two consecutive integers have squares that differ by 63. Find the integers.

39. Find b such that the trinomial $x^2 + bx + 49$ is a square.

40. Solve: $x \quad 4\sqrt{x} - 5 = 0$.

41. A square with sides of length s has the same area as a circle with radius of 5 in. Find s.

Test: Chapter 11

Solve.

1. $7x^2 = 35$

2. $7x^2 + 8x = 0$

3. $48 = t^2 + 2t$

4. $3y^2 - 5y = 2$

5. $(x - 8)^2 = 13$

6. $x^2 = x + 3$

7. $m^2 - 3m = 7$

8. $10 = 4x + x^2$

9. $3x^2 - 7x + 1 = 0$

10. $x - \dfrac{2}{x} = 1$

11. $\dfrac{4}{x} - \dfrac{4}{x + 2} = 1$

12. $\sqrt{6x + 13} = x + 3$

13. Solve by completing the square. Show your work.
$$x^2 - 4x - 10 = 0$$

14. Use Table 2 or your calculator to approximate the solutions to the nearest tenth.
$$x^2 - 4x - 10 = 0$$

15. Solve for n: $d = an^2 + bn$.

16. a) Without graphing, tell whether the graph of the equation $y = -x^2 + x + 5$ opens upward or downward.

    b) Graph the equation.

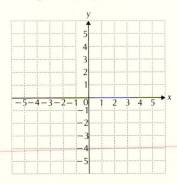

17. Approximate the solutions of $-x^2 + x + 5 = 0$ by graphing.

ANSWERS
1.
2.
3.
4.
5.
6.
7.
8.
9.
10.
11.
12.
13.
14.
15.
16. a)
b)
17.

Solve.

18. $4000 is invested at interest rate r, compounded annually. In 2 years it grows to $6250. What is the interest rate?

19. The width of a rectangle is 4 m less than the length. The area is 16.25 m^2. Find the length and the width.

20. The current in a stream moves at a speed of 2 km/h. A boat travels 44 km upstream and 52 km downstream in a total of 4 hr. What is the speed of the boat in still water?

 SKILL MAINTENANCE

21. Subtract: $\sqrt{240} - \sqrt{60}$.

22. Solve: $3x - 5y = -33$
$2x - 3y = -20$.

23. Multiply and simplify: $\sqrt{7xy}\ \sqrt{14x^2y}$.

24. Pipe A can fill a tank in 8 hr. Pipe B can fill the same tank in 11 hr. How long would it take them running together to fill the tank?

⭐ **EXTENSION**

25. Find the side of a square whose diagonal is 5 ft longer than a side.

26. Solve this system for x. Use the substitution method.

$$x - y = 2$$
$$xy = 4$$

1. What is the meaning of x^3?

2. Evaluate $(x - 3)^2 + 5$ for $x = 10$.

3. Find the prime factorization of 256.

4. Find the LCM of 15 and 48.

5. Find the absolute value of $|-7|$.

Compute and simplify.

6. $-6 + 12 + (-4) + 7$

7. $2.8 - (-12.2)$

8. $-\dfrac{3}{8} \div \dfrac{5}{2}$

9. $-9(7)$

10. Remove parentheses and simplify: $4m + 9 - (6m + 13)$.

Solve.

11. $3x = -24$

12. $3x + 7 = 2x - 5$

13. $3(y - 1) - 2(y + 2) = 0$

14. $x^2 - 8x + 15 = 0$

15. $y - x = 1$
$y = 3 - x$

16. $x + y = 17$
$x - y = 7$

17. $4x - 3y = 3$
$3x - 2y = 4$

18. $x^2 - x - 6 = 0$

19. $x^2 + 3x = 5$

20. $3 - x = \sqrt{x^2 - 3}$

21. $5 - 9x \leq 19 + 5x$

22. $-\dfrac{7}{8}x + 7 = \dfrac{3}{8}x - 3$

23. $0.6x - 1.8 = 1.2x$

24. $-3x > 24$

25. $23 - 19y - 3y \geq -12$

26. $3y^2 = 30$

27. $(x - 3)^2 = 6$

28. $\dfrac{6x - 2}{2x - 1} = \dfrac{9x}{3x + 1}$

29. $\dfrac{2x}{x + 1} = 2 - \dfrac{5}{2x}$

30. $\dfrac{2x}{x + 3} + \dfrac{6}{x} + 7 = \dfrac{18}{x^2 + 3x}$

31. $\sqrt{x + 9} = \sqrt{2x - 3}$

Solve each formula for the given letter.

32. $A = \dfrac{4b}{t}$, for b.

33. $\dfrac{1}{t} = \dfrac{1}{m} - \dfrac{1}{n}$, for m.

34. $r = \sqrt{\dfrac{A}{\pi}}$, for A.

35. $y = ax^2 - bx$, for x.

Simplify.

36. $x^{-6} \cdot x^2$

37. $\dfrac{y^3}{y^{-4}}$

38. $(2y^6)^2$

39. Collect like terms and arrange in descending order: $2x - 3 + 5x^3 - 2x^3 + 7x^3 + x$.

Compute and simplify.

40. $(4x^3 + 3x^2 - 5) + (3x^3 - 5x^2 + 4x - 12)$

41. $(6x^2 - 4x + 1) - (-2x^2 + 7)$

42. $-2y^2(4y^2 - 3y + 1)$

43. $(2t - 3)(3t^2 - 4t + 2)$

44. $\left(t - \dfrac{1}{4}\right)\left(t + \dfrac{1}{4}\right)$

45. $(3m - 2)^2$

46. $(15x^2y^3 + 10xy^2 + 5) - (5xy^2 - x^2y^2 - 2)$

47. $(x^2 - 0.2y)(x^2 + 0.2y)$

48. $(3p + 4q^2)^2$

49. $\dfrac{4}{2x - 6} \cdot \dfrac{x - 3}{x + 3}$

50. $\dfrac{3a^4}{a^2 - 1} \div \dfrac{2a^3}{a^2 - 2a + 1}$

51. $\dfrac{3}{3x - 1} + \dfrac{4}{5x}$

52. $\dfrac{2}{x^2 - 16} - \dfrac{x - 3}{x^2 - 9x + 20}$

Factor.

53. $8x^2 - 4x$

54. $25x^2 - 4$

55. $6y^2 - 5y - 6$

56. $m^2 - 8m + 16$

57. $x^3 - 8x^2 - 5x + 40$

58. $3a^4 + 6a^2 - 72$

59. $16x^4 - 1$

60. $49a^2b^2 - 4$

61. $9x^2 + 30xy + 25y^2$

62. $2ac - 6ab - 3db + dc$

63. $15x^2 + 14xy - 8y^2$

Simplify.

64. $\dfrac{\dfrac{3}{x} + \dfrac{1}{2x}}{\dfrac{1}{3x} - \dfrac{3}{4x}}$

65. $\sqrt{49}$

66. $-\sqrt{625}$

67. $\sqrt{64x^2}$, $x \geq 0$

68. Multiply: $\sqrt{a + b}\,\sqrt{a - b}$.

69. Multiply and simplify: $\sqrt{32ab}\,\sqrt{6a^4b^2}$.

Simplify.

70. $\sqrt{150}$ **71.** $\sqrt{243x^3y^2}$ **72.** $\sqrt{\dfrac{100}{81}}$ **73.** $\sqrt{\dfrac{64}{x^2}}$ **74.** $4\sqrt{12} + 2\sqrt{12}$

75. Divide and simplify: $\dfrac{\sqrt{72}}{\sqrt{45}}$.

76. In a right triangle, $a = 9$ and $c = 41$. Find b.

Graph on a plane.

77. $y = \dfrac{1}{3}x - 2$

78. $2x + 3y = -6$

79. $y = -3$

80. $4x - 3y > 12$

81. $y = x^2 + 2x + 1$

82. Solve $9x^2 - 12x - 2 = 0$ by completing the square. Show your work.

83. Approximate the solutions of $4x^2 = 4x + 1$ to the nearest tenth.

Solve.

84. What percent of 52 is 13?

85. 12 is 20% of what?

86. The speed of a boat in still water is 8 km/h. It travels 60 km upstream and 60 km downstream in a total time of 16 hr. What is the speed of the stream?

87. The length of a rectangle is 7 m more than the width. The length of a diagonal is 13 m. Find the length.

88. Three-fifths of the automobiles entering the city each morning will be parked in city parking lots. There are 3654 such parking spaces filled each morning. How many cars enter the city each morning?

89. A candy shop mixes nuts worth $1.10 per pound with another variety worth $0.80 per pound to make 42 lb of a mixture worth $0.90 per pound. How many pounds of each kind of nuts should be used?

90. In checking records a contractor finds that crew A can pave a certain length of highway in 8 hr. Crew B can do the same job in 10 hr. How long would they take if they worked together?

91. A student's paycheck varies directly as the number of hours worked. The pay was $242.52 for 43 hr of work. What would the pay be for 80 hr of work? Explain the meaning of the variation constant.

92. Find this intersection: $\{a, b, c, d, e\} \cap \{g, h, c, d, f\}$.

93. Find this union: $\{a, b, c, d, e\} \cup \{g, h, c, d, f\}$.

94. Find the slope and y-intercept: $-6x + 3y = -24$.

95. Find the slope of the line containing the points $(-5, -6)$ and $(-4, 9)$.

Find an equation of variation where:

96. y varies directly as x and $y = 100$ when $x = 10$.

97. y varies inversely as x and $y = 100$ when $x = 10$.

 EXTENSION

98. Solve: $|x| = 12$.

99. Simplify: $\sqrt{\sqrt{\sqrt{81}}}$.

100. Find b such that the trinomial $x^2 - bx + 225$ is a square.

101. Find x.

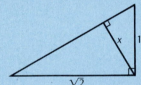

Determine whether each pair of expressions is equivalent.

102. $x^2 - 9$, $(x - 3)(x + 3)$

103. $\dfrac{x + 3}{3}$, x

104. $(x + 5)^2$, $x^2 + 25$

105. $\sqrt{x^2 + 16}$, $x + 4$

106. $\sqrt{x^2}$, $|x|$

Final Examination

1. Evaluate $x^3 + 5$ for $x = -10$.

2. Find the LCM of 16 and 24.

3. Find the absolute value of $|-9|$.

Compute and simplify.

4. $-6.3 + (-8.4) + 5$

5. $-8 - (-3)$

6. $\dfrac{3}{11} \cdot \left(-\dfrac{22}{7}\right)$

7. Remove parentheses and simplify:

$$4y - 5(9 - 3y).$$

8. Simplify:

$$2^3 - 14 \cdot 10 + (3 + 4)^3.$$

Solve.

9. $x + 8 = 13.6$

10. $4x = -28$

11. $5x + 3 = 2x - 27$

12. $5(x - 3) - 2(x + 3) = 0$

13. $x^2 - 2x - 24 = 0$

14. $y = x - 7$
$2x + y = 5$

15. $5x - 3y = -1$
$4x + 2y = 30$

16. $\dfrac{1}{x} - 2 = 8x$

17. $\sqrt{x^2 - 11} = x - 1$

ANSWERS
1.
2.
3.
4.
5.
6.
7.
8.
9.
10.
11.
12.
13.
14.
15.
16.
17.

18. $x^2 = 7 - 3x$

19. $2 - 3x \leq 12 - 7x$

18. _____

19. _____

20. _____

21. _____

22. _____

23. _____

24. _____

25. _____

26. _____

27. _____

28. _____

29. _____

30. _____

31. _____

32. _____

33. _____

34. _____

Solve each formula for the given letter.

20. $A = \dfrac{Bw + 1}{w}$, for w.

21. $K = MT + 2$, for M.

Simplify.

22. $\dfrac{x^8}{x^{-2}}$

23. $(x^{-5})^2$

24. $x^{-5} \cdot x^{-7}$

25. Collect like terms and arrange in descending order:

$$2y^3 - 3 + 4y^3 - 3y^2 + 12 - y.$$

Compute and simplify.

26. $(2x^2 - 6x + 3) - (4x^2 + 2x - 4)$

27. $-3t^2(2t^4 + 4t^2 + 1)$

28. $(4x - 1)(x^2 - 5x + 2)$

29. $(x - 8)(x + 8)$

30. $(2m - 7)^2$

31. $(3ab^2 + 2c)^2$

32. $(3x^2 - 2y)(3x^2 + 4y)$

33. $\dfrac{x}{x^2 - 9} \cdot \dfrac{x - 3}{x^3}$

34. $\dfrac{3x^5}{4x - 4} \div \dfrac{x}{x^2 - 2x + 1}$

Final Examination

35. $\dfrac{2}{3x-1}+\dfrac{1}{4x}$

36. $\dfrac{3}{x-3}-\dfrac{x-1}{x^2-2x-3}$

Factor.

37. $3x^3-15x$

38. $16x^2-25$

39. $6x^2-13x+6$

40. $x^2-10x+25$

41. $2ax+6bx-ay-3by$

42. x^8-81y^4

Simplify.

43. $\sqrt{72}$

44. $\dfrac{\sqrt{54}}{\sqrt{45}}$

45. $2\sqrt{8}+3\sqrt{18}$

46. $\sqrt{24a^2b}\,\sqrt{a^3b^2}$

Graph on a plane.

47. $3x+2y=-4$

48. $x=-2$

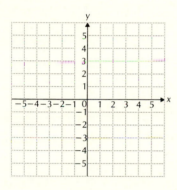

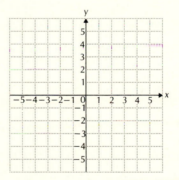

49. $3x-2y<6$

50. $y=x^2-2x+1$

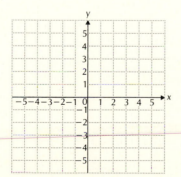

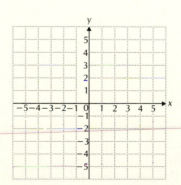

ANSWERS

35. _____

36. _____

37. _____

38. _____

39. _____

40. _____

41. _____

42. _____

43. _____

44. _____

45. _____

46. _____

47. _____

48. _____

49. _____

50. _____

Solve.

51. The sum of the squares of two consecutive odd integers is 74. Find the integers.

52. Solution A is 75% alcohol and solution B is 50% alcohol. How much of each is needed to make 60 L of a solution that is $66\frac{2}{3}$% alcohol?

53. An airplane flew for 6 hr with a 10-km/h tailwind. The return flight against the same wind took 8 hr. Find the speed of the plane in still air.

54. The width of a rectangle is 3 m less than the length. The area is 88 m². Find the length and the width.

55. Find the slope of the line containing the points $(-2, 3)$ and $(4, -5)$.

56. Find this union:
$\{1, 3, 5\} \cup \{1, 2, 3, 4\}$.

Find an equation of variation where:

57. y varies directly as x and $y = 200$ when $x = 25$.

58. y varies inversely as x and $y = 200$ when $x = 25$.

⭐ **EXTENSION**

59. A side of a square is five less than a side of an equilateral triangle. The perimeter of the square is the same as the perimeter of the triangle. Find the length of a side of the square and the length of a side of the triangle.

60. Find c such that the trinomial $x^2 - 24x + c$ is a square.

Final Examination

Table 1

Fractional Notation	Decimal Notation	Percent Notation
$\frac{1}{10}$	0.1	10%
$\frac{1}{8}$	0.125	12.5% or $12\frac{1}{2}$%
$\frac{1}{6}$	$0.16\overline{6}$	$16.6\overline{6}$% or $16\frac{2}{3}$%
$\frac{1}{5}$	0.2	20%
$\frac{1}{4}$	0.25	25%
$\frac{3}{10}$	0.3	30%
$\frac{1}{3}$	$0.333\overline{3}$	$33.3\overline{3}$% or $33\frac{1}{3}$%
$\frac{3}{8}$	0.375	37.5% or $37\frac{1}{2}$%
$\frac{2}{5}$	0.4	40%
$\frac{1}{2}$	0.5	50%
$\frac{3}{5}$	0.6	60%
$\frac{5}{8}$	0.625	62.5% or $62\frac{1}{2}$%
$\frac{2}{3}$	$0.666\overline{6}$	$66.6\overline{6}$% or $66\frac{2}{3}$%
$\frac{7}{10}$	0.7	70%
$\frac{3}{4}$	0.75	75%
$\frac{4}{5}$	0.8	80%
$\frac{5}{6}$	$0.83\overline{3}$	$83.3\overline{3}$% or $83\frac{1}{3}$%
$\frac{7}{8}$	0.875	87.5% or $87\frac{1}{2}$%
$\frac{9}{10}$	0.9	90%
$\frac{1}{1}$	1	100%

Table 2

S Q U A R E   R O O T S

N	$\sqrt{N}$	N	$\sqrt{N}$	N	$\sqrt{N}$	N	$\sqrt{N}$
2	1.414	27	5.196	52	7.211	77	8.775
3	1.732	28	5.292	53	7.280	78	8.832
4	2	29	5.385	54	7.348	79	8.888
5	2.236	30	5.477	55	7.416	80	8.944
6	2.449	31	5.568	56	7.483	81	9
7	2.646	32	5.657	57	7.550	82	9.055
8	2.828	33	5.745	58	7.616	83	9.110
9	3	34	5.831	59	7.681	84	9.165
10	3.162	35	5.916	60	7.746	85	9.220
11	3.317	36	6	61	7.810	86	9.274
12	3.464	37	6.083	62	7.874	87	9.327
13	3.606	38	6.164	63	7.937	88	9.381
14	3.742	39	6.245	64	8	89	9.434
15	3.873	40	6.325	65	8.062	90	9.487
16	4	41	6.403	66	8.124	91	9.539
17	4.123	42	6.481	67	8.185	92	9.592
18	4.243	43	6.557	68	8.246	93	9.644
19	4.359	44	6.633	69	8.307	94	9.695
20	4.472	45	6.708	70	8.367	95	9.747
21	4.583	46	6.782	71	8.426	96	9.798
22	4.690	47	6.856	72	8.485	97	9.849
23	4.796	48	6.928	73	8.544	98	9.899
24	4.899	49	7	74	8.602	99	9.950
25	5	50	7.071	75	8.660	100	10
26	5.099	51	7.141	76	8.718		

Table 3

Table 3

G E O M E T R I C . F O R M U L A S

PLANE GEOMETRY

Rectangle
Area: $A = lw$
Perimeter: $P = 2l + 2w$

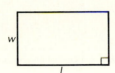

Square
Area: $A = s^2$
Perimeter: $P = 4s$

Triangle

Area: $A = \dfrac{1}{2} bh$

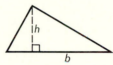

Sum of Angle Measures:
$A + B + C = 180°$

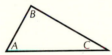

Right Triangle
Pythagorean Theorem
(Equation): $a^2 + b^2 = c^2$

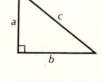

Parallelogram
Area: $A = bh$

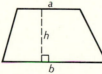

Trapezoid

Area: $A = \dfrac{1}{2} h(a + b)$

Circle
Area: $A = \pi r^2$
Circumference:
$C = \pi D = 2\pi r$

$\left(\dfrac{22}{7}\right.$ and 3.14 are different

approximations for $\pi\Big)$

SOLID GEOMETRY

Rectangular Solid
Volume: $V = lwh$

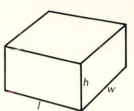

Cube
Volume: $V = s^3$

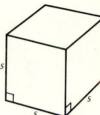
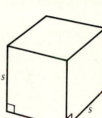

Right Circular Cylinder
Volume: $V = \pi r^2 h$
Lateral Surface Area: $L = 2\pi rh$
Total Surface Area:
$S = 2\pi rh + 2\pi r^2$

Right Circular Cone

Volume: $V = \dfrac{1}{3}\pi r^2 h$

Lateral Surface Area: $L = \pi rs$
Total Surface Area: $S = \pi r^2 + \pi rs$
Slant Height: $s = \sqrt{r^2 + h^2}$

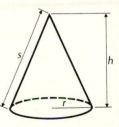
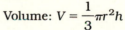
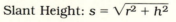

Sphere

Volume: $V = \dfrac{4}{3}\pi r^3$

Surface Area: $S = 4\pi r^2$

Answers

CHAPTER 1

Margin Exercises, Section 1.1, pp. 2–4

1. $2128 + x = 2866$; 738 **2.** 64 **3.** 28 **4.** 25 **5.** 60
6. 192 sq ft **7.** 16 **8.** $y - 12$ **9.** $y + 12$ **10.** $t - 4$
11. $\frac{1}{2}y$ **12.** $8n + 6$ **13.** $x - y$, or $y - x$
14. $59\% y$, or $0.59y$ **15.** $ab - 200$ **16.** $a + b$

Exercise Set 1.1, pp. 5–6

1. 10, 25, 48 **3.** 72 sq yd **5.** 56 **7.** 3 **9.** 2 **11.** 20
13. $m + 6$ **15.** $c - 9$ **17.** $6 + q$ **19.** $b + a$ **21.** $y - x$
23. $98\% x$ **25.** $r + s$ **27.** $2x$ **29.** $5t$
31. $3 - b$, or $b - 3$ **33.** $y + 6$ **35.** $x - 4$ **37.** $x + 3y$
39. $2l + 2w$ **41.** 2

Margin Exercises, Section 1.2, pp. 7–10

1. $3 \cdot 3$; answers may vary
2. $2 \cdot 2 \cdot 2 \cdot 2$; answers may vary
3. $2 \cdot 9, 3 \cdot 6, 3 \cdot 3 \cdot 2, 1 \cdot 18$; answers may vary
4. $1 \cdot 20, 2 \cdot 10, 4 \cdot 5, 2 \cdot 2 \cdot 5$ **5.** 13
6. $2 \cdot 2 \cdot 2 \cdot 2 \cdot 3$ **7.** $2 \cdot 5 \cdot 5$ **8.** $2 \cdot 5 \cdot 7 \cdot 11$
9. 15, 30, 45, 60, . . . **10.** 40 **11.** 360 **12.** 2520
13. 18 **14.** 24 **15.** 36 **16.** 210

Exercise Set 1.2, pp. 11–12

1. $3 \cdot 7$; answers can vary **3.** $12 \cdot 12$; answers can vary
5. $2 \cdot 7$ **7.** $3 \cdot 11$ **9.** $3 \cdot 3$ **11.** $7 \cdot 7$ **13.** $3 \cdot 3 \cdot 2$
15. $5 \cdot 2 \cdot 2 \cdot 2$ **17.** $2 \cdot 3 \cdot 3 \cdot 5$ **19.** $2 \cdot 3 \cdot 5 \cdot 7$
21. $7 \cdot 13$ **23.** $7 \cdot 17$ **25.** 21, 42, 63 **27.** 48, 96, 144
29. 36 **31.** 360 **33.** 150 **35.** 120 **37.** 72 **39.** 315
41. 30 **43.** 72 **45.** 60 **47.** 36 **49.** 294
51. Every 60 yr **53.** Every 420 yr

Margin Exercises, Section 1.3, pp. 13–17

Answers can vary in Exercises 1–7.
1. $\frac{4}{6}, \frac{10}{15}, \frac{6}{9}$ **2.** $\frac{5}{10}, \frac{2}{4}, \frac{3}{6}$ **3.** $\frac{6}{10}, \frac{9}{15}, \frac{12}{20}$
4. $\frac{2}{2}, \frac{5}{5}, \frac{14}{14}$ **5.** $\frac{12}{3}, \frac{4}{1}, \frac{20}{5}$ **6.** $\frac{8}{10}, \frac{12}{15}, \frac{24}{30}$
7. $\frac{16}{14}, \frac{40}{35}, \frac{24}{21}$ **8.** $\frac{2}{3}$ **9.** $\frac{19}{9}$ **10.** $\frac{8}{7}$ **11.** $\frac{1}{2}$ **12.** 4
13. $\frac{5}{2}$ **14.** $\frac{35}{16}$ **15.** $\frac{23}{15}$ **16.** $\frac{7}{12}$ **17.** $\frac{2}{15}$ **18.** $\frac{7}{36}$

Something Extra—Factors and Sums, p. 18

56	63	36	72	140	96	48	168	110	90	432	63
7	7	18	36	14	12	6	21	11	9	24	3
8	9	2	2	10	8	8	8	10	10	18	21
15	16	20	38	24	20	14	29	21	19	42	24

Exercise Set 1.3, pp. 19–20

1. $\frac{8}{6}$, etc. **3.** $\frac{12}{22}$, etc. **5.** $\frac{4}{22}$, etc. **7.** $\frac{25}{5}, \frac{20}{4}$, etc.
9. $\frac{4}{3}$ **11.** $\frac{1}{2}$ **13.** 2 **15.** $\frac{10}{3}$ **17.** $\frac{1}{8}$ **19.** $\frac{51}{8}$ **21.** 1
23. $\frac{7}{6}$ **25.** $\frac{5}{6}$ **27.** $\frac{1}{2}$ **29.** $\frac{5}{18}$ **31.** $\frac{31}{60}$ **33.** $p + q$
35. $3 \cdot 2 \cdot 2 \cdot 2 \cdot 2$ **37.** $\frac{3}{4}$ **39.** $\frac{12}{5}$

Margin Exercises, Section 1.4, pp. 21–24

1. $\frac{162}{100}$ **2.** $\frac{35,431}{1000}$ **3.** 0.875 **4.** 0.8 **5.** $0.\overline{81}$ **6.** $2.\overline{5}$
7. 10^4 **8.** $5 \cdot 5 \cdot 5 \cdot 5$ **9.** $x \cdot x \cdot x \cdot x \cdot x$ **10.** 4 **11.** 1
12. 125 **13.** 1519.76 cm^2 **14.** 119 **15.** (a) 40;
(b) 1000; (c) no

Exercise Set 1.4, pp. 25–26

1. $\dfrac{291}{10}$ 3. $\dfrac{467}{100}$ 5. $\dfrac{362}{100}$ 7. $\dfrac{18,789}{1000}$ 9. 0.5 11. 0.6
13. $0.\overline{2}$ 15. 0.125 17. $0.\overline{45}$ 19. $0.08\overline{3}$ 21. $5 \cdot 5$
23. $m \cdot m \cdot m$ 25. xxx 27. $yyyy$ 29. 1 31. p 33. M
35. 1 37. 27 39. 19 41. 256 43. 1 45. 32
47. 576 m^2 49. 256, 32 51. $x - y$ 53. 54 55. 24
57. x^3y^4

Margin Exercises, Section 1.5, pp. 27–30

1. 22 2. 30 3. 26 4. 30 5. 27 6. 27 7. 27
8. 1820 9. 1820 10. 23 11. 25 12. Commutative
law of multiplication 13. Associative law of addition
14. Associative law of addition
15. Associative law of multiplication 16. (a) 28; (b) 28
17. (a) 77; (b) 77 18. 23 19. 15 20. (a) 240; (b) 240
21. $4(x + y)$ 22. $5(a + b)$ 23. $7(p + q + r)$ 24. 228 ft
25. $5(x + y)$; 35 26. $7(x + y)$; 49

Exercise Set 1.5, pp. 31–32

1. 22 3. 89 5. 41 7. 59 9. Commutative law of
addition 11. Associative law of addition 13. 52
15. $9(x + y)$ 17. $\dfrac{1}{2}(a + b)$ 19. $1.5(x + z)$
21. $4(x + y + z)$ 23. $\dfrac{4}{7}(a + b + c + d)$ 25. $9(x + y)$;
135 27. $10(x + y)$; 150 29. $5(a + b)$; 45
31. $20(a + b)$; 180 33. 120 ft 35. 16, 64
37. $2 \cdot 2 \cdot 2 \cdot 2 \cdot 2 \cdot 3$ 39. (b), (c), (d) 41. (a), (b), (c)

Margin Exercises, Section 1.6, pp. 33–34

1. $5y + 15$ 2. $4x + 8y + 20$ 3. $8m + 24n + 32p$
4. $5(x + 2)$ 5. $3(4 + x)$ 6. $3(2x + 4 + 3y)$
7. $5(x + 2y + 5)$ 8. $Q(1 + ab)$ 9. $8y$ 10. $5x$
11. $1.03x$ 12. $18p + 9q$ 13. $11x + 8y$ 14. $16y$
15. $7s + 13w$ 16. $9x + 10y$ 17. $6a + 1.07b$

Exercise Set 1.6, pp. 35–36

1. $3x + 3$ 3. $4 + 4y$ 5. $36t + 27z$ 7. $7x + 28 + 42y$
9. $15x + 45 + 35y$ 11. $2(x + 2)$ 13. $6(x + 4)$
15. $3(3x + y)$ 17. $7(2x + 3y)$ 19. $5(1 + 2x + 3y)$
21. $8(a + 2b + 8)$ 23. $3(x + 6y + 5z)$ 25. $5x + 12$
27. $11a$ 29. $8x + 9z$ 31. $43a + 150c$
33. $1.09x + 1.2t$ 35. $24u + 5t$ 37. $50 + 6t + 8y$
39. $b + \dfrac{4}{3}$ 41. $\dfrac{13}{4}y$ 43. $\dfrac{5}{6}$ 45. $5x + 15y + 25$;
$5(x + 3y + 5)$

Margin Exercises, Section 1.7, pp. 37–40

1. $\dfrac{4}{7}$ 2. 1 3. $\dfrac{2}{3}$ 4. 1 5. $\dfrac{11}{4}$ 6. $\dfrac{7}{15}$ 7. $\dfrac{1}{5}$ 8. 3
9. $\dfrac{21}{20}$ 10. $\dfrac{10}{33}$ 11. $\dfrac{45}{28}$ 12. $\dfrac{8}{21}$ 13. $\dfrac{5}{6}$ 14. $\dfrac{14}{45}$
15.
16.

Exercise Set 1.7, pp. 41–42

(continued top right)

17. True 18. True 19. False 20. < 21. = 22. >
23. > 24. < 25. > 26. >

Exercise Set 1.7, pp. 41–42

1. $\dfrac{4}{3}$ 3. 8 5. 1 7. $\dfrac{35}{18}$ 9. $\dfrac{10}{3}$ 11. $\dfrac{1}{2}$ 13. $\dfrac{5}{36}$
15. 500 17. $\dfrac{3}{40}$
19. 21.
23. = 25. > 27. > 29. = 31. < 33. 408 cm^2
35. 2475 37. 3.2 39. <

Margin Exercises, Section 1.8, pp. 43–46

Answers may vary in Exercises 1 through 3.
1. $2 + 3 = 5$; $8 - 5 = 3$; $4 + 5.8 = 9.8$ 2. $2 + 3 = 6$;
$8 - 5 = 4$; $4 + 5.8 = 8.8$ 3. $x + 2 = 5$; $y - 1 = 8$;
$10 - y = 3$
4. 4, 8, 10; any three numbers other than 7 5. 7 6. 6
7. 4 8. 5 9. 8 10. 273 11. 4.87 12. $\dfrac{9}{8}$ 13. 7
14. $\dfrac{9}{2}$, or 4.5 15. 5.6 16. Yes 17. No 18. Yes
19. No 20. No 21. No

Exercise Set 1.8, pp. 47–48

1. 2 3. 9 5. 5 7. 20 9. 7 11. 5 13. 19 15. 5.66
17. 2818 19. $\dfrac{5}{12}$ 21. $\dfrac{1}{6}$ 23. 4 25. $\dfrac{5}{4}$, or 1.25
27. 0.24 29. 3.1 31. 8.5 33. $\dfrac{140}{3}$ 35. No 37. No
39. Yes 41. No 43. $\frac{15}{16}$ 45. $27u + 4t + 28$ 47. 8424
49. $0 \cdot x = 0$; answers may vary

Margin Exercises, Section 1.9, pp. 49–50

1. $x + 1,325,034 = 2,148,688$; 823,654 2. $\dfrac{2}{3}x = 44$; 66
3. $145 = p + 132$; 13 million
4. $1.16x = 14,500$; $12,500

Exercise Set 1.9, pp. 51–52

1. $\frac{2}{3}x = 48$; 72 3. $y + 5 = 22$; 17 5. $w = 5 + 4$; 9
7. $78,114 = 4A$; 19,528.5 km^2
9. $\frac{2}{5}t = 35$; $87\frac{1}{2}$ words per min
11. $78.3 = 13.5 + m$; 64.8°C 13. $640 = 1.6s$; 400 kWh
15. $1175 = 1.8c$; $652.78 17. 39.37 in.
19. Salary now

Margin Exercises, Section 1.10, pp. 53–56

1. 0.462 2. 1 3. $\dfrac{67}{100}$ 4. $\dfrac{456}{1000}$ 5. $\dfrac{1}{400}$ 6. 677%
7. 99.44% 8. 25% 9. 37.5% 10. $66.\overline{6}$%, or $66\dfrac{2}{3}$%

11. 11.04 **12.** 10 **13.** 32% **14.** 25% **15.** 225 **16.** 50
17. $x = 19\% \cdot 586{,}400$; 111,416 mi^2
18. $x + 7\%x = 8988$; \$8400
19. $x + 12\%x = 20{,}608$; \$18,400

Exercise Set 1.10, pp. 57–58

1. 0.76 **3.** 0.547 **5.** $\dfrac{20}{100}$ **7.** $\dfrac{786}{1000}$ **9.** 454%

11. 99.8% **13.** 12.5% **15.** 68%
17. $x = 65\% \cdot 840$; 546 **19.** $24\% \cdot x = 20.4$; 85
21. $x\% \cdot 80 = 100$; 125%
23. $76 = x\% \cdot 88$; approx. 86.36%
25. $208 = 26\% \cdot x$; \$800
27. $x = 8\% \cdot 428.86$; \$34.31; \$463.17
29. $x + 9\%x = 8502$; \$7800
31. $(8\% - 7.4\%) \cdot 9600 = x$; \$57.60 **33.** 24%
35. $3\frac{1}{3}\%$, or 3.3%

Exercise Set 1.10A, pp. 59–60

1. 0.38 **3.** 0.654 **5.** 0.0824 **7.** 0.00012 **9.** 0.0073

11. 1.25 **13.** $\dfrac{30}{100}$ **15.** $\dfrac{135}{1000}$ **17.** $\dfrac{32}{1000}$ **19.** $\dfrac{120}{100}$

21. $\dfrac{35}{10{,}000}$ **23.** $\dfrac{42}{100{,}000}$ **25.** 62% **27.** 62.3%

29. 720% **31.** 200% **33.** 7.2% **35.** 0.57% **37.** 17%

39. 70% **41.** 35% **43.** 50% **45.** 60% **47.** $33\frac{1}{3}\%$

49. 95 **51.** 25% **53.** 64

Summary and Review: Chapter 1, pp. 61–62

1. [1.1, ■■] 20 **2.** [1.1, ■■] 6 **3.** [1.1, ■■] 5
4. [1.1, ■■] 4 **5.** [1.4, ■■] 9 **6.** [1.4, ■■] 1
7. [1.4, ■■] 3215.36 mi^2 **8.** [1.1, ■■] $z - 8$
9. [1.1, ■■] $3x$ **10.** [1.1, ■■] $19\%m$ **11.** [1.1, ■■] $x - 1$
12. [1.2, ■■] $2 \cdot 2 \cdot 23$ **13.** [1.2, ■■] $2 \cdot 2 \cdot 2 \cdot 5 \cdot 5 \cdot 7$
14. [1.2, ■■] 96 **15.** [1.2, ■■] 90 **16.** [1.3, ■■] $\frac{5}{12}$
17. [1.3, ■■] $\frac{9}{5}$ **18.** [1.3, ■■■] $\frac{31}{36}$ **19.** [1.7, ■■] $\frac{1}{4}$
20. [1.3, ■■■] $\frac{3}{5}$ **21.** [1.3, ■■■] $\frac{72}{25}$ **22.** [1.7, ■■] >
23. [1.7, ■■] < **24.** [1.4, ■■] $\frac{1798}{100}$ **25.** [1.4, ■■] $\frac{347}{1000}$
26. [1.4, ■■] 0.8125 **27.** [1.4, ■■] 1.571428
28. [1.4, ■■■] $t \cdot t \cdot t \cdot t \cdot t$ **29.** [1.4, ■■] 1
30. [1.4, ■■] t **31.** [1.5, ■■] 32 **32.** [1.5, ■■] 75
33. [1.5, ■■■] Distributive **34.** [1.5, ■■] Commutative
law of addition **35.** [1.5, ■■] Associative law of
multiplication **36.** [1.5, ■■] Commutative law of
multiplication **37.** [1.6, ■■] $18x + 30y$
38. [1.6, ■■] $40x + 24y + 16$ **39.** [1.6, ■■] $6(3x + y)$
40. [1.6, ■■] $4(9x + 4 + y)$ **41.** [1.6, ■■■] $30y + 10a$
42. [1.6, ■■■] $60x + 12b$ **43.** [1.10, ■■] 0.047
44. [1.10, ■■] $\frac{60}{100}$ **45.** [1.10, ■■■] 88.6%
46. [1.10, ■■] 62.5% **47.** [1.10, ■■] 116%
48. [1.8, ■■] 39 **49.** [1.8, ■■■] $\frac{1}{8}$ **50.** [1.8, ■■■] 8
51. [1.8, ■■■] 26 **52.** [1.9, ■■] 67 **53.** [1.9, ■■] 50
54. [1.9, ■■] 32 **55.** [1.10, ■■] 250 **56.** [1.10, ■■] 30
57. [1.10, ■■] \$14,200 **58.** [1.3, ■■] $\dfrac{5a}{9b}$
59. [1.4, ■■] 37 **60.** [1.5] (a), (b), (c)
61. [1.10, ■■] 0.0000006%

CHAPTER 2

Margin Exercises, Section 2.1, pp. 66–69

1. 8 yd, −5 yd **2.** 134°, −76° **3.** −10 sec, 148 sec
4. −\$137, \$289 **5.** > **6.** > **7.** < **8.** > **9.** 18 **10.** 9
11. 29 **12.** 0 **13.** −1 **14.** 2 **15.** 0 **16.** 4 **17.** 1
18. −2 **19.** −5 **20.** 4 **21.** 13 **22.** −28 **23.** 0

Something Extra: Calculator Corner, p. 70

1. 1, 9, 36, 100; 225, 441 **2.** 2^n:2, 4, 8, 16, 32, 64, 128;
$\dfrac{1}{2^n}$:0.5, 0.25, 0.125, 0.0625, 0.0313, 0.0156, 0.0078

Exercise Set 2.1, pp. 71–72

1. −1286 ft, 29,028 ft **3.** −3 sec, 128 sec **5.** > **7.** <
9. < **11.** < **13.** > **15.** < **17.** 3 **19.** 10 **21.** 0
23. 24 **25.** 53 **27.** 8 **29.** 6 **31.** −6 **33.** −7 **35.** 1
37. 12 **39.** −70 **41.** 0 **43.** −34 **45.** 1 **47.** −7
49. 14 **51.** 0 **53.** $\frac{3}{5}$ **55.** Distributive **57.** −7, 7
59. x, $-x$, x

Margin Exercises, Section 2.2, pp. 73–74

1. −3 **2.** −8 **3.** 4 **4.** 0 **5.** $4 + (-5) = -1$
6. $-2 + (-4) = -6$ **7.** $-3 + 8 = 5$ **8.** −11 **9.** −12
10. −34 **11.** −22 **12.** 2 **13.** −4 **14.** −2 **15.** 3
16. 0 **17.** 0 **18.** 0 **19.** 0 **20.** −12

Exercise Set 2.2, pp. 75–76

1. −7 **3.** −4 **5.** 0 **7.** −13 **9.** −16 **11.** −11 **13.** 0
15. 0 **17.** 8 **19.** −8 **21.** −25 **23.** −17 **25.** 0 **27.** 0
29. 3 **31.** −9 **33.** −5 **35.** 7 **37.** −3 **39.** 0 **41.** −5
43. −21 **45.** 2 **47.** −26 **49.** −22 **51.** 32 **53.** 0
55. 45 **57.** −198 **59.** 52 **61.** $21z + 7y + 14$ **63.** 3.5
65. 30,937 **67.** All positive

Margin Exercises, Section 2.3, pp. 77–79

1. −10 **2.** 3 **3.** −5 **4.** −2 **5.** −11 **6.** 4 **7.** −2
8. −6 **9.** −16 **10.** 7 **11.** 3 **12.** −6
13. Three minus eleven; −8 **14.** Twelve minus five; 7
15. Negative twelve minus negative nine; −3
16. Negative twelve minus ten; −22
17. Negative fourteen minus negative fourteen; 0
18. −9 **19.** 511 mL **20.** 792 ft

Something Extra—Calculator Corner: Number Patterns, p. 80

1. 1; 121; 12,321; 1,234,321; 123,454,321
2. 24; 2904; 295,704; 29,623,704; 2,962,903,704;
296,295,703,704 **3.** 48; 408; 4008; 40,008; 400,008
4. 81; 9801; 998,001; 99,980,001; 9,999,800,001

Exercise Set 2.3, pp. 81–82

1. −4 **3.** −7 **5.** −6 **7.** 0 **9.** −4 **11.** −7 **13.** −6
15. 0 **17.** 0 **19.** 14 **21.** 11 **23.** −14 **25.** 5 **27.** −7
29. −1 **31.** 18 **33.** −5 **35.** −3 **37.** −21 **39.** 5
41. −8 **43.** 12 **45.** −23 **47.** −68 **49.** −73 **51.** 116
53. 0 **55.** 37 **57.** −62 **59.** −139 **61.** 62°F

63. 432 ft² **65.** −309,882 **67.** False; 5 − 0 = 5, 0 − 5 = −5

Margin Exercises, Section 2.4, pp. 83–84

1. 20, 10, 0, −10, −20, −30 **2.** −18 **3.** −100 **4.** −80
5. −10, 0, 10, 20, 30 **6.** 12 **7.** 32 **8.** 35 **9.** −2
10. 5 **11.** −3 **12.** 8 **13.** −6

Exercise Set 2.4, pp. 85–86

1. −16 **3.** −42 **5.** −24 **7.** −72 **9.** 16 **11.** 42
13. 24 **15.** 72 **17.** −120 **19.** 1000 **21.** 90 **23.** 200
25. −6 **27.** −13 **29.** −2 **31.** 4 **33.** −8 **35.** 2
37. −12 **39.** −8 **41.** $\frac{9}{8}$ **43.** 125 **45.** 16 **47.** −32
49. −1

Margin Exercises, Section 2.5, pp. 87–88

1. −8.7 **2.** $\frac{8}{9}$ **3.** 7.74 **4.** $\frac{10}{3}$ **5.** 8.32 **6.** $-\frac{5}{4}$ **7.** 12
8. $\frac{5}{6}$ **9.** −17.2 **10.** 4.1 **11.** $\frac{8}{3}$ **12.** 3.5 **13.** −6.2
14. $-\frac{2}{9}$ **15.** $-\frac{19}{20}$ **16.** 510.8 mL **17.** −9 **18.** $-\frac{1}{5}$
19. −7.5 **20.** $-\frac{1}{24}$

Exercise Set 2.5, pp. 89–90

1. 4.7 **3.** $-\frac{7}{2}$ **5.** 7 **7.** 26.9 **9.** $\frac{1}{3}$ **11.** $-\frac{7}{6}$ **13.** 9.3
15. −90.3 **17.** −12.4 **19.** $\frac{9}{10}$ **21.** 34.8 **23.** −567
25. 19.2 **27.** $\frac{2}{3}$ **29.** 89.3 **31.** $\frac{14}{3}$ **33.** −1.8
35. −8.1 **37.** $-\frac{1}{5}$ **39.** $-\frac{8}{7}$ **41.** $-\frac{3}{8}$ **43.** $-\frac{29}{35}$
45. $-\frac{11}{15}$ **47.** −6.3 **49.** −4.3 **51.** −4 **53.** $-\frac{1}{4}$
55. $\frac{1}{12}$ **57.** $-\frac{17}{12}$ **59.** $\frac{1}{8}$ **61.** 19.9 **63.** −9
65. −0.01 **67.** 1 **69.** 0.76 **71.** (a) >; (b) > **73.** −3.4
75. $-\frac{1}{2}$

Margin Exercises, Section 2.6, pp. 91–94

1. −30 **2.** $-\frac{10}{27}$ **3.** $-\frac{7}{10}$ **4.** −30.033 **5.** 64 **6.** $\frac{20}{63}$
7. $\frac{2}{3}$ **8.** 13.455 **9.** −30 **10.** −30.75 **11.** $-\frac{5}{3}$
12. 120 **13.** $\frac{3}{2}$ **14.** $-\frac{4}{5}$ **15.** $-\frac{1}{3}$ **16.** −5
17. $-\frac{2}{3}, \frac{3}{2}; \frac{5}{4}, -\frac{4}{5};$ 0, does not exist; −1, 1;
4.5, $-\frac{1}{4.5}$ or $-\frac{10}{45}$ **18.** $-\frac{20}{21}$ **19.** $-\frac{12}{5}$ **20.** $\frac{16}{7}$
21. −7 **22.** $\frac{19}{20}$ **23.** $\frac{8}{5}$ **24.** $-\frac{10}{3}$ **25.** $-\frac{5}{6}$

26. $-\frac{5}{6}, \frac{-5}{6}$ **27.** $\frac{-8}{7}, \frac{8}{-7}$ **28.** $\frac{-10}{3}, -\frac{10}{3}$

Exercise Set 2.6, pp. 95–96

1. −72 **3.** −12.4 **5.** 24 **7.** 21.7 **9.** $-\frac{2}{5}$ **11.** $\frac{1}{12}$
13. −17.01 **15.** $-\frac{5}{12}$ **17.** −90 **19.** 420 **21.** $\frac{2}{7}$
23. −60 **25.** 150 **27.** $-\frac{1}{5}$ **29.** 4 **31.** $-\frac{5}{7}$ **33.** $-\frac{11}{4}$
35. $-\frac{9}{8}$ **37.** $\frac{5}{3}$ **39.** $\frac{9}{14}$ **41.** $\frac{9}{64}$ **43.** −1 **45.** −2
47. $\frac{11}{13}$ **49.** $-\frac{23}{14}$ **51.** 706.5 cm² **53.** 8(3x + 4y + 8)
55. 1 **57.** −8

Margin Exercises, Section 2.7, pp. 97–98

1. (a) 8; (b) 8 **2.** (a) −4; (b) −4 **3.** (a) −25; (b) −25
4. (a) −20; (b) −20 **5.** 5x, −4y, 3 **6.** −4y, −2x, 3z
7. 3x − 15 **8.** 5x − 5y + 20 **9.** −2x + 6
10. bx − 2by + 4bz **11.** 6(x − 2) **12.** 3(x − 2y + 3)
13. b(x + y − z) **14.** 3x **15.** 6x **16.** 0.59x
17. 3x + 3y **18.** −4x − 5y − 7

Exercise Set 2.7, pp. 99–100

1. 12 **3.** −8 **5.** 8x, −1.4y **7.** −5x, 3y, −14z
9. 8(x − 3) **11.** 4(8 − x) **13.** 2(4x + 5y − 11)
15. a(x − 7) **17.** a(x + y − z) **19.** 7x − 14
21. −7y + 14 **23.** −21 + 3t **25.** −4x − 12y
27. −14x − 28y + 21 **29.** 8x **31.** 16y **33.** −11x
35. 0.17x **37.** 4x + 2y **39.** 7x + y − 11
41. 0.8x + 0.5y **43.** $\frac{3}{5}x + \frac{3}{5}y$ **45.** 78 ft **47.** 6
49. $\frac{1}{2}h(a + b)$ **51.** 2x

Margin Exercises, Section 2.8, pp. 101–103

1. −x − 2 **2.** −5x + 2y + 8 **3.** −6 + t **4.** −x + y
5. 4a − 3t + 10 **6.** −18 + m + 2n − 4z **7.** 2x − 9
8. 3y + 2 **9.** −9x − 8y **10.** −16a + 18
11. −26a + 41b − 48c **12.** 6 **13.** 4 **14.** 5x − y − 8

Something Extra—Volumes of Rectangular Solids, p. 104

1. 200 m³ **2.** 151.2 ft³

Exercise Set 2.8, pp. 105–106

1. −2x − 7 **3.** −5x + 8 **5.** −4a + 3b − 7c
7. −6x − 8y − 5 **9.** −3x − 5y + 6 **11.** 8x + 6y + 43
13. 5x − 3 **15.** 7a − 9 **17.** 5x − 6 **19.** −5x − 2y
21. y − 5z **23.** −19x + 2y **25.** −21a + 34b − 40c
27. −16 **29.** 40 **31.** 19 **33.** 30 **35.** 12x − 2
37. 16x − 4 **39.** 3x + 30 **41.** 9x − 18 **43.** −4x − 64
45. 6y − (−2x + 3a − c) **47.** −2x − a

Margin Exercises, Section 2.9, pp. 107–108

1. -1237 2. 381 3. -135 4. $(1/(4*5))\char94 2$

5. $A\char94 2+B\char94 2-2*A*B$ 6. $X/Y-T/S$ 7. $\dfrac{2}{a+3}$

8. $a^2 - 2ab + b^2$

Exercise Set 2.9, pp. 109–110

1. 1 3. -40 5. 11 7. 4 9. -334 11. -176
13. 1880 15. 64 17. 4682.688 19. $-6{,}902{,}384{,}368$
21. $A\char94 2+2*A*B+B\char94 2$ 23. $2*(3-B)/C$
25. $A/B-C/D$ 27. $2a + 7$ 29. $3a^2 - 5$ 31. $(a + b)^2$
33. $\$5500$ 35. $2y + 3x$ 37. True,
$(-x)^2 = (-x)(-x) = (-1 \cdot x)(-1 \cdot x) = (-1)(-1)xx = x^2$

Summary and Review: Chapter 2, pp. 111–112

1. [2.1, ■●] $-\$45, \$72$ 2. [2.1, ●●●] 38 3. [2.1, ●●●] 7
4. [2.5, ●●] $\frac{5}{2}$ 5. [2.5, ■●] 4.78 6. [2.1, ●●] $<$
7. [2.1, ●●] $>$ 8. [2.1, ●●] $>$ 9. [2.5, ■●] -3.8
10. [2.5, ■●] $\frac{3}{4}$ 11. [2.1, ■■] 34 12. [2.1, ■■] 5
13. [2.6, ●●●] $\frac{8}{3}$ 14. [2.6, ●●●] $-\frac{1}{7}$ 15. [2.6, ●●●] -4
16. [2.6, ●●●] $\dfrac{1}{1.6}$ or $\dfrac{10}{16}$ or 0.625 17. [2.2, ■●] -3
18. [2.5, ●●●] $-\frac{7}{12}$ 19. [2.2, ■●] -4 20. [2.5, ●●●] -10.7
21. [2.3, ●●●] 4 22. [2.5, ●●●] $-\frac{7}{5}$ 23. [2.5, ●●●] -7.9
24. [2.4, ■●] 54 25. [2.6, ●●] -9.18 26. [2.6, ●●] $-\frac{2}{7}$
27. [2.6, ●●●] -210 28. [2.4, ●●●] -7 29. [2.6, ●●●] -3
30. [2.6, ●●●] $\frac{3}{4}$ 31. [2.7, ●●●] $15x - 35$
32. [2.7, ●●●] $-8x + 10$ 33. [2.7, ●●●] $4x + 15$
34. [2.7, ●●●] $-24 + 48x$ 35. [2.7, ■■] $2(x - 7)$
36. [2.7, ■■] $6(x - 1)$ 37. [2.7, ■■] $5(x + 2)$
38. [2.7, ■■] $3(4 - x)$ 39. [2.7, ■■] $7a - 3b$
40. [2.7, ■■] $-2x + 5y$ 41. [2.7, ■■] $5x - y$
42. [2.7, ■■] $-a + 8b$ 43. [2.7, ■■] $6x + 3$
44. [2.7, ■■] $11y - 16$ 45. [2.8, ●●●] -27
46. [2.8, ●●●] -1 47. [2.8, ●●●] 6 48. [2.9, ■●] -66
49. [2.8, ●●●] $-18a + 36$ 50. [2.8, ●●●] $-2b + 21$
51. [2.8, ●●●] $12y - 34$ 52. [2.8, ●●●] $5x + 21$
53. [2.8, ●●●] $-15x + 25$ 54. [2.9, ■●] $A\char94 2-B\char94 3$
55. [2.9, ●●●] $A-C/D$ 56. [2.9, ●●●] $3 - 4b$
57. [2.9, ●●●] $(a - b)^3$ 58. [1.1, ■●] 210 cm^2
59. [1.2, ■■] 270 60. [1.2, ●●] $2 \cdot 2 \cdot 2 \cdot 3 \cdot 3 \cdot 3 \cdot 3$
61. [1.1, ●●] $x - y$ 62. [1.6, ●●] $3(5x + 10y + 2)$
63. [1.10, ■■] 36% 64. [2.5] $-\frac{5}{8}$ 65. [2.5] No solution
66. [2.1, ■■] $-x$
67. [2.5, ■●] All negative rational numbers
68. [2.7] $1.06P$

CHAPTER 3

Margin Exercises, Section 3.1, pp. 116–117

1. -5 2. 13.2 3. -6.5 4. -2 5. $\frac{31}{8}$

Something Extra—Calculator Corner: Number Patterns, p. 118

1. 9; 1089; $110{,}889$; $11{,}108{,}889$; $1{,}111{,}088{,}889$
2. 54; 6534; $665{,}334$; $66{,}653{,}334$; $6{,}666{,}533{,}334$

3. 111; 1221; $12{,}321$; $123{,}321$; $1{,}233{,}321$
4. 111; 222; 333; 444; 555

Exercise Set 3.1, pp. 119–120

1. 4 3. -20 5. $\frac{1}{3}$ 7. $\frac{41}{24}$ 9. 4 11. $\frac{1}{2}$ 13. -5.1

15. 10.9 17. $-\frac{7}{4}$ 19. -4 21. 16 23. -5 25. $1\frac{5}{6}$

27. $6\frac{2}{3}$ 29. $-7\frac{6}{7}$ 31. -11 33. $-\frac{5}{12}$ 35. 342.246

37. No solution

Margin Exercises, Section 3.2, pp. 121–122

1. 5 2. -12 3. $-\frac{7}{4}$ 4. 7800 5. -3 6. 10 7. 28

Exercise Set 3.2, pp. 123–124

1. 6 3. -4 5. -6 7. 63 9. $\frac{3}{5}$ 11. -20 13. 36

15. $\frac{3}{2}$ 17. 10 19. -2 21. 8 23. 13.38 25. $\frac{9}{2}$

27. 15.9 29. -100 31. 54 33. $-7x$ 35. 3.2
37. All rational numbers 39. $-12, 12$

Margin Exercises, Section 3.3, pp. 125–128

1. 5 2. 4 3. 4 4. 39 5. $-\frac{3}{2}$ 6. -4.3 7. -3 8. 800

9. 1 10. 2 11. 2 12. $\frac{17}{2}$ 13. $\frac{8}{3}$ 14. -4.3

Exercise Set 3.3, pp. 129–130

1. 5 3. 10 5. -8 7. 6 9. 5 11. -20 13. 6
15. 8400 17. 7 19. 7 21. 3 23. 5 25. 2 27. 10

29. 4 31. 8 33. $-\frac{3}{5}$ 35. -4 37. $\frac{10}{7}$ 39. $\frac{5}{6}$

41. -14 43. 4.4233464 45. $-\frac{7}{2}$

Margin Exercises, Section 3.4, p. 131

1. 2 2. 3 3. -2 4. $-\frac{1}{2}$

Something Extra—Handling Dimension Symbols (Part 1), p. 132

1. $5\dfrac{\text{mi}}{\text{hr}}$ 2. $34\dfrac{\text{km}}{\text{hr}}$ 3. $2.2\dfrac{\text{m}}{\text{sec}}$ 4. $19\dfrac{\text{ft}}{\text{min}}$ 5. 52 ft

6. 102 sec 7. 31 m 8. 32 hr 9. $\dfrac{23}{20}\text{lb}$ 10. 12 km

11. 22 g 12. $105\dfrac{\text{m}}{\text{sec}}$

Exercise Set 3.4, pp. 133–134

1. 6 3. 2 5. 6 7. 8 9. 4 11. 1 13. 17 15. -8

17. $-\frac{5}{3}$ 19. -3 21. 2 23. 5 25. 22 ft

3

27. $7(x - 3 - 2y)$ **29.** 434.08657 **31.** -0.000036365

Margin Exercises, Section 3.5, pp. 135–138

1. 3 ft, 5 ft **2.** 5 **3.** $5700 **4.** 18, 20 **5.** 240.9 mi
6. Length is 20 m; width is 10 m **7.** 30°, 90°, 60°
8. $11,500

Exercise Set 3.5, pp. 139–144

1. 28° **3.** 19 **5.** -10 **7.** 40 **9.** 20 m; 40 m; 120 m
11. 37; 39 **13.** 56; 58 **15.** 35; 36; 37 **17.** 61; 63; 65
19. Length is 90 m; width is 65 m **21.** Length is 49 m;
width is 27 m **23.** 22.5° **25.** $4400 **27.** $16
29. 412.6 mi **31.** 28°; 84°; 68° **33.** Approx. 4.71 billion
35. 1985 **37.** 20 **39.** $9.17, not $9.10

Margin Exercises, Section 3.6, pp. 145–146

1. 2.8 mi **2.** $I = \dfrac{E}{R}$ **3.** $\dfrac{C}{\pi} = D$ **4.** $4A - a - b - d = c$

5. $I = \dfrac{9R}{A}$

Exercise Set 3.6, pp. 147–148

1. $b = \dfrac{A}{h}$ **3.** $r = \dfrac{d}{t}$ **5.** $P = \dfrac{I}{rt}$ **7.** $a = \dfrac{F}{m}$

9. $w = \dfrac{1}{2}(P - 2l)$ **11.** $r^2 = \dfrac{A}{\pi}$ **13.** $b = \dfrac{2A}{h}$

15. $m = \dfrac{E}{c^2}$ **17.** $3A - a - c = b$ **19.** $t = \dfrac{3k}{v}$

21. $b = \dfrac{2A - ah}{h}$ **23.** $D^2 = \dfrac{2.5H}{N}$ **25.** $S = \dfrac{360A}{\pi r^2}$

27. $t = \dfrac{R - 3.85}{-0.0075}$ **29.** $33\frac{1}{3}\%$, or $33.\overline{3}\%$ **31.** -13.2

33. A quadruples **35.** A increases by $2h$ units

Margin Exercises, Section 3.7, pp. 149–152

1. $\dfrac{1}{4^3}$, or $\dfrac{1}{4 \cdot 4 \cdot 4}$, or $\dfrac{1}{64}$ **2.** $\dfrac{1}{5^2}$, or $\dfrac{1}{5 \cdot 5}$, or $\dfrac{1}{25}$

3. $\dfrac{1}{2^4}$, or $\dfrac{1}{2 \cdot 2 \cdot 2 \cdot 2}$, or $\dfrac{1}{16}$ **4.** 3^{-2} **5.** 5^{-4} **6.** 7^{-3}

7. $\dfrac{1}{5^3}$ **8.** $\dfrac{1}{7^5}$ **9.** $\dfrac{1}{10^4}$ **10.** 3^8 **11.** 5^2 **12.** 6^{-7} **13.** x^{-4}

14. y^{-2} **15.** x^{-8} **16.** 4^3 **17.** 7^{-5} **18.** a^7 **19.** b
20. x^4 **21.** x^7 **22.** 3^{20} **23.** x^{-12} **24.** y^{15} **25.** x^{-32}
26. $16x^{20}y^{-12}$ **27.** $25x^{10}y^{-12}z^{-6}$ **28.** $27y^{-6}x^{-15}z^{24}$
29. $3121.79

Exercise Set 3.7, pp. 153–154

1. $\dfrac{1}{3^2}$, or $\dfrac{1}{3 \cdot 3}$, or $\dfrac{1}{9}$

3. $\dfrac{1}{10^4}$, or $\dfrac{1}{10 \cdot 10 \cdot 10 \cdot 10}$, or $\dfrac{1}{10,000}$ **5.** 4^{-3}

7. x^{-3} **9.** a^{-4} **11.** p^{-n} **13.** $\dfrac{1}{7^3}$ **15.** $\dfrac{1}{a^3}$ **17.** $\dfrac{1}{y^4}$

19. $\dfrac{1}{z^n}$ **21.** 2^7 **23.** 3^3 **25.** x^{-1} **27.** x^7 **29.** x^{-13}
31. 1 **33.** 7^3 **35.** x^2 **37.** x^9 **39.** z^{-4} **41.** x^3 **43.** 1
45. 2^6 **47.** 5^{-6} **49.** x^{12} **51.** $x^{-12}y^{-15}$ **53.** $x^{24}y^8$
55. $9x^6y^{-16}z^{-6}$ **57.** $2508.80 **59.** $22,318.40
61. 130.9 **63.** False **65.** True

Margin Exercises, Section 3.8, pp. 155–158

1. 4.6×10^{11} **2.** 9.3×10^7 **3.** 1.235×10^{-8}
4. 1.7×10^{-24} **5.** 3.14×10^{-4} **6.** 2.18×10^8
7. 789,300,000,000 **8.** 0.0000567 **9.** 5.6×10^{-15}
10. 7.462×10^{-13} **11.** 2.0×10^3 **12.** 5.5×10^2
13. $3.\overline{3} \times 10^{-2}$ **14.** 2.3725×10^9

Exercise Set 3.8, pp. 159–160

1. 4.2×10^6 **3.** 4.8×10^9 **5.** 7.8×10^{10}
7. 9.07×10^{17} **9.** 3.74×10^{-6} **11.** 1.8×10^{-8}
13. 10^7 **15.** 10^{-9} **17.** 784,000,000
19. 0.0000000008764 **21.** 100,000,000 **23.** 0.0001
25. 6×10^9 **27.** 3.38×10^4 **29.** 8.1477×10^{-13}
31. 2.5×10^{13} **33.** 5.0×10^{-4} **35.** 3.0×10^{-21}
37. $6.\overline{6} \times 10^{-2}$ **39.** $-\frac{1}{12}$ **41.** 24 **43.** 3.5×10^{-10}
45. 2.5×10^{-11}

Summary and Review: Chapter 3, pp. 161–162

1. [3.1, ▪] -22 **2.** [3.2, ▪] 7 **3.** [3.2, ▪] -192
4. [3.1, ▪] 1 **5.** [3.2, ▪] $-\frac{7}{3}$ **6.** [3.1, ▪] 25
7. [3.1, ▪] $\frac{1}{2}$ **8.** [3.2, ▪] $-\frac{15}{64}$ **9.** [3.1, ▪] 9.99
10. [3.3, ▪] -8 **11.** [3.3, ▪▪] -5 **12.** [3.3, ▪▪] $-\frac{1}{3}$
13. [3.3, ▪] 4 **14.** [3.3, ▪▪] 3 **15.** [3.3, ▪▪] 4
16. [3.3, ▪▪] 16 **17.** [3.4, ▪] 6 **18.** [3.4, ▪] -3
19. [3.4, ▪] 12 **20.** [3.4, ▪] 4 **21.** [3.6, ▪] $D = \dfrac{C}{\pi}$

22. [3.6, ▪] $B = \dfrac{3V}{h}$ **23.** [3.6, ▪] $a = 2A - b$

24. [3.5, ▪] 9 m, 7 m **25.** [3.5, ▪] 9
26. [3.5, ▪] 57, 59 **27.** [3.5, ▪] width = 11 cm,
length = 17 cm **28.** [3.5, ▪] $220
29. [3.5, ▪] $26,087 **30.** [3.5, ▪] 35°, 85°, 60°

31. [3.7, ▪] y^{-4} **32.** [3.7, ▪] $\dfrac{1}{5^3}$ **33.** [3.7, ▪▪] x^{-2}

34. [3.7, ▪▪▪] t^9 **35.** [3.7, ▪] 7^{-10} **36.** [3.7, ▪▪▪] 4^{-15}
37. [3.7, ▪▪] 8^9 **38.** [3.7, ▪▪] $81a^{-24}$
39. [3.7, ▪▪] $x^{10}y^{-5}z^{-35}$ **40.** [3.7, ▪▪] $5180.12
41. [3.8, ▪] 2.78×10^{-5} **42.** [3.8, ▪] 3.9×10^9
43. [3.8, ▪] 0.00000005 **44.** [3.8, ▪] 12,800
45. [3.8, ▪] 2.09×10^4 **46.** [3.8, ▪] 5.12×10^{-5}
47. [3.8, ▪▪▪] 6.205×10^{10} **48.** [3.8, ▪▪] 2.65×10^2
49. [1.3, ▪▪▪] $\frac{11}{8}$ **50.** [1.3, ▪▪▪] $\frac{15}{32}$ **51.** [1.3, ▪▪▪] $\frac{1}{8}$
52. [1.7, ▪▪] $\frac{6}{5}$ **53.** [2.5, ▪▪] 45.78 **54.** [2.5, ▪▪] $-\frac{1}{24}$
55. [2.5, ▪▪▪] 5.5 **56.** [2.6, ▪▪▪] 2.3
57. [2.8, ▪▪] $-43x + 8y$ **58.** [3.3, ▪] $-23, 23$
59. [3.2, ▪▪] No solution **60.** [3.3, 3.4] 4
61. [3.8, ▪▪], [3.2, ▪] 2.0×10^3
62. [3.5, ▪] Amazon: 6437 km; Nile: 6671 km
63. [3.5, ▪] $14,150

CUMULATIVE REVIEW: CHAPTERS 1–3, pp. 165–166

1. [1.1, ▩] $\frac{3}{2}$ **2.** [1.1, ▩] $\frac{15}{4}$ **3.** [1.4, ▩] 24
4. [1.5, ▩] 52 cm **5.** [1.1, ▩] $2w - 4$
6. [1.1, ▩] $34\%t$ **7.** [1.2, ▩] $2 \cdot 2 \cdot 2 \cdot 3 \cdot 3 \cdot 3 \cdot 3$
8. [1.2, ▩] 120 **9.** [1.3, ▩] $\frac{5}{9}$ **10.** [1.10, ▩] 0.026
11. [1.10, ▩] $\frac{80}{100}$ or $\frac{4}{5}$ **12.** [1.10, ▩] 190%
13. [1.10, ▩] 87.5% **14.** [2.1, ▩] > **15.** [2.1, ▩] >
16. [2.1, ▩] < **17.** [2.5, ▩], [2.6, ▩] $-\frac{2}{5}, \frac{2}{5}, \frac{5}{2}$
18. [2.1, ▩] -10 **19.** [2.5, ▩] -4.4 **20.** [2.5, ▩] $-\frac{5}{2}$
21. [2.6, ▩] $\frac{5}{6}$ **22.** [2.6, ▩] -105 **23.** [2.4, ▩] -9
24. [2.6, ▩] -0.3 **25.** [2.6, ▩] $\frac{32}{125}$
26. [2.7, ▩] $15x + 25y + 10z$ **27.** [2.7, ▩] $-12x - 8$
28. [2.7, ▩] $-12y + 24x$
29. [1.6, ▩], [2.7, ▩] $2(32 + 9x + 12y)$
30. [2.7, ▩] $8(2y - 7)$ **31.** [2.7, ▩] $5(a - 3b + 5)$
32. [1.6, ▩] $15b + 22y$ **33.** [1.6, ▩] $9y + 6z + 4$
34. [2.7, ▩] $-3a - 9d + 1$ **35.** [2.7, ▩] $-2.6x - 5.2y$
36. [2.8, ▩] $3x - 1$ **37.** [2.8, ▩] $-2x - y$
38. [2.8, ▩] $-7x + 6$ **39.** [2.8, ▩] $8x$
40. [2.8, ▩] $5x - 13$ **41.** [3.1, ▩] 4.5 **42.** [3.2, ▩] $\frac{4}{25}$
43. [3.1, ▩] 10.9 **44.** [3.1, ▩] $\frac{23}{6}$ **45.** [3.2, ▩] -48
46. [3.2, ▩] 12 **47.** [3.2, ▩] -6.2 **48.** [3.3, ▩] -3
49. [3.3, ▩] $-\frac{12}{5}$ **50.** [3.3, ▩] 8 **51.** [3.4, ▩] 7
52. [3.3, ▩] $-\frac{4}{5}$ **53.** [3.3, ▩] $-\frac{10}{3}$
54. [3.6, ▩] $h = \dfrac{2A}{b + c}$ **55.** [1.10, ▩] 30%
56. [1.10, ▩] 50 **57.** [3.5, ▩] 154 **58.** 3.5, ▩] $45
59. [3.5, ▩] $1500 **60.** [3.5, ▩] 50 m, 53 m, 40 m
61. [3.7, ▩] x^6 **62.** [3.7, ▩] $16x^{-8}y^{20}$
63. [3.7, ▩] x^{-7} **64.** [2.9, ▩] -15
65. [2.8, ▩] $-4x - 20$ **66.** [1.10, ▩] 6 ft, 7 in.
67. [3.3, ▩] $-4, 4$ **68.** [3.4, ▩] All rational numbers
69. [3.4, ▩] No solution **70.** [3.3, 3.4] 3
71. [3.4, ▩] All rational numbers
72. [3.6, ▩] $Q = \dfrac{2 - PM}{P}$, or $\dfrac{2}{P} - M$

CHAPTER 4

Margin Exercises, Section 4.1, pp. 168–170

1. $x^2 + 2x - 8$; $x^3 + x^2 - 2x - 5$; $x - 7$, answers may vary **2.** -19 **3.** -104 **4.** -13 **5.** 8 **6.** 90 **7.** 360 ft
8. $-9x^3 + (-4x^5)$ **9.** $-2y^3 + 3y^7 + (-7y)$
10. $3x^2, 6x, \frac{1}{2}$ **11.** $-4y^5, 7y^2, -3y, -2$
12. $4x^3$ and $-x^3$ **13.** $4t^4$ and $-7t^4$; $-9t^3$ and $10t^3$
14. $8x^2$ **15.** $2x^3 + 7$ **16.** $-\frac{1}{4}x^5 + 2x^2$ **17.** $-4x^3$
18. $5x^3$ **19.** $25 - 3x^5$ **20.** $6x$ **21.** $4x^3 + 4$
22. $-\frac{1}{4}x^3 + 4x^2 + 7$ **23.** $3x^2 + x^3 + 9$

Exercise Set 4.1, pp. 171–172

1. -18 **3.** 19 **5.** -12 **7.** Approx. 449 **9.** 2 **11.** 4
13. 11 **15.** 2, $-3x$, x^2 **17.** $6x^2$ and $-3x^2$
19. $2x^4$ and $-3x^4$, $5x$ and $-7x$ **21.** $-3x$ **23.** $-8x$
25. $11x^3 + 4$ **27.** $x^3 - x$ **29.** $4b^5$ **31.** $\frac{3}{4}x^5 - 2x - 42$
33. x^4 **35.** $\frac{15}{16}x^3 - \frac{7}{6}x^2$ **37.** $3s + 3t + 24$ **39.** $5x$
41. $3x^2 + 2x + 1$ **43.** $5x^9 + 4x^8 + x^2 + 5x$

Margin Exercises, Section 4.2, pp. 173–174

1. $6x^7 + 3x^5 - 2x^4 + 4x^3 + 5x^2 + x$
2. $7x^5 - 5x^4 + 2x^3 + 4x^2 - 3$
3. $14t^7 - 10t^5 + 7t^2 - 14$ **4.** $-2x^2 - 3x + 2$
5. $10x^4 - 8x - \frac{1}{2}$ **6.** 4, 2, 1, 0,; 4 **7.** 5, 6, 1, -1, 4
8. x **9.** x^3, x^2, x, x^0 **10.** x^2, x **11.** x^3 **12.** Monomial
13. None of these **14.** Binomial **15.** Trinomial

Exercise Set 4.2, pp. 175–176

1. $x^5 + 6x^3 + 2x^2 + x + 1$
3. $15x^9 + 7x^8 + 5x^3 - x^2 + x$
5. $-5y^8 + y^7 + 9y^6 + 8y^3 - 7y^2$ **7.** $x^6 + x^4$
9. $13x^3 - 9x + 8$ **11.** $-5x^2 + 9x$ **13.** $12x^4 - 2x + \frac{1}{4}$
15. 3, 2, 1, 0; 3 **17.** 2, 1, 6, 4; 6 **19.** $-3, 6$
21. 6, 7, -8, -2 **23.** x^2, x **25.** x^3, x^2, x^0
27. None missing **29.** Trinomial **31.** None of these
33. Binomial **35.** Monomial **37.** 6 **39.** 30 **41.** 10

Margin Exercises, Section 4.3, pp. 177–178

1. $x^2 + 7x + 3$ **2.** $-4x^5 + 7x^4 + 3x^3 + 2x^2 + 4$
3. $24x^4 + 5x^3 + x^2 + 1$ **4.** $2x^3 + \frac{10}{3}$ **5.** $2x^2 - 3x - 1$
6. $8x^3 - 2x^2 - 8x + \frac{5}{2}$ **7.** $-8x^4 + 4x^3 + 12x^2 + 5x - 8$
8. $-x^3 + x^2 + 3x + 3$ **9.** 224 **10.** $\frac{7}{2}x^2$ **11.** 224

Exercise Set 4.3, pp. 179–182

1. $-x + 5$ **3.** $x^2 - 5x - 1$ **5.** $3y^5 + 13y^2 + 6y - 3$
7. $-4x^4 + 6x^3 + 6x^2 + 2x + 4$ **9.** $12x^2 + 6$
11. $5x^4 - 2x^3 - 7x^2 - 5x$
13. $9x^8 + 8x^7 - 3x^4 + 2x^2 - 2x + 5$
15. $-\frac{1}{2}x^4 + \frac{2}{3}x^3 + x^2$
17. $0.01x^5 + x^4 - 0.2x^3 + 0.2x + 0.06$
19. $-3t^4 + 3t^2 + 4t$ **21.** $3x^5 - 3x^4 - 3x^3 + x^2 + 3x$
23. $5x^3 - 9x^2 + 4x - 7$ **25.** $\frac{1}{4}x^4 -$
$\frac{1}{4}x^3 + \frac{3}{2}x^2 + 6\frac{3}{4}x + \frac{1}{4}$ **27.** $-x^4 + 3x^3 + 2x + 1$
29. $x^4 + 4x^2 + 12x - 1$ **31.** $x^5 - 6x^4 + 4x^3 - x^2 + 1$
33. $7p^4 - 2p^3 + 7p^2 + 4p + 9$

35. $3x^5 + x^4 + 10x^3 + x^2 + 3x - 6$
37. $1.05x^4 + 0.36x^3 + 14.22x^2 + x + 0.97$
39. (a) $5x^2 + 4x$; **(b)** $57{,}352$ **41.** x^{10} **43.** 9400
45. (a) Compare $(ax + b) + (cx + d)$ and
$(cx + d) + (ax + b)$. $(ax + b) + (cx + d) =$
$ax + b + cx + d = (a + c)x + (b + d) =$
$(c + a)x + (d + b)$; by the commutative law of
addition; $(cx + d) + (ax + b) = cx + d + ax + b =$
$(c + a)x + d + b$; **(b)** similar to (a).

Margin Exercises, Section 4.4, pp. 183–184

1. $-(12x^4 - 3x^2 + 4x)$; $-12x^4 + 3x^2 - 4x$
2. $-(-4x^4 + 3x^2 - 4x)$; $4x^4 - 3x^2 + 4x$
3. $-(-13x^6 + 2x^4 - 3x^2 + x - \frac{5}{13})$;
$13x^6 - 2x^4 + 3x^2 - x + \frac{5}{13}$
4. $-(-7y^3 + 2y^2 - y + 3)$; $7y^3 - 2y^2 + y - 3$
5. $-4x^3 + 6x - 3$ **6.** $-5x^4 - 3x^2 - 7x + 5$
7. $-14x^{10} + \frac{1}{2}x^5 - 5x^3 + x^2 - 3x$ **8.** $2x^3 + 2x + 8$
9. $x^2 - 6x - 2$ **10.** $-8x^4 - 5x^3 + 8x^2 - 1$
11. $x^3 - x^2 - \frac{4}{3}x - 0.9$ **12.** $2x^3 + 5x^2 - 2x - 5$
13. $-x^5 - 2x^3 + 3x^2 - 2x + 2$

Exercise Set 4.4, pp. 185–186

1. $-(-5x)$, $5x$ **3.** $-(-x^2 + 10x - 2)$, $x^2 - 10x + 2$
5. $-(12x^4 - 3x^3 + 3)$, $-12x^4 + 3x^3 - 3$ **7.** $-3x + 7$
9. $-4x^2 + 3x - 2$ **11.** $4x^4 + 6x^2 - \frac{3}{4}x + 8$
13. $2x^2 + 14$ **15.** $-2x^5 - 6x^4 + x + 2$
17. $9x^2 + 9x - 8$ **19.** $\frac{3}{4}x^3 - \frac{1}{2}x$
21. $0.06x^3 - 0.05x^2 + 0.01x + 1$ **23.** $3x + 6$
25. $4x^3 - 3x^2 + x + 1$
27. $11x^4 + 12x^3 - 9x^2 - 8x - 9$
29. $-4x^5 + 9x^4 + 6x^2 + 16x + 6$ **31.** $x^4 - x^3 + x^2 - x$
33. $8x - 16$ **35.** 1 **37.** $569.607x^3 - 15.168x$
39. $-3y^4 - y^3 + 5y - 2$

Margin Exercises, Section 4.5, pp. 187–189

1. $-15x$ **2.** $-x^2$ **3.** x^2 **4.** $-x^5$ **5.** $12x^7$ **6.** $-8y^{11}$
7. $7y^5$ **8.** 0 **9.** $8x^2 + 16x$ **10.** $-15t^3 + 6t^2$
11. $x^2 + 13x + 40$ **12.** $x^2 + x - 20$
13. $5x^2 - 17x - 12$ **14.** $6x^2 - 19x + 15$
15. $x^4 + 3x^3 + x^2 + 15x - 20$
16. $6y^5 - 20y^3 + 15y^2 + 14y - 35$
17. $3x^3 + 13x^2 - 6x + 20$
18. $20x^4 - 16x^3 + 32x^2 - 32x - 16$
19. $6x^4 - x^3 - 18x^2 - x + 10$

Something Extra—Expanded Notation and Polynomials, p. 190

1. $8 \cdot 10^3 + 7 \cdot 10^2 + 6 \cdot 10 + 2$; $8x^3 + 7x^2 + 6x + 2$
2. $7 \cdot 10^2 + 8 \cdot 10 + 6$; $7x^2 + 8x + 6$

3. $1 \cdot 10^4 + 6 \cdot 10^3 + 4 \cdot 10^2 + 3 \cdot 10 + 2$;
$x^4 + 6x^3 + 4x^2 + 3x + 2$ **4.** $7 \cdot 10^3 + 6 \cdot 10 + 3$;
$7x^3 + 6x + 3$

Exercise Set 4.5, pp. 191–192

1. $-12x$ **3.** $42x^2$ **5.** $30x$ **7.** $-2y^3$ **9.** x^6 **11.** $6x^6$
13. 0 **15.** $-0.02x^{10}$ **17.** $8x^2 - 12x$ **19.** $-6x^4 - 6x^3$
21. $4x^6 - 24x^5$ **23.** $-x^2 - 4x + 12$
25. $2y^2 - 15y + 25$ **27.** $9x^2 - 25$ **29.** $2x^2 + \frac{5}{2}x - \frac{3}{4}$
31. $x^3 + 7x^2 + 7x + 1$ **33.** $-10q^3 - 19q^2 - q + 3$
35. $3x^4 - 6x^3 - 7x^2 + 18x - 6$
37. $6t^4 + t^3 - 16t^2 - 7t + 4$ **39.** $x^4 - 1$
41. $x = \dfrac{y - b}{m}$ **43.** t^{-20} **45.** $2x^2 + 18x + 36$
47. $v = 4x^3 - 48x^2 + 144x$, $S = -4x^2 + 144$

Margin Exercises, Section 4.6, pp. 193–194

1. $8x^3 - 12x^2 + 16x$ **2.** $10y^6 + 8y^5 - 10y^4$
3. $x^2 + 7x + 12$ **4.** $x^2 - 2x - 15$ **5.** $2x^2 + 9x + 4$
6. $2x^3 - 4x^2 - 3x + 6$ **7.** $12x^5 + 6x^2 + 10x^3 + 5$
8. $y^6 - 49$ **9.** $-2x^7 + x^5 + x^3$ **10.** $x^2 - \frac{16}{25}$
11. $x^5 + 0.5x^3 - 0.5x^2 - 0.25$ **12.** $8 + 2x^2 - 15x^4$
13. $30x^5 - 3x^4 - 6x^3$

Exercise Set 4.6, pp. 195–196

1. $4x^2 + 4x$ **3.** $-3x^2 + 3x$ **5.** $x^5 + x^2$
7. $6x^3 - 18x^2 + 3x$ **9.** $x^3 + 3x + x^2 + 3$
11. $x^4 + x^3 + 2x + 2$ **13.** $x^2 - x - 6$
15. $9x^2 + 15x + 6$ **17.** $15x^2 + 4x - 12$ **19.** $9x^2 - 1$
21. $4x^2 - 6x + 2$ **23.** $x^2 - \frac{1}{16}$ **25.** $t^2 - 0.01$
27. $2x^3 + 2x^2 + 6x + 6$ **29.** $-2x^2 - 11x + 6$
31. $a^2 + 14a + 49$ **33.** $1 - x - 6x^2$
35. $x^5 - x^2 + 3x^3 - 3$ **37.** $x^3 - x^2 - 2x + 2$
39. $3q^6 - 6q^2 - 2q^4 + 4$ **41.** $6x^7 + 18x^5 + 4x^2 + 12$
43. $8x^6 + 65x^3 + 8$ **45.** $4x^3 - 12x^2 + 3x - 9$
47. $4x^6 + 4x^5 + x^4 + x^3$ **49.** 0.000047
51. $8y^3 + 72y^2 + 160y$ **53.** -7

Margin Exercises, Section 4.7, pp. 197–200

1. $x^2 - 25$ **2.** $4x^2 - 9$ **3.** $x^2 - 4$ **4.** $x^2 - 49$
5. $36 - 16y^2$ **6.** $4x^6 - 1$ **7.** $x^2 + 16x + 64$
8. $x^2 - 10x + 25$ **9.** $x^2 + 4x + 4$ **10.** $a^2 - 8a + 16$
11. $4x^2 + 20x + 25$ **12.** $16x^4 - 24x^3 + 9x^2$
13. $49 + 14y + y^2$ **14.** $9x^4 - 30x^2 + 25$
15. $x^2 + 11x + 30$ **16.** $t^2 - 16$
17. $-8x^5 + 20x^4 + 40x^2$ **18.** $81x^4 + 18x^2 + 1$
19. $4a^2 + 6a - 40$ **20.** $25x^2 + 5x + \frac{1}{4}$
21. $4x^2 - 2x + \frac{1}{4}$

Something Extra—Factors and Sums, p. 200

56	63	−36	−72	140	−96	48	168	−110	90	−432	63
−7	−7	−18	36	−14	−12	−6	−21	−11	−9	−24	−3
−8	−9	2	−2	−10	8	−8	−8	10	−10	18	−21
−15	−16	−16	34	−24	−4	−14	−29	−1	−19	−6	−24

Exercise Set 4.7, pp. 201–202

1. $x^2 - 16$ 3. $4x^2 - 1$ 5. $25m^2 - 4$ 7. $4x^4 - 9$
9. $9x^8 - 16$ 11. $x^{12} - x^4$ 13. $a^8 - 9a^2$ 15. $x^{24} - 9$
17. $4x^{16} - 9$ 19. $x^2 + 4x + 4$ 21. $9x^4 + 6x^2 + 1$
23. $x^2 - x + \frac{1}{4}$ 25. $9 + 6x + x^2$ 27. $y^4 + 2y^2 + 1$
29. $4 - 12x^4 + 9x^8$ 31. $25 + 60t^2 + 36t^4$
33. $9 - 12x^3 + 4x^6$ 35. $4x^3 + 24x^2 - 12x$
37. $4x^4 - 2x^2 + \frac{1}{4}$ 39. $-1 + 9p^2$
41. $15t^5 - 3t^4 + 3t^3$ 43. $36x^8 + 48x^4 + 16$
45. $12x^3 + 8x^2 + 15x + 10$ 47. $64 - 96x^4 + 36x^8$
49. TV: 50 watts; lamps: 500 watts; AC: 2000 watts
51. $4567.0564x^2 + 435.891x + 10.400625$
53. $625t^8 - 450t^4 + 81$

Summary and Review: Chapter 4, pp. 203–204

1. [4.1, ■■] -17 2. [4.1, ■■] 10 3. [4.1, ■■] $3x^2, 6x, \frac{1}{2}$
4. [4.1, ■■] $-4y^5, 7y^2, -3y, -2$ 5. [4.2, ■■] x^2, x^0
6. [4.2, ■■] $6, 17$ 7. [4.2, ■■] $4, 6, -5, 0.43$
8. [4.2, ■■] $3, 1, 0; 3$ 9. [4.2, ■■] $0, 4, 9, 6, 3; 9$
10. [4.2, ■■] Binomial 11. [4.2, ■■] None
12. [4.2, ■■] Monomial 13. [4.2, ■■] $-x^2 + 9x$
14. [4.2, ■■] $-2t^3 + 4t^2 + 7$ 15. [4.2, ■■] $-3x^5 + 25$
16. [4.2, ■■] $-2y^2 - 3y + 2$
17. [4.2, ■■] $10x^4 - 7x^2 - x - \frac{1}{2}$
18. [4.3, ■■] $x^5 - 2x^4 + 6x^3 + 3x^2 - 9$
19. [4.3, ■■] $2t^5 - 6t^4 + 2t^3 - 2t^2 + 2$
20. [4.3, ■■] $\frac{1}{4}x^3 - \frac{1}{3}x^2 - \frac{7}{4}x + \frac{3}{8}$
21. [4.4, ■■] $2x^2 - 4x - 6$
22. [4.4, ■■] $x^5 - 3x^3 - 2x^2 + 8$
23. [4.4, ■■] $-y^5 + y^4 - 5y^3 - 2y^2 + 2y$
24. [4.5, ■■] $-12x^3$ 25. [4.7, ■■] $49x^2 + 14x + 1$
26. [4.6, ■■] $y^2 + \frac{7}{6}y + \frac{1}{3}$
27. [4.6, ■■] $0.3t^2 + 0.65t - 8.45$
28. [4.5, ■■] $12x^3 - 23x^2 + 13x - 2$
29. [4.7, ■■] $x^2 - 18x + 81$
30. [4.6, ■■] $15x^7 - 40x^6 + 50x^5 + 10x^4$
31. [4.6, ■■] $x^2 - 3x - 28$
32. [4.6, ■■] $x^2 - 1.05x + 0.225$
33. [4.5, ■■] $x^7 + x^5 - 3x^4 + 3x^3 - 2x^2 + 5x - 3$
34. [4.7, ■■] $9y^4 - 12y^3 + 4y^2$
35. [4.6, ■■] $2t^4 - 11t^2 - 21$
36. [4.5, ■■] $4x^5 - 5x^4 - 8x^3 + 22x^2 - 15x$
37. [4.7, ■■] $9x^4 - 16$ 38. [4.7, ■■] $4 - m^2$
39. [4.6, ■■] $13x^2 - 172x + 39$ 40. [1.10, ■■] 74.1

41. [1.10, ■■] 23% 42. [2.7, ■■] $16(4t - 2m + 1)$
43. [2.7, ■■] $-24x + 40y - 32$
44. [2.7, ■■] $7x + y - 11$ 45. [3.3, ■■] 13
46. [3.3, ■■] $\frac{172}{13}$ 47. [3.5, ■■] $100°, 25°, 55°$
48. [4.6, ■■] $\frac{94}{13}$ 49. [4.7, ■■, ■■] $1 - 128t^2 + 4096t^4$
50. [4.2, ■■] $-28x^8$ 51. [4.4, ■■] $4x^3 + 2x^2 + x + 2$
52. [4.4, ■■], [4.6, ■■] $-6x$

CHAPTER 5

5

Margin Exercises, Section 5.1, pp. 208–210

1. (a) $12x^2$; (b) $3x \cdot 4x$, $2x \cdot 6x$, answers may vary
2. (a) $16x^3$; (b) $(2x)(8x^2)$, $(4x)(4x^2)$, answers may vary
3. $8x \cdot x^3$; $4x^2 \cdot 2x^2$; $2x^3 \cdot 4x$, answers may vary
4. $7x \cdot 3x$, $(-7x)(-3x)$, $(21x)(x)$, answers may vary
5. $6x^4 \cdot x$, $(-2x^3)(-3x^2)$, $(3x^3)(2x^2)$, answers may vary
6. (a) $3x + 6$; (b) $3(x + 2)$
7. (a) $2x^3 + 10x^2 + 8x$; (b) $2x(x^2 + 5x + 4)$ 8. $x(x + 3)$
9. $y^2(3y^4 - 5y + 2)$ 10. $3x^2(3x^2 - 5x + 1)$
11. $\frac{1}{4}(3t^3 + 5t^2 + 7t + 1)$ 12. $7x^3(5x^4 - 7x^3 + 2x^2 - 9)$
13. $(x + 2)(x + 5)$ 14. $(y + 3)(y - 4)$
15. $(2x^2 - 3)(2x - 3)$ 16. $(2t^2 + 3)(4t + 1)$
17. $(3m^3 + 2)(m^2 - 5)$

Exercise Set 5.1, pp. 211–212

1. $6x^2 \cdot x$, $3x^2 \cdot 2x$, $(-3x^2)(-2x)$, answers may vary
3. $(-9x^4) \cdot x$, $(-3x^2)(3x^3)$, $(-3x)(3x^4)$, answers may vary
5. $(8x^2)(3x^2)$, $(-8x^2)(-3x^2)$, $(4x^3)(6x)$, answers may vary
7. $x(x - 4)$ 9. $x^2(x + 6)$ 11. $8x^2(x^2 - 3)$
13. $17x(x^4 + 2x^2 + 3)$ 15. $5(2x^3 + 5x^2 + 3x - 4)$
17. $\frac{x^3}{3}(5x^3 + 4x^2 + x + 1)$ 19. $(y + 1)(y + 4)$
21. $(x - 2)(x + 5)$ 23. $(4 - x)(4 - 3x)$
25. $(2x^2 + 1)(x + 3)$ 27. $(x^2 - 3)(x + 8)$
29. $(4t^2 + 1)(3t - 4)$ 31. $(2x^3 + 3)(2x^2 + 3)$
33. $x^2 - 16$ 35. -2 37. $(x^4 + 1)(x^2 + 1)$ 39. No
41. Yes

Margin Exercises, Section 5.2, pp. 213–214

1. Yes 2. No 3. No 4. No 5. Yes 6. Yes 7. Yes
8. $(x - 3)(x + 3)$ 9. $(t + 8)(t - 8)$
10. $8y^2(2 + y^2)(2 - y^2)$ 11. $x^4(8 + 5x)(8 - 5x)$
12. $5(1 + 2t^3)(1 - 2t^3)$ 13. $(9x^2 + 1)(3x + 1)(3x - 1)$
14. $m^4(7 + 5m^3)(7 - 5m^3)$

Exercise Set 5.2, pp. 215–216

1. Yes 3. No 5. Yes 7. No 9. $(x - 2)(x + 2)$
11. $(t + 3)(t - 3)$ 13. $(4a - 3)(4a + 3)$
15. $(2x - 5)(2x + 5)$ 17. $2(2x - 7)(2x + 7)$
19. $x(6 - 7x)(6 + 7x)$ 21. $y^2(4 - 5y)(4 + 5y)$
23. $(7a^2 - 9)(7a^2 + 9)$ 25. $a^2(a^5 + 2)(a^5 - 2)$
27. $(9y^3 - 5)(9y^3 + 5)$ 29. $(x^2 + 1)(x - 1)(x + 1)$
31. $4(x^2 + 4)(x + 2)(x - 2)$
33. $(1 + y^4)(1 + y^2)(1 + y)(1 - y)$
35. $(x^6 + 4)(x^3 + 2)(x^3 - 2)$ 37. $(\frac{1}{4} + y)(\frac{1}{4} - y)$
39. $(5 - \frac{1}{7}y)(5 + \frac{1}{7}y)$ 41. $(4 + t^2)(2 + t)(2 - t)$
43. $t^2 - 18t + 81$ 45. 80 47. $(1.8x + 0.9)(1.8x - 0.9)$
49. $2(0.8t + 1)(0.8t - 1)$ 51. $(x^2 + 1)(x - 1)^2(x + 1)^2$

Margin Exercises, Section 5.3, pp. 217–218

1. Yes 2. No 3. No 4. Yes 5. No 6. Yes 7. No
8. Yes 9. $(x + 1)^2$ 10. $(x - 1)^2$ 11. $(t + 2)^2$
12. $(5x - 7)^2$ 13. $(7 - 4y)^2$ 14. $3(4m + 5)^2$

Exercise Set 5.3, pp. 219–220

1. Yes 3. No 5. No 7. No 9. $(x - 7)^2$ 11. $(y + 8)^2$
13. $(m - 1)^2$ 15. $(x + 2)^2$ 17. $2(x - 1)^2$ 19. $x(x - 9)^2$
21. $5(2t + 5)^2$ 23. $(7 - 3x)^2$ 25. $5(y^2 + 1)^2$
27. $(y^3 + 13)^2$ 29. $(4x^5 - 1)^2$ 31. $x^2 + 2x - 24$
33. 3.125 L 35. $0.1(9x + 8)(9x - 8)$ 37. $(2a + 15)^2$

Margin Exercises, Section 5.4, pp. 221–224

1. $(x + 4)(x + 3)$ 2. $(x + 9)(x + 4)$ 3. $(x - 5)(x - 3)$
4. $(t - 5)(t - 4)$ 5. $(x + 6)(x - 2)$ 6. $(y - 6)(y + 2)$
7. $(t + 7)(t - 2)$ 8. $(x - 6)(x + 5)$ 9. Not factorable
10. $(y + 4)(y + 4)$, or $(y + 4)^2$

Something Extra—Calculator Corner: Nested Evaluation, p. 224

1. $x(x(x(x(x + 3) - 1) + 1) - 1) + 9$; 6432.0122;
170,616.6106; −5,584,092
2. $x(x(x(5x - 17) + 2) - 1) + 11$; 1476.2115;
43,101.696; 1,607,136
3. $x(x(x(2x - 3) + 5) - 2) + 18$; 1279.3152;
22,235.0912; 598,892
4. $x(x(x(x(-2x + 4) + 8) - 4) - 3) + 24$; −4120.9615;
−208,075.1491; 13,892,691

Exercise Set 5.4, pp. 225–226

1. $(x + 5)(x + 3)$ 3. $(x + 1)(x - 2)$ 5. $(x + 3)(x + 4)$
7. $(x + 5)(x - 3)$ 9. $(y + 8)(y + 1)$
11. $(x^2 + 2)(x^2 + 3)$ 13. $(x - 4)(x + 7)$
15. $(4 + x)(4 + x)$ 17. $(a - 11)(a - 1)$
19. $\left(x - \frac{1}{5}\right)\left(x - \frac{1}{5}\right)$ 21. $(y - 0.4)(y + 0.2)$
23. $(y + 7)(y + 4)$ 25. $(a + 6)(a + 5)$
27. $(x - 7)(x + 6)$ 29. $(x - 11)(x + 9)$
31. $(x + 12)(x - 6)$ 33. $(x + 10)^2$
35. $16x^3 - 48x^2 + 8x$ 37. 2 39. $(x + \frac{1}{4})(x - \frac{3}{4})$
41. $15, -15, 27, -27, 51, -51$

Margin Exercises, Section 5.5, pp. 227–231

1. $(2x + 5)(x - 3)$ 2. $(4x + 1)(3x - 5)$
3. $(3x - 4)(x - 5)$ 4. $2(5x - 4)(2x - 3)$
5. $(2x + 1)(3x + 2)$ 6. $3(2x + 3)(x + 1)$
7. $2(y + 3)(y - 1)$ 8. $2(2t + 3)(t - 1)$
9. $(2x - 1)(3x - 1)$

Something Extra—Calculator Corner: A Number Pattern, p. 232

1. 1, 1, 1, 1, 1, 1 2. 1, 1, 1 3. $x^2 - (x + 1)(x - 1) = 1$;
$x^2 - [x^2 - 1] = x^2 - x^2 + 1 = 1$

Exercise Set 5.5, pp. 233–234

1. $(3x + 1)(x + 1)$ 3. $4(3x - 2)(x + 3)$
5. $(2x + 1)(x - 1)$ 7. $(3x - 2)(3x + 8)$
9. $5(3x + 1)(x - 2)$ 11. $(3x + 4)(4x + 5)$
13. $(7x - 1)(2x + 3)$ 15. $(3x^2 + 2)(3x^2 + 4)$
17. $(3x - 7)(3x - 7)$ 19. $2x(3x + 5)(x - 1)$
21. $(3a - 1)(3a + 5)$ 23. $(5x + 4)^2$
25. $(1 + 3m)(7 + 2m)$ 27. $2(3x + 5)(x - 1)$
29. $x(3x + 4)(4x + 5)$ 31. $x^2(2x + 3)(7x - 1)$
33. Not factorable 35. y^{-4} 37. 2
39. $(3x + 7)(3x - 7)(3x - 7)$ 41. $(x^{3a} - 1)(3x^{3a} + 1)$

Margin Exercises, Section 5.6, pp. 235–236

1. $3(m^2 + 1)(m - 1)(m + 1)$ 2. $(x^3 + 4)^2$
3. $2x^2(x + 1)(x + 3)$ 4. $(3x^2 - 2)(x + 4)$
5. $8x(x - 5)(x + 5)$

Exercise Set 5.6, pp. 237–238

1. $2(x - 8)(x + 8]$ 3. $(a - 5)^2$ 5. $(2x - 3)(x - 4)$
7. $x(x + 12)^2$ 9. $(x + 3)(x + 2)$ 11. $6(2x + 3)(2x - 3)$
13. $4x(5x + 9)(x - 2)$ 15. Not factorable
17. $(x^2 + 7)(x^2 - 3)$ 19. $x^3(x - 7)^2$
21. $2(2 - x)(5 + x)$ 23. Not factorable
25. $4(x^2 + 4)(x + 2)(x - 2)$
27. $(1 + y^4)(1 + y^2)(1 + y)(1 - y)$ 29. $x^3(x^2 - 4x - 3)$
31. $\left(6a - \frac{5}{4}\right)^2$ 33. $(a - 1)^2(a + 1)^2$
35. About 7,397,895 37. $\frac{22}{9}$ 39. $(3.5x - 1)^2$
41. $5(x + 1.8)(x + 0.8)$ 43. $(y - 1)^3$

Margin Exercises, Section 5.7, pp. 239–241

1. 3, −4 2. 7, 3 3. $-\frac{1}{4}, \frac{2}{3}$ 4. 0, $\frac{17}{3}$ 5. −2, 3 6. 7, −4
7. 3 8. 0, 4 9. 4, −4

Something Extra—Calculator Corner: Number Patterns, p. 242

1. 1, 4, 9, 16, 25, 36; 49, 64 2. $\frac{8 \cdot 9}{2}$ or 36,
$\frac{9 \cdot 10}{2}$ or 45; $\frac{n(n + 1)}{2}$

Exercise Set 5.7, pp. 243–244

1. −8, −6 3. 3, −5 5. 12, 11 7. 0, 13 9. 0, −21

11. $-\dfrac{5}{2}$, -4 13. $\dfrac{1}{3}$, -2 15. 0, $\dfrac{2}{3}$ 17. 2, 6 19. $\dfrac{1}{3}$, 20

21. -1, -5 23. 0, 5 25. -3 27. $\dfrac{5}{3}$, -1 29. 4, $-\dfrac{5}{3}$

31. $\dfrac{2}{3}$, $-\dfrac{1}{4}$ 33. 0, $\dfrac{3}{5}$ 35. 7, -2 37. $\dfrac{9}{8}$, $-\dfrac{9}{8}$ 39. -144

41. 4, -5 43. 5, 3 45. -2000, -51.546 47. -3, $\dfrac{5}{4}$, 7

Margin Exercises, Section 5.8, pp. 245–248

1. 5, -5 2. 7, 8 3. -4, 5
4. Length is 5 cm; width is 3 cm 5. (a) 342; (b) 9
6. 21 and 22; -22 and -21

Exercise Set 5.8, pp. 249–252

1. $-\dfrac{3}{4}$, 1 3. 13 and 14, -13 and -14 5. 13 and 15,
-13 and -15 7. 5 9. 6 m 11. 506 13. 12 15. 120

17. 20 19. $h = \dfrac{S}{2\pi r}$ 21. -100.1 23. 37 25. 5 in.

Margin Exercises, Section 5.9, pp. 253–258

1. -7940 2. -176 3. 32 4. -3, 3, -2, 1, 2
5. 3, 7, 1, 1, 0; 7 6. $2x^2y + 3xy$ 7. $5pq + 4$
8. $-4x^3 + 2x^2 - 4x + 2$ 9. $14x^3y + 7x^2y - 3xy - 2y$
10. $-5p^2q^4 + 2p^2q^2 + 3p^2q + 6pq^2 + 3q + 5$
11. $-8s^4t + 6s^3t^2 + 2s^2t^3 - s^2t^2$
12. $-9p^4q + 10p^3q^2 - 4p^2q^3 - 9q^4$
13. $x^5y^5 + 2x^4y^2 + 3x^3y^3 + 6x^2$
14. $p^5q - 4p^3q^3 + 3pq^3 + 6q^4$
15. $3x^3y + 6x^2y^3 + 2x^3 + 4x^2y^2$
16. $2x^2 - 11xy + 15y^2$ 17. $16x^2 + 40xy + 25y^2$
18. $9x^4 - 12x^3y^2 + 4x^2y^4$ 19. $4x^2y^4 - 9x^2$
20. $16y^2 - 9x^2y^4$ 21. $9y^2 + 24y + 16 - 9x^2$
22. $4a^2 - 25b^2 - 10bc - c^2$ 23. $x^2y(x^2y + 2x + 3)$
24. $2p^4q^2(5p^2 - 2pq + q^2)$ 25. $(a - b)(2x + 5 + y^2)$
26. $(a + b)(x^2 + y)$ 27. $(x^2 + y^2)^2$ 28. $-(2x - 3y)^2$
29. $(xy + 4)(xy + 1)$ 30. $2(x^2y^3 + 5)(x^2y^3 - 2)$
31. $(5xy^2 - 2a)(5xy^2 + 2a)$

Exercise Set 5.9, pp. 259–264

1. -1 3. -7 5. \$12,597.12 7. 44.4624 in^2
9. Coefficients: 1, -2, 3, -5; degrees: 4, 2, 2, 0; 4
11. Coefficients: 17, -3, -7; degrees: 5, 5, 0; 5
13. $-a - 2b$ 15. $3x^2y - 2xy^2 + x^2$ 17. $8u^2v - 5uv^2$
19. $-8au + 10av$ 21. $x^2 - 4xy + 3y^2$ 23. $3r + 7$
25. $-x^2 - 8xy - y^2$ 27. $2ab$
29. $-2a + 10b - 5c + 8d$ 31. $6z^2 + 7zu - 3u^2$
33. $a^4b^2 - 7a^2b + 10$
35. $a^4 + a^3 - a^2y - ay + a + y - 1$ 37. $a^6 - b^2c^2$
39. $y^6x + y^4x + y^4 + 2y^2 + 1$ 41. $12x^2y^2 + 2xy - 2$
43. $12 - c^2d^2 - c^4d^4$ 45. $m^3 + m^2n - mn^2 - n^3$
47. $x^9y^9 - x^6y^6 + x^5y^5 - x^2y^2$ 49. $x^2 + 2xh + h^2$
51. $r^6t^4 - 8r^3t^2 + 16$ 53. $p^8 + 2m^2n^2p^4 + m^4n^4$
55. $4a^6 - 2a^3b^3 + \frac{1}{4}b^6$ 57. $3a^3 - 12a^2b + 12ab^2$
59. $4a^2 - b^2$ 61. $c^4 - d^2$ 63. $a^2b^2 - c^2d^4$
65. $2\pi r(h + r)$ 67. $(a + b)(2x + 1)$

69. $(x + 1)(x - 1 - y)$ 71. $(n + p)(n + 2)$
73. $(2x + z)(x - 2)$ 75. $(x - y)^2$ 77. $(3c + d)^2$
79. $(7m^2 - 8n)^2$ 81. $(y^2 + 5z^2)^2$ 83. $(\frac{1}{2}a + \frac{1}{3}b)^2$
85. $(a + b)(a - 2b)$ 87. $(m + 20n)(m - 18n)$
89. $(mn - 8)(mn + 4)$ 91. $a^3(ab + 5)(ab - 2)$
93. $a^3(a - b)(a + 5b)$ 95. $(x^3 - y)(x^3 + 2y)$
97. $(x - y)(x + y)$ 99. $(ab - 3)(ab + 3)$
101. $(3x^2y - b)(3x^2y + b)$ 103. $3(x - 4y)(x + 4y)$
105. $(8z - 5cd)(8z + 5cd)$
107. $7(p^2 + q^2)(p + q)(p - q)$
109. $(9a^2 + b^2)(3a - b)(3a + b)$ 111. $2m^2(3m + 1)^2$
113. $(x + 3)(y - 2)(y + 2)$ 115. $p(p + t)(p - 2t)$
117. $-(a + 3b)(a - 2b)$ 119. $ab(b^2 - b - 1)$
121. $4xy - 4y^2$; $4y(x - y)$ 123. $2xy + \pi x^2$; $x(2y + \pi x)$
125. $A^3 + 3A^2B + 3AB^2 + B^3$

Summary and Review: Chapter 5, pp. 265–266

1. [5.1, ■] $-10x \cdot x$, $-5x \cdot 2x$, $5x(-2x)$;
answers may vary 2. [5.1, ■] $6x \cdot 6x^4$, $4x^2 \cdot 9x^3$,
$2x^4 \cdot 18x$; answers may vary
3. [5.2, ■■■] $5(1 + 2x^3)(1 - 2x^3)$ 4. [5.1, ■■■] $x(x - 3)$
5. [5.2, ■■■] $(3x + 2)(3x - 2)$ 6. [5.4, ■■] $(x + 6)(x - 2)$
7. [5.3, ■■■] $(x + 7)^2$ 8. [5.1, ■■] $3x(2x^2 + 4x + 1)$
9. [5.1, ■■■] $(x^2 + 3)(x + 1)$
10. [5.5, ■] $(3x - 1)(2x - 1)$
11. [5.2, ■■■] $(x^2 + 9)(x + 3)(x - 3)$
12. [5.5, ■] $3x(3x - 5)(x + 3)$
13. [5.2, ■■■] $2(x + 5)(x - 5)$
14. [5.1, ■■■] $(x^3 - 2)(x + 4)$
15. [5.2, ■■■] $(4x^2 + 1)(2x + 1)(2x - 1)$
16. [5.1, ■■] $4x^4(2x^2 - 8x + 1)$ 17. [5.3, ■■] $3(2x + 5)^2$
18. [5.2, ■] Not factorable 19. [5.4, ■■] $x(x - 6)(x + 5)$
20. [5.2, ■■■] $(2x + 5)(2x - 5)$ 21. [5.3, ■■] $(3x - 5)^2$
22. [5.5, ■■] $2(3x + 4)(x - 6)$ 23. [5.3, ■■] $(x - 3)^2$
24. [5.5, ■] $(2x + 1)(x - 4)$ 25. [5.3, ■■] $2(3x - 1)^2$
26. [5.2, ■■■] $3(x + 3)(x - 3)$ 27. [5.4, ■■] $(x - 5)(x - 3)$
28. [5.3, ■■] $(5x - 2)^2$ 29. [5.7, ■■] 1, -3
30. [5.7, ■■] -7, 5 31. [5.7, ■■] -4, 3
32. [5.7, ■■] $\frac{2}{3}$, 1 33. [5.7, ■■] $\frac{3}{2}$, -4
34. [5.7, ■■] 8, -2 35. [5.8, ■] 3 and -2
36. [5.8, ■] -18 and -16, 16 and 18
37. [5.8, ■] -19 and -17, 17 and 19
38. [5.8, ■] $\frac{5}{2}$ and -2 39. [5.9, ■] 49
40. [5.9, ■■] Coefficients: 1, -7, 9, -8; degrees:
6, 2, 2, 0; 6 41. [5.9, ■■] Coefficients:
1, -1, 1; degrees: 16, 40, 23; 40
42. [5.9, ■■] $9w - y - 5$
43. [5.9, ■■] $m^6 - 2m^2n + 2m^2n^2 + 8n^2m - 6m^3$
44. [5.9, ■■] $-9xy - 2y^2$
45. [5.9, ■■] $11x^3y^2 - 8x^2y - 6x - 6x^2 + 6$
46. [5.9, ■■] $p^3 - q^3$ 47. [5.9, ■■] $9a^8 - 2a^4b^3 + \frac{1}{9}b^6$
48. [5.9, ■■] $(xy + 4)(xy - 3)$
49. [5.9, ■■] $3(2a + 7b)^2$ 50. [5.9, ■] $(m + t)(m + 5)$
51. [5.9, ■■] $32(x^2 - 2y^2z^2)(x^2 + 2y^2z^2)$
52. [2.6, ■■] -27.73 53. [2.6, ■■] $\frac{10}{7}$ 54. [2.6, ■■] $\frac{8}{35}$
55. [2.6, ■■] -12 56. [3.4, ■] 2 57. [3.4, ■] $\frac{4}{3}$

58. $[3.6,$ ■ $]$ $b = \dfrac{A - a}{2}$ **59.** $[3.6,$ ■ $]$ $g = \dfrac{2S}{t^2}$

60. $[4.7,$ ■ $]$ $4a^2 + 12a + 9$ **61.** $[4.7,$ ■ $]$ $4a^2 - 9$

62. $[4.7,$ ■ $]$ $10a^2 - a - 21$ **63.** $[5.7,$ ■ $]$ $2, -3, \frac{5}{2}$

64. $[5.7,$ ■ $]$ $3, -11$

65. $[5.3,$ ■ $], [5.1,$ ■ $]$ $(y + 1)(y - 2)^2$

66. $[5.2,$ ■ $]$ $(x - 1.5)(x + 1.5)$

67. $[5.4,$ ■ $]$ $(x - \frac{1}{2})(x + \frac{1}{4})$

68. $[5.6,$ ■ $]$ Not factorable

69. $[5.2,$ ■ $]$ $(x^n - y^m)(x^n + y^m)$

70. $[5.8,$ ■ $]$ 0 and 2

CHAPTER 6

Margin Exercises, Section 6.1, pp. 270–271

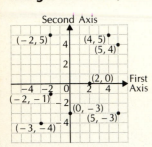

9. Both are negative numbers   **10.** First positive; second negative   **11.** I   **12.** III   **13.** IV   **14.** II

15. $B(-3, 5)$; $C(-4, -3)$; $D(2, -4)$; $E(1, 5)$; $F(-2, 0)$; $G(0, 3)$

Something Extra—An Application: Coordinates, p. 272

1. Latitude 32.5° north, longitude 64.5° west

2. Latitude 27° north, longitude 81° west

Exercise Set 6.1, pp. 273–274

1.

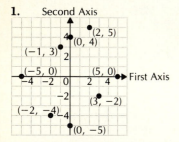

3. II   **5.** IV   **7.** III   **9.** I   **11.** Negative; negative

13. $A(3, 3)$; $B(0, -4)$; $C(-5, 0)$; $D(-1, -1)$; $E(2, 0)$

15.

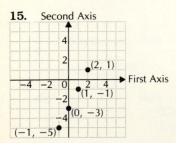

17. $-\frac{13}{6}$ **19.** -5

21. I, IV   **23.** I, III

Margin Exercises, Section 6.2, pp. 275–279

1. No   **2.** Yes

3.

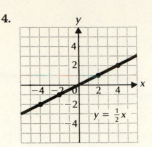

4.

5.

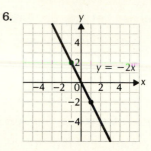

6.

7.

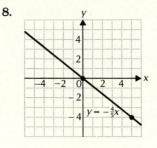

8.

9. $y = x + 3$ looks like $y = x$ moved *up* 3 units.

10. $y = x - 1$ looks like $y = x$ moved *down* 1 unit.

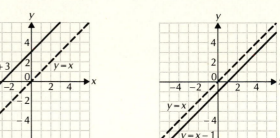

11. $y = 2x + 3$ looks like $y = 2x$ moved *up* 3 units.

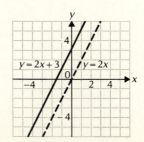

12.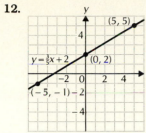

13.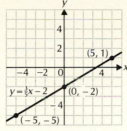

19.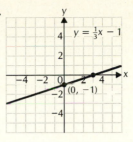

21.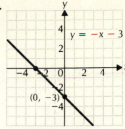

14.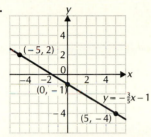

15.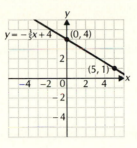

23.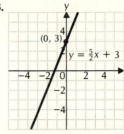

25.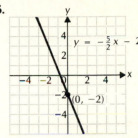

Something Extra—Applications, p. 280

1. $D = \dfrac{1}{2}L - \dfrac{1}{4}C_1 - \dfrac{1}{4}C_2$

Exercise Set 6.2, pp. 281–282

1. Yes   **3.** No   **5.** No

7.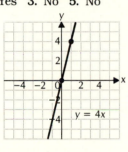

9.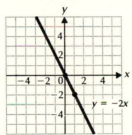

11.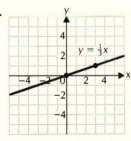

13.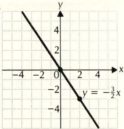

15.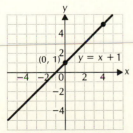

17.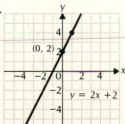

27. $(4 + t^2)(2 - t)(2 + t)$ **29.** $x^3(x - 7)(x + 5)$
31. $(0, 6), (1, 5), (2, 4), (3, 3), (4, 2), (5, 1) (6, 0)$

Margin Exercises, Section 6.3, pp. 283–285

1. (a) $(4, 0)$; (b) $(0, 3)$
2. x-intercept is $(3, 0)$;   **3.** x-intercept is $(-3, 0)$;
y-intercept is $(0, 2)$ y-intercept is $(0, 4)$

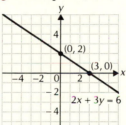

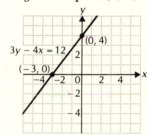

4.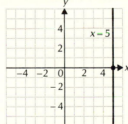

5.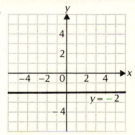

6.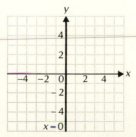

7.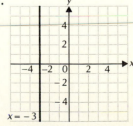

6

Something Extra—Depreciation: The Straight-Line Method, p. 286

1. $3250, $2762.50, $2275, $1787.50, $1300

Exercise Set 6.3, pp. 287–288

1. x-intercept is (3, 0);
y-intercept is (0, −5)

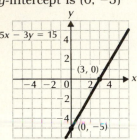

3. x-intercept is (2, 0);
y-intercept is (0, 4)

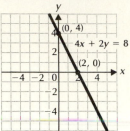

5. x-intercept is (1, 0);
y-intercept is (0, −1)

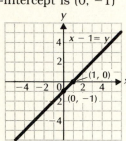

7. x-intercept is $\left(\frac{1}{2}, 0\right)$;
y-intercept is (0, −1)

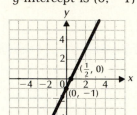

9. x-intercept is (3, 0);
y-intercept is (0, −4)

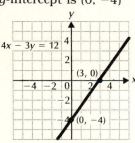

11. x-intercept is $\left(\frac{6}{7}, 0\right)$;
y-intercept is (0, 3)

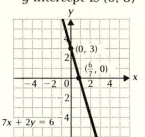

13. x-intercept is (−1, 0);
y-intercept is (0, −4)

15.

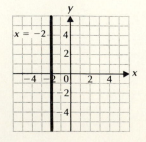

17.

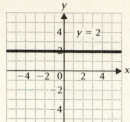

19.

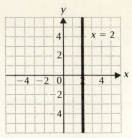

21.

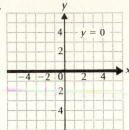

23.

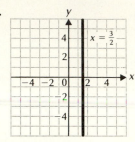

25. $2x^2 + 16$ **27.** $-\frac{1}{2}x^2 - x + 6$ **29.** $y = 0$ **31.** $(-3, 6)$

Margin Exercises, Section 6.4, pp. 289–290

1. $y = 7x$ **2.** $y = \frac{5}{8}x$ **3.** $0.4667; $0.0194 **4.** 79.2 kg

Exercise Set 6.4, pp. 291–292

1. $y = 4x$ **3.** $y = 1.75x$ **5.** $y = 3.2x$ **7.** $y = \frac{2}{3}x$
9. $183.75 **11.** $22\frac{6}{7}$ **13.** $16\frac{2}{3}$ kg **15.** 68.4 kg **17.** 31
19. 11, −11
21. $C = 2\pi r$, $k = 2\pi$ **23.** $A = \pi r^2$, π

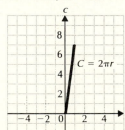

Margin Exercises, Section 6.5, pp. 293–298

1. $\frac{2}{5}$ **2.** $-\frac{5}{3}$ **3.** $\frac{2}{5}$ **4.** $-\frac{5}{3}$ **5.** 0 **6.** No slope **7.** No slope
8. 0 **9.** −8 **10.** $-\frac{4}{5}$ **11.** $\frac{1}{4}$ **12.** $\frac{5}{4}$ **13.** 5, (0, 0) **14.** $-\frac{3}{2}$,
(0, −6) **15.** 2, $\left(0, -\frac{17}{2}\right)$ **16.** $-\frac{3}{4}$, $\left(0, \frac{15}{4}\right)$ **17.** $-\frac{7}{5}$,
$\left(0, -\frac{22}{5}\right)$ **18.** $y = 5x - 18$ **19.** $y = -3x - 5$
20. $y = 6x - 13$ **21.** $y = -\frac{2}{3}x + \frac{14}{3}$ **22.** $y = x + 2$
23. $y = 2x + 4$

Exercise Set 6.5, pp. 299–300

1. 0 **3.** $-\frac{4}{5}$ **5.** 7 **7.** 2 **9.** 0 **11.** No slope
13. No slope **15.** 0 **17.** No slope **19.** 0 **21.** $-\frac{3}{2}$
23. $-\frac{1}{4}$ **25.** 2 **27.** −4, (0, −9) **29.** 1.8, (0, 0)
31. $-\frac{8}{7}$, (0, −3) **33.** 3, $\left(0, -\frac{5}{3}\right)$ **35.** $-\frac{3}{2}$, $\left(0, -\frac{1}{2}\right)$
37. 0, (0, −17) **39.** $y = 5x - 5$ **41.** $y = \frac{3}{4}x + \frac{5}{2}$
43. $y = x - 8$ **45.** $y = -3x - 9$ **47.** $y = \frac{1}{4}x + \frac{5}{2}$
49. $y = -\frac{1}{2}x + 4$ **51.** $y = -\frac{3}{2}x + \frac{13}{2}$ **53.** $y = \frac{3}{4}x - \frac{5}{2}$
55. 33 ft **57.** −5 **59.** $y = 3x - 9$ **61.** $y = \frac{3}{2}x - 2$

Summary and Review: Chapter 6, pp. 301–302

1. 2. 3. [6.1, 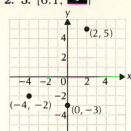]

4. [6.1, ■] IV
5. [6.1, ■] III
6. [6.1, ■] I
7. [6.1, ■] (−5, −1)
8. [6.1, ■] (−2, 5)
9. [6.1, ■] (3, 0)
10. [6.2, ■] No
11. [6.2, ■] Yes

12. [6.2, ■]

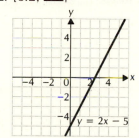

13. [6.2, ■]

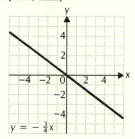

14. [6.2, ■]

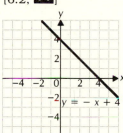

15. [6.3, ■]

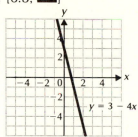

16. [6.3, ■]

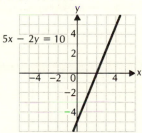

17. [6.3, ■]

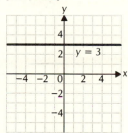

18. [6.3, ■]

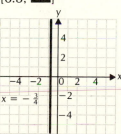

19. [6.3, ■]

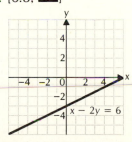

20. [6.4, ■] $y = 3x$ 21. [6.4, ■] $y = 0.8x$
22. [6.4, ■] $247.50   23. [6.5, ■] $\frac{3}{2}$ 24. [6.5, ■] 0

25. [6.5, ■] No slope   26. [6.5, ■] 2   27. [6.5, ■] 0
28. [6.5, ■] No slope   29. [6.5, ■] $-\frac{4}{3}$
30. [6.5, ■] −9, (0, 46)   31. [6.5, ■] −1, (0, 9)
32. [6.5, ■] $\frac{3}{5}$, $(0, -\frac{4}{5})$ 33. [6.5, ■] $y = 3x − 1$
34. [6.5, ■] $y = \frac{2}{3}x − \frac{11}{3}$ 35. [6.5, ■] $y = −2x + 4$
36. [6.5, ■] $y = x + 2$ 37. [6.5, ■] $y = \frac{1}{2}x − 1$
38. [3.5, ■] $1108
39. [3.5, ■] Length = 125 m; Width = 38 m
40. [4.4, ■] $17r^4 − 7r^3 + 16r^2 − 5r + 1$
41. [4.4, ■] $-2.43x^2 − 13.4x + 56$
42. [5.6, ■] $8y(5 − y)(5 + y)$
43. [5.6, ■] $2(3x^2 − 2)(x + 4)$
44. [5.6, ■] $10x^2(x + 3)(x + 1)$ 45. [5.6, ■] $x(x^3 + 4)^2$
46. [2.9, ■] 172,134.4   47. [2.9, ■] 608
48. [6.2, ■] −1   49. [6.2, ■] (0, 4), (1, 3), (−2, 2), answers may vary   50. [6.3, ■] $y = −5$
51. [6.5, ■] $y = −2x − 5$

CUMULATIVE REVIEW: CHAPTERS 4–6, pp. 305–306

1. [4.1, ■] −13   2. [4.1, ■] −15
3. [4.2, ■] 1, −2, 1, −1   4. [4.2, ■] 3, 2, 1, 0; 3
5. [4.2, ■] None   6. [5.7, ■] 0, 4   7. [5.7, ■] −5, 4
8. [5.7, ■] 0, 10   9. [5.7, ■] $\frac{1}{2}$, −4
10. [4.2, ■] $2x^3 − 3x^2 − 2$
11. [4.2, ■] $-4x^3 − \frac{1}{7}x^2 − 2$
12. [4.3, ■] $2x^5 + 6x^4 + 2x^3 − 10x^2 + 3x − 9$
13. [4.4, ■] $-y^3 − 2y^2 − 2y + 7$
14. [4.6, ■] $12x^3 + 16x^2 + 4x$ 15. [4.6, ■] $a^2 − 9$
16. [4.5, ■] $2x^5 + x^3 − 6x^2 − x + 3$
17. [4.6, ■] $2 − 10x^2 + 12x^4$
18. [4.6, ■] $6x^7 − 12x^5 + 9x^2 − 18$
19. [4.7, ■] $4x^6 − 1$ 20. [4.7, ■] $36x^2 − 60x + 25$
21. [4.7, ■] $64 − \frac{1}{9}x^2$ 22. [5.1, ■] $9(4 − 9y)$
23. [5.9, ■] $-2(3 + x + 6y)$ 24. [5.4, ■] $(x − 6)(x − 4)$
25. [5.5, ■] $(2x + 1)(4x + 3)$
26. [5.1, ■] $3x^2(2x^3 − 12x + 3)$
27. [5.2, ■] $2(x + 3)(x − 3)$ 28. [5.3, ■] $(4x + 5)^2$
29. [5.5, ■] $(3x − 2)(x + 4)$
30. [5.1, ■] $(m^3 − 3)(m + 2)$
31. [5.1, ■] $4t(3 − t − 12t^3)$
32. [5.2, ■] $(4y^2 + 9)(2y + 3)(2y − 3)$
33. [5.5, ■] $2(3x − 2)(x − 4)$
34. [5.2, ■] $3(1 + 2x^3)(1 − 2x^3)$
35. [5.9, ■] $(2x^2 − 3y)^2$
36. [5.9, ■] $xy^3 − 2xy^2 − 4x^2y$
37. [5.9, ■] $9x^4 − 16y^2$
38. [5.9, ■] $4a^4b^2 − 20a^3b^3 + 25a^2b^4$
39. [6.2, ■]   40. [6.3, ■]

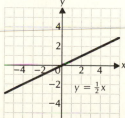

41. [6.2, 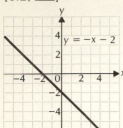]

42. [6.3, 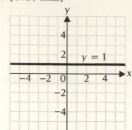]

43. [6.3, ⬛⬛]

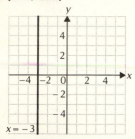

44. [6.4, ⬛] $y = \frac{2}{3}x$ **45.** [6.4, ⬛] $y = 0.2x$
46. [6.5, ⬛⬛] No slope **47.** [6.5, ⬛] $-\frac{3}{7}$ **48.** [6.5, ⬛⬛] 0
49. [6.5, ⬛⬛] $\frac{4}{3}$, $(0, -2)$ **50.** [6.5, ⬛⬛] $y = -4x + 5$
51. [6.5, ⬛⬛] $y = 6x - 3$ **52.** [6.5, ⬛⬛] $y = \frac{1}{6}x - \frac{17}{6}$
53. [6.5, ⬛⬛] $y = -\frac{10}{7}x - \frac{8}{7}$ **54.** [5.8, ⬛] 4, -5
55. [5.8, ⬛] 14 ft **56.** [6.4, ⬛⬛] \$78 **57.** [4.4, 4.6] 12
58. [4.7, ⬛⬛] $16y^6 - y^4 + 6y^2 - 9$
59. [5.2, ⬛⬛] $2(a^{16} + 81b^{20})(a^8 + 9b^{10})(a^4 + 3b^5) \times$
$(a^4 - 3b^5)$ **60.** [5.7, ⬛] 4, -7, 12 **61.** [6.5, ⬛⬛] $y = \frac{2}{3}x$

CHAPTER 7

Margin Exercises, Section 7.1, pp. 308–309

1. $x + y = 115$, $x - y = 21$, where x is one number and
y is the other **2.** $31.95 + 0.33m = c$,
$34.95 + 0.29m = c$, where m = mileage and c = cost
3. $2l + 2w = 76$, $l = w + 17$, where l = length and
w = width

Something Extra—Calculator Corner, p. 310

AEDG of length 893

Exercise Set 7.1, pp. 311–312

1. $x + y = 58$, $x - y = 16$, where x = one number and
y = the other number **3.** $2l + 2w = 400$, $w = l - 40$,
where l = length and w = width **5.** $53.95 + 0.30m = c$,
$54.95 + 0.20m = c$, where m = mileage and c = cost
7. $x - y = 16$, $3x = 7y$, where x = the larger number
and y = the smaller **9.** $x + y = 180$, $y = 3x + 8$, where
x and y are the angles **11.** $x + y = 90$, $x - y = 34$,
where x and y are the angles **13.** $x + y = 820$,
$y = x + 140$, where x = number of hectares of Riesling
and y = hectares of Chardonnay **15.** 52 **17.** $108x^{-13}$
19. $x = 0.2y$, $x + 20 = 0.52(y + 20)$, x = Patrick's age,
y = his father's age **21.** $\frac{1}{3}(b + 2) = h - 1$,
$\frac{1}{2}(b + 2)(h - 1) = 24$, b = base, h = height

Margin Exercises, Section 7.2, pp. 313–314

1. Yes **2.** No **3.** $(2, -3)$
4. No solution; lines are parallel.

Exercise Set 7.2, pp. 315–316

1. Yes **3.** No **5.** Yes **7.** $(-12, 11)$ **9.** $(4, 3)$
11. No solution **13.** $(5, 3)$ **15.** $\frac{1}{625}x^8y^{-20}$
17. $25x^2 - 60x + 36$ **19.** $A = 2$, $B = 2$ **21.** Three
lines intersecting in one point; the solution is one point,
or ordered pair.

Margin Exercises, Section 7.3, pp. 317–318

1. $(3, 2)$ **2.** $(3, -1)$ **3.** $\left(\dfrac{24}{5}, -\dfrac{8}{5}\right)$

Exercise Set 7.3, pp. 319–320

1. $(1, 3)$ **3.** $(1, 2)$ **5.** $(4, 3)$ **7.** $(-2, 1)$ **9.** $(-1, -3)$
11. $\left(\dfrac{17}{3}, \dfrac{16}{3}\right)$ **13.** $\left(\dfrac{25}{8}, -\dfrac{11}{4}\right)$ **15.** $(-3, 0)$ **17.** $(6, 3)$
19. No solution **21.** $49x^4 - 16$ **23.** $-\frac{5}{3}$
25. $(4.3821792, 4.3281211)$ **27.** $(10, -2)$

Margin Exercises, Section 7.4, pp. 321–324

1. $(3, 2)$ **2.** $(1, -1)$ **3.** $(1, 4)$ **4.** $(1, 1)$ **5.** $(1, -1)$
6. $\left(\dfrac{17}{13}, -\dfrac{7}{13}\right)$ **7.** No solution **8.** $(1, -1)$

Exercise Set 7.4, pp. 325–326

1. $(9, 1)$ **3.** $(5, 3)$ **5.** $\left(3, -\dfrac{1}{2}\right)$ **7.** $\left(-1, \dfrac{1}{5}\right)$
9. $(-3, -5)$ **11.** No solution **13.** $(2, -2)$ **15.** $(1, -1)$
17. $(-2, 3)$ **19.** $(8, 6)$ **21.** $(50, 18)$ **23.** $(2, -1)$
25. $14x^4 + 9x^3 - 19x^2$ **27.** 180 **29.** $(5, 2)$ **31.** $(4, 0)$

Margin Exercises, Section 7.5, pp. 327–332

1. 75 miles **2.** A is 47; B is 21 **3.** Length is
27.5 cm; width is 10.5 cm **4.** 7 quarters, 13 dimes
5. 125 adults, 41 children **6.** 22.5 L of 50%,
7.5 L of 70% **7.** 30 lb of A, 20 lb of B

Exercise Set 7.5, pp. 333–338

1. 10 miles **3.** Sammy is 44; his daughter is 22
5. 28 and 12 **7.** 43° and 137° **9.** 62° and 28°
11. 480 hectares Chardonnay; 340 hectares Riesling
13. Length is 120 m; width is 80 m
15. 70 dimes; 33 quarters **17.** 300 nickels; 100 dimes
19. 203 adults; 226 children
21. 130 adults; 70 students **23.** 40 mL of A;
60 mL of B **25.** 300 gal of A; 100 gal of B
27. 80 L of 30%; 120 L of 50%
29. 6 kg of cashews; 4 kg of pecans

31.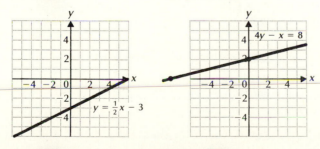

33.

35. About 89.7 L **37.** 54
39. Tweedledum: 120; Tweedledee: 121

Margin Exercises, Section 7.6, pp. 339–342

1. 168 km **2.** 275 km/h **3.** 324 mi **4.** 3 hr

Exercise Set 7.6, pp. 343–346

1. 2 hr **3.** 4.5 hr **5.** $7\frac{1}{2}$ hr after the first train leaves, or $4\frac{1}{2}$ hr after the second train leaves **7.** 14 km/h
9. 384 km **11.** 330 km/h **13.** 15 mi
15. 317.02702 km/h **17.** 180 mi, 96 mi

Summary and Review: Chapter 7, pp. 347–348

1. [7.2, ■] No **2.** [7.2, ■] Yes **3.** [7.2, ■] Yes
4. [7.2, ■] No **5.** [7.2, ■] $(6, -2)$ **6.** [7.2, ■] $(6, 2)$
7. [7.2, ■] $(0, 5)$ **8.** [7.2, ■] No solution; lines are parallel **9.** [7.3, ■] $(0, 5)$ **10.** [7.3, ■] $(-2, 4)$
11. [7.3, ■] $(1, -2)$ **12.** [7.3, ■] $(-3, 9)$
13. [7.3, ■] $(1, 4)$ **14.** [7.3, ■] $(3, -1)$
15. [7.4, ■] $(3, 1)$ **16.** [7.4, ■] $(1, 4)$
17. [7.4, ■] $(5, -3)$ **18.** [7.4, ■] $(-4, 1)$
19. [7.4, ■] $(-2, 4)$ **20.** [7.4, ■] $(-2, -6)$
21. [7.4, ■] $(3, 2)$ **22.** [7.4, ■] $(2, -4)$
23. [7.5, ■] $10, -2$ **24.** [7.5, ■] 12, 15
25. [7.5, ■] Length = 38.5 cm, width = 10.5 cm
26. [7.6, ■] 135 km/h **27.** [7.5, ■] 297 orchestra,
211 balcony **28.** [7.5, ■] 40 L of each
29. [7.5, ■] Jeff: 27; son: 9 **30.** [3.7, ■] t^8
31. [3.7, ■] t^{-18} **32.** [3.7, ■] $x^{12}y^{-15}$
33. [4.7, ■] $9t^2 - 64$
34. [4.7, ■] $25y^4 + 100y^2 + 100$
35. [4.7, ■] $20y^2 - 3y - 56$ **36.** [5.7, ■] $-\frac{8}{5}, \frac{7}{4}$
37. [5.7, ■] $\frac{8}{3}, -\frac{8}{3}$
38. [6.3, ■]   **39.** [6.3, ■]

40. [7.2, ■] $C = 1, D = 3$ **41.** [7.3, ■] $(2, 0)$
42. [7.5, ■] 24 **43.** [7.5, ■] $96

CHAPTER 8

Margin Exercises, Section 8.1, pp. 352–354

1. (a) Yes; **(b)** yes; **(c)** yes; **(d)** no; **(e)** no
2. (a) Yes; **(b)** no; **(c)** no; **(d)** yes; **(e)** no **3.** $\{x|x > 2\}$
4. $\{x|x < 13\}$ **5.** $\{x|x < -3\}$ **6.** $\{x|x \leq \frac{2}{15}\}$ **7.** $\{y|y \leq -3\}$

Exercise Set 8.1, pp. 355–356

1. (a) Yes; **(b)** yes; **(c)** no; **(d)** yes
3. (a) No; **(b)** no; **(c)** yes; **(d)** yes
5. (a) No; **(b)** no; **(c)** yes; **(d)** no **7.** $\{x|x > -5\}$
9. $\{y|y > 3\}$ **11.** $\{x|x \leq -18\}$ **13.** $\{x|x \leq 16\}$
15. $\{y|y > -5\}$ **17.** $\{x|x > 3\}$ **19.** $\{x|x \geq 13\}$
21. $\{x|x < 4\}$ **23.** $\{c|c > 14\}$ **25.** $\left\{y|y \leq \frac{1}{4}\right\}$
27. $\left\{x|x > \frac{7}{12}\right\}$ **29.** $\{x|x > 0\}$ **31.** $\{r|r < -2\}$
33. $\{x|x \geq 1\}$ **35.**   **37.** -7
39. $\{x|x \leq -18{,}058{,}999\}$ **41.** Yes

Margin Exercises, Section 8.2, pp. 357–358

1. $\{x|x < 8\}$ **2.** $\{y|y \geq 32\}$ **3.** $\{x|x \geq -6\}$
4. $\left\{y|y < -\frac{13}{5}\right\}$ **5.** $\left\{x|x > -\frac{1}{4}\right\}$ **6.** $\{x|x \leq -1\}$
7. $\left\{y|y \leq \frac{19}{9}\right\}$

Exercise Set 8.2, pp. 359–360

1. $\{x|x < 7\}$ **3.** $\{y|y \leq 9\}$ **5.** $\left\{x|x < \frac{13}{7}\right\}$ **7.** $\{x|x > -3\}$
9. $\left\{y|y \geq -\frac{2}{5}\right\}$ **11.** $\{x|x \geq -6\}$ **13.** $\{y|y \leq 4\}$
15. $\left\{x|x > \frac{17}{3}\right\}$ **17.** $\left\{y|y < -\frac{1}{14}\right\}$ **19.** $\left\{x|\frac{3}{10} \geq x\right\}$
21. $\{x|x < -1\}$ **23.** $\{x|x \leq 3\}$ **25.** $\left\{t|t \leq \frac{9}{2}\right\}$
27. $\{c|c < -2\}$ **29.** $\{x|x \geq 4\}$ **31.** 17 **33.** $(8, 1)$
35. $\{t|t \leq 0\}$ **37.** All nonzero rational numbers
39. $\{m|m > 50\}$

Margin Exercises, Section 8.3, pp. 361–364

1.   **2.**

3.   **4.**

5.   **6.**

7.   **8.**

9. No

10. $y < x$

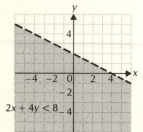

11. $y \geq x + 2$

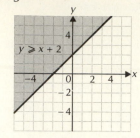

12. $2x + 4y < 8$

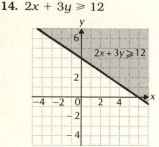

13. $3x - 5y < 15$

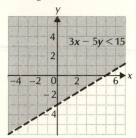

14. $2x + 3y \geq 12$

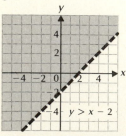

Exercise Set 8.3, pp. 365–366

1.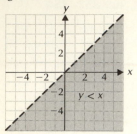

3.

5.

7.

9.

11.

13.

15. $y > x - 2$

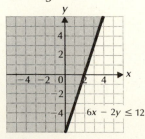

17. $6x - 2y \leq 12$

19. $3x - 5y \geq 15$

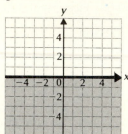

21. $y - 2x < 4$

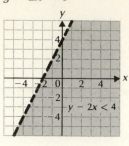

23. Base $= 7$ m, height $= 2$ m

25. $x \geq 3$

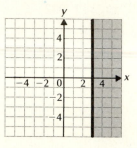

27. $y \leq 0$

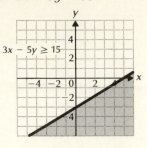

Margin Exercises, Section 8.4, pp. 367–368

1. {2, 3, 4, 5, 6, 7, 8, 9, 10} **2.** {32, 34, 36, 38} **3.** True
4. False **5.** True **6.** True **7.** {a, 1, 9} **8.** {2} **9.** $\emptyset$
10. $\emptyset$ **11.** {0, 1, 2, 3, 4, 5, 6, 7} **12.** The set of integers

Exercise Set 8.4, pp. 369–370

1. {3, 4, 5, 6, 7, 8} **3.** {41, 43, 45, 47, 49} **5.** {3, −3}
7. False **9.** True **11.** True **13.** {c, d, e} **15.** {1, 10}
17. $\emptyset$ **19.** {a, e, i, o, u, q, c, k} **21.** {0, 1, 2, 5, 7, 10}
23. {a, e, i, o, u, m, n, f, g, h} **25.** $(\frac{1}{8}, \frac{9}{8})$ **27.** 9
29. 0.6 km **31.** The set of positive even integers

Summary and Review: Chapter 8, pp. 371–372

1. [8.1, ▪•▪] Yes **2.** [8.1, ▪•▪] No **3.** [8.1, ▪•▪] Yes
4. [8.1, ▪•▪] {$y|y \geq -\frac{1}{2}$} **5.** [8.2, ▪•▪] {$x|x \geq 7$}
6. [8.2, ▪•▪] {$y|y > 2$} **7.** [8.2, ▪•▪] {$y|y \leq -4$}
8. [8.1, ▪•▪] {$x|x < -11$} **9.** [8.2, ▪•▪] {$y|y > -7$}
10. [8.2, ▪•▪] {$x|x > -6$} **11.** [8.2, ▪•▪] {$x|x > -\frac{9}{11}$}
12. [8.2, ▪•▪] {$y|y \leq 7$} **13.** [8.2, ▪•▪] {$x|x \geq -\frac{1}{12}$}

14. [8.3, ▪•▪]

15. [8.3, ▪•▪]

16. [8.3, ▪•▪]

17. [8.3, 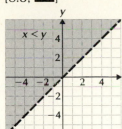]

18. [8.3, ■■■]

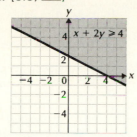

19. [8.4, ■■] {32, 36, 40, 44, 48, 52, 56}
20. [8.4, ■■] False **21.** [8.4, ■■] False
22. [8.4, ■■] [2, 4] **23.** [8.4, ■■] ∅
24. [8.4, ■■] {0, 1, 2, 3} **25.** [8.4, ■■] {A, W, R, E, F, B}
26. [2.1, ■■] 10 **27.** [2.1, ■■] −33 **28.** [2.1, ■■] 56
29. [2.1, ■■] 0 **30.** [2.1, ■■] 76 **31.** [3.6, ■] $t = \dfrac{d}{r}$
32. [3.6, ■] $p = \dfrac{Q + 7q}{5}$ **33.** [7.4, ■] (1, −1)
34. [7.4, ■] (2, 14) **35.** [5.8, ■] 24 m
36. [5.8, ■] 18 and 20, −20 and −18
37. [8.2, ■■] $\{y | y > \frac{8}{3}\}$ **38.** [8.2, ■■] $\{x | x \leq \frac{15}{2}\}$
39. [8.2, ■■] 86 **40.** [8.2, ■■] $\{w | w > 17 \text{ cm}\}$
41. [8.3, ■■] $x < -3$ **42.** [8.3, ■■] $y \geq -2$

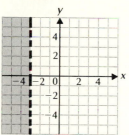

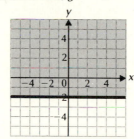

43. [8.4, ■■] {7, 9, 11, 13}

CHAPTER 9

Margin Exercises, Section 9.1, pp. 376–380

1. $\dfrac{(x + 3)(x + 2)}{5(x + 4)}$ **2.** $\dfrac{-3 \cdot 4}{(2x + 1)(2x - 1)}$ **3.** $\dfrac{x(2x + 1)}{x(3x - 2)}$
4. $\dfrac{(x + 1)(x + 2)}{(x - 2)(x + 2)}$ **5.** $\dfrac{-1(x - 8)}{-1(x - y)}$ **6.** 5 **7.** $\dfrac{x}{3}$ **8.** $\dfrac{2x + 1}{3x + 2}$
9. $\dfrac{x + 1}{2x + 1}$ **10.** $x + 2$ **11.** $\dfrac{y + 2}{4}$ **12.** $\dfrac{a - 2}{a - 3}$
13. $\dfrac{x - 5}{2}$

Exercise Set 9.1, pp. 381–382

1. $\dfrac{(x - 2)(x - 2)}{(x - 5)(x + 5)}$ **3.** $\dfrac{(c - 3d)(c + 3d)}{(c + d)(c - d)}$
5. $\dfrac{(2a - 1)(3a - 1)}{(2a - 1)(3a + 2)}$ **7.** $\dfrac{3x + 2}{3x - 2}$ **9.** $\dfrac{a + b}{a - b}$ **11.** $\dfrac{t - 5}{t - 4}$

13. $\dfrac{x + 2}{2(x - 4)}$ **15.** $\dfrac{a - 3}{a - 4}$ **17.** $\dfrac{6}{x - 3}$ **19.** $\dfrac{a^2 + 1}{a + 1}$
21. $\dfrac{t}{t + 2}$ **23.** $\dfrac{12a}{a - 2}$ **25.** $\dfrac{1}{a}$ **27.** $\dfrac{1}{16}$ **29.** $\dfrac{9}{10}$
31. $x + 2y$ **33.** 1 **35.** 0, 2, 7

Margin Exercises, Section 9.2, pp. 383–384

1. $\dfrac{2}{7}$ **2.** $\dfrac{2x^3 - 1}{x^2 + 5}$ **3.** $\dfrac{1}{x - 5}$ **4.** $x^2 - 3$ **5.** $\dfrac{6}{35}$
6. $\dfrac{(x - 3)(x - 2)}{(x + 5)(x + 5)}$ **7.** $\dfrac{x - 3}{x + 2}$ **8.** $\dfrac{(x - 3)(x - 2)}{x + 2}$
9. $\dfrac{y + 1}{y - 1}$

Exercise Set 9.2, pp. 385–386

1. $\dfrac{x}{4}$ **3.** $\dfrac{1}{x^2 - y^2}$ **5.** $\dfrac{x^2 - 4x + 7}{x^2 + 2x - 5}$ **7.** $\dfrac{3}{10}$ **9.** $\dfrac{1}{4}$
11. $\dfrac{y^2}{x}$ **13.** $\dfrac{(a + 2)(a + 3)}{(a - 3)(a - 1)}$ **15.** $\dfrac{(x - 1)^2}{x}$ **17.** $\dfrac{1}{2}$ **19.** $\dfrac{3}{2}$
21. $\dfrac{(x + y)^2}{x^2 + y}$ **23.** $\dfrac{x + 3}{x - 5}$ **25.** $\dfrac{1}{(c - 5)^2}$ **27.** $\dfrac{t + 5}{t - 5}$ **29.** $\frac{5}{2}$
31. $(x + 1)(x + 2)$ **33.** $\dfrac{4}{x + 7}$ **35.** $\dfrac{3(y + 2)^3}{y(y - 1)}$

Margin Exercises, Section 9.3, pp. 387–388

1. $\dfrac{7}{9}$ **2.** $\dfrac{3 + x}{x - 2}$ **3.** $\dfrac{6x + 4}{x - 1}$ **4.** $\dfrac{x - 5}{4}$ **5.** $\dfrac{x - 1}{x - 3}$ **6.** $\dfrac{4}{11}$
7. $\dfrac{x^2 + 2x + 1}{2x + 1}$ **8.** $\dfrac{3x - 1}{3}$ **9.** $\dfrac{4x - 3}{x - 2}$

Exercise Set 9.3, pp. 389–392

1. 1 **3.** $\dfrac{6}{3 + x}$ **5.** $\dfrac{2x + 3}{x - 5}$ **7.** $\dfrac{1}{4}$ **9.** $-\dfrac{1}{t}$ **11.** $\dfrac{-x + 7}{x - 6}$
13. $y + 3$ **15.** $\dfrac{2b - 14}{b^2 - 16}$ **17.** $-\dfrac{1}{y + z}$ **19.** $\dfrac{5x + 2}{x - 5}$
21. −1 **23.** $\dfrac{-x^2 + 9x - 14}{(x - 3)(x + 3)}$ **25.** $\dfrac{1}{2}$ **27.** 1 **29.** $\dfrac{4}{x - 1}$
31. $\dfrac{8}{3}$ **33.** $\dfrac{13}{a}$ **35.** $\dfrac{4x - 5}{4}$ **37.** $\dfrac{x - 2}{x - 7}$ **39.** $\dfrac{2x - 16}{x^2 - 16}$
41. $\dfrac{2x - 4}{x - 9}$ **43.** $\dfrac{-9}{2x - 3}$ **45.** $\dfrac{18x + 5}{x - 1}$ **47.** 0
49. $\dfrac{20}{2y - 1}$ **51.** $(2x - 1)(x - 1)$ **53.** 60 **55.** $\dfrac{x}{3x + 1}$
57. 0

Margin Exercises, Section 9.4, pp. 393–394

1. 144 **2.** 12 **3.** 10 **4.** 120 **5.** $\dfrac{35}{144}$ **6.** $\dfrac{1}{4}$ **7.** $\dfrac{11}{10}$
8. $\dfrac{9}{40}$ **9.** $60x^3y^2$ **10.** $(y + 1)^2(y + 4)$

9

11. $7(t^2 + 16)(t - 2)$
12. $3x(x + 1)^2(x - 1)$, or $3x(x + 1)^2(1 - x)$

Exercise Set 9.4, pp. 395–396

1. 108 **3.** 72 **5.** 126 **7.** 360 **9.** 420 **11.** $\dfrac{59}{300}$

13. $\dfrac{71}{120}$ **15.** $\dfrac{23}{180}$ **17.** $8a^2b^2$ **19.** c^3d^2

21. $8(x - 1)$, or $8(1 - x)$ **23.** $(a + 1)(a - 1)^2$
25. $(3k + 2)(3k - 2)$ or $(3k + 2)(2 - 3k)$
27. $18x^3(x - 2)^2(x + 1)$ **29.** $10x^3(x - 1)(x + 1)^2$
31. $2x(3x + 2)$ **33.** 2, 3
35. One expression is a multiple of the other.

Margin Exercises, Section 9.5, pp. 397–398

1. $\dfrac{x^2 + 6x - 8}{(x + 2)(x - 2)}$ **2.** $\dfrac{4x^2 - x + 3}{x(x - 1)(x + 1)^2}$

3. $\dfrac{8x + 88}{(x + 16)(x + 1)(x + 8)}$ **4.** $\dfrac{2a^2 - 5a + 15}{(a + 2)(a - 9)}$

5. $\dfrac{x - 3}{(x + 1)(x + 3)}$

Exercise Set 9.5, pp. 399–400

1. $\dfrac{2x + 5}{x^2}$ **3.** $\dfrac{x^2 + 4xy + y^2}{x^2y^2}$ **5.** $\dfrac{4x}{(x - 1)(x + 1)}$

7. $\dfrac{x^2 + 6x}{(x + 4)(x - 4)}$ **9.** $\dfrac{3x - 1}{(x - 1)^2}$ **11.** $\dfrac{x^2 + 5x + 1}{(x + 1)^2(x + 4)}$

13. $\dfrac{2x^2 - 4x + 34}{(x - 5)(x + 3)}$ **15.** $\dfrac{3a + 2}{(a + 1)(a - 1)}$

17. $\dfrac{2x + 6y}{(x + y)(x - y)}$ **19.** $\dfrac{3x^2 + 19x - 20}{(x + 3)(x - 2)^2}$ **21.** 25,704 ft^2

23. $\{t|t < 112\}$ **25.** $\dfrac{16y + 28}{15}$, $\dfrac{y^2 + 2y - 8}{15}$

27. $\dfrac{(z + 6)(2z - 3)}{z^2 - 4}$

Margin Exercises, Section 9.6, p. 401

1. $\dfrac{-x - 7}{15x}$ **2.** $\dfrac{6x^2 - 2x - 2}{3x(x + 1)}$

Something Extra—An Application: Handling Dimension Symbols (Part 2), p. 402

1. 4 yd **2.** 96 oz **3.** 27 km **4.** 60 m **5.** 3 g **6.** 18 mi
7. 2.347 km **8.** 550 mm **9.** 0.7 kg^2/m^2
10. 720 lb-mi^2/ft-hr^2 **11.** 14 m-kg/sec^2

Exercise Set 9.6, pp. 403–404

1. $\dfrac{-(x + 4)}{6}$ **3.** $\dfrac{7z - 12}{12z}$ **5.** $\dfrac{4x^2 - 13xt + 9t^2}{3x^2t^2}$

7. $\dfrac{2x - 40}{(x + 5)(x - 5)}$ **9.** $\dfrac{3 - 5t}{2t(t - 1)}$ **11.** $\dfrac{2s - st - s^2}{(t + s)(t - s)}$

13. $\dfrac{2}{y(y - 1)}$ **15.** $\dfrac{z - 3}{2z - 1}$ **17.** $\dfrac{1 - 3x}{(2x - 3)(x + 1)}$

19. $\dfrac{1}{2c - 1}$ **21.** x **23.** $30x^{12}$

25. $\dfrac{-3xy - 3a + 6x}{(a + 2x)(a - 2x)(y - 3)^2}$

27. $\dfrac{11z^4 - 22z^2 + 6}{(2z^2 - 3)(z^2 + 2)(z^2 - 2)}$

Margin Exercises, Section 9.7, pp. 405–408

1. $\dfrac{33}{2}$ **2.** 3 **3.** $\dfrac{3}{2}$ **4.** $-\dfrac{1}{8}$ **5.** 1 **6.** 2 **7.** 4

Exercise Set 9.7, pp. 409–412

1. $\dfrac{12}{5}$ **3.** -2 **5.** 3 **7.** 10 **9.** 5 **11.** 3 **13.** $\dfrac{17}{2}$

15. No solution **17.** $-4, -1$ **19.** -1 **21.** 3 **23.** 12

25. $\dfrac{5}{3}$ **27.** $\dfrac{2}{9}$ **29.** $\dfrac{1}{2}$ **31.** No solution

33.

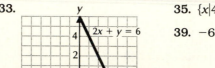

35. $\{x|4 < x\}$ **37.** $-\dfrac{1}{6}$

39. -6

Margin Exercises, Section 9.8, pp. 413–417

1. -3 **2.** 40 km/h, 50 km/h **3.** $\dfrac{24}{7}$, or $3\dfrac{3}{7}$ hr

4. 58 km/L **5.** 0.280 **6.** 124 km/h **7.** 2.4 fish/yd^2
8. 3.45 **9.** 42 **10.** 2074

Something Extra—An Application, p. 418

1. 57.848% **2.** 62.6% **3.** 66.56% **4.** 70.52%

Exercise Set 9.8, pp. 419–422

1. $\dfrac{20}{9}$, or $2\dfrac{2}{9}$ **3.** 20 and 15 **5.** 30 km/h, 70 km/h

7. 20 mph **9.** Passenger: 80 km/h, freight: 66 km/h

11. $2\dfrac{2}{9}$ hr **13.** $5\dfrac{1}{7}$ hr **15.** 9 **17.** 2.3 km/h

19. $581\dfrac{9}{11}$ **21.** 702 km **23.** 1.92 g **25.** 287

27. (a) 1.92 tons; **(b)** 14.4 kg **29.** 2 mph **31.** $\frac{3}{4}$

Margin Exercises, Section 9.9, pp. 423–424

1. $M = \dfrac{fd^2}{km}$ **2.** $a^2 = \dfrac{6V - \pi h^3}{3\pi h}$ **3.** $p = \dfrac{n}{2 - m}$

4. $f = \dfrac{pq}{p + q}$

Exercise Set 9.9, pp. 425–426

1. $r = \dfrac{S}{2\pi h}$ **3.** $b = \dfrac{2A}{h}$ **5.** $n = \dfrac{S + 360}{180}$

7. $b = \dfrac{3V - kB - 4kM}{k}$ **9.** $r = \dfrac{S - a}{S - l}$ **11.** $h = \dfrac{2A}{b_1 + b_2}$

13. $a = \dfrac{v^2 pL}{r}$ **15.** $p = \dfrac{qf}{q - f}$ **17.** $A = P(1 + r)$

19. $R = \dfrac{r_1 r_2}{r_1 + r_2}$ **21.** $D = \dfrac{BC}{A}$ **23.** $h_2 = \dfrac{p(h_1 - q)}{q}$

25. $a = \dfrac{b}{K - C}$ **27.** $23x^4 + 50x^3 + 23x^2 - 163x + 41$

29. $30y^4 + 9y^2 - 12$ **31.** $T = \dfrac{FP}{u + EF}$ **33.** $-40°$

Margin Exercises, Section 9.10, pp. 427–428

1. $\dfrac{20}{21}$ **2.** $\dfrac{2(6 + x)}{5}$ **3.** $\dfrac{7x^2}{3(2 - x^2)}$ **4.** $\dfrac{x}{x - 1}$

Exercise Set 9.10, pp. 429–430

1. $\dfrac{25}{4}$ **3.** $\dfrac{1}{3}$ **5.** $\dfrac{1 + 3x}{1 - 5x}$ **7.** -6 **9.** $\dfrac{5}{3y^2}$ **11.** 8

13. $-\dfrac{1}{a}$ **15.** $\dfrac{x + y}{x}$ **17.** $\dfrac{x - 2}{x - 3}$ **19.** 0.0000347

21. $(10m - 9)(10m + 9)$ **23.** $\dfrac{(x - 1)(3x - 2)}{5x - 3}$

25. $\dfrac{5x + 3}{3x + 2}$

Margin Exercises, Section 9.11, pp. 431–432

1. $x^2 + 3x + 2$ **2.** $2x^2 + x - \dfrac{2}{3}$ **3.** $4x^2 - \dfrac{3}{2}x + \dfrac{1}{2}$

4. $2x^2 - 3x + 5$ **5.** $x - 2$ **6.** $x + 4$ **7.** $x + 4$,

R $- 2$; or $x + 4 + \dfrac{-2}{x + 3}$ **8.** $x^2 + x + 1$

Exercise Set 9.11, pp. 433–434

1. $1 - 2u - u^4$ **3.** $5t^2 + 8t - 2$ **5.** $-4x^4 + 4x^2 + 1$

7. $1 - 2x^2y + 3x^4y^5$ **9.** $x - 5 + \dfrac{-50}{x - 5}$; or $x - 5$, R -50

11. $x + 2$ **13.** $x - 2 + \dfrac{-2}{x + 6}$; or $x - 2$, R -2

15. $x^4 + x^3 + x^2 + x + 1$ **17.** $t^2 + 1$ **19.** $x^3 - 6$

21. $(4, 2)$ **23.** $\{t | t > -8\}$ **25.** $a + 3$, R 5

27. $y^3 - ay^2 + a^2y - a^3$, R $a^2(a^2 + 1)$ **29.** -5

Margin Exercises, Section 9.12, pp. 435–436

1. $y = \dfrac{63}{x}$ **2.** $y = \dfrac{900}{x}$ **3.** 8 hr **4.** $7\frac{1}{2}$ hr

Exercise Set 9.12, pp. 437–438

1. $y = \dfrac{75}{x}$ **3.** $y = \dfrac{80}{x}$ **5.** $y = \dfrac{1}{x}$ **7.** $y = \dfrac{1050}{x}$

9. $y = \dfrac{0.06}{x}$ **11.** $5\frac{1}{3}$ hr **13.** 320 cm³ **15.** 54 min

17. 2.4 ft **19.** 4 **21.** $(3x^2 + 2)(x + 7)$ **23.** No **25.** No

Summary and Review: Chapter 9, pp. 439–440

1. [9.1, ▪▪▪] $\dfrac{x - 2}{x + 1}$ **2.** [9.1, ▪▪▪] $\dfrac{7x + 3}{x - 3}$

3. [9.1, ▪▪▪] $\dfrac{y - 5}{y + 5}$ **4.** [9.1, ▪▪] $\dfrac{a - 6}{5}$

5. [9.1, ▪▪] $\dfrac{6}{2t - 1}$ **6.** [9.2, ▪▪▪] $-20t$

7. [9.2, ▪▪] $\dfrac{2x^2 - 2x}{x + 1}$ **8.** [9.4, ▪▪▪] $30x^2y^2$

9. [9.4, ▪▪▪] $4(a - 2)$, or $4(2 - a)$

10. [9.4, ▪▪▪] $(y - 2)(y + 2)(y + 1)$

11. [9.3, ▪▪] $\dfrac{-3x + 18}{x + 7}$ **12.** [9.5, ▪] -1

13. [9.3, ▪▪▪] $\dfrac{4}{x - 4}$ **14.** [9.6, ▪] $\dfrac{x + 5}{2x}$

15. [9.3, ▪▪] $\dfrac{2x + 3}{x - 2}$ **16.** [9.5, ▪] $\dfrac{2a}{a - 1}$

17. [9.3, ▪▪] $d + c$

18. [9.6, ▪▪] $\dfrac{-x^2 + x + 26}{(x - 5)(x + 5)(x + 1)}$

19. [9.6, ▪▪] $\dfrac{2(x - 2)}{x + 2}$ **20.** [9.7, ▪] 8

21. [9.7, ▪▪] 3, -5 **22.** [9.8, ▪] $5\frac{1}{7}$ hr

23. [9.8, ▪▪] 240 km/h, 280 km/h **24.** [9.8, ▪] -2

25. [9.8, ▪▪▪] 160 **26.** [9.9, ▪] $s = \dfrac{rt}{r - t}$

27. [9.9, ▪] $C = \frac{5}{9}(F - 32)$, or $C = \dfrac{5F - 160}{9}$

28. [9.10, ▪] $\dfrac{z}{1 - z}$ **29.** [9.10, ▪] $c - d$

30. [9.11, ▪] $5x^2 - \frac{1}{2}x + 3$

31. [9.11, ▪] $3x^2 - 7x + 4$, R 1

32. [9.12, ▪] $y = \dfrac{30}{x}$ **33.** [9.12, ▪] $y = \dfrac{0.65}{x}$

34. [9.12, ▪▪] 1 hr **35.** [5.6, ▪] $(5x^2 - 3)(x + 4)$

36. [8.2, ▪▪] $\{x | x \leq -2\}$

37. [4.4, ▪▪] $-2x^3 + 3x^2 + 12x - 18$

38. [7.5, ▪] 5670 **39.** [9.1] 0, 5, 3

CUMULATIVE REVIEW: CHAPTERS 7–9, pp. 443–444

1. [7.3, ▪▪] $(3, -3)$ **2.** [7.4, ▪▪] $(-2, 2)$

3. [7.3, ▪▪] $(4, 1)$ **4.** [7.4, ▪▪] $(3, -8)$

5. [8.1, ▪▪] $\{x | x \geq 7\}$ **6.** [8.2, ▪▪] $\{x | x \leq -6\}$

7. [8.2, ▪▪] $\{x | x > -6\}$ **8.** [8.2, ▪▪] $\{x | 20 \geq x\}$

9. [9.7, ▪▪] 2 **10.** [9.7, ▪▪] No solution

11. [9.9, ▪] $p = \dfrac{mn}{T - Rn}$ **12.** [9.9, ▪] $R = \dfrac{r^2}{E - r}$

13. [8.3, ▪]

```
←—+——+——+——+——+——+——+——+——+——+——o—+—→
  -6 -5 -4 -3 -2 -1  0  1  2  3  4  5  6
```

14. [8.3, ▪▪]

```
←—+——+——+——●——+——+——+——+——●——+——+——+—→
  -6 -5 -4 -3 -2 -1  0  1  2  3  4  5  6
```

15. [8.3, 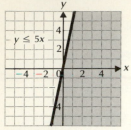] $y \leq 5x$

16. [8.3, ■■■] $2y - 3x > -6$

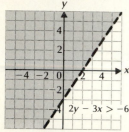

17. [8.4, ■■] {1, 3, 4, 6, 7, 9, 11, 12, 15, 18, 19, 21, 23, 24, 25, 27, 30, 31} **18.** [8.4, ■■■] {0, 1}

19. [9.12, ■■] $y = \dfrac{15}{x}$ **20.** [9.12, ■■] $y = \dfrac{0.05}{x}$

21. [7.5, ■■] 38°, 76°, 66°

22. [7.5, ■■] Hamburger: $1.30; milkshake: $0.90

23. [7.5, ■■] $L = 80$ ft, $w = 30$ ft **24.** [7.5, ■■] 300 L

25. [9.12, ■■] 72 ft, $k = 360$ **26.** [9.8, ■■] 35, 28

27. [9.8, ■■] 35 mph, 25 mph **28.** [9.1, ■■■] $\dfrac{x + 3}{2x - 1}$

29. [9.1, ■■] $\dfrac{t - 4}{t + 4}$ **30.** [9.1, ■■] $\dfrac{2}{3(y + 2)}$

31. [9.2, ■■] 2 **32.** [9.3, ■■] $x + 4$

33. [9.5, ■■] $\dfrac{-5x - 28}{5(x - 5)}$, or $\dfrac{5x + 28}{5(5 - x)}$

34. [9.3, ■■] $\dfrac{4x - 1}{x - 2}$ **35.** [9.6, ■■] $\dfrac{2x - 6}{(x + 2)(x - 2)}$

36. [9.10, ■■] $\dfrac{8x - 12}{17x}$

37. [9.11, ■■] $3x^2 + 4x + 9$, R 13

38. [9.11, ■■] $a^2 + ab + b^2$

39. [9.10, ■■] LCM is $(a - b)^2(a + b)^2$: $\dfrac{y^2 - x^2}{y^2 + x^2} =$

$\dfrac{a^2b^2(a + b)^2 - a^2b^2(a - b)^2}{a^2b^2(a + b)^2 + a^2b^2(a - b)^2} =$

$\dfrac{(a + b)^2 - (a - b)^2}{(a + b)^2 + (a - b)^2} = \dfrac{4ab}{2a^2 + 2b^2} = \dfrac{2ab}{a^2 + b^2}$

40. [8.2] $\emptyset$ **41.** [8.3, ■■■] $x < 3$

42. [7.2, ■■] $A = -4$, $B = -\frac{7}{5}$

43. [7.4, ■■] (0, −1)

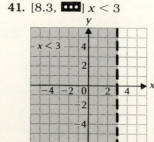

CHAPTER 10

Margin Exercises, Section 10.1, pp. 446–449

1. 6, −6 **2.** 8, −8 **3.** 15, −15 **4.** 10, −10 **5.** 4 **6.** 7
7. 10 **8.** 21 **9.** −7 **10.** −13 **11.** Irrational
12. Rational **13.** Irrational **14.** Irrational **15.** $\frac{7}{128}$
16. −0.6781 **17.** 5.69895 **18.** Rational **19.** Rational
20. Rational **21.** Irrational **22.** 2.828 **23.** 7.874
24. 20, 16

Something Extra—Calculator Corner: Finding Square Roots on a Calculator, p. 450

1. 4.123 **2.** 8.944 **3.** 10.488 **4.** 8.307 **5.** 14.142
6. 3.240 **7.** 29.833 **8.** 16.303 **9.** 1.414 **10.** 0.484
11. 1.772 **12.** 1.932

Exercise Set 10.1, pp. 451–452

1. 1, −1 **3.** 4, −4 **5.** 10, −10 **7.** 13, −13 **9.** 2
11. −3 **13.** −8 **15.** 15 **17.** 19 **19.** 18 **21.** Irrational
23. Irrational **25.** Rational **27.** Irrational **29.** Rational
31. Rational **33.** Rational **35.** Irrational **37.** Rational
39. Irrational **41.** Rational **43.** 2.236 **45.** 4.123
47. 6.557 **49.** 13, 24 **51.** 5 · 5 **53.** 8 **55.** 2
57. −5 and −6 **59.** 65.038

Margin Exercises, Section 10.2, pp. 453–454

1. $45 + x$ **2.** $\dfrac{x}{x + 2}$ **3.** No **4.** Yes **5.** Yes **6.** No

7. Yes **8.** No **9.** $a \geq 0$ **10.** $x \geq 3$ **11.** $x \geq \frac{5}{2}$
12. All real numbers **13.** $|xy|$ **14.** $|xy|$ **15.** $|x - 1|$
16. $|x + 4|$ **17.** xy **18.** xy **19.** $x - 1$ **20.** $x + 4$
21. $5y$ **22.** $\frac{1}{2}t$

Exercise Set 10.2, pp. 455–456

1. $a - 4$ **3.** $t^2 + 1$ **5.** $\dfrac{3}{x + 2}$ **7.** Yes **9.** No **11.** Yes

13. No **15.** $x \geq 0$ **17.** $t \geq 5$ **19.** $y \geq -8$ **21.** $x \geq -20$
23. $y \geq \frac{7}{2}$ **25.** All real numbers **27.** t **29.** $3x$ **31.** ab
33. $34d$ **35.** $x + 3$ **37.** $a - 5$ **39.** $2x - 5$ **41.** 3
43. $\dfrac{3x^2 - 1}{x^2(x + 1)}$ **45.** 6, −6 **47.** 7, −7

49. $[t | t \leq -2$ or $t \geq 2]$

Margin Exercises, Section 10.3, pp. 457–458

1. (a) 8; (b) 8 **2.** $\sqrt{21}$ **3.** 5 **4.** $\sqrt{x^2 + x}$ **5.** $\sqrt{x^2 - 1}$
6. $4\sqrt{2}$ **7.** $x + 7$ **8.** $5x$ **9.** $6m$ **10.** $2\sqrt{19}$ **11.** $x - 4$
12. $8t$ **13.** $10a$ **14.** 16.585 **15.** 10.097

Exercise Set 10.3, pp. 459–460

1. $\sqrt{6}$ **3.** 3 **5.** $\sqrt{14}$ **7.** $\sqrt{\dfrac{3}{10}}$ **9.** $\sqrt{2x}$ **11.** $\sqrt{x^2 - 3x}$

13. $\sqrt{x^2 + 3x + 2}$ **15.** $\sqrt{x^2 - y^2}$ **17.** $\sqrt{86x}$ **19.** $2\sqrt{3}$
21. $5\sqrt{3}$ **23.** $10\sqrt{2x}$ **25.** $4a$ **27.** $7t$ **29.** $x\sqrt{x - 2}$
31. $2x - 1$ **33.** $3(a - b)$ **35.** 11.180 **37.** 18.972
39. 17.320 **41.** 11.043 **43.** $\frac{35}{12}$ **45.** 2 · 5 · 5 **47.** 0.1

49. x^2 **51.** 7, $7\sqrt{10}$, 70, $70\sqrt{10}$, 700; each is $\sqrt{10}$ times the preceding.

Margin Exercises, Section 10.4, pp. 461–462

1. $4\sqrt{2}$ **2.** $5h\sqrt{2}$ **3.** $\sqrt{3}(x-1)$ **4.** x^5 **5.** $(x+2)^7$
6. $x^7\sqrt{x}$ **7.** $3\sqrt{2}$ **8.** 10 **9.** $4x^3y^2$ **10.** $5xy^2\sqrt{2xy}$

Exercise Set 10.4, pp. 463–464

1. $2\sqrt{6}$ **3.** $2\sqrt{10}$ **5.** $5\sqrt{7}$ **7.** $4\sqrt{3x}$ **9.** $2x\sqrt{7}$
11. $\sqrt{2}(2x+1)$ **13.** t^3 **15.** $x^2 \cdot \sqrt{x}$ **17.** $(y-2)^4$
19. $6m\sqrt{m}$ **21.** $8x^3y\sqrt{7y}$ **23.** $3\sqrt{6}$ **25.** $6\sqrt{7}$ **27.** 10
29. $5b\sqrt{3}$ **31.** $a\sqrt{bc}$ **33.** $6xy^3\sqrt{3xy}$ **35.** $10ab^2\sqrt{5ab}$
37. -2 **39.** $y = \frac{7}{32}x$ **41.** $2x^3\sqrt{5x}$ **43.** $0.2x^{2n}$

Margin Exercises, Section 10.5, pp. 465–466

1. $\frac{4}{3}$ **2.** $\frac{1}{5}$ **3.** $\frac{1}{3}$ **4.** $\frac{3}{4}$ **5.** $\frac{15}{16}$ **6.** $\frac{1}{5}\sqrt{15}$ or $\frac{\sqrt{15}}{5}$
7. $\frac{1}{4}\sqrt{10}$ or $\frac{\sqrt{10}}{4}$ **8.** 0.535 **9.** 0.791

Exercise Set 10.5, pp. 467–468

1. $\frac{3}{7}$ **3.** $\frac{1}{6}$ **5.** $\frac{4}{9}$ **7.** $\frac{8}{17}$ **9.** $\frac{13}{14}$ **11.** $\frac{6}{a}$ **13.** $\frac{3a}{25}$
15. $\frac{1}{5}\sqrt{10}$ or $\frac{\sqrt{10}}{5}$ **17.** $\frac{1}{4}\sqrt{6}$ or $\frac{\sqrt{6}}{4}$ **19.** $\frac{1}{2}\sqrt{2}$ or $\frac{\sqrt{2}}{2}$
21. $\frac{1}{x}\sqrt{3x}$ or $\frac{\sqrt{3x}}{x}$ **23.** 0.655 **25.** 0.577 **27.** 0.592
29. 1.549 **31.** $-8x^3 - 3x^2 + 7$ **33.** $-2, -3$ **35.** $\frac{\sqrt{5}}{40}$
37. $\frac{\sqrt{5x}}{5x^2}$ **39.** 1.57 sec, 3.14 sec, 8.880 sec, 11.101 sec

Margin Exercises, Section 10.6, pp. 469–470

1. $\frac{1}{3}$ **2.** $\frac{\sqrt{3}}{3}$ **3.** $x\sqrt{6x}$ **4.** $\frac{\sqrt{35}}{7}$ **5.** $\frac{8y\sqrt{7y}}{7}$ **6.** $\frac{\sqrt{xy}}{y}$

Exercise Set 10.6, pp. 471–472

1. 3 **3.** 2 **5.** $\sqrt{5}$ **7.** $\frac{2}{5}$ **9.** 2 **11.** $3y$ **13.** $x^2\sqrt{5}$ **15.** 2
17. $\frac{\sqrt{10}}{5}$ **19.** $\sqrt{2}$ **21.** $\frac{\sqrt{6}}{2}$ **23.** 5 **25.** $\frac{\sqrt{3x}}{x}$
27. $\frac{4y\sqrt{3}}{3}$ **29.** $\frac{a\sqrt{2a}}{4}$ **31.** $\frac{\sqrt{2}}{4a}$ **33.** 3 **35.** 4
37. $\frac{3\sqrt{30}}{40}$ **39.** $\frac{2}{3}$

Margin Exercises, Section 10.7, pp. 473–474

1. $12\sqrt{2}$ **2.** $5\sqrt{5}$ **3.** $-12\sqrt{10}$ **4.** $5\sqrt{6}$ **5.** $\sqrt{x+1}$
6. $\frac{3}{2}\sqrt{2}$ **7.** $\frac{2}{15}\sqrt{15}$

Something Extra—An Application: Wind Chill Temperature, p. 474

1. $0°$ **2.** $-10°$ **3.** $-22°$ **4.** $-64°$

Exercise Set 10.7, pp. 475–476

1. $7\sqrt{2}$ **3.** $-8\sqrt{a}$ **5.** $8\sqrt{3}$ **7.** $\sqrt{3}$ **9.** $13\sqrt{2}$
11. $-24\sqrt{2}$ **13.** $(2+9x)\sqrt{x}$ **15.** $3\sqrt{2x+2}$
17. $(3xy - x^2 + y^2)\sqrt{xy}$ **19.** $\frac{2}{3}\sqrt{3}$ **21.** $\frac{13}{2}\sqrt{2}$
23. $\frac{1}{18}\sqrt{3}$ **25.** $3(x-5)$ **27.** $(6, 2)$
29. $\frac{(2+x^2)\sqrt{1+x^2}}{1+x^2}$ **31.** $-\sqrt{3} - \sqrt{5}$
33. Any pairs of numbers a, b such that $a = 0$, $b \geq 0$; or $a \geq 0$, $b = 0$.

Margin Exercises, Section 10.8, pp. 477–478

1. $c = \sqrt{65} \approx 8.062$ **2.** $a = \sqrt{75} \approx 8.660$
3. $b = \sqrt{10} \approx 3.162$ **4.** $a = \sqrt{175} \approx 13.229$
5. $\sqrt{325} \approx 18.028$

Exercise Set 10.8, pp. 479–480

1. $c = 17$ **3.** $c = \sqrt{32} \approx 5.657$ **5.** $b = 12$ **7.** $b = 4$
9. $c = 26$ **11.** $b = 12$ **13.** $a = 2$ **15.** $b = \sqrt{2} \approx 1.414$
17. $a = 5$ **19.** $\sqrt{75} \approx 8.660$ m **21.** $\sqrt{208} \approx 14.422$ ft
23. $\sqrt{7200} \approx 84.853$ ft **25.** 3 **27.** $\sqrt{181} \approx 13.454$ cm
29. $12 - 2\sqrt{6} \approx 7.101$ **31.** 6 **33.** $A = \frac{a^2\sqrt{3}}{4}$

Margin Exercises, Section 10.9, pp. 481–482

1. $\frac{64}{3}$ **2.** 2 **3.** 66 **4.** Approx. 313.050 km
5. Approx. 15.652 km **6.** 676 m

Exercise Set 10.9, pp. 483–484

1. 25 **3.** 38.44 **5.** 397 **7.** $\frac{621}{2}$ **9.** 5 **11.** 3 **13.** $\frac{17}{4}$
15. No solution **17.** No solution
19. 346.48 km, approx. **21.** 11,236 m **23.** 125 ft,
245 ft **25.** $E = \frac{180A}{\pi r^2}$ **27.** \$7200 **27.** 0 **31.** 9

Summary and Review: Chapter 10, pp. 485–486

1. [10.1, ■] 8, -8 **2.** [10.1, ■] 5, -5
3. [10.1, ■] 14, -14 **4.** [10.1, ■] 20, -20
5. [10.1, ■] 6 **6.** [10.1, ■] -9 **7.** [10.1, ■] 7
8. [10.1, ■] -13 **9.** [10.1, ■] Irrational
10. [10.1, ■] Rational **11.** [10.1, ■] Irrational
12. [10.1, ■] Rational **13.** [10.1, ■] Rational
14. [10.1, ■] Rational **15.** [10.1, ■] Rational
16. [10.1, ■] Irrational **17.** [10.1, ■] 1.732
18. [10.1, ■] [10.3, ■] 10.392 **19.** [10.5, ■] 0.354
20. [10.5, ■] 0.742 **21.** [10.2, ■] $x^2 + 4$
22. [10.2, ■] $5ab^3$ **23.** [10.2, ■] Yes
24. [10.2, ■] No
25. [10.2, ■] Yes **26.** [10.2, ■] Yes
27. [10.2, ■] $x \geq -7$ **28.** [10.2, ■] $y \geq 10$
29. [10.2, ■] m **30.** [10.2, ■] $7t$ **31.** [10.2, ■] p
32. [10.2, ■] $x - 4$ **33.** [10.3, ■] $\sqrt{21}$
34. [10.3, ■] $\sqrt{at}$ **35.** [10.3, ■] $\sqrt{x^2 - 9}$

10

36. [10.3, ■] $\sqrt{6xy}$ 37. [10.3, ■], [10.4, ■] $-4\sqrt{3}$
38. [10.4, ■] $4t\sqrt{2}$ 39. [10.4, ■] $x + 8$
40. [10.4, ■] $\sqrt{t} - 7\sqrt{t} + 7$ 41. [10.4, ■] x^4
42. [10.4, ■] $m^7\sqrt{m}$ 43. [10.4, ■] $2\sqrt{15}$
44. [10.4, ■] $2x\sqrt{10}$ 45 [10.4, ■] $5xy\sqrt{2}$
46. [10.4, ■] $10a^2b\sqrt{ab}$ 47. [10.5, ■] $\frac{5}{8}$
48. [10.5, ■] $\frac{2}{3}$ 49. [10.5, ■] $\frac{7}{t}$ 50. [10.5, ■] $\frac{\sqrt{2}}{2}$
51. [10.5, ■] $\frac{\sqrt{2}}{4}$ 52. [10.5, ■] $\frac{\sqrt{5y}}{y}$
53. [10.6, ■] $\frac{2\sqrt{3}}{3}$ 54. [10.6, ■] $\frac{\sqrt{15}}{5}$
55. [10.6, ■] $\frac{x\sqrt{30}}{6}$ 56. [10.7, ■] $13\sqrt{5}$
57. [10.7, ■] $\sqrt{5}$ 58. [10.7, ■] $\frac{1}{2}\sqrt{2}$
59. [10.8, ■] $b = 20$ 60. [10.8, ■] $c = \sqrt{3} \approx 1.732$
61. [10.8, ■] $7\sqrt{2}$ cm 62. [10.9, ■] 52
63. [10.9, ■] No solution 64. [10.9, ■] 405 ft
65. [9.7, ■] 15 66. [9.6, ■] $\frac{-4x + 3}{(x - 3)(x + 3)(x - 2)}$
67. [6.4, ■] 100 68. [7.6, ■] $1\frac{1}{3}$ hr
69. [10.8, ■] 50 ft^2
70. [10.1, ■] $\sqrt{5}$ 71. [10.9, ■] No solution

CHAPTER 11

Margin Exercises, Section 11.1, pp. 490–492

1. $x^2 - 7x = 0$; $a = 1$, $b = -7$, $c = 0$
2. $x^2 + 9x - 3 = 0$; $a = 1$, $b = 9$, $c = -3$
3. $4x^2 + 2x + 4 = 0$; $a = 4$, $b = 2$, $c = 4$ 4. $\sqrt{5}$, $-\sqrt{5}$
5. 0 6. $\frac{\sqrt{6}}{2}$, $-\frac{\sqrt{6}}{2}$ 7. $\frac{3}{2}$, $-\frac{3}{2}$ 8. 9.3 sec

Exercise Set 11.1, pp. 493–494

1. $x^2 - 3x - 2 = 0$; $a = 1$, $b = -3$, $c = -2$
3. $7x^2 - 4x + 3 = 0$; $a = 7$, $b = -4$, $c = 3$
5. $3x^2 - 2x + 8 = 0$; $a = 3$, $b = -2$, $c = 8$ 7. 2, -2
9. 7, -7 11. $\sqrt{7}$, $-\sqrt{7}$ 13. $\sqrt{10}$, $-\sqrt{10}$
15. $2\sqrt{2}$, $-2\sqrt{2}$ 17. $\frac{5}{2}$, $-\frac{5}{2}$ 19. $\frac{7\sqrt{3}}{3}$, $-\frac{7\sqrt{3}}{3}$
21. $\sqrt{3}$, $-\sqrt{3}$ 23. 7.9 sec 25. 3.3 sec 27. 0, 6
29. $2\sqrt{5}$ 31. Approx. 50, -50 33. $\sqrt{21}$, $-\sqrt{21}$

Margin Exercises, Section 11.2, pp. 495–496

1. 0, $-\frac{5}{3}$ 2. 0, $\frac{3}{5}$ 3. $\frac{2}{3}$, -1 4. 4, 1 5. (a) 14; (b) 11

Exercise Set 11.2, pp. 497–498

1. 0, -7 3. 0, $-\frac{2}{3}$ 5. 0, -1 7. 0, $\frac{1}{5}$ 9. 4, 12
11. 3, -7 13. -5 15. $\frac{3}{2}$, 5 17. 4, $-\frac{5}{3}$ 19. -2, 7
21. 4, -5 23. 9 25. 7 27. $9x^2 + 6x + 1$ 29. 4.123
31. $-\frac{1}{3}$, 1 33. 0; $-28,160,000$ 35. $\sqrt{2}$, $-\sqrt{2}$

Margin Exercises, Section 11.3, pp. 499–501

1. 7, -1 2. $-3 \pm \sqrt{10}$ 3. $1 \pm \sqrt{5}$
4. $x^2 - 8x + 16 = (x - 4)^2$ 5. $x^2 - 10x + 25 = (x - 5)^2$
6. $x^2 + 7x + \frac{49}{4} = \left(x + \frac{7}{2}\right)^2$
7. $x^2 - 3x + \frac{9}{4} = \left(x - \frac{3}{2}\right)^2$ 8. $1299.60 9. 12.5%
10. 73.2%

Something Extra—Calculator Corner: Compound Interest, p. 502

1. $1360.49 2. $1425.76
3. (a) $1160.00; (b) $1166.40; (c) $1169.86;
(d) $1173.47; (e) $1173.51 4. (a) $2; (b) $2.25;
(c) $2.44; (d) $2.7146; (e) $2.7181

Exercise Set 11.3, pp. 503–504

1. -7, 3 3. $-1 \pm \sqrt{6}$ 5. $3 \pm \sqrt{6}$
7. $x^2 - 2x + 1 = (x - 1)^2$ 9. $x^2 + 18x + 81 = (x + 9)^2$
11. $x^2 - x + \frac{1}{4} = \left(x - \frac{1}{2}\right)^2$
13. $x^2 + 5x + \frac{25}{4} = \left(x + \frac{5}{2}\right)^2$ 15. 10% 17. 18.75%
19. 8% 21. 20% 23. (4, -1) 25. $3\frac{1}{3}$ hr 27. 8, -10
29. 12.6%

Margin Exercises, Section 11.4, pp. 505–506

1. -2, -6 2. $5 \pm \sqrt{3}$ 3. $-3 \pm \sqrt{10}$ 4. 5, -2
5. -7, 2 6. $\frac{-3 \pm \sqrt{33}}{4}$ 7. $\frac{1 \pm \sqrt{10}}{3}$

Exercise Set 11.4, pp. 507–508

1. -2, 8 3. -21, -1 5. $1 \pm \sqrt{6}$ 7. $9 \pm \sqrt{7}$ 9. -9, 2
11. -3, 2 13. $\frac{7 \pm \sqrt{57}}{2}$ 15. $\frac{-3 \pm \sqrt{17}}{4}$
17. $\frac{-2 \pm \sqrt{7}}{3}$ 19. $-\frac{7}{2}$, $\frac{1}{2}$ 21. $-\frac{5}{2}$, $\frac{2}{3}$ 23. $\sqrt{6}$
25. $(-3, 5)$ 27. $\pm 2\sqrt{55}$ 29. $\frac{-1 \pm \sqrt{1 - c}}{2}$

Margin Exercises, Section 11.5, pp. 509–511

1. $\frac{1}{2}$, -4 2. $\frac{4 \pm \sqrt{31}}{5}$ 3. -0.3, 1.9

Something Extra—An Application: Handling Dimension Symbols (Part 3), p. 512

1. 6 ft 2. 1020 min 3. 172,800 sec 4. 0.1 hr
5. 600 g/cm 6. 30 mi/hr 7. 2,160,000 cm^2
8. 0.81 ton/yd^3 9. 150¢/hr 10. 60 person-days
11. Approx. 5,865,696,000,000 mi/yr
12. 6,570,000 mi/yr

Exercise Set 11.5, pp. 513–514

1. $-3, 7$ **3.** 3 **5.** $-\dfrac{4}{3}, 2$ **7.** $-3, 3$ **9.** $5 \pm \sqrt{3}$

11. $1 \pm \sqrt{3}$ **13.** No real-number solutions

15. $\dfrac{-1 \pm \sqrt{10}}{3}$ **17.** $\dfrac{-7 \pm \sqrt{61}}{2}$ **19.** $0, 2$

21. $-1.3, 5.3$ **23.** $-0.2, 6.2$ **25.** $-1.7, 0.4$

27. $3x^2\sqrt{3x}$ **29.** $\frac{4}{3}\sqrt{3}$ **31.** $0.25, -0.45$ **33.** $a \geq -\dfrac{1}{3}$

Margin Exercises, Section 11.6, pp. 515–516

1. $13, 5$ **2.** 2 **3.** 7

Exercise Set 11.6, pp. 517–518

1. $6, -\dfrac{2}{3}$ **3.** $6, -4$ **5.** 1 **7.** $5, 2$ **9.** No solution

11. $\sqrt{7}, -\sqrt{7}$ **13.** $-2 \pm \sqrt{3}$ **15.** $2 \pm \sqrt{34}$ **17.** 9

19. 7 **21.** $1, 5$ **23.** 3 **25.** $25, 13$ **27.** No solution

29. 3 **31.** $-4, 3$ **33.** 0

Margin Exercises, Section 11.7, pp. 519–520

1. $L = \dfrac{r^2}{20}$ **2.** $L = \dfrac{T^2 g}{4\pi^2}$ **3.** $m = \dfrac{E}{c^2}$ **4.** $r = \sqrt{\dfrac{A}{\pi}}$

5. $d = \sqrt{\dfrac{C}{P}} + 1$ **6.** $n = \dfrac{1 \pm \sqrt{1 + 4N}}{2}$

7. $t = \dfrac{-v \pm \sqrt{v^2 + 32h}}{16}$

Exercise Set 11.7, pp. 521–522

1. $A = \dfrac{N^2}{6.25}$ **3.** $T = \dfrac{cQ^2}{a}$ **5.** $c = \sqrt{\dfrac{E}{m}}$

7. $d = \dfrac{c \pm \sqrt{c^2 + 4aQ}}{2a}$ **9.** $a = \sqrt{c^2 - b^2}$ **11.** $t = \sqrt{\dfrac{2S}{g}}$

13. $r = \dfrac{-\pi h \pm \sqrt{\pi^2 h^2 + \pi A}}{\pi}$ **15.** $r = 6\sqrt{\dfrac{10A}{\pi S}}$

17. $a = \sqrt{c^2 - b^2}$ **19.** $a = \dfrac{2h\sqrt{3}}{3}$ **21.** $\dfrac{2x - 14}{x^2 - 9}$

23. $4x^3 y^2$ **25.** $T = \dfrac{2 \pm \sqrt{4 - a(m - n)}}{a}$

27. $x = \dfrac{-b \pm \sqrt{b^2 - 4a(c - y)}}{2a}$

Margin Exercises, Section 11.8, pp. 523–524

1. 20 m **2.** 2.3 cm; 3.3 cm **3.** 3 km/h

Exercise Set 11.8, pp. 525–530

1. 3 cm **3.** 7 ft; 24 ft **5.** Width 8 cm; length 10 cm

7. Length 20 cm; width 16 cm

9. Width 5 m; length 10 m **11.** 4.6 m; 6.6 m

13. Width 3.6 in.; length 5.6 in.

15. Width 2.2 m; length 4.4 m **17.** 7 km/h **19.** 8 mph

21. 4 km/h **23.** 36 mph **25.** $1 + \sqrt{2} \approx 2.41$

27. $d = 10\sqrt{2} = 14.14$; two 10-inch pizzas

Margin Exercises, Section 11.9, pp. 531–536

1. (a) Upward;
(b)

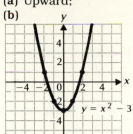

2. (a) Upward;
(b)

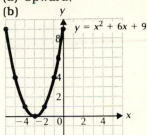

3. (a) Downward;
(b)

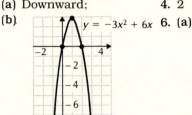

4. 2 **5.** $0.2, -2.2$ **6. (a)**

x	y	
0	8	← y-intercept
-2	0	⎱
-4	0	⎰ x-intercepts
-3	-1	← vertex
-6	8	

(b)

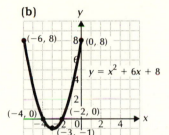

7.

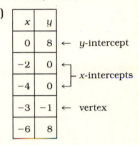

8.

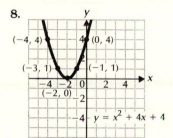

Exercise Set 11.9, pp. 537–538

1. $y = x^2$ Upward

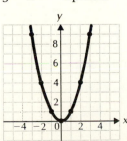

3. $y = -1 \cdot x^2$ Downward

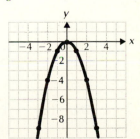

5. $y = -x^2 + 2x$
Downward

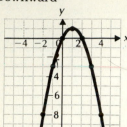

7. $y = 8 - x - x^2$
Downward

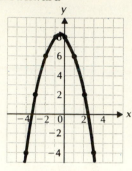

9. $y = x^2 - 2x + 1$
Upward

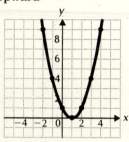

11. $y = x^2 + 2x - 3$
Upward

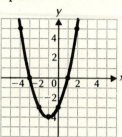

13. $y = -2x^2 - 4x + 1$
Downward

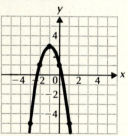

15. $2.2, -2.2$ **17.** $2.4, -3.4$ **19.** 4 **21.** No solution
23. No solution **25.** $2t\sqrt{5}$ **27.** $(0, c)$

Summary and Review: Chapter 11, pp. 539–540

1. [11.1, ■■] $\sqrt{3}, -\sqrt{3}$ **2.** [11.2, ■■] $\frac{3}{5}, 1$
3. [11.5, ■] $1 \pm \sqrt{11}$ **4.** [11.2, ■■] $\frac{1}{3}, -2$
5. [11.3, ■] $-8 \pm \sqrt{13}$ **6.** [11.1, ■■] 0

7. [11.2, ■■] $0, \frac{7}{5}$ **8.** [11.5, ■] $\dfrac{1 \pm \sqrt{10}}{3}$

9. [11.5, ■■] $-3 \pm 3\sqrt{2}$ **10.** [11.5, ■] $\dfrac{2 \pm \sqrt{3}}{2}$

11. [11.5, ■■] $\dfrac{3 \pm \sqrt{33}}{2}$

12. [11.5, ■■] No real solution **13.** [11.2, ■■] $0, \frac{4}{3}$
14. [11.1, ■■] $-2\sqrt{2}, 2\sqrt{2}$ **15.** [11.4, ■] $\frac{5}{3}, -1$

16. [11.4, ■■] $\dfrac{5 \pm \sqrt{17}}{2}$ **17.** [11.5, ■■] $4.6, 0.4$

18. [11.5, ■■] $-1.9, -0.1$ **19.** [11.6, ■■] $3, -5$
20. [11.6, ■■] 1 **21.** [11.6, ■■] 4 **22.** [11.6, ■■] $0, \frac{1}{4}$

23. [11.7, ■] $T = \dfrac{L}{4V^2 - 1}$

24. [11.9, ■]

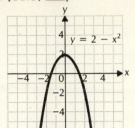

25. [11.9, ■]

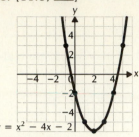

$y = x^2 - 4x - 2$

26. [11.9, ■■] $-0.5, 4.5$ **27.** [11.8, ■■] 1.7 m, 4.7 m
28. [11.3, ■■■] 30% **29.** [11.8, ■■] $W = 7$ m, $L = 10$ m
30. [11.8, ■] 30 km/h **31.** [7.4, ■■] $(3, 5)$
32. [7.4, ■] $(-4, 7)$ **33.** [10.4, ■■■] $6\sqrt{a}$
34. [10.4, ■■■] $2xy\sqrt{15y}$ **35.** [10.7, ■] $12\sqrt{11}$
36. [10.7, ■■] $4\sqrt{10}$ **37.** [9.8, ■■] $15\frac{5}{9}$ days
38. [11.8, ■■] $31, 32; -32, -31$ **39.** [11.3, ■■] $14, -14$
40. [11.6, ■■] 25 **41.** [11.8, ■■] $s = 5\sqrt{\pi}$

CUMULATIVE REVIEW: CHAPTERS 1–11, pp. 543–546

1. [1.4, ■■■] $x \cdot x \cdot x$ **2.** [1.4, ■■] 54
3. [1.2, ■■] $2 \cdot 2 \cdot 2 \cdot 2 \cdot 2 \cdot 2 \cdot 2 \cdot 2$ **4.** [1.2, ■■] 240
5. [2.1, ■■■] 7 **6.** [2.2, ■■] 9 **7.** [2.5, ■■■] 15
8. [2.6, ■■■] $-\frac{3}{20}$ **9.** [2.4, ■■] -63
10. [2.8, ■■] $-2m - 4$ **11.** [3.2, ■■] -8
12. [3.3, ■■] -12 **13.** [3.4, ■■] 7 **14.** [5.7, ■■] $3, 5$
15. [7.3, ■■] $(1, 2)$ **16.** [7.4, ■■] $(12, 5)$
17. [7.4, ■■] $(6, 7)$ **18.** [5.7, ■■] $3, -2$

19. [11.5, ■■] $\dfrac{-3 \pm \sqrt{29}}{2}$ **20.** [11.6, ■■] 2

21. [8.2, ■■] $\{x | x \geq -1\}$ **22.** [3.3, ■■] 8
23. [3.3, ■■] -3 **24.** [8.2, ■■] $\{x | x < -8\}$
25. [8.2, ■■] $\{y | y \leq \frac{35}{22}\}$ **26.** [11.1, ■■] $\sqrt{10}, -\sqrt{10}$
27. [11.3, ■■] $3 \pm \sqrt{6}$ **28.** [9.7, ■] $\frac{2}{9}$
29. [9.7, ■■] [11.6, ■■] -5
30. [9.7, ■■] [11.6, ■■] No solution **31.** [10.9, ■] 12

32. [9.9, ■■] $b = \dfrac{At}{4}$ **33.** [9.9, ■] $m = \dfrac{tn}{t + n}$

34. [11.7, ■] $A = \pi r^2$

35. [11.7, ■] $x = \dfrac{b \pm \sqrt{b^2 + 4ay}}{2a}$ **36.** [3.7, ■■] x^{-4}

37. [3.7, ■■■] y^7 **38.** [3.7, ■■] $4y^{12}$
39. [4.1, ■■] $10x^3 + 3x - 3$
40. [4.3, ■■] $7x^3 - 2x^2 + 4x - 17$
41. [4.4, ■■] $8x^2 - 4x - 6$
42. [4.6, ■■] $-8y^4 + 6y^3 - 2y^2$
43. [4.5, ■■] $6t^3 - 17t^2 + 16t - 6$ **44.** [4.7, ■] $t^2 - \frac{1}{16}$
45. [4.7, ■■] $9m^2 - 12m + 4$
46. [5.9, ■■] $15x^2y^3 + x^2y^2 + 5xy^2 + 7$
47. [5.9, ■■] $x^4 - 0.04y^2$

48. [5.9, ■■] $9p^2 + 24pq^2 + 16q^4$ **49.** [9.1, ■■] $\dfrac{2}{x + 3}$

50. [9.2, ■■] $\dfrac{3a(a - 1)}{2(a + 1)}$ **51.** [9.5, ■] $\dfrac{27x - 4}{5x(3x - 1)}$

52. [9.6, ■] $\dfrac{-x^2 + x + 2}{(x + 4)(x - 4)(x - 5)}$

53. [5.1,] $4x(2x - 1)$ **54.** [5.2, ▦] $(5x - 2)(5x + 2)$
55. [5.5, ▪] $(3y + 2)(2y - 3)$ **56.** [5.3, ▦] $(m - 4)^2$
57. [5.1, ▦] $(x^2 - 5)(x - 8)$
58. [5.4, ▪], [5.2, ▦] $3(a^2 + 6)(a + 2)(a - 2)$
59. [5.2, ▦] $(2x + 1)(2x - 1)(4x^2 + 1)$
60. [5.9, ▦] $(7ab - 2)(7ab + 2)$
61. [5.9, ▦] $(3x + 5y)^2$ **62.** [5.9, ▦] $(2a + d)(c - 3b)$
63. [5.9, ▦] $(5x - 2y)(3x + 4y)$ **64.** [9.10, ▪] $-\frac{4}{5}$
65. [10.1, ▪] 7 **66.** [10.1, ▪] -25 **67.** [10.2, ▦] $8x$
68. [10.3, ▪] $\sqrt{a^2 - b^2}$ **69.** [10.4, ▦] $8a^2b\sqrt{3ab}$
70. [10.4, ▪], [10.3, ▦] $5\sqrt{6}$ **71.** [10.4, ▪] $9xy\sqrt{3x}$
72. [10.5, ▪] $\frac{10}{9}$ **73.** [10.5, ▪] $\frac{8}{x}$

74. [10.7, ▪] $6\sqrt{12}$, or $12\sqrt{3}$ **75.** [10.6, ▪] $\frac{2\sqrt{10}}{5}$

76. [10.8, ▪] 40

77. [6.2, ▪]

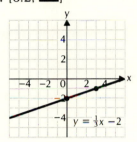

78. [6.3, ▪]

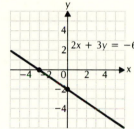

79. [6.3, ▪]

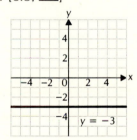

80. [8.3, ▦]

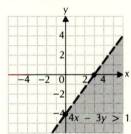

81. [11.9, ▪, ▪]

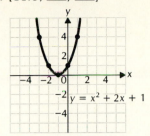

82. [11.4, ▪] $\frac{2 \pm \sqrt{6}}{3}$ **83.** [11.5, ▪] $-0.2, 1.2$

84. [1.10, ▦] 25% **85.** [1.10, ▦] 60
86. [11.8, ▪] 2 km/h   **87.** [11.8, ▪] $12m$
88. [3.5, ▪] 6090
89. [7.5, ▪] 14 lb of $\$1.10$; 28 lb of $\$0.80$
90. [9.8, ▪] $4\frac{4}{9}$ hr
91. [6.4, ▪] $\$451.20$; variation constant is the amount earned per hour.
92. [8.4, ▦] $\{c, d\}$ **93.** [8.4, ▦] $\{a, b, c, d, e, f, g, h\}$
94. [6.5, ▦] 2, $(0, -8)$ **95.** [6.5, ▪] 15

96. [6.4, ▪] $y = 10x$ **97.** [9.12, ▪] $y = \dfrac{1000}{x}$

98. [2.1] $12, -12$ **99.** [10.1, ▪] $\sqrt{3}$

100. [11.4] $-30, 30$ **101.** [10.8, ▪] $\dfrac{\sqrt{6}}{3}$

102. [5.2, ▪] Yes   **103.** [9.3, ▪] No
104. [5.3, ▪] No   **105.** [10.7, ▪] No
106. [10.2, ▦] Yes

Index

O

One
dividing by, 37
as exponent, 23
multiplying by, 14, 37, 377, 466
Operations, order of, 107
Opposite of a number, *see* Additive inverse
Order
ascending, 173
descending, 173
on number line, 39, 67
of operations, 107
Ordered pairs, 270
as solutions of equations, 275, 313
Origin, 270

P

Pairs, ordered, 270
Parabola, 531. *See also* Graphing, equations, quadratic.
intercepts, 534
line of symmetry, 431
vertex, 534
Parallel lines, 314, 324
Parallelogram, area of, 147
Parentheses, 27
in equations, 131
within parentheses, 103
removing, 102
Parking, 449
Pendulum, period of, 468
Percent notation, 53
converting from decimal notation, 54
converting to decimal notation, 53
converting from fractional notation, 54
converting to fractional notation, 53
and problem solving, 54–56
Perfect square, 454
Perimeter, rectangle, 30, 147
Pi (π), 448
Planet orbits, 12
Plotting points, 270
Point–slope equation, 297, 298
Points
coordinates of, 271
plotting, 270
Polygon, number of diagonals, 496
Polynomials, 168
addition of, 177, 254
additive inverse of, 183
in ascending order, 173
binomials, 174
coefficients, 174, 254
collecting like terms, 170, 173, 254
degree of, 173, 254
in descending order, 173
division of, 431
equivalent, 169, 183
evaluating, 168, 224, 253
and expanded notation, 190
factoring, *see* Factoring, polynomials
missing terms in, 174
monomials, 168, 174
multiplication of, 187–189, 255
special products, 193, 197–200
relatively prime, 212
in several variables, 253–258
subtraction of, 184, 255
terms of, 169
trinomials, 174
Population growth, 49
Positive numbers, 66, 87
Positive square root, 446
Powers, 22. *See also* Exponents.
raising to a power, 151
square roots of, 461
Prime factorization, 8
and LCM, 9
Prime numbers, 7
table of, 8
Principal, *see* Interest
Principal square root, 446
Principle
addition, 116
multiplication, 121
of squaring, 481
of zero products, 239
Problem-solving tips, 135, 245. *See also* Applied problems; Problems.
Problems, *see also* Applied problems
age, 327
board cutting, 135
earned run average, 417
falling object, distance traveled by, 492
geometric
area, 523
diagonals of a polygon, 496
perimeter, 137, 328
triangle, 138
gravitational force, 423
horizon, distance to, 482
integer, 136, 268. *See also* Problems, number.
interest
compound, 152, 500, 502
simple, 34
I.Q. (intelligence quotient), 146
mixture, 329–332
motion, 339–342, 413, 524
number, 136, 245, 413
parking-lot arrival spaces, 449
pendulum, period of, 468
percent, 54–56
population growth, 49
price reduction, 138
rent-a-car, 137, 327
right triangles, 477, 478, 523
skidding car, speed of, 460
Torricelli's theorem, 519
wildlife population, estimating, 417
work, 414
Products, *see also* Multiplication
of polynomials, 187–189, 193, 197–200, 255
raising to a power, 151
of square roots, 457, 462
Property, Pythagorean, 477
Proportion, 416
extremes, 422
means, 422
Proportional, 416
Pythagorean property, 477

Q

Quadrants in graphs, 271
Quadratic equations, 490
approximating solutions
using calculator or table, 511
by graphing, 534
discriminant, 510
graphs of, 531–536
solving, 491
by completing the square, 505
by factoring, 495
by quadratic formula, 509–511
in standard form, 490
Quadratic formula, 509
Quotient
square root of, 465
of square roots, 469

R

Radical equations, 481, 516
Radical expressions, 453. *See also* Square roots.

ANSWER KEYS FOR INTRODUCTORY ALGEBRA CHAPTER TESTS

Test: Chapter 1

1. [1.1, ·] 6   2. [1.1, ·] 59   3. [1.1, ·] 8

4. [1.1, · ·] n – 8   5. [1.1, ·] 240 ft²

6. [1.3, · · ·] $\frac{2}{3}$ 7. [1.3, · · ·] $\frac{23}{15}$ 8. [1.3, · · ·] $\frac{9}{32}$

9. [1.7, · ·] $\frac{4}{3}$ 10. [1.4, · · ·] yyyy

11. [1.4, ·] $\frac{569}{100}$ 12. [1.4, · ·] 0.375

13. [1.5, ·] 62   14. [1.5, · ·] Associative law of multiplication   15. [1.5, :·:] 5(x + y); 50

16. [1.6, · ·] 5(3y + 1)   17. [1.6, · ·] 8(3x + 2 + y)

18. [1.2, · ·] 2·2·3·5·5   19. [1.2, : :] 120

20. [1.3, · ·] $\frac{2}{3}$ 21. [1.6, ·] 90x + 30y

22. [1.6, ·] 63m + 14x + 7   23. [1.6, · · ·] 26x + 102

24. [1.6, · · ·] 22y + 39a   25. [1.10, ·] 0.007

26. [1.10, · ·] $\frac{91}{100}$ 27. [1.10, ·] 44%

28. [1.8, · ·] 8.4   29. [1.8, · · ·] 6

30. [1.8, · ·] $\frac{3}{5}$ 31. [1.9, ·] 17

32. [1.9, ·] 14   33. [1.9, ·] 30

34. [1.10, :·:] $6\frac{2}{3}$% or 6.$\overline{6}$%   35. [1.10, :·:] 64

36. [1.10, :·:] 77.5   37. [1.10, :·:] $25,000

38. [1.7, : :] <   39. [1.3, · ·] $\frac{33}{100}$

40. [1.1, ·] 15

Test: Chapter 2

1. [2.1, · ·] <   2. [2.1, · ·] >   3. [2.1, · ·] <

4. [2.1, · · ·] 9   5. [2.5, ·] $\frac{9}{4}$ 6. [2.5, ·] $-\frac{2}{3}$

7. [2.5, ·] 1.4   8. [2.1, : :] 8   9. [2.6, · · ·] $-\frac{1}{2}$

10. [2.6, · · ·] $\frac{7}{4}$ 11. [2.5, · ·] –1.6

12. [2.3, ·] –8   13. [2.3, ·] 10

14. [2.5, · · ·] –2.5   15. [2.5, · · ·] $\frac{7}{8}$

16. [2.4, ·] –48   17. [2.4, · ·] –9

18. [2.6, · · ·] $\frac{3}{4}$ 19. [2.6, · · ·] –9.728

20. [2.7, · · ·] 18 – 3x   21. [2.7, · · ·] –5y + 5

22. [2.7, : :] 2(6 – 11x)

23. [2.7, : :] 7(x + 3 + 2y)   24. [2.3, ·] 12

25. [2.8, · ·] 2x + 7

26. [2.7, · · ·], [:·:] 9a – 12b – 7

27. [2.8, · · ·] 84y + 12   28. [2.8, · · ·] 6

29. [2.9, ·] 166   30. [2.9, · ·] 2 * A * B + B^3

31. [2.9, · ·] $\frac{a - b}{a^2}$ 32. [1.1, ·] 480 yd²

33. [1.10, :·:] 48%   34. [1.2, · ·] 2·2·2·7·5

35. [1.2, : :] 240   36. [2.1, · · ·] No solution

37. [2.8, · · ·] 4a

Test: Chapter 3

1. [3.1, ·] 8   2. [3.1, ·] 26   3. [3.2, ·] –6

4. [3.2, ·] 49   5. [3.3, · ·] –12   6. [3.3, ·] 2

7. [3.3, ·] –8   8. [3.1, ·] $-\frac{7}{20}$ 9. [3.4, ·] 7

10. [3.4, ·] –5   11. [3.3, · ·] 2.5

12. [3.5, ·] 7 cm, 11 cm   13. [3.5, ·] 6

14. [3.5, ·] 81, 83, 85   15. [3.5, ·] $750

16. [3.6, ·] r = $\frac{A}{2\pi h}$ 17. [3.6, ·] l = $\frac{P - 2w}{2}$

18. [3.7, ·] t⁻⁵   19. [3.7, ·] $\frac{1}{y^4}$

20. [3.7, · · ·] x⁻¹²   21. [3.7, · ·] 6⁻¹³

22. [3.7, : :] 16a⁻¹²   23. [3.8, ·] 3.28 × 10⁻⁵

24. [3.8, ·] 8,300,000   25. [3.7, :·:] $1442.90

26. [3.8, · ·] 2.3 × 10⁻⁵   27. [2.5, · · ·] $-\frac{2}{9}$

28. [1.3, · · ·] $\frac{11}{20}$ 29. [2.5, · ·] –139.1

30. [2.6, · ·] 1440   31. [2.8, · ·] –18x + 37y

32. [3.6, ·] d = $\frac{ca - 1}{c}$, or a $- \frac{1}{c}$

33. [3.3, ·], [2.1, · · ·] –15, 15

34. [1.5, · ·] All rational numbers

Test: Chapter 4

1. [4.1, ·] –7   2. [4.2, : :] $\frac{1}{3}$, –1, 7

3. [4.2, · · ·] 4, 0, 1, 6; 6   4. [4.2, :::] Trinomial

5. [4.1, : :] 5a² – 6   6. [4.1, : :] $\frac{7}{4}$y² – 4y

7. [4.2, · ·] x⁵ + 2x³ + 4x² – 8x + 3

8. [4.3, ·] 4x⁵ + x⁴ + 2x³ – 8x² + 2x – 7

9. [4.3, ·] 2y⁴ + 5y² + y + 5

10. [4.4, · ·] –4x⁴ + x³ – 8x – 3

11. [4.4, · ·] –t⁵ + 0.7t³ – 0.8t² – 21

12. [4.6, ·] –12x⁴ + 9x³ + 15x²

13. [4.7, · ·] x² $- \frac{2}{3}$x + $\frac{1}{9}$ 14. [4.7, ·] 100 – 9y²

15. [4.6, · ·] 3b² – 4b – 15

16. [4.6, · ·] x¹⁴ – 4x⁸ + 4x⁶ – 16

17. [4.6, · ·] 48 + 34y – 5y²

18. [4.5, : :], [:·:] 6x³ – 7x² – 11x – 3

19. [4.7, · ·] 25t² + 20t + 4   20. [1.10, :·:] 32%

21. [3.3, · ·] $\frac{9}{4}$ 22. [2.7, : :] 25(t – 2 + 4m)

23. [3.5, ·] Width = 125.5 m, length = 144.5 m

24. [4.7, ·], [· ·], [3.3, · ·] $-\frac{61}{12}$

25. [1.4, ∴] $\pi r^2 - 9\pi$

Test: Chapter 5

1. [5.1, ·] (4x)(x²); (2x²)(2x); (-2x)(-2x²)

2. [5.4, ·] (x - 5)(x - 2)   3. [5.3, · ·] (x - 5)²

4. [5.1, · ·] 2y²(2y² - 4y + 3)

5. [5.1, ···] (x² + 2)(x + 1)   6. [5.1, · ·] x(x - 5)

7. [5.6, ·] x(x + 3)(x - 1)

8. [5.6, ·] 2(5x - 6)(x + 4)

9. [5.2, · ·] (2x - 3)(2x + 3)

10. [5.4, ·] (x - 4)(x + 3)

11. [5.6, ·] 3m(2m + 1) × (m + 1)

12. [5.6, ·] 3(w + 5)(w - 5)

13. [5.6, ·] 5(3x + 2)²

14. [5.6, ·] 3(x² + 4)(x + 2) × (x - 2)

15. [5.3, · ·] (7x - 6)²

16. [5.5, ·] (5x - 1)(x - 5)

17. [5.1, ···] (x³ - 3)(x + 2)

18. [5.6, ·] 5(4 + x²)(2 + x) × (2 - x)

19. [5.5, ·] (2x + 3)(2x - 5)

20. [5.6, ·] 3t(2t + 5)(t - 1)   21. [5.7, · ·] 5, -4

22. [5.7, · ·] $\frac{3}{2}$, -5   23. [5.7, · ·] 7, -4

24. [5.8, ·] 8, -3   25. [5.8, ·] l = 10 m, w = 4 m

26. [5.9, ···] -5x³y - y³ + xy³ - x²y² + 19

27. [5.9, ∴] 8a²b² + 6ab² + 6ab + ab³ - 4b³

28. [5.9, ∴] 9x¹⁰ - 16y¹⁰

29. [5.9, ∴] 3(m + 2n) × (m - 5n)

30. [2.6, ···] $-\frac{10}{11}$ 31. [3.4, ·] $\frac{19}{3}$

32. [3.6, ·] T = $\frac{I}{PR}$ 33. [4.7, · ·] 25t⁴ - 70t² + 49

34. [5.8, ·] l = 15, w = 3

35. [4.7, · ·], [4.2, · ·], [5.4, ·] (a - 4)(a + 8)

Test: Chapter 6

1. [6.1, · ·] II   2. [6.1, · ·] III

3. [6.1, ···] (3, 4)   4. [6.1, ···] (0, -4)

5. [6.2, ·] Yes

6. [6.2, · ·]

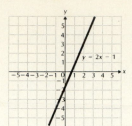

7. [6.3, ·]

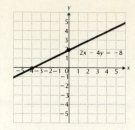

8. [6.3, · ·]

9. [6.2, · ·]

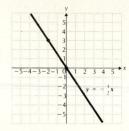

10. [6.4, ·] y = 2x   11. [6.4, ·] y = 0.5x

12. [6.4, · ·] 240 km   13. [6.5, ·] No slope

14. [6.5, ·] $\frac{7}{12}$ 15. [6.5, · ·] 0

16. [6.5, · ·] No slope   17. [6.5, ∶∶] 2, $\left(0, -\frac{1}{4}\right)$

18. [6.5, ∶∶] $\frac{4}{3}$, (0, -2)   19. [6.5, ∴] y = x + 2

20. [6.5, ∴] y = -3x - 6   21. [6.5, ∴] y = -3x + 4

22. [6.5, ∴] y = $\frac{1}{4}$x - 2

23. [4.4, · ·] -3x³ + 3x² - 3x - $\frac{1}{4}$ 24. [2.9, ·] 269

25. [5.6, ·] x³(x² - 7) × (x² + 4)

26. [3.5, ·] 139, 141

27. [6.2, · ·] answers may vary (0, 1), (-2, 3), (4, 5)

28. [6.5, ···] and [∶∶] y = $\frac{2}{3}$x + $\frac{11}{3}$

Test: Chapter 7

1. [7.2, ·] No   2. [7.2, · ·] (2, -1)

3. [7.3, ·] (8, -2)

4. [7.3, · ·] (-1, 3)

5. [7.4, ·] (1, -5)

6. [7.4, · ·] (12, -6)

7. [7.4, · ·] (0, 1)

8. [7.4, · ·] (5, 1)

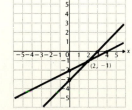

9. [7.5, ·] 5 and -3   10. [7.5, ·] 40 km/h

11. [7.5, ·] 40 L of A, 20 L of B

12. [3.7, ∶] x⁶y⁻²¹   13. [4.7, · ·] 16 - 40y + 25y²

14. [5.7, · ·] 5, -2   15. [6.3, ·]

16. [7.5, ·] 5

17. [7.2, ·], [3.3, ·]

$A = -\frac{9}{5}$, $B = 23$

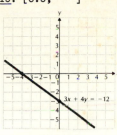
$3x + 4y = -12$

Test: Chapter 8

1. [8.1, ·] No   2. [8.1, ·] Yes

3. [8.1, ·] ~~Yes~~   4. [8.1, · ·] {x|x ⩽ -4}

5. [8.1, · ·] {x|x > -13}   6. [8.2, ·] {x|x ⩽ 5}

7. [8.2, ·] {y|y ⩽ -13}   8. [8.2, ·] {y|y ⩾ 8}

9. [8.2, ·] $\left\{x \mid x \leqslant -\frac{1}{20}\right\}$ 10. [8.2, · ·] {x|x < -6}

11. [8.2, · ·] {x|x ⩽ -1}

12. [8.3, ·]

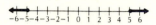

13. [8.3, · ·]

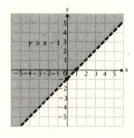

14. [8.3, · · ·]   15. [8.3, · · ·]

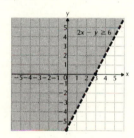
$y > x - 1$ $2x - y \geq 6$

16. [8.4, ·] {17, 19, 23}   17. [8.4, · · ·] {0}

18. [8.4, · · ·] {r, s, t}

19. [8.4, : :] {-4, -2, -1, 0, 1, 2, 4}

20. [8.4, : :] {a, b}   21. [3.6, ·] c = 4A + d

22. [7.4, · ·] (2, -2)   23. [2.1, · · ·] 37

24. [2.1, : :] -19

25. [5.8, ·] 17 and 18, -18 and -17

26. [8.1, ·] All rational numbers

27. [8.2, ·] $\left\{w \mid w < 21\frac{1}{2} \text{ cm}\right\}$

Test: Chapter 9

1. [9.1, · · ·] $\frac{3x + 7}{x + 3}$ 2. [9.1, : :] $\frac{a + 5}{2}$

3. [9.2, · ·] $\frac{(5x + 1)(x + 1)}{3x(x + 2)}$

4. [9.4, · · ·] (y - 3) × (y + 3)(y + 7)

5. [9.3, ·] $\frac{23 - 3x}{x^3}$ 6. [9.3, · · ·] $\frac{8 - 2t}{t^2 + 1}$

7. [9.3, · ·] $\frac{-3}{x - 3}$ 8. [9.3, : :] $\frac{2x - 5}{x - 3}$

9. [9.5, ·] $\frac{8t - 3}{t(t - 1)}$

10. [9.6, ·] $\frac{-x^2 - 7x - 15}{(x + 4)(x - 4)(x + 1)}$

11. [9.6, · ·] $\frac{x^2 + 2x - 7}{(x - 1)^2(x + 1)}$ 12. [9.7, ·] 12

13. [9.7, ·] 5, -3   14. [9.8, ·] 4

15. [9.8, · · ·] 16   16. [9.8, ·] 45 km/h, 65 km/h

17. [9.9, ·] $t = \frac{g}{M - L}$

18. [9.11, ·] $4x^2 + 3x - 5$

19. [9.11, · ·] $2x^2 - 4x - 2$, R 17

20. [9.10, ·] $\frac{3y + 1}{y}$ 21. [9.12, ·] $y = \frac{12}{x}$

22. [9.12, · ·] $1\frac{1}{5}$ hr   23. [5.2, · ·] (4a - 7)(4a + 7)

24. [8.2, · ·] $\left\{x \mid x > \frac{9}{2}\right\}$

25. [4.4, · ·] $13x^2 - 29x + 76$

26. [7.5, ·] $789\frac{1}{4}$ yd²

27. [9.8, ·] Team A: 4 hr; team B: 10 hr

28. [9.10, ·] $\frac{3a + 2}{2a + 1}$

Test: Chapter 10

1. [10.1, ·] 9, -9   2. [10.1, ·] 8

3. [10.1, ·] -5   4. [10.1, · ·] Irrational

5. [10.1, · ·] Rational   6. [10.1, · ·] Rational

7. [10.1, · ·] Irrational   8. [10.1, · ·] Rational

9. [10.1, : :] 10.770   10. [10.1, : :] 9.327

11. [10.1, : :] 1.732   12. [10.2, ·] $4 - y^3$

13. [10.2, · ·] No   14. [10.2, · ·] Yes

15. [10.2, · ·] x ⩽ 8   16. [10.2, · · ·] a

17. [10.2, · · ·] 6y   18. [10.3, ·] $\sqrt{30}$

19. [10.3, ·] $\sqrt{x^2 - 9}$ 20. [10.3, · ·] $3\sqrt{3}$

21. [10.3, · ·] $5\sqrt{x - 1}$ 22. [10.4, · ·] $t^2\sqrt{t}$

23. [10.4, · · ·] $5\sqrt{2}$ 24. [10.4, · · ·] $3ab^2\sqrt{2}$

25. [10.5, ·] $\frac{3}{2}$ 26. [10.5, ·] $\frac{12}{a}$

27. [10.5, · ·] $\frac{\sqrt{10}}{5}$ 28. [10.5, · ·] $\frac{\sqrt{2xy}}{y}$

29. [10.6, · ·] $\frac{3\sqrt{6}}{8}$ 30. [10.6, ·] $\frac{\sqrt{7}}{4y}$

31. [10.7, ·] $-2\sqrt{18}$, or $-6\sqrt{2}$

32. [10.7, ·] $\frac{4}{5}\sqrt{5}$

33. [10.8, ·] $c = 4\sqrt{5} \approx 8.944$ 34. [10.9, ·] 48

35. [10.9, · ·] About 5000 m   36. [6.4, · ·] 44 kg

37. [9.7, ·] -5   38. [9.6, ·] $\frac{-x^2 + x + 17}{(x^2 - 16)(x + 1)}$

39. [7.6, ·] 261 mph **40.** [10.4, ··] y^{8n}

41. [10.8, ··] 8 ft

Test: Chapter 11

1. [11.1, ··] $\pm\sqrt{5}$ **2.** [11.2, ·] 0, $-\frac{8}{7}$

3. [11.2, ··] -8, 6 **4.** [11.2, ··] $-\frac{1}{3}$, 2

5. [11.3, ·] $8\pm\sqrt{13}$ **6.** [11.5, ·] $\frac{1\pm\sqrt{13}}{2}$

7. [11.5, ·] $\frac{3\pm\sqrt{37}}{2}$ **8.** [11.5, ·] $-2\pm\sqrt{14}$

9. [11.5, ·] $\frac{7\pm\sqrt{37}}{6}$ **10.** [11.6, ·] 2, -1

11. [11.6, ·] 2, -4 **12.** [11.6, ··] -2, 2

13. [11.4, ·] $2\pm\sqrt{14}$ **14.** [11.5, ··] 5.7, -1.7

15. [11.7, ·] $n = \frac{-b\pm\sqrt{b^2+4ad}}{2a}$

16. [11.9, ·] a) Downward b) [11.9, ·]

17. [11.9, ··] -1.8, 2.8

18. [11.3, ···] 25%

19. [11.8, ·] Width = 2.5 m, length = 6.5 m

20. [11.8, ·] 24 km/h

21. [10.7, ·] $\sqrt{60}$, or $2\sqrt{15}$ **22.** [7.4, ··] (-1, 6)

23. [10.4, ···] $7xy\sqrt{2x}$ **24.** [9.8, ·] $4\frac{12}{19}$ hr

25. [11.8, ·] $5+5\sqrt{2}$

26. [7.3, ·], [11.5, ·] $1\pm\sqrt{5}$

Final Examination

1. [1.1, ·] -995 **2.** [1.2, ∶∶] 48

3. [2.1, ···] 9 **4.** [2.5, ··] -9.7

5. [2.2, ·] -5 **6.** [2.6, ·] $-\frac{6}{7}$

7. [2.8, ··] 19y - 45 **8.** [2.9, ·] 211

9. [3.1, ·] 5.6 **10.** [3.2, ·] -7

11. [3.3, ··] -10 **12.** [3.4, ·] 7

13. [5.7, ··] 6, -4 **14.** [7.3, ·] (4, -3)

15. [7.4, ··] (4, 7) **16.** [9.7, ·] $\frac{1}{4}$, $-\frac{1}{2}$

17. [10.9, ·] 6 **18.** [5.7, ··] $\frac{-3\pm\sqrt{37}}{2}$

19. [8.2, ··] $\left\{x \mid x \leqslant \frac{5}{2}\right\}$ **20.** [9.9, ·] $w = \frac{1}{A-B}$

21. [3.6, ·] $M = \frac{K-2}{T}$ **22.** [3.7, ···] x^{10}

23. [3.7, ∶∶] x^{-10} **24.** [3.7, ··] x^{-12}

25. [4.2, ··] $6y^3 - 3y^2 - y + 9$

26. [4.4, ··] $-2x^2 - 8x + 7$

27. [4.6, ·] $-6t^6 - 12t^4 - 3t^2$

28. [4.5, ∶·∶] $4x^3 - 21x^2 + 13x - 2$

29. [4.7, ·] $x^2 - 64$ **30.** [4.7, ··] $4m^2 - 28m + 49$

31. [5.9, ∶∶] $9a^2b^4 + 12ab^2c + 4c^2$

32. [5.9, ∶∶] $9x^4 + 6x^2y - 8y^2$

33. [9.1, ∶∶] $\frac{1}{x^2(x+3)}$ **34.** [9.2, ··] $\frac{3x^4(x-1)}{4}$

35. [9.5, ·] $\frac{11x-1}{4x(3x-1)}$

36. [9.6, ·] $\frac{2x+4}{(x-3)(x+1)}$

37. [5.1, ··] $3x(x^2 - 5)$

38. [5.2, ··] $(4x - 5)(4x + 5)$

39. [5.5, ·] $(3x - 2)(2x - 3)$

40. [5.3, ··] $(x - 5)^2$

41. [5.9, ∶·∶] $(2x - y)(a + 3b)$

42. [5.9, ∶·∶] $(x^4 + 9y^2)(x^2 - 3y) \times (x^2 + 3y)$

43. [10.4, ·] $6\sqrt{2}$ **44.** [10.6, ··] $\frac{\sqrt{30}}{5}$

45. [10.7, ·] $13\sqrt{2}$ **46.** [10.4, ···] $2a^2b\sqrt{6ab}$

47. [6.3, ·] **48.** [6.3, ··]

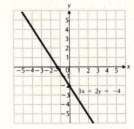 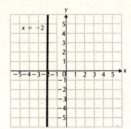

49. [8.3, ···] **50.** [11.9, ···]

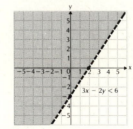 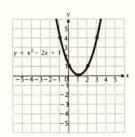

51. [3.5, ·] 5, 7; -7, -5

52. [7.5, ·] 40 L of 75%; 20 L of 50%

53. [7.6, ·] 70 km/h

54. [5.8, ·] Length = 11 m, width = 8 m

55. [6.5, ·] $-\frac{4}{3}$ **56.** [8.4, ∶∶] {1, 2, 3, 4, 5}

57. [6.4, ·] y = 8x **58.** [9.12, ·] $y = \frac{5000}{x}$

59. [3.5, ·] Side of square is 15; side of triangle is 20 **60.** [11.3, ··] 144